W9-BBF-370

DISCARD
BE...MY
CO...
LIBRARY

EUTROPHICATION:

CAUSES,
CONSEQUENCES,
CORRECTIVES

PROCEEDINGS OF A SYMPOSIUM

NATIONAL ACADEMY OF SCIENCES

Washington, D. C.
1969

Standard Book Number 309-01700-9

Available from

PRINTING AND PUBLISHING OFFICE
NATIONAL ACADEMY OF SCIENCES
2101 CONSTITUTION AVENUE
WASHINGTON, D.C. 20418

Library of Congress Catalog Card Number 68-62704

Preface

Events leading to the International Symposium on Eutrophication were as follows:

1. In April 1965 the National Academy of Sciences–National Research Council, in recognition of growing concern over problems associated with eutrophication of lakes, streams, and estuaries, appointed a Planning Committee on Eutrophication. Members were: Gerard A. Rohlich (Chairman), University of Wisconsin (Madison); A. M. Beeton, University of Wisconsin (Milwaukee); David C. Chandler, University of Michigan; W. T. Edmondson, University of Washington; Gordon M. Fair, Harvard University; David G. Frey, Indiana University; Arthur D. Hasler, University of Wisconsin (Madison); F. Ronald Hayes, Fisheries Research Board of Canada; Kenneth M. Mackenthun, Robert A. Taft Sanitary Engineering Center, U.S. Public Health Service; Charles E. Renn, Johns Hopkins University; and Jacob Verduin, Southern Illinois University.

2. In May 1965 the Committee submitted a report containing a recommendation that

... an international symposium on eutrophication be held in order that the present worldwide state of knowledge and understanding of this phenomenon can be discussed in open forum and recommendations developed for the effective management of problems and for the course of future research.

3. In October 1965 the National Academy of Sciences–National Research Council appointed a Planning Committee for the International Symposium on

628.168
Jn8
1967

104485

Eutrophication. Five members of the Planning Committee on Eutrophication made up the new committee: Gerard A. Rohlich (Chairman), A. M. Beeton, David C. Chandler, W. T. Edmondson, and Arthur D. Hasler.

The symposium was held at the University of Wisconsin, Madison, June 11–15, 1967, under the sponsorship of the National Academy of Sciences–National Research Council, the U.S. Atomic Energy Commission, the U.S. Department of the Interior, the National Science Foundation, and the Office of Naval Research, U.S. Department of the Navy. Almost 600 persons, representing 11 foreign countries and the United States, attended.

It is hoped that these proceedings will serve as a useful reference on eutrophication, both for scientists and for persons responsible for managing water resources.

Gerard A. Rohlich, Chairman
Planning Committee for the
International Symposium on
Eutrophication

Contents

VI. CONTRIBUTIONS TO SCIENCE FROM THE STUDY OF EUTROPHICATION

I.

INTRODUCTION,

SUMMARY,

AND RECOMMENDATIONS

Introduction,

Summary,

and Recommendations

INTRODUCTION

Man's activities, which introduce excess nutrients, along with other pollutants, into lakes, streams, and estuaries, are causing significant changes in aquatic environments. The excess nutrients greatly accelerate the process of eutrophication.

The pollution problem is critical because of increased population, industrial growth, intensification of agricultural production, river-basin development, recreational use of waters, and domestic and industrial exploitation of shore properties. Accelerated eutrophication causes changes in plant and animal life—changes that often interfere with use of water, detract from natural beauty, and reduce property values. Occasionally the changes threaten the destruction of water resources. A common change is excessive growth of algae and larger aquatic plants. Such growth chokes the open water, may make it nonpotable, and may greatly increase the cost of filtration. The excessive vegetation decomposes, fouls the air, and consumes the deep-water oxygen vital for fish and other animal life.

The term "eutrophic" means well-nourished; thus, "eutrophication" refers to natural or artificial addition of nutrients to bodies of water and to the *effects* of added nutrients.

Eutrophication of lakes is a natural process that can be greatly accelerated by man. Eutrophication is an aspect of aging; it increases the rate at which lakes disappear. Some disagreement exists as to the applicability of the term to other bodies of water. Streams do not age in the same sense as lakes, although added nutrients increase their productivity. Consequently, a distinction should be made between (1) eutrophication *sensu stricto,* the increase in nutrient supply, and (2) the effects of eutrophication, which may be expressed in various ways. When the effects are undesirable, eutrophication may be considered a form of pollution.

It is not certain that the biotic changes accompanying culturally accelerated eutrophication of a lake parallel those accompanying natural maturation. Some limnologists doubt that eutrophication is a reversible process on any practical time scale, although some lakes have shown definite improvement in response to diversion of sewage or other remedial action.

Human sewage and industrial wastes are significant sources of nutrients that contribute to eutrophication of lakes. Drainage from farmland is also an important source, most of the nutrients coming from farm manure that is spread on frozen ground and is subsequently flushed into streams during spring thaws and rains. Runoff from urban areas is rich in phosphate and nitrate. Substantial quantities of nitrates from the ignition of fossil fuels augment these sources of nutrients in rain.

Substances other than inorganic phosphorus and nitrogen compounds contribute to eutrophication. Examples are vitamins, growth hormones, amino acids, and trace elements. Some of these substances are synthesized in the biological treatment of sewage.

The abundance and species composition of planktonic, bacterial, benthic, and fish populations change as eutrophication progresses, and changes of this nature may be used to detect and measure the degree and rate of eutrophication. In some stages of enrichment, benthic invertebrates may be eliminated, as in Lake Erie. Enriched lakes develop dense populations of planktonic algae commonly dominated by a few species of blue-green algae.

The notion that chemical enrichment can be desirable is seldom discussed, but this enrichment is precisely what is needed to increase the productivity of marine bays and estuaries. As we consider its effects, we are appalled by the waste involved in sewage disposal. Our first knowledge of eutrophication was derived from efforts to increase production of fish ponds through fertilization. Fish production does increase with fertilization, and in many parts of the world, greater yields are achieved by fertilizing ponds with sewage effluent. However, because of the species changes that typically accompany enrichment, these increases are a major benefit only in areas where the value of fish is determined by weight rather than by species.

A technology efficient in harvesting macrophytes and plankton and marketing them as useful products would turn eutrophic waters to assets. Almost certainly, however, success would depend on controlling species composition by controlling the nutrient regime.

SUMMARY

Participants in the International Symposium on Eutrophication agreed that we need greater knowledge of the processes involved in eutrophication. Furthermore, we should study eutrophication on an ecosystem basis. The limits of the system cannot be drawn at the water's edge but must include the waters of lakes, their sediments, and their drainage basins. Such a multifaceted approach will benefit from computer simulation models.

Many of the participants called for a more thorough understanding of algal physiology and ecology. More algal culture studies, coupled with experiments in chemical alteration of lake waters, are needed to better understand the interaction between organisms and nutrients. Such studies require more sensitive analytical techniques for plant nutrients, especially techniques that distinguish between available and nonavailable forms of the elements. Furthermore, we need better documentation of the numbers and species of algae that are considered a nuisance.

Considering the beneficial aspects of eutrophication, our concern must not be limited to environmental factors that favor high primary production; we must also learn more about conditions that favor efficient conversion of phytoplankton into zooplankton and of these into fish.

Documentation of the histories of enrichment in lakes could provide a basis for timing management programs in the future. Such documentation may come from advances in paleolimnology or may, to a large extent, already be available in the form of data that have not been analyzed from this point of view.

Several participants thought we should have more basic biological studies of organisms at all trophic levels, to establish their relation to the trophy of the environment. This is especially true for bacteria. What changes in numbers, kinds, and activity of bacteria are associated with eutrophication?

Speakers agreed that prevention of further damage to water resources is a matter of great urgency and that reversal of deterioration already under way should be attempted wherever an attempt is practicable.

The most ecologically sound approach to the problem is to prevent the introduction of nutrients resulting from man's activities. As one speaker said, "Prophylactic action should be preferable to therapeutic."

Models of simple ecosystems suggest that a reduction in nutrient supply is the only management program that has the cooperation of the system. The eradication of nuisance organisms is less effective because those that are eliminated are replaced by other types that may present an even greater problem of control.

DIVERTING NUTRIENTS FROM LAKES

Diverting sewage, sewage-treatment-plant effluents, and other sources of nutrients away from lakes is practicable in certain situations. Studies of Lake Washington, the Madison lakes, and others have demonstrated the benefits to be derived from diversion. Lake Michigan undoubtedly would show greater eutrophication if Chicago's sewage were not diverted into the Mississippi River drainage basin. Canals or conduits carrying sewage from other cities on or near the lake could contribute greatly to reducing eutrophication.

Diversion from lakes to rivers can be justified in part by the fact that rivers have a greater ability to mix and aerate pollutants, and thus can handle more effectively an effluent with a high biochemical oxygen demand and nutrient content. Rivers, unlike lakes, are more readily rescued from a state of overfertilization. Estuaries, because of tidal flushing, have a greater potential than rivers for disposal of organic waste.

REMOVING NUTRIENTS FROM SEWAGE

Where diversion of sewage from lakes is impracticable, the sewage should be treated for removal of nutrients.

Difficulty in determining the maximum amount of wastes and nutrients that a specific body of water can assimilate without undergoing deterioration in water quality is a barrier to efficient removal of nutrients from sewage.

By testing sewage with bioassay methods, we can determine the elements that should be removed in order to control nuisance growths of algae and macrophytes.

The symposium reflected widespread concern about removal of phosphorus from sewage. Most speakers who discussed this point thought that removal of phosphorus would aid substantially in controlling eutrophication. Use of nonphosphate detergents was also recommended.

Many growth stimulators are found in sewage, but little has been done to develop a technology for removing them.

IMPROVING AGRICULTURAL PRACTICES

The widespread use of commercial fertilizers may make it necessary to modify agricultural processes to reduce enrichment of surface waters. Aerial application of fertilizers to forests may inadvertently include direct fertilization of streams.

Even the old methods of applying fertilizer may require modification. Manure spread in winter may contribute large amounts of phosphorus as it runs off the frozen soil. The remedy is to spread the manure on thawed soil.

In some areas, wastes from domestic animals make up a considerable part of agricultural wastes washed into streams. Serious attention should be given to improved handling and treatment of these wastes.

More consideration should be given to the use of waste to increase the fertility of the land. Forest and agricultural soils, if not eroded, have a remarkable capacity for accumulating and retaining mineral ions, especially phosphorus.

CONTROLLING AVAILABILITY OF NUTRIENTS WITHIN LAKES

In situations where it is impractical or impossible to control nutrient influx, it may be possible to make nutrients unavailable within lakes. As a last resort, flocculants could be used to precipitate algae and nutrients out of the trophogenic zone. This procedure would be especially effective where lakes could be made meromictic (nutrients would be permanently locked in the hypolimnion) and would be appropriate where runoff laden with silt or salts tends to develop a density stratification.

Another way of controlling the availability of nutrients is to prevent thermocline formation and overturning in a lake in summer. Deliberate alteration of the chemistry of lakes is also proposed.

REMOVING NUTRIENTS FROM LAKES

There has been considerable interest in the possibility of changing the trophic state of a lake by removing nutrients. One method would be to flush the lake with large volumes of water low in nutrients. Flushing action of this kind would be applicable only to small lakes.

Another method would be to dredge sediments that are high in nutrients. An added benefit of dredging would be deepening of a lake, which would reduce the region available for growth of aquatic macrophytes.

Removing macrophytes and large quantities of fish would remove nutrients from the ecosystem and have some economic value.

RELIEVING SYMPTOMS OF EUTROPHICATION

Several methods may be used to control plankton blooms, large growths of macrophytes, large populations of midges, and coarse fish—all symptoms of eutrophication.

Chemical toxicants have been used extensively for controlling the symptoms named here, but we have inadequate information on residues of toxicants, rates of breakdown, and long-term effects on aquatic organisms. Chemicals, however, distort the structure of multispecies aquatic communities and hence are less useful in lakes than they are in agriculture, where weeds for a single crop (e.g., wheat) have to be eradicated.

In addition, management of chemicals is more difficult in lakes than on land. The development of selective toxicants should continue to be investigated.

Removing plants by mechanical harvesting has advantages over using toxicants. Relief is immediate, and decomposition of plants in the water is avoided.

Biological control of algae and macrophytes was suggested as an important area for further research.

Fish management practices have been developed to control large populations of coarse fish, such as carp, which flourish in some eutrophic waters. Coarse fish can be harvested efficiently, and they can be marketed at a profit. Selective toxicants to eradicate a species have been tried, but it is difficult to prevent fish from decomposing on the bottom and thus aggravating the eutrophication problem.

MEASURING THE EXTENT OF EUTROPHICATION

A major recommendation was for nationwide or worldwide surveillance and monitoring networks to assess man's impact on the productivity of aquatic systems. To do this we need some indices of eutrophication that will measure small changes in the ecosystem.

The criteria of a good index are sensitivity, specificity, ubiquity, and ease of measurement and interpretation.

Organisms, populations of organisms, or communities may meet the criteria of good indices. When nutrients increase, the diversity of the community decreases, and new species invade the lake and may become abundant.

Indicator organisms are useful in characterizing eutrophic waters. The presence or absence of certain species, or the relative abundance of groups of species, relieves the biologist from continuous monitoring and surveillance of chemical conditions. Many of the difficulties in using indicator species result from a lack of specific information about how these organisms are related to the enrichment process. The relation between species diversity and nutrient levels has great potential as an index. Other disturbances to natural systems, however, are likely to reduce diversity, and this criterion alone is not diagnostic of eutrophication. Fish may provide one of the most sensitive indices because an early response to eutrophication is enhancement in growth.

The turnover time of ^{32}P may assess the entire phosphate pool and therefore may be a useful indicator. Certain definite concentrations of phosphorus may indicate critical changes in estuarine environments, because when phosphorus concentrations reach these levels, the oxygen demand of the organic matter that is produced by photosynthesis equals the available oxygen supply.

POLITICAL AND SOCIOLOGICAL ASPECTS

Jacob H. Beuscher, professor of law at the University of Wisconsin, now deceased, pointed out that the racing processes of deterioration could not tolerate delay until all the facts were verified. He asserted that circumstantial evidence justifies immediate action. Enough is known today about eutrophication to justify expensive diversion of effluents in order to cope with the problem until later technology can solve the problem of nutrient control. Among the actions that can be taken are the establishment, by ordinance, of minimum lot size and setback from lakes and the protection of shorelines from excessive deforestation, marshland destruction, landfills, and erosive scarring.

Professor Beuscher, who helped draft the 1966 Wisconsin Water Resources Act, stated that man's presence near lakes has resulted in water pollution, destruction of scenery, and destruction of wildlife habitats. Because of these abuses, a program of shoreland control is necessary. He said that such a program requires a close working relation between state and local units of government, more technically trained personnel, effective and fair administrative procedures operating at local levels, and keeping abreast of research.

RECOMMENDATIONS

The following recommendations are based on papers presented at the symposium and on discussions by members of the Planning Committee for the International Symposium on Eutrophication.

EDUCATION AND INFORMATION

General goals Information on the nature of eutrophication and on prevention and control should be made available to the general public, to government agencies (federal, state, and local), and to professional and business groups concerned with land and water management (e.g., conservationists, hydrologists, engineers, and manufacturers).

Media should include newspapers, periodicals, booklets for distribution to schools and civic organizations, educational films, and videotapes.

Professional science writers should be asked to aid in carrying out the program.

Scientists and engineers who have been active in the eutrophication field should provide the background information needed by media representatives.

Specific recommendations

1. An authoritative booklet should be prepared that will show the legal, governmental, social, recreational, conservational, scientific, and economic implications of eutrophication.

2. A newsletter (probably a quarterly) should be distributed widely to scientific, professional, and civic organizations.

3. Materials for magazines, newspapers, and television and radio documentaries should be prepared.

4. Subsidies should be obtained from appropriate government agencies for a visiting-lecturer program on problems of eutrophication. This program would be directed to waste-treatment design engineers, university and secondary schools, operators of waste-treatment plants, and various civic organizations.

5. Summer institutes should be held to bring university and secondary school people up to date on the problems of eutrophication. At least 12 such programs should be held throughout the country.

6. Federal support should be provided to educational institutions for adult-education clinics and for extension courses dealing with problems of eutrophication and pollution. These would be directed to operators of waste-water treatment plants, schoolteachers, clergymen, lake association officials, district sanitation officials, county board members, city aldermen and councilmen, realtors, and other interested groups.

7. Appropriate agencies should sponsor summer workshops at which policy-makers, decision-makers, and experts could exchange information and views on problems of eutrophication and pollution.

8. The National Academy of Sciences–National Research Council (NAS-NRC) should sponsor a second international symposium on eutrophication in 1972 to examine progress in dealing with the problem and to recommend application of promising techniques investigated since the last international symposium.

9. The NAS-NRC should establish a panel of seven experts on eutrophication who would be prepared to appear before congressional committees and provide information on the subject.

10. Means should be sought to encourage the inclusion of questions dealing with eutrophication in civil-service examinations for professional positions related to management of water resources.

11. The importance of the engineering aspects of eutrophication problems should be called to the attention of the National Academy of Engineering, which should be asked to consider the dissemination of pertinent information to engineers and the possibility of developing a national or international symposium on engineering aspects of eutrophication problems.

12. Cross-disciplinary seminars, similar in concept to those sponsored by the Federal Water Pollution Control Administration in Cincinnati, Ohio, should be encouraged. Their purposes should be to identify problems, find possible solutions, and strengthen the qualifications of persons working in contributory disciplines.

RESEARCH

1. A cost–benefit analysis of waste-disposal systems should be initiated with the purpose of finding a system that will involve our waterways to the least possible extent. All present nonwater techniques should be evaluated.

2. Long-term multidisciplinary studies should be established to improve understanding of the mechanisms and processes of eutrophication and to formulate feasible management procedures, both preventive and corrective. Such studies should be started immediately on bodies of water in which eutrophication has led to critical socioeconomic conditions (e.g., Lakes Erie and Michigan, Lake Washington, and some estuaries).

3. Research should be undertaken in the following areas:

a. The role of various sources as contributors of nutrients, including drainage basins or floodplains, municipalities, industries, agricultural lands, and forests.

b. The recycling of nutrients in aquatic systems.

c. The effects of impounding free-flowing water on production and nutrient availability.

d. The socioeconomic aspects of accelerated eutrophication (e.g., the impact of accelerated eutrophication on recreational use of a body of water, on commercial fishing, on public health, and on cost of water treatment).

e. Detailed documentation of eutrophic changes that have occurred in natural waters and a critical evaluation of the available data.

f. Projections of the consequences of unchecked eutrophication in various natural waters and of what may happen with the initiation of various corrective measures. Such information would be essential to any socio-economic approaches involving models for systems analysis.

4. Experiments on a number of small lakes should be initiated, with the recognition that each aquatic ecosystem is unique. In these experiments, investigators would seek ways of dealing with eutrophication; for example, ways of:

a. Limiting nutrient input.

b. Accelerating nutrient outgo.

c. Impairing nutrient availability.

d. Reducing the volume of water participating in production of plant material.

e. Altering stratification.

f. Modifying ecological systems to provide for accelerated consumption of plant material by an appropriate array of animal populations.

5. In the United States, at least four centers should be established at which scientists and engineers could develop interdisciplinary research on eutrophication. These centers would be responsible for carrying out research recommendations 1 and 2. In addition, they would provide for laboratory studies such as:

a. Enhancing the effectiveness of fungi that can parasitize algae and so destroy populations.

b. Finding an array of viruses that can destroy select algal populations.

c. Developing algicides and herbicides that can be used safely, specifically, and effectively in restricted areas.

d. Developing a system for measuring and monitoring eutrophication that could be used on a nationwide or worldwide basis.

e. Ascertaining environmental requirements of algae and aquatic macrophytes.

6. Aspects of eutrophication that have a potential for increasing the world's food supply should be realistically evaluated. The following are examples of studies that would be useful:

a. More developmental effort on engineering techniques for harvesting and processing aquatic products.

b. Development of information on the potential value of algae and rooted aquatic plants as crops, as food, or as sources of pharmaceuticals or other biochemical products.

c. Development of food technology for the utilization of "trash" species of fish (e.g., alewife, carp, and goldfish) and other marine organisms.

7. Ecosystem analysis and research on models for simulating trends in eutrophication should be strengthened.

8. Information should be obtained (a) on the nature and performance of conduits constructed around small lakes and constructed or proposed for large lakes in the U.S. and abroad and (b) on the performance of schemes for disposing of waste water by discharging it onto agricultural and forested areas and onto natural wastelands and wastelands formed by soil-denuding operations such as strip-mining.

IMPLEMENTING RECOMMENDATIONS

The National Academy of Sciences–National Research Council should organize a committee on eutrophication and charge it with working out the means for implementing the above recommendations. This committee should work closely with the NAS-NRC, the government agencies that cosponsored the first international symposium, the Public Affairs Committee of the Ecological Society of America, the U.S. National Committee for the International Biological Program, and the American Society of Limnologists and Oceanographers.

II.

EUTROPHICATION,

PAST

AND

PRESENT

G. E. HUTCHINSON
Yale University, New Haven, Connecticut

Eutrophication,

Past and Present

It would be well, at the beginning of this symposium, to try to find out exactly what we are about to discuss.

The terms "eutrophic," "mesotrophic," and "oligotrophic" in their German forms—*"nährstoffreichere* (eutrophe) *dan mittelreiche* (mesotrophe) *und zuletzt nährstoffarme* (oligotrophe)"—were introduced into science by Weber in 1907 to describe the general nutrient conditions determined by the chemical nature of the soil solution in German bogs. In Weber's case, the succession proceeded in this direction from eutrophic to oligotrophic as a raised bog, subject to continual leaching, was built up. The plant association characterizing the eutrophic low bog was described as eutraphent, or well-nourished,* while that on the oligotrophic raised bog was oligotraphent. These terms in *-traphent* have been used to a limited extent by plant ecologists, mainly in northern Europe, but the environmental terms in *-trophic* have proliferated along with some of the phenomena that they describe.

The introduction of Weber's terms into limnology was due primarily to Einar Naumann (1919), who combined an extraordinary eye for lakes with a passion for a succinct, classificatory style of presentation. This style led him into difficulties. Wesenberg-Lund (1926) stated,

Naumann has tried to press nature into a series of highly artificial schemata which are unquestionably very valuable for all those scientists whose time is just as scanty as his own, whereas from a purely scientific point of view, as far as I can see, they have very little value.

*In contemporary Greek, εὐτραφής still means thick or corpulent. The etymological aspects of all these terms are discussed by Holmberg and Naumann (1927).

Looking back, we can now see, as is usually the case, how far both of these great men were right and how both are among the founders of modern limnology. Naumann's contribution was to provide a theoretical classificatory scheme that enabled a large number of casual observations to be coordinated, although parts of this scheme are probably of limited application. The fundamental concepts of oligotrophy and eutrophy have survived well enough for us to have, 60 years after Weber's initial paper and 48 years after Naumann's first contribution, a conference on a very important subject that we call eutrophication.

It is important to notice that Naumann originally considered his terminology to refer to water types. These types were characterized in theory by content of nutrient materials and in practice were recognized by the capacity to support poor or rich communities of phytoplankton. At the time the classification was developed, the criteria for nutrient contents given by Naumann (1927) were either completely uncertain or wildly unrealistic, and, admittedly, were based on quite inadequate analyses. The practical criterion was that when eutrophic water filled a lake, the phytoplankton rendered the water turbid or colored for much of the year, whereas in a lake filled with oligotrophic water, this rarely or never happened.

The considerable amount of knowledge available about the phytoplankton of the lakes of northwestern and central Europe had led Teiling (1916) to the concept of an extreme Caledonian type of phytoplankton, characterized by desmids, in the unproductive lakes of Scotland, Wales, and the mountainous parts of Scandinavia, in contrast to the Baltic type of southern Sweden, Denmark, and northern Germany, in which myxophycean water blooms were commonplace. This concept gave a geographical complexion to Naumann's distinction and soon led to the recognition of various regional water types. Earlier, Thienemann (1913), as a result of his studies of the *maare* of the Eifel, had recognized two lake types, basing the distinction on hypolimnetic oxygen concentrations and on correlated differences in the benthic chironomid fauna. A classification into oligotrophic, eutrophic, and dystrophic (Thienemann, 1925) synthetic lake types now followed. The extreme oligotrophic type was deep and nutrient-deficient; it did not produce a water bloom; it had what we now call an orthograde distribution of oxygen; and it had a stenoxybiont benthic fauna and *Coregonus* in the hypolimnion. The eutrophic type was shallow and nutrient-rich; it produced a water bloom; it had a clinograde oxygen distribution; it had a euryoxybiont *Chironomus* fauna; and it had no stenotopic hypolimnetic fishes. At this point, the terminology began to proliferate. At the same time, difficulties began to appear, particularly in relation to the effect on oxygen distribution of the largely independent variation of depth and nutrient concentration (Lundbeck, 1934). Even if we restrict ourselves to the original concept of water types, as Järnefelt (1956) heroically attempted to do, we are still faced with considerable conceptual problems.

In 1931, Juday and Birge, commenting on the very large body of data that they had accumulated in their study of the lakes of northeastern Wisconsin, pointed out that often there was no evidence of the loss of phosphorus from the water as algal populations were maintained or even somewhat increased. About the same time, Pearsall (1932) commented on the curious fact that the blue-green blooms, which usually give rise to the *sehr stark getrübt oder sogar vollständig verfärbt* appearance of markedly eutrophic lakes, actually tended to appear when there was a most striking nutrient deficiency in the water.

This type of situation led to a renewed series of investigations. It became apparent (Hutchinson, 1941) that in many small lakes the nutrient elements were undergoing very rapid cyclical changes, passing from the sediments into the free water and back, in dying plankton or littoral vegetation, over and over again. The easy availability of artificial radioisotopes after 1945 made the detailed investigation of this kind of cycle possible (Hutchinson and Bowen, 1947, 1950; Coffin *et al.,* 1949; Hayes *et al.,* 1952) and culminated in Rigler's extraordinary discovery (1956, 1964) that the turnover time of ionic phosphorus in the epilimnion of a lake in summer can be of the order of 1 minute.*

It is now quite apparent that we should think not of oligotrophic or eutrophic water types, but of lakes and their drainage basins and sediments as forming oligotrophic or eutrophic systems. This concept is somewhat different from that of the synthetic lake types in which the biological characters of the lake appear partly, but only partly, as the results of the nutrient supply. By a eutrophic system, I mean one in which the total *potential* concentration of nutrients is high; there may happen to be an extremely low concentration in the water because the whole supply at that moment is locked up somewhere else in the system—in sediments or in the bodies of organisms. This is exemplified by what evidently happens in many fairly shallow lakes in the Temperate Zone. After considerable plankton production in the spring, *Dinobryon,* which seems to be an extreme oligotraphent poisoned by quite low phosphorus concentrations, appears. Later, after increasing temperature has promoted decomposition in the shallow-water sediments, the eutraphent blue-green algae become dominant, taking up soluble phosphate as fast as it is produced into the water. The stationary concentration of the assimilable form of any nutrient thus will be

---

*I should like to pay tribute to the energy of my former colleague Dr. E. C. Pollard, who in 1941 attempted to provide Dr. W. T. Edmondson and me with ^{32}P, manufactured with a cyclotron, to do an experiment on Linsley Pond. The amount of the radioisotope available and the sensitivity of the counters then existing were not great enough for us to achieve success.

of little interest in such a system; what is important is the total available supply in all forms and the rate at which it undergoes circulation.

The differences between the more extreme oligotrophic and eutrophic water types as understood originally by Weber and by Naumann were, as has been indicated, based largely on guesses as to phosphorus and nitrogen contents. It was also evident that hard waters were more likely to be eutrophic than soft waters, although initially it was impossible to sort out the interrelations between the various possible nutrients. Today the situation is quite different; we have experimental evidence that under certain circumstances a number of elements (e.g., C, N, P, Fe, Mo, Mg, Na) may be limiting, and we have some idea of the dynamics of some of the processes involved. If the results at times still seem confusing, this is because we now have a vast amount of partially assimilated information. Clarification of a great many of the details of our current problems may be expected from the papers to be presented here. I want, therefore, to confine my remarks to two rather general and interrelated topics.

We have spoken of eutrophic or oligotrophic systems rather than waters, emphasizing that, although such a system may be a lake basin and its contents, the concept is a little different from that of the synthetic lake type. A shallow stratified lake at a great altitude may develop a clinograde oxygen curve and a stenoxybiont benthic fauna and still be relatively oligotrophic; an example is Yaye Tso in Indian Tibet (Hutchinson 1937a). A deep lake may be very productive, but if the hypolimnion is large enough, it may show until late summer a fairly orthograde oxygen curve and support *Mysis relicta* at intermediate depths; an example is Green Lake in Wisconsin (Birge and Juday, 1911; Juday and Birge, 1927). As systems, these two lakes seem to be, respectively, moderately oligotrophic and quite eutrophic, even though the old Thienemannian classification would tend to put them in the opposite categories. Looking at the lake as a dynamic trophic system rather than as a member of a static type, we can immediately ask not merely about its reserves of nutrients, but also about how fast they circulate. In attempting to answer such questions, it became clear, as Einsele (1936) and Hasler and Einsele (1948) first showed in the case of phosphorus, that the chemical dynamics within the lake are as important as the quantities of material available. Moreover, as Ohle (1954) found in the case of sulfur, input of an element not primarily involved as a limiting factor can, by its modification of the chemical dynamics, have a profound effect on what is happening. As an illustration of these points, providing confirmation of the brilliant work of Einsele and Ohle, I should like to mention the results of a comparison (Cowgill, 1968) of some of the chemical conditions in Linsley Pond as I observed them in 1936–1939 with the present condition studied by my colleague Dr. Ursula Cowgill in 1966. In the years between the investigations, the number of houses in the drainage basin increased from about 10 to more than 100. The new analyses are not complete, but it is clear that sulfur has increased considerably, probably as a result of a general increase in settlement and industrialization and, consequently, a greater sulfate content in the local rain. The sulfur tends to be reduced in the hypolimnion and to precipitate iron as a black cloud of ferrous sulfide; the iron content of the deep water is less than it was 30 years ago; the manganese and phosphorus contents are markedly greater. These observations accord so well with what we might have expected from the earlier European work that they may appear common-place. The fact that they do all fit together shows, however, that we now have some real understanding of what happens and can make significant generalizations that can be used in the study and possibly in eventual control of undesirable events in lakes.

The second general aspect that I want to consider concerns the problem of succession. Weber clearly showed in his original work on peat bogs that the more eutrophic conditions were antecedent to the more oligotrophic ones. But it soon became usual to regard the succession as going the other way in lakes (from oligotrophic to eutrophic), at least as long as a lake remained a drainage lake. When for any reason the discrete influents disappear and the lake becomes a seepage lake, as has evidently happened often in northeastern Wisconsin, the influent water, now largely derived directly from rain or from groundwater that has passed through peat or other sorbtive materials, may become oligotrophic, and the succession is clearly from a moderately hard and fairly productive locality to one that is soft and often very unproductive. Assuming an initial transitory clear-water (ortho-oligotrophic) period, the succession would be from this condition to eutrophy and then back to oligotrophy, often with brown water.

Early in our work in Connecticut we came to suspect that as a drainage lake developed, it went through a process of eutrophication until the full trophic potential of the environment was reached, and that the lake then remained in *trophic equilibrium* for a long period (Hutchinson and Wollack, 1940) unless some external change altered the trophic potential. Although the interpretation of the early and apparently oligotrophic stages in the process have been questioned by Livingstone (1957), largely on valid grounds, the general concept of a trophic equilibrium under relatively constant hydrographic, and so climatic, conditions seems to accord with the facts.

It is, therefore, of some interest to inquire whether the very extreme types of eutrophication that Wetzel (1966) calls hypereutrophication, which are often noxious, are always man-made or whether the heavy water bloom is ever a natural phenomenon corresponding to an extreme type of eutrophic equilibrium. The first really scientific account of a water bloom is probably that of de Candolle (1826), who, largely on the basis of observations by

Engelhardt and Trechsel, described a bloom of *Oscillatoria rubescens* in the Lac de Morat. This alga has a special biology, and although it became a major problem in Switzerland during this century, owing to cultural eutrophication, its occurrence in massive populations under probably natural conditions may not bear on the general problem of other myxophycean blooms.

Griffiths (1939) has examined a number of very early records of water blooms in lakes and rivers. Some of these records are too vague to be of much scientific interest. The earliest one that might refer to a quite natural water bloom had to do with Lake Llangors, a lake in Brechnocshire in Wales, 3 km long and 2 km wide, which was said by that learned, credulous, and most attractive writer Giraldus Cambrensis* to have become very green (*viridissimi coloris inveniretur*) on two occasions prior to 1188. The event was clearly unusual because in both cases it was supposed to have foretold devastation of the area by war, which though frequent was not an annual event in the region. The lake appears now to have flat low shores with much littoral vegetation, and, no doubt, in some years there is a diversion of nutrients into the water from decaying vegetation in the littoral zone. A modern study would be interesting.

A record of a bloom of *Gloeotrichia echinulata* in an undesignated lake in Anglesea (Smith, 1804) may refer to a completely natural locality. But in the other cases that Griffiths considers, there seem to be hints of external disturbance, although in one case the disturbance was by birds rather than by man. A more reliable source of information is provided by paleolimnological studies, notably those of Korde (1960). Some of her cases from the Central Russian Plain indicated marked variations in the proportions of myxophycean remains, but many of these are benthic. The conditions for the production of striking populations of such benthic species and for the production of striking water blooms are certainly quite different. On the whole, the hypereutrophic condition must have been quite rare in the humid temperate regions before the time of significant human disturbance.

There is one type of natural locality in which a condition may exist that is as hypereutrophic as in any polluted basin, namely, shallow, closed, mineralized lakes, such as those of central Africa, where the lesser flamingo *Phoeniconaias* strains out the *Arthrospira* and *Spirulina* filaments from the

---

*Giraldus Cambrensis (?1146–1220) was a Welsh priest of considerable learning who had a vivid literary style. In 1188 he accompanied Archbishop Baldwin, as a local expert, on a tour of Wales undertaken to raise enthusiasm for the Third Crusade. This resulted in his *Itinerarium Kambriae,* from which the material discussed above is taken (Lib. 1, Cap. II, page 21 of J. F. Dimock's edition, London, 1868). He also wrote a *Topographia Hibernica* and a number of theological and polemical works, some of which relate to his disputed election to the Bishopric of Saint David's. Though he lacked objectivity where his country or his own interests were concerned, his writing provides a very entertaining account of how the world looked to a medieval man.

soupy alkaline waters (Beadle, 1932; Jenkin, 1936, 1957). In general, closed and somewhat mineralized lakes are likely to acquire phosphorus along with other elements. In some such lakes—for example, Pyramid Lake, in Nevada, which contains 0.07 to 0.11 mg of P. PO_4 per liter (Hutchinson, 1937b)—considerable soluble phosphorus may occur without any evidence of high productivity. Iron may well be limiting in these lakes; in the shallow, closed lakes, some suspension of sediment is likely to provide the phytoplankton with a source of iron and so permit full utilization of phosphorus and, in some cases, perhaps, fixation of nitrogen also.

Finally, I would like to say something about a locality on which a great deal of work has been done, at Yale and elsewhere, in the past few years. I am referring to the charming little crater lake, on the road from Rome to Siena, known as the Lago di Monterosi.* This lake was formed by a volcanic explosion, probably about 26,000 years ago. It appears initially to have acquired sediment fairly rapidly and was probably moderately productive. But for the greater part of its history, from about 22,000 B.C. to 200 B.C., the sedimentation rate was exceedingly low and the organic productivity apparently very small. All through this time the water was clearly dilute but probably on the alkaline side of neutrality. Chrysophyceae and soft-water diatoms such as *Eunotia* were characteristic. This phase of dilute—and, I think, legitimately oligotrophic—water persisted from the height of the last glacial peak in the Alps (Würm III, about 20,000 to 18,000 B.C.) until Roman times. It is, I think, reasonable to suppose that a condition of trophic equilibrium had been established at a quite low level of productivity and that this persisted in spite of dramatic changes in vegetation from *Artemisia* steppe through grassland to *Abies* and finally mixed oak forest. Then suddenly something happened: erosion and sedimentation increased, and the whole biological association altered.

Archaeologically, the area around the lake is barren of early remains, but in or about 171 B.C. the Via Cassia, which the modern highway still largely follows, was built. The result seems to have been a sudden eutrophication of the lake. The present influents, which are springs just above the margin of the lake, supply a moderately hard water, which has drained through tuff or volcanic ash. The water entering the pre-Cassian lake must, from the nature of the microflora, have been quite different. In this case, the eutrophication process seems to have involved a change in the way in which the water entered the lake. In pre-Cassian times, when the basin was undisturbed, presumably the source of water was largely precipitation on the lake surface, and such superficial runoff as was not taken up and transpired by the plants of the drainage basin. We have independent evidence that the lake was

*The final report on this locality will be published shortly under the title *Ianula*, with contributions by a number of investigators.

shallower at this time than it is today. When the road was put through, a good deal of clearing undoubtedly occurred, and water began to move through the ash, picking up calcium as bicarbonate and probably dissolving some apatite. The changes initiated by the building of the Via Cassia at first seem to have been accompanied by very little settlement. In postimperial times the appearance of much pollen of Urticaceae, coincident with a further temporary increase in sedimentation rate, both inorganic and organic, suggests the spread of cultivation in the basin. As there was no increase in large graminean pollen until medieval times, we suspect the presence of vegetable gardens and pig farming by that time; documents exist indicating that this was to be expected in the region. Later the lake became rather less eutrophic. The full chemical history of the basin is exceedingly complicated and full of small geochemical surprises. In broad outline, however, we seem to have a case in which cultural eutrophication was initiated not by the artificial liberation of specific nutrient elements into the water, but by a rather subtle change in hydrographic regime. As such, the history of the Lago di Monterosi may have a specific lesson for us today, providing a picturesque beginning for our program.

REFERENCES

Beadle, L. C. 1932. Scientific results of the Cambridge expedition to the East African lakes 1930-1. 4. The waters of some East African lakes in relation to their fauna and flora. J. Linn. Soc. (Zool.) 38:157–211.

Birge, E. A., and C. Juday. 1911. The inland lakes of Wisconsin. The dissolved gases and their biological significance. Bull. 22, Wis. Geol. Nat. Hist. Surv. 259 pp.

Candolle, L. de. 1826. Notice sur la matière qui a coloré en rouge le Lac de Morat. Première partie. Sur la matière rouge considérée sous le rapporte de l'histoire naturelle. Mem. Soc. Phys. Hist. Nat. Genève. 3, pt. 2:29–37.

Coffin, C. C., F. R. Hayes, L. H. Jodrey, and S. C. Whiteway. 1949. Exchange of materials in a lake as studied by the addition of radioactive phosphorus. Can. J. Res. D27:207–222.

Cowgill, U. M. 1968. A comparative study in eutrophication. Developments in Applied Spectroscopy 6:229–321.

Einsele, W. 1936. Ueber die Beziehungen des Eisenkreislaufs zum Phosphatkreislauf in eutrophen See. Arch. Hydrobiol. 29:664–686.

Griffiths, B. M. 1939. Early references to waterbloom in British lakes. Proc. Linn. Soc. Lond. 151:12–19.

Hasler, A. D., and W. Einsele. 1948. Fertilization for increasing productivity of natural inland waters. Trans. 13th N. Am. Wildlife Conf. 527–554.

Hayes, F. R., J. A. McCarter, M. L. Cameron, and D. A. Livingstone. 1952. On the kinetics of phosphorus exchange in lakes. J. Ecology 40:202–216.

Holmberg, O. R., and E. Naumann. 1927. Die Trophie-Begriffe in sprachlicher Hinsicht. Bot. Notiser 1927:211–214.

Hutchinson, G. E. 1937a. Limnological studies in Indian Tibet. Int. Revue Ges. Hydrobiol. Hydrogr. 35:134–176.

Hutchinson, G. E. 1937b. A contribution to the limnology of arid regions primarily founded on observations made in the Lahontan Basin. Trans. Conn. Acad. Arts Sci. 33:47–132.

Hutchinson, G. E. 1941. Limnological studies in Connecticut. IV. Mechanisms of intermediary metabolism in stratified lakes. Ecol. Monogr. 11:21–60.

Hutchinson, G. E., and V. T. Bowen, 1947. A direct demonstration of the phosphorus cycle in a small lake. Proc. Nat. Acad. Sci. U.S. 33:148–153.

Hutchinson, G. E., and V. T. Bowen. 1950. A quantitative radiochemical study of the phosphorus cycle in Linsley Pond. Ecology 31:194–203.

Hutchinson, G. E., and A. Wollack. 1940. Studies on Connecticut lake sediments. II. Chemical analyses of a core from Linsley Pond. Am. J. Sci. 238:493–517.

Järnefelt, H. 1956. Zur Limnologie einiger Gewässer Finnlands. XVI Mit besonderer Berucksichtigung des Planktons. Suomal. eläin-ja kasvit seur. van eläin. Julk. 17 (No. 7):201 p.

Jenkin, P. M. 1936. Reports on the Percy Sladen expedition to some Rift Valley lakes in Kenya in 1929. VII. Summary of the ecological results with special reference to the alkaline lakes. Ann. Mag. Nat. Hist. Ser. 10, 18:161–181.

Jenkin, P. M. 1957. The filter feeding and food of Flamingoes (Phoenicopteri). Phil. Trans. Roy. Soc. 240B:401–493.

Juday, C., and E. A. Birge. 1927. Pontoporeia and Mystis in Wisconsin lakes. Ecology 8:445–452.

Juday, C., and E. A. Birge. 1931. A second report on the phosphorus content of Wisconsin lake waters. Trans. Wis. Acad. Sci. Arts Lett. 26:353–382.

Korde, N. 1960. Biostratifikatsiya i tipologiya Russkikh sapropelei. Izdatel'stvo Akad. Nauk. SSSR. Moscow. (Trans., J. E. S. Bradley; ed., K. E. Marshall.) National Lending Library for Science and Technology, Boston Spa, Yorkshire, England. 165 p.

Livingstone, D. A. 1957. On the sigmoid growth phase in the history of Linsley Pond. Am. J. Sci. 255:304–373.

Lundbeck, J. 1934. Über den "primar oligotrophen" Seetypus und den Wollingster See als dessen mitteleuropäischen Vertreter. Arch. Hydrobiol. 27:221–250.

Naumann, E. 1919. Några synpunkter angående planktons ökologi. Med särskild hänsyn till fytoplankton. Svensk. bot. Tidskr. 13:129–158.

Naumann, E. 1927. Ziel und Hauptprobleme der regionale Limnologie. Bot. Notiser 1927:81–103.

Ohle, W. 1954. Sulfat als "Katalysator" des limnischen Stoffkreislaufes. Vom Wasser 21:13–32.

Pearsall, W. H. 1932. Phytoplankton in the English Lakes. II. The composition of the phytoplankton in relation to dissolved substances. J. Ecology 20:241–262.

Rigler, F. H. 1956. A tracer study of the phosphorus cycle in lake water. Ecology 37:550–562.

Rigler, F. H. 1964. The phosphorus fractions and the turnover time of inorganic phosphorus in different types of lakes. Limnol. Oceanogr. 9:511–518.

Smith, J. E. 1804. English Botany XXI. pl 1378. London (cited in Griffiths 1939).

Teiling, E. 1916. En Kaledonisk fytoplanktonformation. Svensk. bot. Tidskr. 10:506–519.

Thienemann, A. 1913. Der Zusammenhang zwischen dem Sauerstoffgehalt des Tiefenwassers und der Zusammensetzung der Tierfauna unserer Seen. Vorläufige Mitteilung. Int. Rev. Hydrobiol 6:243–249.

Thienemann, A. 1925. Die Binnengewässer mitteleuropas. Binnengewässer 1. E. Schweizerbart'sche Verlagsbuchhandlung, Stuttgart. 225 p.

Weber, C. A. 1907. Aufbau und Vegetation der Moore Norddeutschlands. Beiblatt zu den Botanischen Jahrbuchern 90:19–34.

Wesenberg-Lund, C. 1926. Contributions to the biology and morphology of the genus Daphnia. K. danske Vidensk. Selsk. Skr. Naturw. Math. Afd. (ser. 8) 11 no. 2:92–250.

Wetzel, R. G. 1966. Variations in productivity of Goose and hypereutrophic Sylvan lakes, Indiana. Invest. Indiana Lakes Streams 7:147–184.

III.

GEOGRAPHICAL CONCEPTS OF EUTROPHICATION

EUGENE A. THOMAS
Zurich University and Cantonal Laboratory, Switzerland

The Process of Eutrophication
in Central European Lakes

In the course of many years, a good number of Central European lakes have been examined with biological, physical, chemical, and general limnological methods. It is difficult to summarize the research reports about these lakes, because the results have been published in many scientific journals and because many of the lakes have changed in character profoundly in recent decades.

For this reason, I questioned about 20 of my colleagues concerning the content of nitrates and phosphates in the surface water of large Central European lakes during the homothermic state and during the summer stagnation period. I also asked about the content of oxygen in the hypolimnion and about the precipitation of lime during the summer stagnation period. This information, part of which has not yet been published, has been very useful and has enabled me to provide the material given in Table 1. The work on the comprehensive data collected here is not yet complete. However, this paper confirms the experiences gained from the Swiss lakes (Thomas, 1953).

GEOLOGICAL AND GEOGRAPHICAL NOTES

The Central European lakes can be divided into the following groups: Bavarian lakes, lakes in the Salzkammergut, lakes of Kärnten, lakes of the southern Alps, Savoyan lakes, lakes of the Swiss midland, and lakes of the higher Alps.

29

TABLE 1 Characteristics of Central European Lakes

Lake	Surface Area (km^2)	Maximum Depth (m)	Oxygen Content (mg per liter) At Depth of 50 m	Oxygen Content (mg per liter) At Bottom	Nitrate Ions as NO_3^- (mg per liter) Homothermic	Nitrate Ions as NO_3^- (mg per liter) Summer Minimum	Phosphate Ions as PO_4^{3-} (mg per m^3) Homothermic[a]	Phosphate Ions as PO_4^{3-} (mg per m^3) Summer Minimum[a]	CaCO$_3$ (mg per liter) Homothermic	CaCO$_3$ (mg per liter) Summer Minimum
Ammersee	47	82	10.4	9.7	–	4.6	15	9	–	185
Starnbergersee	57	127	9.1	6.8	–	3.0	–	(40)	–	126
Chiemsee	80	73	7.3	5.7	–	1.0	–	(80)	–	134
Tegernsee	9.1	71	4.4	0	1.8		70		136	
Zellersee (at Salzach)	4.6	68	5.9	0.3	2.2	0.9	6	0	50	35
Obertrumersee	4.7	35	0.0[b]	0.0	3.1	0.1	3	0	145	115
Fuschlsee	2.6	67	9.4	9.1	3.2	0.6	([c])	0	152	145
Mondsee	14	68	5.6	2.9	2.1	0.8	2	0	150	135
Attersee	47	170	11.1	11.3[d]	2.2	0.8	0	0	125	119
Traunsee	25	190	8.9	6.0	2.0	1.2	0(12)	0	110	105
Wolfgangsee	13	114	10.3	3.8[d]	1.7	1.3	2	0	124	118
Hallstättersee	8.6	125	9.5	1.1	1.8	1.6	0	0	96.5	95
Weissensee[e]	6.6	99	0.0	0.0	0.4	0.05	0	0	155	137
Millstättersee	13.3	140	7.6	1.9	2.0	0.4	3	3	104	83
Ossiachersee	10.6	45	–	0.3	2.2	0.09	3	0	104	84
Wörthersee[e]	19.4	84	0.0	0.0	0.7	0.0	3	0	128	110
Lago di Garda	370	346	7.8	6.6	0.9	0.7	24	9	106	–
Lago d'Iseo	65	251	([f])	([f])	–	1.9	–	<15	–	81.5
Lago di Como	146	410	8.6	5.8	2.2	1.7	6		65	55
Lago Maggiore	212	370	8.7	6.9	2.9	1.0	20	3	40	30
Lago d'Orta	18	143	2.7	0.9	22.1	–	<15	<15	–	–
Lago di Lugano	49	288	3.5	1.3	1.7	<0.2	90	<7	120	975
Lac d'Annecy	27	81	5.4	0.1	0.8	0.1	(60)	(45)	132	123

Lake										
Lac de Bourget	44	145	8.8	0.6			—	148	(45)	125
Lac Léman	589	309	9.0	4.6	2.6	0.3	(87)	88	(18)	82
Lac de Joux	8.7	33.5	4.2	2.2	1.8	0.8	15	143	4	137.5
Lac de Neuchatel	214	153	5.7	8.3	0.9	0.2	(31)	130	(45)	106
Brienzersee	29	260	[f]	[f]	3.1	1.4	<10	75	<10	60
Thunersee	47	218	8.8	6.1	0.9	0.4	50	105	<10	81
Bielersee	40	74	0	0	1.4	0.4	200	140	<10	107
Urnersee	20	205	8.0	8.4	4.2	0.9	10	95	<5	75
Vierwaldstättersee	114	214	7.8	3.5	1.7	0.4	21	99	1	72
Aegerisee	7.2	83	6.0	4.0	1.6	0.09	<10	112	0	100
Zugersee	38	198	3.5	0	1.4	0.1	150	128	0	107
Sempachersee	14	86	7.0	0.3	1.5	0.2	20	125	0	100
Baldegersee	5.2	67	0.1	0	1.6	0	420	202	0	120
Hallwilersee	10.2	48	0.1	0	1.6	0	200	162	0	115
Walensee	24	145	9.5	6.8	1.8	0.5	<10	111	0	87
Zürich-Obersee	20	48	1.9	0	2.9	0	45	132	0	105
Zürichsee	68	138	4.7	0	3.2	0–0.4	290	125	0	90
Pfäffikersee	3.3	34	0[g]	0	5.0	0	700	190	0	102
Greifensee	8.6	32	0[g]	0	8.0	0–1.5	1,100	215	0	105
Bodensee	474	252	9.5	4.3	3.3	2.9	45	130	<6	105
Untersee	63	46	3.4[g]	0	2.2	1.1	20	120	<14	106
Silsersee	4.1	71	[f]	[f]	0.4	<0.2	<20	37	<20	31
St. Moritzersee	0.8	44	—	0	0.4	0.4	<20	40	<20	37

[a] Figures in parentheses refer to total P as PO_4^{3-} (mg per m³).
[b] At depth of 12 m.
[c] Traces.
[d] At depth of 100 m.
[e] Meromictic.
[f] Sufficient.
[g] At depth of 20 m.

SOURCES: Ambühl (1964), Bachofen (1960), Bianucci (1966), Bosset (1962), Elster and Lehn (1963), Eschmann (1964), Findenegg (1959, 1965, 1966), Gessner (1942/45), Grim (1955), Halbfass (1923), Hubault (1947), Jaag (1952), Kiefer (1955), Laurent and Monod (1964), Liebmann (1959, 1966), Liepolt (1957, 1958), H. Liiönd (unpublished data), Merlo and Mozzi (1963), Monod (1966), Morton (1930/31), Näher (1963), Nümann (1963), Picotti (1964), Schürmann (1964), H. Sollberger (unpublished data), Suchet (1954), Tiso (1962), Thomas (1955, 1965a, 1968), Vollenweider (1963, 1965).

BAVARIAN LAKES

The Bavarian lakes are situated on the Swabian-Bavarian plateau, more than 500 m above sea level. Their catchment area is rich in calcareous minerals, as is evident from the hardness of the water.

LAKES IN THE SALZKAMMERGUT

These lakes are in a mountainous region. Minerals in this tract of country consist of dolomite and calcium carbonate (Trias and Jura). The carbonate hardness of these lakes is similar to that of the Bavarian lakes.

LAKES OF KÄRNTEN

Protected by the mountains, many of these lakes are meromictic. Whereas the Millstätter lake, the Ossiacher lake, and the Wörther lake receive their water partly from a crystalline area, the Weissensee is situated in an area that is rich in dolomite, which can be recognized from the hardness of the water.

LAKES OF THE SOUTHERN ALPS

The catchment areas of these beautiful lakes are large and include parts of different geological areas. The Lago di Garda receives water from two areas that are rich in calcium carbonate: the crystalline areas of the Alps and the lower parts of the Alps. In contrast, the catchment area of the Lago Maggiore is, in general, poor in calcium carbonate.

SAVOYAN LAKES

The Savoyan lakes are situated in a rich calcony area. Two lakes from this area are mentioned. Their carbonate hardness is not extremely high.

LAKES OF THE SWISS MIDLAND

The Swiss midland covers the area from Jura to the northern part of the Alps and from the Bodensee to Lac Léman. There are 19 lakes with an expanse ranging from 581.4 km^2 to 3.0 km^2. If you calculate that this area covers 25,000 km^2, then 7 percent of the area consists of lakes. Lac Léman, the Brienzersee, the Thunersee, the Urnersee, and the Vierwaldstättersee receive most of their water from the crystalline Alps region. They therefore have a lower hardness than the other lakes in this area. Lac de Joux, which is completely in the Jura region, has a lower hardness during total circulation than the lakes in the low-lying plane (for example, the Baldeggersee, the Hallwilersee, the Pfäffikersee, and the Greifensee).

LAKES OF THE HIGHER ALPS

As would be expected, the higher-altitude Alpine lakes are characterized by their low content of salts. Many of these little lakes are in the region of primitive rocks and are poor in calcony. (For examples, see Table 1.) Few are meromictic. Since morphological and hydrological conditions vary considerably in these lakes, they cannot be presented as a homogeneous group. These lakes are interesting with respect to fishery.

THE CONCEPT OF LAKE EUTROPHICATION

There are no generally available definitions of the words "eutrophic" and "oligotrophic." We are not going to discuss the possible definitions of these words, but it is necessary to point out what we mean in this paper by "eutrophication."

Every observer of Central European lakes knows that most of these lakes have changed greatly during the last few years or decades. The most remarkable change consists of an extraordinarily rapid growth of algae in the surface water. Planktonic algae cause turbidity and floating films. Shore algae cause ugly muddying and, very often, floating films and damage to reeds. Decay of these algae causes oxygen depletion in the deep water, in the thermic layer, and in shallow water near the beach (Thomas, 1960, 1964a, 1964b).

This rapid growth of algae gives rise to numerous undesirable effects on treatment of water for drinking purposes and on fisheries, bathing sanitation, and recreation (tourism). Undoubtedly, this increase in algal growth is caused by increased fertilization of the lakes through the influence of man. Overfertilization results from the increase of waste water in the lakes.

By "eutrophication," therefore, we mean the nutrients that are connected with the augmentation of algal production in the lake.

Harmful algae are best controlled by limiting inflow of phosphates, for the following reasons:

1. Phosphate is present only in traces in oligotrophic lakes.
2. Natural tributaries running into these lakes contain very little phosphate as long as they are not subjected to pollution by the influence of man, but contain large quantities of nitrates.
3. Fewer phosphates than nitrogenous compounds are washed out of agricultural land.
4. Rainwater often contains large quantities of nitrogenous compounds that can be utilized by plants.

5. Bacteria and blue-green algae living in lake water are able to bind gaseous nitrogen organically or to produce the growth factors for algae.

6. Addition of phosphate alone to lake water is sufficient to increase the growth of bacteria and blue-green algae.

7. Some blue-green algae produce toxins that are very toxic to warm-blooded animals.

8. Out of putrefied parts of organisms and sludge, nitrogenous compounds return in larger quantities than phosphate compounds in the biochemical cycle.

9. In eutrophic lakes, nitrates are eliminated from time to time by the process of denitrification.

10. It is cheap and easy to eliminate phosphates from sewage water (by $FeCl_3$ in the activated sludge process).

LAKE MORPHOLOGY AS AN INDICATOR OF SUSCEPTIBILITY TO THE EUTROPHICATION PROCESS

Forel (1901), Thienemann (1928), Grote (1934), and other authors have already pointed out that lake morphology greatly influences both the life cycle and the eutrophication of a lake. To assess the eutrophication process, we must compare the present chemical and biological conditions of a lake with its primitive state. For lakes that have not been studied or for which no data are available regarding primitive conditions, paleolimnological research of lake sediments can yield fruitful results as to the primitive conditions. Considering the susceptibility to eutrophication, Central European lakes can be divided morphologically into three groups: small lakes, medium-sized lakes, and large lakes (Thomas, 1949, 1964a, 1964b).

SMALL LAKES

This group includes lakes having a surface area of less than 0.5 km^2 and depth less than 20 m. Such lakes receive, by action of tributaries and the wind, quantities of leaves, pollen, insects, and other organisms that are much greater (per square meter of surface and per cubic meter of water) than those received by larger lakes. The dead organisms decompose in the lake water and set free all the nutrients for growth of planktonic and shore algae without any influence of man. In these little lakes, oxygen depletion—as well as the appearance of much ferric sulfide, manganese sulfide, and hydrogen sulfide—is expected in the deep waters during the summer stagnation period. Lakes having depths between 10 and 20 m are very susceptible to their oxygen balance because of the lower metalimnion, very close to the bottom sediment, that is exposed to oxygen depletion. Moreover, the possibility of oxygen transport into the hypolimnion is not as good during summer stagnation in these lakes as in big lakes because of the lesser influence of storms. In small lakes that are less than 10 m deep, the bad-weather period in summer again frees nutrients from the hypolimnion into the surface water where the nutrients become available to plankton. The nutrient cycle becomes intensive. Small lakes are sensitive to the inflow of sewage and should be protected against it.

MEDIUM-SIZED LAKES

Central European lakes having a depth of 20 to 50 m and a surface area of more than 50 ha can be included in this group of lakes. Years ago, these lakes belonged to the *Coregonenseen,* having rich oxygen content in the deeper water during the whole year. But because they do not have large volumes of water, these lakes are sensitive to sewage inflow. The addition of sewage causes extreme development of algae, which destroy the balance of the food chain in the lake. One of the worst effects of this is that the oxygen is used up in all areas below a depth of 5 to 10 m. Depletion of oxygen destroys the living conditions of the salmonids. Treatment of water from such lakes for drinking purposes is costly and difficult. On the shores of these lakes, masses of algae collect and create trouble and complaints. The plankton algae develop in immense numbers in the highly fertilized lakes of this size, and from time to time these algae float on the surface in masses several centimeters thick, giving a very unpleasant appearance.

LARGE LAKES

These lakes are more than 50 m deep. Centuries ago they were oligotrophic. Agricultural activities of man for thousands of years had no influence on their degree of trophy. The materials transported to the lakes by rain have been beneficial to the fishes. Only a few of these lakes have retained their primitive conditions. The blue or greenish-blue water of these oligotrophic lakes is clear and transparent. During the summer their phosphate content, especially phosphate ions, is low; the nitrate-ion content is high. Even in the surface water, phytoplankton development is scant, and there is little production of littoral plants. Because of the lack of phosphates, plants cannot fully utilize the nitrates that are present. Throughout the year in clean, large lakes, calcium carbonate content and oxygen content are about the same from the surface down to the greatest depths. The bottom fauna is well developed. Salmonids are dominant.

INDICATIONS OF INCREASED PHOSPHATE TROPHY IN SOME CENTRAL EUROPEAN LAKES

Some lakes in Central Europe have been regularly analyzed during the last few years regarding the increase in phosphate. We are going to consider three lakes with different degrees of eutrophication: Bodensee, Zürichsee, and Greifensee.

BODENSEE

Wagner (1967, page 100) has summarized with a graphic illustration the phosphate trophy of the Bodensee. Following are values, expressed as mg per m^3 of PO_4 ion, taken from that illustration:

Year	PO_4-Ion Content	Year	PO_4-Ion Content
1940	0	1960	30
1950	4.5	1964	50
1955	12		

Wagner states that phosphates (PO_4 ions) increase 3 to 6 mg per m^3 per year. There was, however, no increase from 1964 to 1965. Wagner attributed this fact to climatic factors. The year 1965 was rainy; only a little phosphate was washed from the ground, but more phosphate was carried out of the lake through its outlet. I made similar observations on Swiss lakes.

Although the phosphate content of the Bodensee has increased considerably, phosphate is still a minimal factor in the epilimnion.

ZÜRICHSEE

A short article about the phosphate trophy of the Zürichsee and a comprehensive work on the subject have been published (Thomas, 1967, 1968). In order to compare the Zürichsee with the Bodensee, I give here some results illustrating the increase of phosphate content (PO_4 ions) of the Zürichsee (annual average of the content for the total volume). Data are in milligrams per cubic meter.

Year	PO_4-Ion Content	Year	PO_4-Ion Content
1946	69.6	1962	189.2
1950	82.4	1963	202.3
1956	126.1	1964	269.0
1960	152.0	1965	231.6
1961	171.7	1966	234.8

After the war, the phosphate content in Zürichsee apparently increased sharply (Figure 1). Algal masses first appeared on the beaches in 1949. A very interesting fact is that phosphate content was highest in a very dry year, 1964, whereas in the very rainy years 1965 and 1966 the phosphate content did not increase (there was no special washout of phosphates from agricultural areas, but a decrease in PO_4^{3-} was observed). Before being sedimented, clay particles washed into the lake from the shore seemed to bind some phosphate.

Nitrates played a role opposite that of phosphates. The nitrate content of the Zürichsee was higher in the summer months of 1965 and 1966, when rains were heavy, than in previous years. Thus, heavy rains caused a decrease in phosphates and an increase in nitrates.

GREIFENSEE

Almost every year, Greifensee has a complete mixing, that is, a period of total circulation during which the phosphate content is uniform from the surface to the bottom. The mean annual phosphate content in relation to the total volume has not been calculated from the analyses that have been made for many years. The figures for phosphate content, therefore, are the values at the time of total circulation. Data are expressed as mg of PO_4 ions per m^3 of water.

Year	PO_4-Ion Content	Year	PO_4-Ion Content
1950	180	1960	900
1951	220	1961	1,000
1953	320	1962	1,000
1954	550	1963	1,100
1955	550	1964	1,300
1956	600	1965	1,300
1957	900	1966	1,000
1958	800	1967	900 (spring)
1959	800		

In this lake also, phosphates did not increase in the years with heavy rains but decreased markedly.

Since 1960 the purification plant at Uster has eliminated about 90 percent of the phosphates from the sewage water. Only a small proportion of the inhabitants in the catchment area of this lake live in Uster. There was a steady increase in phosphates between 1950 and 1960, but not between 1960 and 1967.

Because of the elimination of phosphates by the iron-return sludge process, today the outflow of the Uster purification plant contains less phosphates than the water of the Greifensee during the total circulation.

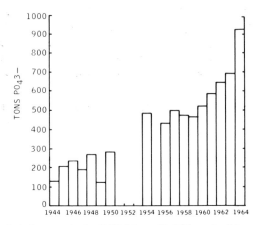

FIGURE 1   Phosphate-ion content of Zürichsee (0–136-m depth); yearly averages, 1944–1964.

PROBLEMS OF EUTROPHICATION THROUGH ADDITION OF PHOSPHATES

THE BASIC VALUE

Because of total circulation in holomictic lakes, all the water mechanisms are in balance from top to bottom.

Lakes having no total circulation are also known, as shown by Bachmann (1924, page 21):

Wie aus den chemischen Auseinandersetzungen hervorgeht, bildete der Ritomsee vor dem Jahre 1917 zwei übereinander geschichtete total verschiedene Wassermassen, die nie durch eine Vollzirkulation durcheinander gemischt wurden. Die Grenzschicht lag bei 12-15 m. Tiefe.*

Findenegg (1934) called such lakes "meromictic lakes."

Some other lakes such as the Türlersee (Thomas, 1948) and the Zürichsee (Thomas and Märki, 1949; Kutsche, 1966, page 69), can be classified as "facultative meromictic." The state of maximum circulation should not be called "total circulation" but "main circulation" (Thomas, 1949, page 479).

In all these types of lakes at the time of spring turnover, the maximum quantities of phosphate ions are available to the phytoplankton of the epilimnion. We characterize this spring maximal value as the basic value. At the time of increasing stratification and increasing water temperature, the available phosphate ions are quickly consumed.

*Translation: As shown by chemical analysis, the Ritomsee consisted of two totally different layers of water, which were never mixed. The depth of the upper layer was 12–15 m.

On analysis of the samples, the values obtained are too high because in the method used for analysis some of the bound phosphates are freed. If phosphate addition from the catchment area of the lake were stopped, the growth of algae would stop as soon as the algae consumed the basic phosphate content.

CONTINUOUS ADDITION OF PHOSPHATE

In the Central European lakes the basic value of phosphate ions decreases to traces or zero in the period of March to June. For the Zürichsee we have proved the significance of the *daily addition of phosphate ions* by sewage. If we assume the addition of 2.5 g of phosphate ions per capita per day (a modest figure), then 375 kg of free phosphates are added per day in the lower part of the lake (Zürich-Untersee). This does not take into consideration the population around the Walensee and the Zürich-Obersee. Phosphates from the industrial waste and the household waste detergents are not included; these will double the value to 750 kg per day, or 22.5 metric tons per month.

As long as the Zürichsee is in total circulation or is being mixed by spring storms (i.e., up to the month of April), all the 22.5 tons of phosphates does not reach the epilimnion. Some of the phosphates will be present in the deeper parts of the lake at that time of year. During summer stagnation, starting in May, vegetation of the epilimnion will consume the total quantity of phosphate of the sewage water. As shown in Figure 2 (dots and black area), this monthly addition is comparatively larger than the basic phosphate content. If you calculate the supply of phosphate to the vegetation zone (Figure 2), you will see that about two thirds of the phosphate used in the epilimnion in 1 year comes directly from sewage water (disregarding summer turbulence, zooplankton, empneuston, and agriculture). This finding is surprising (Thomas, 1968).

It is also possible for a lake to have a low basic value of phosphate ions, but because of high daily additions of phosphate, it may, nevertheless, show a trend toward eutrophication. We often find this phenomenon in large lakes to which a lot of phosphates have recently been added.

PHOSPHATE IONS AS STIMULANT FOR GROWTH OF BACTERIA

After Bosset (1965) had shown that the bacterial content of drinking water increases greatly because of addition of phosphates, we investigated the effect of phosphate addition on bacteria in the waters of the Zürich-Obersee and the Zürich-Untersee. Water samples were collected at the surface and at depths of 20 m and 50 m. Analytical sodium phosphate was added to the unsterilized lake water, which was kept in darkness, along with control water, at 20°C for 20 days. After 20 days the samples were examined for bacterial densities.

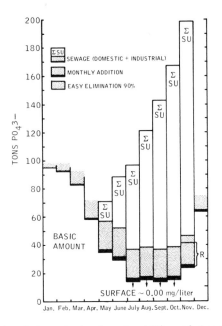

FIGURE 2   Basic phosphate content and yearly additions of phosphate through sewage in the Zürichsee, calculated for the 0–10-m depth of water (1964). The quantity of phosphate additions to the lake through the sewage during the vegetation period is almost double the value of the maximum basic phosphate content. R = phosphate that remains (10 percent) in spite of phosphate elimination at the sewage-treatment plants (equal to the sum of all black areas in the figure). ΣSU = successive additions from May to November (period of vegetation).

Plate-count agar was used for a total count of bacteria. The results of the experiments (Table 2) show that in the Zürichsee increase in bacterial density is directly proportional to the phosphate content of the water.

PHOSPHATE AS A STIMULANT FOR BACTERIA FOR PRODUCTION
OF GROWTH FACTORS FOR ALGAE

It is well known that most freshwater algae use growth factors for thriving. We have shown that for *Cladophora glomerata* there are numerous water bacteria that are able to produce such growth factors (Thomas, 1963). With our bacterial pure culture S_2 we sought to determine whether bacteria that produce growth factors are also stimulated by the addition of phosphate ions. Preliminary experiments show that the bacteria are so stimulated. Further tests are in progress to determine whether there is a simultaneous increase in production of growth factors and in growth of bacteria, as presumed.

Judging from our studies, increased addition of phosphates in the lake

TABLE 2   Increase in Bacterial Content of the Zürichsee 20 Days after Addition of 2 mg per Liter of PO_4^{3-} (average of two samples)

Sampling Location	PO_4^{3-} Content (mg per liter)	Bacterial Colonies (total per milliliter)
Zürich-Obersee[a]		
Surface		
Without addition	<0.02	7,315
With addition	1.7	25,025
20-m depth		
Without addition	<0.02	1,365
With addition	1.65	5,313
Zürich-Untersee[b]		
Surface		
Without addition	<0.02	3,653
With addition	1.65	32,300
50-m depth		
Without addition	0.25	1,773
With addition	2.1	5,225

[a]Middle of the lake at Bollingen.
[b]Middle of the lake at Thalwil.

would bring about the following eutrophication mechanism:

Increase in bacterial content
Increase in oxygen demand
Increase in production of growth factors for the algae
Increase in growth of algae

The phosphate hypertrophy of lakes allows the increased production of growth factors, and this production promotes the growth of algae. To avoid algal damage in lakes, we need to reduce the supply of phosphates.

NITRATE AND PHOSPHATE CONTENTS OF LARGE CENTRAL EUROPEAN LAKES

In my research on the production of lakes, an interesting question arises: If only the supply of phosphorous and nitrogenous compounds in lake water increases, is the supply of phosphorous and nitrogenous compounds and other essential substances already in the water enough to produce an equal increase of plankton? To determine this, we placed water from nearly 40 lakes in flasks and added sufficient phosphate and nitrate to increase the phosphate to 2 mg per liter of PO_4^{3-} and the nitrate to 20 mg per liter of NO_3^- (compare Thomas, 1953). Results of the experiment with summer and winter water show that addition of nitrate and phosphate alone is enough to increase production of phytoplankton in Swiss lakes. It is certain, moreover, that oligotrophic lakes on which man has had little or no influence all have phosphate as the limiting factor; free nitrate ions were present in these lakes throughout the year. Elimination of sewage nitrate in these cases, therefore, would be useless.

Observations on numerous Swiss lakes led us to draw up the scheme shown in Figure 3. The scheme is concerned with the process of lake eutrophication and the behavior of nitrates and phosphates in the epilimnion.

Through the cooperation of my colleagues in central Europe, I was able to prepare the summary in Table 1. Because of different methods of analysis there may be some differences in these values; therefore, the values are not to be compared absolutely. The table is useful for finding major trends.

The three Bavarian lakes, the Ammersee, the Starnbergersee, and the Chiemsee, may be considered as oligotrophic because of their oxygen content. Their high nitrate content indicates their primitive stage.

Conduit pipes have just been constructed around the Tegernsee and the Zellersee. Their effectiveness will be of great interest. It is possible that the presence of free nitrates in the epilimnion of the Zellersee during the summer stagnation period can be considered as the first sign of oligotrophication.

The Fuschlsee, near Salzburg, is still oligotrophic, but there is already a slight decrease of nitrates in this lake during the summer. That is possible

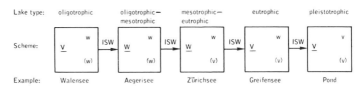

FIGURE 3   Scheme of eutrophication expressed through maximal values of surface water in summer.

V = more than 0.2 mg/liter of NO_3^-

W = less than 0.2 mg/liter of NO_3^-

v = more than 20 mg/m^3 of PO_4^{3-}

w = less than 20 mg/m^3 of PO_4^{3-}

(v) = more than 40 mg/m^3 of PO_4^{3-} after ignition

(w) = less than 40 mg/m^3 of PO_4^{3-} after ignition

ISW = incoming sewage water

only through the daily inflow of phosphates. We hope the inflow of phosphates through the sewage will be stopped so the rest of the nitrates will not be consumed.

The Austrian lakes Mondsee, Attersee, Traunsee, Wolfgangsee, Hallstätter-see, and Millstättersee still show good oxygen conditions. The inflow of phosphates in all these lakes is so small that the reserve of nitrate is not consumed in summer. The Ossiachersee, which is only 45 m deep, is already very susceptible to a small addition of phosphates. But Weissensee and Wörthersee, meromictic lakes, have to be specially classified.

In 1948 Professor Dr. Edgardo Baldi, who died in 1949, suggested that the large lakes in the southern part of the Alps are only slightly subject to evolution because of morphometric factors, and that these (Lago Maggiore is included; Lago di Lugano was not named) are very resistant to eutrophication (Baldi, 1949). Lago di Garda, Lago d'Iseo, and Lago di Como are still oligotrophic; during summer they contain much nitrate in the epilimnion. That is their primitive character. But Lago di Lugano shows serious signs of eutrophication: decrease of oxygen in the hypolimnion and mass production of algae. It is apparent that this eutrophication was caused only by the inflow of phosphates through sewage.

Even at Lago Maggiore there is a remarkable daily inflow of phosphates by sewage. This causes a big decrease of nitrate in the epilimnion during the summer while there is an abundant growth of plankton algae. The lake has to be inspected regularly.

Because of inflow of industrial waste water, Lago d'Orta shows a large supply of nitrate in winter. It will be interesting to follow carefully the changing of nitrate and phosphate during the year.

In Lac d'Annecy and Lac de Bourget, eutrophication can be recognized by the decrease in nitrate content during summer, which happens in many Central European lakes. In Lac Léman, Lac de Joux, Lac de Neuchâtel, the Brienzersee, the Thunersee, the Urnersee, the Sempachersee, the Walensee, and the Bodensee, the primitive amount of nitrate in the epilimnion is never consumed, or is consumed only in special climatic circumstances. Fortunately the inflow of phosphates here is only moderate. In the Bielersee, on the other hand, as well as in the Greifensee, there is a big inflow of nitrates, so we sometimes find nitrates in the epilimnion in the summer (September 23, 1965, summer minimum 1.5 mg per liter NO_3^-). This presence is attributed to large inflows of sewage and rainwater. Even in the Zürichsee the nitrate minimum did not sink lower than 0.4 mg per liter of NO_3^- in 1965 (August 19).

As examples of high-altitude Alpine lakes, the lakes Silsersee and St. Moritzersee, 1,800 m and 1,771 m above sea level, respectively, are mentioned in Table 1. The water from the oligotrophic Silsersee passes through the Silvaplanersee and the Campfersee before reaching the St. Moritzersee, a lake that is only 1.5 km long. This lake receives sewage from the village of St. Moritz Bad. Eutrophication of this lake is indicated by growth of algae and by oxygen depletion in the deep water. During homothermic conditions the phosphate content is not high; eutrophication apparently is caused by the daily inflow of phosphates through the sewage. In Table 1 we can see that signs of eutrophication must be expected in lakes having a high basic phosphate content during homothermic conditions, but the daily phosphate inflow in a lake is certainly more important for maintaining eutrophication.

The lime content of the Central European lakes varies greatly from lake to lake; in each case too great a supply of phosphates causes eutrophication. The degree of eutrophication is indicated not only by oxygen depletion in the hypolimnion but also by precipitation of lime during the summer stagnation period.

PRECAUTIONS AGAINST EUTROPHICATION OF LAKES IN CENTRAL EUROPE

CIRCUMFERENTIAL PIPELINES

Circumferential pipelines are used to collect waste water from the catchment area of a lake and to carry it to the central purification plant. Purified water then is diverted straight to the outflow of the lake. These pipes are especially suited for small lakes, where installation does not involve extending the pipeline over too long a distance.

Such pipelines were installed some years ago around the Tegernsee (Germany), the Zellersee (Austria; advised by Professor Dr. H. Liepolt), and the Hallwilersee (Switzerland). As early as 1912, two pipes, each about 6 km long, were installed around the lower part of the Zürichsee, one from Kilchberg and the other from Zollikon, to the outflow of the Zürichsee; both of these cover a part of the city of Zürich. These pipes were installed to protect the drinking-water supply from this part of the lake. Undoubtedly, installation of these pipes is a most effective precaution.

High-water level during rainy weather should be considered when such pipes are being installed. Whether pipes can be installed around big lakes depends on many local factors. Reports about the first results are given by Liepolt (1966) and I. Findenegg (personal communication).

ELIMINATING PHOSPHATES FROM SEWAGE

There are many publications on elimination of phosphates from sewage. I have 10 years of experience with phosphate elimination through the activated-sludge process. Ferric chloride (10 mg per liter of Fe^{3+}) is continuously added to the inflow or outflow of the aeration tank. In this way about 90 percent of the phosphates can be eliminated. Volume of sludge does not increase, and the ferric phosphate sludge does not dissolve during decay.

In Switzerland the cost of installations for eliminating phosphates is 1 to 2 percent of the total cost of mechanical and biological purification. The cost of chemicals used is 2 to 3 dollars per 1,000 m^3, or about 50 cents per person per year (Thomas, 1965b, 1966).

Since September 1966, there have been standards in Switzerland for the condition of water running out of purification plants. This cleaned sewage water, measured daily, must not contain more than 2 mg per liter of PO_4^{3-} In the purification plants of Uster (20,000 inhabitants) and Stäfa (10,000 inhabitants), the phosphates are precipitated continuously, and the required effect is achieved. On February 2, 1967, the administrative advisor (Regierungsrat) of the state of Zürich requested all communities in the catchment areas of the Zürichsee, the Greifensee, and the Pfäffikersee to eliminate phosphates from sewage. This request is now being implemented.

Elimination of nitrates or other ions is out of the question.

PROHIBITING ADDITION OF PHOSPHATES TO DRINKING WATER

Phosphates are often added to drinking water, but this greatly increases the content of bacteria in the distribution system and impairs the quality of drinking water (research by Dr. E. Bosset, Lausanne, 1965). For this reason, the Swiss Public Health Department in Bern has forbidden addition of phosphates to drinking water (Kreisschreiben Nr. 16 betreffend die Verwendung von Phosphaten zur Wasseraufbereitung, 1966). This precaution is very important for the protection of water. In this way it is possible to stop the extra increase of phosphates in sewage.

PREVENTING DISPOSAL OF ANIMAL LIQUID MANURE AND SILO WASTE

In some Central European countries it is forbidden to discharge animal liquid manure and silo waste into bodies of water. Such agricultural sewage is better used in the field.

TWO ASPECTS OF LAKE SANITATION

THE IMMEDIATE EFFECT

In many lakes the area between 0 and 5 m in depth may be regarded as a special zone for phytoplankton activity. In the Zürichsee this zone covers an area of about 350×10^6 m^3. Added to this zone are 750×10^6 mg of phosphate per day or 2 μg of PO_4^{3-} per liter per day, or 0.02 mg per liter of PO_4^{3-} in 10 days. These are easily detectable quantities but are continuously consumed by the phytoplankton of the lake. In other words, the daily addition of phosphate, especially through sewage, allows the increased phytoplankton to be continuously maintained during the summer. Because the winter supply of the surface water is used up in May, the further addition of phosphate to the lake causes plankton growth in the remaining half of the summer.

These facts are of great importance for repressing eutrophication. If phosphate from purification plants is decreased 90 percent by eliminating the phosphates, the growth of algae in the surface water must decrease in a similar manner from about the middle of May. This means an improvement for the oxygen content of the deep water. If phosphate reduction for a lake is very well organized, there will be an immediate reduction of algal production the following summer.

THE LONG-TERM EFFECT

With a sufficient decrease in the phosphate addition, the basic value of phosphate in a lake will decrease, not increase. After many years, smaller quantities of phosphates will be offered to the algae before the beginning of the vegetative period, and the trophic content of the lake will nearly approach its original state.

ACKNOWLEDGMENTS

I should like to acknowledge the help I received from many of my colleagues in gathering the data presented in Table 1. In particular, the following limnologists furnished data: Dr. W. Nümann (Langenargen/Bodensee); Dr. H.-J. Elster (Falkau); Dr. F. Kiefer (Konstanz); Dr. H. Liebmann (München); Dr. E. Danecker (Scharfling/Mondsee); J. Kopecky (Salzburg); Dr. R. Liepolt (Wien); Dr. I. Findenegg (Klagenfurt), who furnished most of the data about the Austrian lakes; Dr. G. Bonomi (Pallanza); Dr. R. A. Vollenweider (Pallanza); Dr. P. Laurent (Thonon); Dr. R. Monod (Lausanne); Dr. E. Bosset (Lausanne); Dr. H. Sollberger (Neuchâtel); Dr. H. Zschaler (Bern); Dr. H. Lüönd (Zürich); Dr. O. Jaag (Zürich); Dr. H. Ambühl (Zürich); and K. H. Eschmann (Zug). I remember with gratitude assistance received from two limnologists who have passed away, Dr. V. Tonolli (Pallanza) and Dr. W. Einsele (Scharfling/Mondsee).

The scientific projects mentioned on pages 39 to 41 were carried out with a grant from the Swiss National Foundation.

REFERENCES

Ambühl, H. 1964. Die Nährstoffelimination aus der Sicht des Limnologen. Schweiz. Z. f. Hydrol. 26:569–594.

Bachmann, H. 1924. Temperaturen des ti Romsees. Schweiz. Z. f. Hydrol., 2 Jahrg., H.3, 21–28.

Bachofen, R. 1960. Stoffhaushalt und Sedimentation im Baldegger- und Hallwilersee. Diss. Universität Zürich, 118 p., Juris-Verlag, Zürich.

Baldi, E. 1949. Alcuni caratteri generali dei laghi marginali sudalpini. Verh. I.V.L. 10:50–69.

Bianucci, G. 1966. L'inquinamento del Lago di Varese. Studio chimico-fisico. Acqua Industriale 43 (May–June).

Bosset, E. 1962. Le lac de Joux, étude hydrologique du bassin, 1953–57. Schweiz. Z. f. Hydrol. 24:90–151.

Bosset, E. 1965. Incidences hygiéniques de la vaccination des eaux de boisson au moyen de polyphosphates. Monatsbull. Schweiz. Ver. Gas-u. Wasserfachm. 45:146–148.

Elster, H. J., and H. Lehn. 1963. Bodensee-Projekt der deutschen Forschungsgemeinschaft. Franz Steiner Verlag, Wiesbaden.

Eschmann, K. H. 1964. Das Wasser des Zuger- und Aegerisees und seine Eignung zur Trinkwasserbereitung. Monatsbull. Schweiz. Ver. Gas-u. Wasserfachm. nr. 1/2.

Findenegg, I. 1934. Beiträge zur Kenntnis des Ossiacher Sees. Carinthia 2.

Findenegg, I. 1953. Kärntner Seen naturkundlich betrachtet. Ferd. Kleinmayr, Klagenfurt. 101 p.

Findenegg, I. 1959. Die Gewässer Oesterreichs. Lunz, 1959.

Findenegg, I. 1965. Limnologische Unterschiede zwischen den österreichischen und ostschweizerischen Alpenseen und ihre Auswirkung auf das Phytoplankton. Vierteljsschr. Naturf. Ges Zürich. 110:289–300.

Findenegg, I. 1966. Phytoplankton und Primärproduktion einiger ostschweizerischer Seen und des Bodensees. Schweiz. Z. f. Hydrol. 28:148–172.

Forel, F. A. 1901. Handbuch der Seenkunde; Allgemeine Limnologie. Stuttgart, Verlag J. Engelhorn. 249 p.

Gessner, F. 1942/45 (Tegernsee) Arch. f. Hydrobiol. 687–732.

Grim, J. 1955. Die chemischen und planktologischen Veränderungen des Bodensee-Obersees in den letzten 20 Jahren. Arch. Hydrobiol. Suppl. 22:310–322.

Grote, A. 1934. Der Sauerstoffhaushalt der Seen. Binnengewässer 14:217 p.

Halbfass, W. 1923. Die Seen der Erd. Peterm. Mitt. 1923.

Hubault, E. 1947. Etudes thermiques, chimiques et biologiques des eaux des lacs de l'Est de La France. Ann. de l'Ecole Nat. Eaux et Forêts 10:115–259.

Jaag, O. 1952. Der derzeitige Zustand der schweizerischen Gewässer. Fisch und Fischerei, Verlag G. Schmid Winterthur, 343–354.

Kiefer, F. 1955. Naturkunde des Bodensees. Jan Thorbecke Verlag, Lindau und Konstanz. 169 p.

Kreisschreiben Nr. 16 betreffend die Verwendung von Phosphaten zur Wasseraufbereitung (sog. "Phosphat-Impfung"). Eidgenöss. Gesundheitsamt, 3011 Bern, 26 April 1966.

Kutschke, I. 1966. Die thermischen Verhältnisse im Zürichsee zwischen 1937 und 1963 und ihre Beeinflussung durch meteorologische Faktoren. Vierteljsschr. Natf. Ges. Zürich 111:47–124.

Laurent, P., and R. Monod. 1964. Rapport sur l'état sanitaire du Léman de 1957–60. L'ère nouvelle S. A., Lausanne. 292 p.

Liebmann, H. 1959 (Tegernsee) -Münchn. Beiträge z. Abwasser- Fischerei-u. Fluss-biologie, Bd. 6.

Liebmann, H. 1966. Untersuchungen zur Biologie des Chiemsees. Allg. Fischerei-Zeitg. 91, H. 12.

Liepolt, R. 1957. Die Verunreinigung des Zellersees. Wasser und Abwasser, Bd. 1957, 9–37.

Liepolt, R. 1958. Zur limnologischen Erforschung des Zellersees in Salzburg. Wasser und Abwasser, Bd. 1958, 18–101.

Liepolt, R. 1966. Die limnologischen Verhältnisse des Zellersees, seine Verunreinigung und Sanierung. Foederation Europäischer Gewässerschutz, Symposium Salzburg, Sept. 1966.

Merlo, S. and C. Mozzi. 1963. Richerche limn. sul. Lago di Garda. Arch. Oceanogr. Limn. (Venezia), 13/a:1–124.

Monod, R. 1966. Rapport concernant l'évolution physico-chimique des eaux du Léman, campagne 1965. Commission internationale pour la protection des eaux du lac Léman et du Rhône contre la pollution, Lausanne.

Morton, F. 1930–31. Thermik und Sauerstoffverteilung im Hallstättersee. Arch. f. Hydrobiol. 23:177.

Näher, W. 1963 (Starnbergersee). Arch. f. Hydrobiol. 59/4:401–466.

Nümann, W. 1963. Endbericht an die Arbeitsgemeinschaft Industrieller Forschungs-vereinigungen über den Zustand des Bodensees und seiner Mündungsgebiete. Staatl. Institut f. Seenforschung, Langenargen.

Picotti, M. 1964 (Lago d'Orta). Bolletino di pesca, piscicoltura e idrobiologia 19/1:5–193.

Schürmann, J. 1964. Untersuchungen über organische Stoffe im Wasser des Zürichsees. Vierteljsschr. Natf. Ges. Zürich 109:409–460.

Suchet, M. 1954. Etude physico-chimique des eaux du lac d'Annecy. Ann. Station Centrale d'Hydrobiol. App. 5:159–184.

Thienemann, A. 1928. Der Sauerstoff im eutrophen und oligotrophen See. Binnenge-wässer, Band 4, 175 p.

Tiso, A. 1962. I sali nutritivi nelle acque del Lago di Garda. Arch. Oceanogr. Limnol. 3:361–378.

Thomas, E. A. 1948. Limnologische Untersuchungen am Türlersee. Schweiz. Z. f. Hydrol. 11:90–177.

Thomas, E. A. 1949. Regionallimnologische Studien an 25 Seen der Nordschweiz. Verh. I. V. L. 10:489–495.

Thomas, E. A. 1953. Zur Bekämpfung der See-Eutrophierung: Empirische und experimentelle Untersuchungen zur Kenntnis der Minimumstoffe in 46 Seen der Schweiz und angrenzender Gebiete. Monatsbull. Schweiz. Ver. Gas-u. Wasserfachm. 33:25–32, 71–79.

Thomas, E. A. 1955. Sedimentation in oligotrophen und eutrophen Seen als Ausdruck der Produktivität. Verh. I. V. L. 12:383–393.

Thomas, E. A. 1960. Sauerstoffminima und Stoffkreisläufe im ufernahen Oberflächen-wasser des Zürichsees (Cladophora- und Phragmites-Gürtel). Monatsbull. Schweiz. Ver. Gas und Wasserfachm. 40:140–167.

Thomas, E. A. 1963. Versuche über die Wachstumsförderung von Cladophora- und Rhizoclonium-Kulturen durch Bakterienstoffe. Ber. Schweiz. Bot. Ges. 73:504–518.

Thomas, E. A. 1964a. Massenentwicklung von Lamprocystis roseo-persicina als tertiäre Verschmutzung am Ufer des Zürichsees. Vierteljsschr. Natf. Ges. Zürich 109:267–276.

Thomas, E. A. 1964b. Seetypen und Gewässerschutz. Vierteljsschr. Natf. Ges. Zürich 109:511–517.

Thomas, E. A. 1965a. Der Verlauf der Eutrophierung des Zürichsees. Mitteil. d. Oesterr. Sanitätsverwaltung 66, H. 5, 11 p.

Thomas, E. A. 1965b. Phosphat-Elimination in der Belebtschlammanlage von Männedorf und Phosphat-Fixation in See- und Klärschlamm. Vierteljsschr. Natf. Ges. Zürich 110:419–434.

Thomas, E. A. 1966. Phosphatfällung in der Kläranlage von Uster und Beseitigung des Eisen-Phosphat-Schlammes (1960 und 1966). Vierteljsschr. Natf. Ges. Zürich 111:309–318.

Thomas, E. A. 1967. Die Phosphat-Hypertrophie der Gewässer. Chemisch Weekblad, Koninklijke Nederlandse Chemische Vereniging.

Thomas, E. A. 1968. Die Phosphattrophierung des Zürichsees und anderer Schweizerseen; Symposium Plön 1965. Mitteilungen, I. V. L. 14:231–242.

Thomas, E. A., and E. Märki. 1949. Der heutige Zustand des Zürichsees. Verh. I. V. L. 10:476–488.

Vollenweider, R. A. 1963. Studi sulla situazione attuale del regime chimico e biologico del Lago d'Orta. Mem. Ist. Ital. Idrobiol. 16:21–125.

Vollenweider, R. A. 1965. Materiali ed idee per una idrochimica delle acque insubriche. Mem. Ist. Ital. Idrobiol. 19:213–286.

Wagner, G. 1967. Beiträge zum Sauerstoff-, Stickstoff- und Phosphorhaushalt des Bodensees. Arch. Hydrobiol. 63:86–103.

WILHELM RODHE

Institute of Limnology, Uppsala, Sweden

Crystallization of Eutrophication Concepts
in Northern Europe

Several of the fundamental concepts that aim to describe the biological status of a lake, its classification, and development originated in Northern Europe. There the process of terminological crystallization started 60 years ago. But there followed a period of disintegration and dissolution, resulting in considerable confusion. More recently, a recondensation of concepts, based on nuclei provided by new methods and new results, seems to have created a simpler and more coherent pattern of ideas than before.

The beginning was connected not with live lakes but with remnants of extinct lakes. Studying the evolution of north German peat bogs, Weber (1907) found that the upper layers, formed during the terrestrial stage of the bog, indicated much poorer conditions than the deeper strata, which reflected a rich supply of nutrients in the lake that had preceded the bog. To distinguish between the two kinds of layers, he called them "oligotrophic" and "eutrophic," respectively. He introduced these new terms in brackets, as synonyms for "poor in nutrients" and "rich in nutrients."

Thus two of our fundamental concepts were born at the border of the limnological realm, in a most unpretentious manner.

A couple of years later, Einar Naumann studied plankton algae and fresh sediments of various lakes in Sweden. Guided more by intuition than by analytical data, he postulated a direct relation between phytoplankton and nutrient conditions, and he anticipated phosphorus and nitrogen as dominant factors for the composition and quantity of phytoplankton. In his first paper on these matters, the dissertation thesis (1917), Naumann applied some of Weber's general views about peat bogs to recent lake deposits. In the second paper (1919, where Weber was not quoted), the terms "oligotrophic" and "eutrophic" were adopted and adapted to characterize different associations (formations) of phytoplankton. To these autotrophic associations he added a heterotrophic category for polluted waters and attempted to coordinate them with the saprobic system of Kolkwitz and Marsson (see further below).

Here it is appropriate to remember that limnology in those early days still lacked most of the basic information that is now commonplace. Lakes were classified according to their geographical sites, such as the Baltic and the sub-Alpine types. Naumann realized the importance of the watershed for the organic production of every lake and used phytoplankton as a yardstick to compare productivities of different lakes. He emphasized the physiological aspects and had an ecological approach long before ecology had been accepted as a science of its own.

Whereas Naumann founded his typological scheme upon plankton algae and the conditions in the photosynthetic (trophogenic) layer of free water, August Thienemann of Germany arrived at very similar conclusions at about the same time, starting from bottom animals in the aphotic (tropholytic) zone. There the oxygen conditions may be of overwhelming importance. Thienemann (1918) defined, to begin with, the Baltic and sub-Alpine (later, eutrophic and oligotrophic) lake types mainly according to their hypolimnic oxygen regime and the corresponding dominance of different chironomid larvae. Somewhat later (1921) he accepted Naumann's terminology and amalgamated their convergent definitions. He also added a third type, the dystrophic lakes with humic water, an expansion that Naumann first refused (1921) but then agreed to.

Thus Naumann and Thienemann laid the foundation stones of the *Seetypenlehre,* the ecological classification of lakes, just as they jointly initiated the foundation of the International Association of Theoretical and Applied Limnology in 1922. Both actions had an immense impact on the progress of limnology, but lake typology was ambiguous and confused. This is already apparent in the extensions of the original ideas elaborated by Thienemann (1925, 1928) and Naumann (1932) themselves.

In general, a concept used for definitions must be logically defined in itself. The trophic concepts for the productivity of lakes, however, were not strictly bound to organic production but included conditions of production (e.g., nutrient supply) as well as some of its effects (e.g., hypolimnic oxygen depletion). In cases where these aspects were altogether congruent, the interpretation of oligotrophy and eutrophy did not cause any trouble, but the terms lost much of their sense as soon as the structure and functions of the ecosystem deviated from accepted rules. For instance, according to the rule, a productive lake was eutrophic beyond dispute if it had a small hypolimnion with oxygen deficiency. But it was said to exhibit a secondary or morphometric oligotrophy if its hypolimnion was bigger than the epilimnion and had plenty of oxygen. The nomenclature was further complicated when Naumann (1932) introduced four other trophic types (alkalitrophy, acidotrophy, argillotrophy, and siderotrophy) and suggested no fewer than 24 "pure" and "combined" types of lakes. Other authors contributed eagerly in the same direction, and the terminology reached such a degree of sophistication that the entire system could only disintegrate.

Here I permit myself a short digression from Northern Europe to North America. Contemporarily with Naumann and Thienemann, E. A. Birge and C. Juday collected an unprecedented amount of qualified data from more than 500 Wisconsin lakes. The difference between their lines of action was almost as vast as the ocean that separated the two teams of scientists. On one side, a synthetic attitude and the endeavor to construct, as fast as possible, a framework with general applicability; on the other side, an analytical approach and the tedious task of bringing together single bricks for the future structure. In limnology these two modes of scientific procedure, though opposed, are complementary and should be combined to warrant full success. Generalized ideas may be extremely stimulating, but they do not offer a shortcut to truth if they are based on premature premises. Mere collection of facts, on the other hand, will fail to provoke due response and progress unless the results are abstracted and transformed into general terms.

Birge and Juday were strikingly economical in their use of general terms, but they did introduce, nevertheless, some fundamental ones. With regard to the origin of dissolved organic matter, they made (1927) a distinction between autotrophic lakes, dependent on internal sources, and allotrophic lakes, which receive a major part of their organic solutes from external sources. A similar division between orthotrophic and paratrophic lakes was suggested earlier by Naumann (1921), but he did not refer only to organic matter, though humic extractives apparently were in his mind for the paratrophic type. It is, therefore, preferable to use the terms "autotrophy" and "allotrophy" for the supply of autochthonous and allochthonous organic matter, respectively.

"Supply" is a concept that brings us from structural and static statements into the wider field of functional and dynamic dimensions. Naumann had already stressed the production of phytoplankton, but he could not assess more than the population and biomass present at the moment of investigation; and Thienemann considered the oxygen profile and its development but calculated only instant relationships. Both of them were thus mainly restricted to parameters that do not include time in their dimensions. "Supply," however, involves the time factor as well as quantities and concentrations. It has to be expressed as a rate, with time in the denominator.

The first attempt to obtain numerical data for the rate of total organic supply to lakes was made by Strøm (1931) in Norway. He introduced the daily loss of oxygen per unit area of the hypolimnion as a measure for comparative purposes. This roundabout method to infer the delivery of organic substances from their rate of destruction has been refined and completed by Hutchinson (1957, p. 639 ff.), who found that the hypolimnic areal deficit develops faster in eutrophic lakes than in oligotrophic ones (0.05 to 0.14 and 0.004 to 0.03 mg of O_2 cm^{-2} day^{-1}, respectively). Ohle (1952) evaluated the rate of hypolimnic carbon dioxide accumulation to be used for similar aims also after complete oxygen depletion.

This line of approach is restricted by difficulties that arise in several ways, one of which is interference of autochthonous and allochthonous organic matter. Regardless of all the inherent methodological obstacles, however, the trophic concepts must be kept free from confusion and contamination. In the beginning, Naumann (1919, 1921) vacillated between nutritional standard and phytoplankton production as criteria for the concepts of oligotrophy and eutrophy. Mainly for practical reasons, he finally chose to let the phytoplankton have the deciding vote in cases where the algal production and the nutrient level appeared to disagree (1932, p. 14). This decision was also quite logical, and its line was pursued by Åberg and Rodhe (1942, p. 230) in the following definitions:

1. The trophy of a lake indicates the intensity and kind of its supply of organic matter.
2. The autotrophy of a lake indicates the intensity of its own total production of organic matter.
3. The allotrophy of a lake indicates the intensity of organic supply from its environment.

According to the first statement, trophy should not be equated with the status of nutrient conditions, as advocated by Findenegg (1955). On the contrary, trophy has the dimension of a rate and refers to the amount of organic matter supplied in or to the lake per unit time. The conditions of organic supply within and around the lake are, of course, of paramount importance, but the trophy has to express their combined effect, which is manifested in the supply of organic matter.

From the two other statements, it follows that there are, as a principle, two different sources of organic matter supply and that their rates of action should be expressed by the degree of autotrophy and allotrophy. This can be visualized, as anticipated by Strøm (1928) in Norway and by Ohle (1934) in Germany, in a simple two-dimensional diagram (cf. Åberg and Rodhe, 1942, p. 232).

Under natural conditions, humic substances represent the most common component of allochthonous organic material. From this point of view,

Thienemann (1921; 1925, p. 199) distinguished between "clear-water lakes" and "brown-water lakes." The former he subdivided into oligotrophic and eutrophic lakes according to their autotrophic standard, whereas he introduced the term "dystrophic" for lakes with a high content of humic (colored) matter. In most cases the productivity of brown-water lakes is rather low, but Järnefelt (1925, p. 310) found some productive lakes in Finland to be rich in humic matter, and he called them mixotrophic. Thus we have four trophic lake types as cornerstones in our scheme (Figure 1).

Looking on the abscissa—the extent of autotrophy—a clear statement is needed as to the "production of organic matter." Åberg and Rodhe (1942) accepted the definition given by Thienemann (1931, p. 619), saying that the production of organic matter of a biotype during a certain period of time is identical with the "total quantity of organisms and their excreta" produced during that period. Such a complex quantity is hardly available to direct determinations, but even if it were, it would not be relevant in the present context. This was pointed out by Elster (1958, p. 109) in an illuminating metaphor: to measure the single processes of organic production and then to add them would be the same as to measure the water consumed in a house by determining (a) the inflow through the main pipe (= primary production), (b) the use of water in the various floors (= secondary production, tertiary production, and so on), and (c) the outflow through the discharge pipe (= excreta) and by leakage (= sedimentation, and so on), and then to regard the sum of these figures as the total water consumption. But the water meter of the house has to be, of course, at only one place, namely, the main intake pipe, before the first tap. Correspondingly, the autotrophy of lakes must be measured at the level of primary production, to be plotted along the abscissa

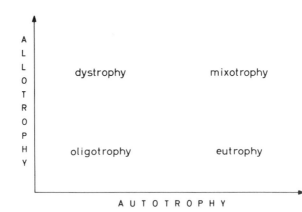

FIGURE 1 The two-dimensional concept of autotrophy and allotrophy and the major trophic types of lakes.

in adequate units of the first assimilation product (organic matter, organic carbon, or fixed energy per unit area and unit time).

An important step toward the direct determination of autotrophy in various waters was taken when Steemann Nielsen (1952) introduced the use of ^{14}C for measuring aquatic primary production. I first applied his method to lakes and found that the daily rate of plankton photosynthesis in different unpolluted lakes may vary within a wide range, from about 20 mg of C/m^2 in Lapland to more than 2,000 mg of C/m^2 in central Sweden (Rodhe, 1958a, Figure 8). A comparison of daily production rates with the corresponding biomass of the producers (fresh weight of phytoplankton) indicated, in general, a direct relationship despite variations within a magnitude of 10 in the individual cases (Rodhe, 1958a, Figure 9). This corroborates the idea that we are entitled to use phytoplankton production as a basis for the trophic classification of lakes instead of phytoplankton quantity, originally suggested by Naumann for that purpose.

That is easy enough to say, but not to do. The primary production at one station may fluctuate considerably from day to day because of phytoplankton patchiness and wind action (Rodhe, 1958b, Figure 3), and the seasonal variations are also very large (Rodhe, 1958b, Figure 2). Thus single data are often insignificant for the entire lake, and a thorough study throughout the year is required to achieve a reliable assessment.

Evidently, not many lakes have been studied in such a way, but from the area considered here a few examples can be given.

In southern Lapland the lakes Ransaren and Kultsjön were investigated continuously from 1954 to 1959 (Rodhe, 1964). Under natural conditions their primary production during the summer months (June, July, August) was found to have a mean level of about 50 mg of C/m^2 per day, with maximum values between 100 and 150 mg of C/m^2 per day. In accordance with more-scattered data from other regions, this seems to be the range one can expect for undisturbed lakes that would be classified from any point of view as oligotrophic. (After the impoundment of Lake Ransaren in 1956, its phytoplankton production increased 2 to 3 times within 2 years.)

Lake Erken in middle Sweden (Rodhe, 1958a, 1958b) and Lake Esrom in Denmark (Jónasson, 1965) are two naturally eutrophic lakes where the seasonal variations have been studied sufficiently to permit estimations of the primary production per year. For both lakes single peaks from 1,000 to about 2,000 mg of C/m^2 per day were noted, but most frequent were daily values of about 500 in Erken and between 500 and 1,000 in Lake Esrom. In Erken the phytoplankton produced about 100 g of C/m^2 in the year 1954, whereas the mean annual gross production of Lake Esrom during 9 years (1955–1963) was 185 g of C/m^2 with 175 and 225 as extreme values.

Results of ^{14}C-testing covering all seasons are also available for several Danish lakes (Steemann Nielsen, 1955; Johnsen et al., 1962; Kristiansen and

Mathiesen, 1964) and for one Swedish lake (Ahlgren and Ahlgren, 1963), all of which are fertilized by domestic and industrial pollution. Their phytoplankton production was high from spring to autumn, with most of the daily rates between 1,500 and 3,000 mg of C/m^2, but some up to 6,000 mg of C/m^2. The annual gross production was between 350 and 650 g of C/m^2.

In addition to phytoplankton, the sessile algae and higher vegetation contribute to the primary production of lakes. We need similar data for their production rates in order to calculate the total supply of autochthonous organic matter in lakes, but we still know very little, if anything, in these respects.

It is risky to generalize, but the results mentioned above seem to justify a tentative connection of phytoplankton production with the autotrophic concepts, as suggested in Figure 2. The "gaps" between the ranges are not real but are due to our lack of precise knowledge. It would, perhaps, be more adequate if we had overlapping ranges to indicate that fixed boundary stones do not exist in the autotrophic continuum. The transition from oligotrophy to eutrophy is better marked by an intermediate range of mesotrophy than by a certain figure or limit.

Transition involves a temporal aspect, and the axis of autotrophy has, in fact, a particular bearing on the development and aging of lakes. In lake evolution the eutrophic period is preceded, generally, by more or less oligotrophic conditions and communities. Consequently, during the life-span of a lake, the position of the lake on the autotrophic axis is moving from left to right. Every permanent change toward a higher degree of autotrophy deserves to be called "eutrophication." Of all the causes of such a trend, natural aging is the most basic and irresistible one.

Whereas the direction of aging processes is the same for all lakes—as it is for all organisms doomed to a limited life—the speed of aging processes is extremely variable and is dependent on the prevailing morphometric, edaphic, and climatic conditions of the individual lake. Moreover, the autotrophic

AUTOTROPHY
(phytoplankton)

	oligotrophic lakes	eutrophic lakes natural	polluted	
mean rates in growing season	30 -100	300 - 1,000	1,500 - 3,000	mg of C/m^2/day
annual rates	7 - 25	75 - 250	350 - 700	g of C/m^2/year

FIGURE 2 Approximate ranges of phytoplankton production in oligotrophic and eutrophic lakes.

increase is not uniform throughout the lifetime of a lake; as a rule it seems to follow a sigmoid (logistic) form of curve (see Lindeman, 1942, and Hasler, 1947). This means that the eutrophication rate of a middle-aged lake is being accelerated by a mechanism similar to a chain reaction and that it is then slowed down by counteracting processes.

The investigations of Mortimer (1941–1942) on the exchange of dissolved substances between mud and water in the English Lake District have provided a key for our understanding of seasonal as well as evolutional changes in lake metabolism. He found that the redox conditions in the mud–water interface largely regulate the flux of dissolved phosphate and other solutes from the mud to the free water, and vice versa. In the oxidized mud surface layer, precipitates of ferric iron and associated colloidal complexes constitute an effective trap for many substances, in particular, phosphate ions, and form a barrier for the exchange of nutrients between mud and water. If the mud surface becomes anaerobic, however, the ferric iron is reduced to ferrous iron and the absorbing colloids are broken down, resulting in the liberation of nutrients from the vast reserves stored in the bottom muds to the supernatant water. This is equivalent to an internal fertilization of the waters to which the phytoplankton responds with an increased rate of production. The more organic matter produced in the photosynthetic layer, the more oxygen is consumed in the mud–water interface and the more nutrients are liberated additionally from the sediments to enhance the primary production to still higher levels. This is the main feedback mechanism that speeds up the eutrophication rate under natural conditions as soon as the mud surface in the deepest part of a lake has lost its oxidized and absorbent barrier.

Mortimer distinguished three phases of lake evolution: First a period (phase I) of slow increase in productivity, with a surface-oxidized mud layer, then a second period (phase II) of accelerated increase, in which the absorbing influence of ferric complexes is destroyed because of reducing conditions, and, finally, possibly a highly deoxygenated sterile period (phase III), in which the iron may be precipitated again as sulfide.

The rates of these changes are determined by geochemical and morphometric factors. Mortimer (1941–1942, p. 192) stated:

Very deep lakes and geochemically oligotrophic lakes may never pass out of phase I, before they become silted up. Shallow and highly eutrophic lakes may pass through all three stages in a very short time after their formation, or they may become arrested in phases I and II, . . . as long as ample nutrient supplies are continuously available from the drainage area.

He concluded that

the greatest rate of change is probably at the end of phase I, and [that] we may expect to observe this at the present day in mesotrophic lakes of moderate depths, or in lakes in which edaphic determinants are being changed by cultural influences.

This view, presented more than 25 years ago, is still valid. But since then, more and more concern has been expressed, even among laymen, because of the trophic changes of lakes due to so-called cultural influences. Strictly speaking, it is not culture but progress of civilization that is the villain in the present tragedy of so many waters. Our modern civilization has done more harm to lakes in a few decades than human culture did during preceding millennia.

In the lakes around Plön in Holstein, Germany, Ohle has carried out long-term investigations of the causes and effects of eutrophication. He found that the acceleration induced by man was greatly exceeding natural aging rates, and he described the total process as *rasante Seenalterung* (1953a) or *rasante Seeneutrophierung* (1955), which means, roughly, "racing aging" or "racing eutrophication" of lakes. Ohle's concept, covering natural and induced effects, has been more readily accepted than the term "auxotrophication," suggested in 1948 by Thunmark in Lund, Sweden, for human effects alone (quoted and translated by Lillieroth, 1950, p. 280).

The lakes in Holstein are situated in an area with intense farming, and Ohle (1953a, 1955, 1965) found that not only domestic sewage but also agriculture must be blamed for loading the lakes with additional nutrients. Increasing use of commercial fertilizers has been one of the causes of rapid eutrophication that in some lakes (e.g., the Grosser Plöner See) started as early as about 1930. These lakes have, by nature, a rather rich supply of nitrogen compounds. Among the various determinants accelerating their aging processes, phosphorus was found to be the prime and initiating factor, whereas a simultaneous increase of sulfate catalyzed the speed of eutrophication (Ohle, 1953b, 1954). An important aspect, repeatedly stressed by Ohle, is the stimulating effect of raised concentrations of phosphate and sulfate on the heterotrophic metabolism in lakes, provided that sufficient organic matter is available. Every increase of bacterial activities shortens the regeneration time of nutrients within the ecosystem and delivers more growth-promoting metabolites. In stratified lakes with anaerobic bottom water, gas bubbles of methane and nitrogen rising from the mud transport nutrient-rich water from the aphotic to the photic layers and thus enhance the internal fertilization still more (Ohle, 1960).

Several Danish lakes, nowadays loaded with domestic sewage, have been studied for periods long enough to bear evidence of drastic changes. The most comprehensive information refers to Lake Fure (Furesø), where Wesenberg-Lund made many of his classic investigations in the early twentieth century.

Berg and associates (1958) could therefore compare their results (1950–1954) with data obtained 50 years before. They found a serious deterioration of environmental conditions, such as a decrease of transparency and of hypolimnic oxygen, but even more radical were the qualitative and quantitative changes of the biological community. It is interesting that the nutritional loading by pollution, which certainly has been a major agent in the recent progress of eutrophication, was comparatively modest. In 1952–1953 the annual contribution from sewage was about 2.6 g of total N and 0.4 g of PO_4-P per m^2 of the lake surface. On the other hand, only small fractions of these increments, namely, 14 percent of the nitrogen and 2 percent of the phosphorus, were found to leave the lake through its outlet. It is the high and continuous accumulation of inflowing nutrients that aggravates pollution effects in standing waters.

The enormous storage of nutrients, mainly bound to the mud, is one of the reasons that an immediate or complete reversion of induced eutrophication processes appears highly improbable in lakes that have been heavily polluted for a long time. In the small Danish lakes Lyngby Sø and Bagsvaerd Sø, however, a substantial decrease of phytoplankton production was observed for the first two years after the diversion of sewage water in 1959 (Johnsen et al., 1962; Mathiesen, 1963). Further experience is strongly needed from cases like these. In general, however, with lakes as with human beings, prophylactic actions should be preferable to therapeutic cures.

My previous discussion started from the autotrophic axis, in order to demonstrate that eutrophication concepts can be based on the autochthonous supply of organic matter. From this platform most other aspects and criteria, abiotic as well as biotic, can be viewed and evaluated. In the present context, we must bring into focus the allotrophic contributions caused by man, in particular, domestic and industrial pollution.

In 1902, Kolkwitz and Marsson, scientific members of the governmental institute of water supply and sewage treatment in Berlin, introduced the terms "saprobionts" (*Saprobien*) and "saprobic," with the subdivisions of polysaprobic, mesosaprobic, and oligosaprobic species, for organisms living in water more or less polluted by putrescible organic matter. They also suggested the use of indicator species and communities for a "biological analysis" of the degree of organic pollution. Later (1908, 1909) they presented detailed lists of organisms regarded as indicative of specific degrees of saprobicity and intended to form the basis of the *Saprobiensystem*. Other authors, in particular Liebmann (1951) and Fjerdingstad (1964), have tried to improve and extend the original system. But, outside of Germany and a few other countries, its practical use has remained rather limited.

The shortcoming of the *Saprobiensystem* depended, to a large extent, on its insufficient foundation on scientific principles and definitions. In recent years, however, the establishment of revised and new saprobic concepts has been vividly discussed in Europe. The most promising line of approach seems to be that of Caspers and Karbe (1966, 1967) in Hamburg. Of their suggestions, which were adopted at a symposium on saprobicity in Prague in 1966, some have immediate relations to the trophic aspects reviewed here.

According to Caspers and Karbe (1967), "saprobicity is the sum total of all those metabolic processes which are the antithesis of primary production." In this extended sense, saprobicity is an expression of all heterotrophic processes within the bioactivity of a body of water. It comprises not only the metabolic processes of secondary producers (i.e., decomposition in the narrower sense) but also the intensities of respiration of primary producers. It is therefore identical with the total respiration of the ecosystem and can be measured, at least as to the order of magnitude, by metabolic-dynamic methods.

Each of the polysaprobic, mesosaprobic, and oligosaprobic zones is now divided into one α- and one β-level, of which the former is more saprobic than the latter. Thus, all together, six levels are distinguished, covering the continuous spectrum of saprobicity from the highest degree of organic pollution (level VI) to oligotrophic conditions (level I).

Caspers and Karbe emphasize that saprobicity is a general phenomenon, of which the reactions induced by man are merely a special case whereby existing heterotrophic processes are intensified. In the energy budget of the ecosystem, these processes, which imply a loss of potential energy, are the opposite of the autotrophic processes, which store potential energy. A new and valuable feature in the extended concept of saprobicity is the possibility of correlating saprobic (i.e., heterotrophic) and autotrophic levels by means of the intensities and balance of total respiration and primary net production in one or various water bodies. Howard T. Odum (1956) made the first attempt in this direction, and Caspers and Karbe (1967) used his autotrophy–heterotrophy coordinates to illustrate their definitions of the saprobicity levels. To me their classification diagram appeals still more, if it is turned 90° and reversed in order to get the autotrophic axis along the abscissa (Figure 3). Remembering our previous diagram (Figure 1), we see that the saprobic levels (shaded for standing waters, unshaded for running waters where the flow acts as an additional trophication factor) in the oligosaprobic zone (I and II) roughly correspond to oligotrophic and moderately eutrophic levels on the autotrophic axis. The mesosaprobic levels (III and IV) belong to the upper eutrophic range, but their high rates of primary production are usually more than counterbalanced by respiratory activities. This imbalance is extreme in the polysabrobic levels (V and VI) where heterotrophy is most pronounced but autotrophy, if present, is low.

Here I would like to put in a reflection of my own. The sequence VI–I may be regarded as a succession, most clearly demonstrated in a river that

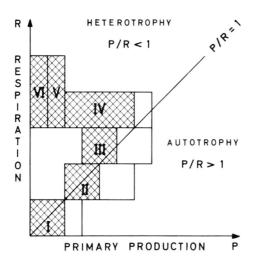

FIGURE 3 Saprobicity levels (heterotrophy) in relation to primary net production (autotrophy). (Redrawn from Caspers and Karbe, 1967.)

recovers downstream from an outlet of organic sewage. But what about the reverse sequence? Let us assume that the levels I to IV mark the course of autotrophic–heterotrophic evolution in lakes that are continuously polluted (levels V and VI would rarely be attained from this direction). Then it sounds reasonable that the development due to man-induced allosaprobization (= decomposition of allochthonous organic matter, cf. Caspers and Karbe, 1966) diverges from the route of naturally aging lakes with prevailing auto-saprobization (= decay of autochthonous organic matter). In the case of naturally aging lakes, the long-term deposition in sediments involves such an immobilization of decomposable material that the evolution route can be expected to proceed within the autotrophic half of the autotrophy–heterotrophy diagram. At the present state of our knowledge, this is no more than a guess, of course.

Caspers and Karbe are anxious to stress that the metabolic-dynamic assessment of saprobicity can be combined with (possibly even be replaced by) qualitative and quantitative determinations of the communities of organisms. The biomass exchange of autotrophs (primary producers) and heterotrophs (consumers and decomposers) can yield useful information concerning the structure of biocenoses and the level of saprobicity. Thus a connection is established between trophic concepts and essential aspects of the *Saprobiensystem,* a connection that has been a challenge ever since Naumann tackled it.

Here the story ends, for the present. I have tried to demonstrate that concepts related to eutrophication are subject to transformation and evolution, as are the lakes themselves. Also, the concepts should be protected, as far as possible, against misuse and aging. To keep them, as well as the lakes, young and active, we need much more information, and much more research.

REFERENCES

Åberg, B., and W. Rodhe. 1942 Über die Milieufaktoren in einigen südschwedischen Seen. Symb. Bot. Ups. 5 (3):1–256.

Ahlgren, G., and I. Ahlgren. 1963. Näringsbalans och primärproduktion i sjön Norrviken. Mimeographed. Institute of Limnology, Uppsala, Sweden.

Berg, K., K. Andersen, T. Christensen, F. Ebert, E. Fjerdingstad, C. Holmquist, K. Korsgaard, G. Lange, J. M. Lyshede, H. Mathiesen, G. Nygaard, S. Olsen, C. V. Otterstrøm, U. Røen, A. Skadhauge, and E. Steemann Nielsen. 1958. Furesøunder-søgelser 1950–1954. Folia Limnologica Scandinavica 10:1–189.

Birge, E. A., and C. Juday. 1927. The organic content of the water of small lakes. Proc. Am. Phil. Soc. 66:357–372.

Caspers, H., and L. Karbe. 1966. Trophie and Saprobität als stoffwechseldynamischer Komplex. Gesichtspunkte für die Definition der Saprobitätsstufen. Arch. Hydrobiol. 61:453–470.

Caspers, H., and L. Karbe. 1967. Vorschläge für eine saprobiologische Typisierung der Gewässer. Int. Rev. Ges. Hydrobiol. 52(2):145–162. (English version edited by World Health Organization: EBL/66.80.)

Elster, H. J. 1958. Das limnologische Seetypensystem, Rückblick und Ausblick. Verh. intern. Ver. Lmnol. 13:101–120.

Findenegg, I. 1955. Trophiezustand und Seetypen. Schweiz. Z. Hydrol. 17:87–97.

Fjerdingstad, E. 1964. Pollution of streams estimated by benthal phytomicro-organisms. I. A system based on communities of organisms and ecological factors. Int. Rev. Ges. Hydrobiol. 49:63–131.

Hasler, A. D. 1947. Eutrophication of lakes by domestic drainage. Ecology 28:383–395.

Hutchinson, G. E. 1957. A treatise on limnology. I. John Wiley & Sons, Inc., New York, 1015 p.

Järnefelt, H. 1925. Zur Limnologie einiger Gewässer Finnlands. Ann. Soc. Zool. Bot. Fennicae Vanamo 2:185–352.

Johnsen, P., H. Mathiesen, and U. Røen. 1962. Sorø-søerne, Lyngby Sø og Bagsvaerd Sø. Dansk Ingeniørforening, Spildevandskomiteen 14:1–135.

Jónasson, P. M. 1965. Factors determining population size of Chironomus anthracinus in Lake Esrom. Mitt. Intern. Verein. Limnol. 13:139–162.

Kolkwitz, R., and M. Marsson. 1902. Grundsätze für die biologische Beurteilung des Wassers nach seiner Flora und Fauna. Mitt. d. kgl. Prüfungsanstalt f. Wasserversorgung u. Abwässerbeseitigung 1:33–72.

Kolkwitz, R., and M. Marsson. 1908. Ökologie der pflanzlichen Saprobien. Ber. Deutsch. Botan. Ges. 26a:505–519.

Kolkwitz, R., and M. Marsson. 1909. Ökologie der tierischen Saprobien. Int. Rev. Ges. Hydrobiol. 2:126–152.

Kristiansen, J., and H. Mathiesen. 1964. Phytoplankton of the Tystrup-Bavelse lakes, primary production and standing crop. Oikos 15:1–43.

Liebmann, H. 1951. Handbuch der Frischwasser- und Abwasserbiologie I. (2. Aufl. 1962). Verlag von R. Oldenburg, München. 539 p.

Lillieroth, S. 1950. Über Folgen kulturbedingter Wasserstandsenkungen für Makrophyten- und Planktongemeinschaften in seichten Seen des südschwedischen Oligotrophie- gebietes. Acta Limnologica 3:1–288.

Lindeman, R. L. 1942. The trophic-dynamic aspect of ecology. Ecology 23:399–418.

Mathiesen, H. 1963. Om planteplanktonets produktion af organisk stof i nogle naeringsrige søer på Sjaelland. Ferskvandsfiskeribladet 61:7–9, 20–25.

Mortimer, C. H. 1941–1942. The exchange of dissolved substances between mud and water in lakes. J. Ecol. 29:280–329, 30:147–201.

Naumann, E. 1917. Undersökningar över fytoplankton och under den pelagiska regionen försiggående gyttje-och dybildningar inom vissa syd- och mellansvenska urbergs- vatten. K. Sv. Vetensk. Akad. Handl. 56 (6):1–165.

Naumann, E. 1919. Några synpunkter angående limnoplanktons ökologi med särskild hänsyn till fytoplankton. Sv. Bot. Tidskr. 13:129–163.

Naumann, E. 1921. Einige Grundlinien der regionalen Limnologie. Lunds Univ. Årsskr. N.F. 2(17):1–22.

Naumann, E. 1932. Grundzüge der regionalen Limnologie. Binnengewässer 11:1–176.

Odum, H. T. 1956. Primary production in flowing waters. Limnol. Oceanogr. 1:102–117.

Ohle, W. 1934. Chemische und physikalische Untersuchungen norddeutscher Seen. Arch. Hydrobiol. 26:386–464, 584–658.

Ohle, W. 1952. Die Kohlendioxyd-Akkumulation als produktionsbiologischer Indikator. Arch. Hydrobiol. 46:153–285.

Ohle, W. 1953a. Der Vorgang rasanter Seenalterung in Holstein. Naturwissenschaften 40:153–162.

Ohle, W. 1953b. Phosphor als Initialfaktor der Gewässereutrophierung. Vom Wasser 20:11–23.

Ohle, W. 1954. Sulfat als "Katalysator" des limnischen Stoffkreislaufes. Vom Wasser 21:13–32.

Ohle, W. 1955. Die Ursachen der rasanten Seeneutrophierung. Verh. Intern. Ver. Limnol. 12:373–382.

Ohle, W. 1960. Fernsehen, Photographie and Schallortung der Sedimentoberfläche in Seen. Arch. Hydrobiol. 57:135–160.

Ohle, W. 1965. Nährstoffanreicherung der Gewässer durch Düngemittel und Meliora- tionen. Münchner Beiträge 12:54–83.

Rodhe, W. 1958a. Primärproduktion und Seetypen. Verh. Intern. Ver. Limnol. 13:121–141.

Rodhe, W. 1958b. The primary production in lakes: some results and restrictions of the 14C method. Rapp. et Proc. Verb. Cons. Intern. Explor. de la Mer 144:122–128.

Rodhe, W. 1964. Effects of impoundment on water chemistry and plankton in Lake Ransaren (Swedish Lappland). Verh. Intern. Ver. Limnol. 15: 437–443.

Steemann Nielsen, E. 1952. The use of radio-active carbon (C^{14}) for measuring organic production in the sea. J. du Cons. 18:117–140.

Steemann Nielsen, E. 1955. The production of organic matter by the phytoplankton in a Danish lake receiving extraordinarily great amounts of nutrient salts. Hydrobiologia 7:68–74.

Strøm, K. M. 1928. Recent advances in limnology. Proc. Linnean Soc. London 140:96–110.

Strøm, K. M. 1931. Feforvatn. A physiographical and biological study of a mountain lake. Arch. Hydrobiol. 22:491–536.

Thienemann, A. 1918. Untersuchungen über die Beziehungen zwischen dem Sauerstoffgehalt des Wassers und der Zusammensetzung der Fauna in norddeutschen Seen. Arch. Hydrobiol. 12:1–65.

Thienemann, A. 1921. Seetypen. Naturwissenschaften 9.

Thienemann, A. 1925. Die Binnengewässer Mitteleuropas. Binnengewässer 1:1–255.

Thienemann, A. 1928. Der Sauerstoff im eutrophen und oligotrophen See. Binnengewässer 4:1–175.

Thienemann, A. 1931. Der Produktionsbegriff in der Biologie. Arch. Hydrobiol. 22:616–622.

Weber, C. A. 1907. Aufbau und Vegetation der Moore Norddeutschlands. Bot. Jahrb. 40. Beibl. 90:19–34.

MILAN STRAŠKRABA AND VĚRA STRAŠKRABOVÁ

Hydrobiological Laboratory, Czechoslovak Academy of Sciences,
Prague, Czechoslovakia

Eastern European Lakes

The word "eutrophication," which has become popular in the West, is not widely used in Eastern Europe. This does not mean there are no problems of eutrophication in this part of the world; on the contrary, many related problems have appeared in recent years in both natural lakes and reservoirs.

The difference between the area reported on in this paper and the areas reported on by Professors Thomas and Rodhe is mainly one of natural eutrophy. The waters of Eastern Europe have a generally high degree of natural eutrophy, but the waters of Northern and Central Europe in their natural state tend to be oligotrophic.

The difference in natural eutrophy might also be responsible for differences in theoretical approach to problems of water quality. Changes in an already eutrophic lake caused by cultural factors will not be as extensive or striking as changes in a clear and oligotrophic lake. Investigators, therefore, will need a much more detailed understanding of the processes to identify and account for this narrower change.

A thorough background in theory also is needed to predict water quality in artificial lakes that are situated in productive, cultivated lowlands. In these areas, oligotrophy can hardly be expected; investigators must be able to distinguish in detail all the factors leading to change from a high degree of eutrophy to a higher degree.

In recent years it has become evident that factors other than the classical increase in nutrient content are at work in the rising of levels of eutrophy. There no longer seems to be a direct relation among nutrient content, amount of phytoplankton, amount of zooplankton, speed of decomposition, and related water quality. This is the main reason, aside from our personal experience and views, why in this summary we have shortened the general introduction, giving some examples of more or less classical processes of eutrophication of lakes in the Eastern European area, and have treated some of the general aspects of eutrophication problems to which studies in the area have contributed most. These general aspects are as follows:

1. Specific problems of artificial lakes
2. Bacteria in bodies of water of different degrees of eutrophy
3. Biotic interrelations in a body of water as a water-quality factor

EXAMPLES OF EUTROPHICATION IN THE AREA

LAKES

Oligotrophic lakes are classic objects of eutrophication studies, and one of these is the Caucasian Lake Sevan. In this lake, however, the eutrophication processes were due to somewhat unusual reasons. During the last 30 years, the lake (1,416 km^2 area, 99 m maximum depth, 41 m mean depth, catchment area 4,891 km^2, 1,916 m above sea level, according to Vladimirova, 1947, and Mechkova, 1961) has been heavily used for irrigation and generation of hydroelectric power, both of which, because of the small catchment area, caused a substantial decrease in the water level.

From January 1938 to the end of 1965, the level dropped 18 m (Mechkova, 1966) and is continuing to drop at a rate of about 1 m per year (Mechkova, 1962, and Slobodchikov, 1955). A comparison of the oxygen conditions during 1926–1930 (Liatti, 1932), 1938–1939 (Vladimirova, 1947), 1947–1948 (Slobodchikov, 1951, 1955), and 1956–1957 (Mechkova, 1962) does not show a continuous decrease in the oxygen conditions. Only during some recent summers was there an exceptionally low value of about 60-percent saturation.

Minimal values of permanganate oxidation almost doubled, and mean values increased 1.5 times. Maximum Secchi disc readings decreased from about 20 m to about 15 m. The amount of phosphate phosphorus, reaching more than 1 mg per liter* in the late twenties, decreased almost 10 times.

Phytoplankton composition and amounts, compared for 1936–1938 (Vladimirova, 1947), 1947–1948 (Stroikina, 1952), and 1956–1958 (Mechkova, 1962), showed an increase in the number of species, appearance of several species of blue-green algae, and the mass occurrence of a few species not previously occurring, for example, *Oocystis submarina, Crucigenia*

---

*According to several published records. Method is not indicated in sufficient detail to allow proof.

quadrata, and *Microcystis pulverea.* The total phytoplankton standing crop
was about doubled, comparing 1947 with 1957–1958. The total standing
crop of zooplankton does not change much; however, there was a decrease in
Cladocera (in relation to other groups) and an increase in the number of eggs
and changed life cycles of the *Daphnia* species (Mechkova, 1947, 1953, and
1962). From all these observations, Mechkova (1961, 1962) concluded that
at present a slow but continuous eutrophication of Lake Sevan is taking
place. This is an example of cultural eutrophication due to a substantial
decrease of the lake volume as a result of man's activity.

Similar phenomena, although they were manifested quite differently,
were observed in the largest "lake" of the area–the Caspian Lake or the
Caspian Sea (area 422,000 km^2, maximum depth 980 m). The water is highly
saline (5 to 13 percent, with several areas between 2 and 3 percent).
(Appropriately, the translation of the Russian name for the lake is
"Lake-Sea.") Regulation of the Volga River, construction of a series of
reservoirs, and extensive use for irrigation brought a decrease in the water
level of about 2 m in the period 1932–1945. Since the mean depth is 184 m,
this is not a very large decrease, and no changes have been observed or are
expected in the deep southern part. However, the northern Caspian Sea is
very shallow, and here the influence is more pronounced. The main change is
the increased salinity, a "sea" phenomenon that will not be discussed here.
[It is also a phenomenon that recurred several times in the interesting
geological history of the Caspian and resulted in the well-known immigration
of a large number of endemic animals from the sea to freshwater
(Mordukhai-Boltovskoi, 1960).]

The decrease in water level brought other changes in the northern Caspian
Sea that are relevant to the scope of the present topic. In areas near the
mouth of the Volga, the decrease in water level caused the rise of an
underwater ridge (Birstein, 1953), which directed the flow of river water with
its increasing amount of easily decomposable organic matter and prevented its
mixing with water of the "Lake-Sea." As a result, an oxygen depletion
appeared at the bottom on areas as large as 5,000 km^2 (about one fifth of the
area of Lake Erie), indicating increased eutrophication. The primary
production of these areas, studied by the dark and light oxygen bottle
method (Vineckaya, 1961), is very high; a seasonal trend was correlated with
the increased amount of water coming from the Volga. An orientative budget
of the organic matter of the Caspian Sea (Datsko, 1961) is shown in Table 1.
The amount of organic matter coming directly from the river is much lower
in proportion to the contribution of nutrients; however, the budget shows
how important the river is. Russian authors tend to discuss the observed
changes in terms of fishery biology, sometimes using reversed criteria for
what is "improvement" or "worsening."

TABLE 1 Budget of Organic Matter in the Caspian Sea[a]

Organic Matter	Amount[b]
INCOMING	
Production of phytoplankton	200,000
Incoming with river water	7,000
Production of phytobenthos	375
Total	207,375
OUTGOING	
Deposition in bottom	8,000
Fishing	135
Mineralization	199,240
Total	207,375

[a]Source: Datsko (1961).
[b]In thousands of tons of dry weight.

Aside from the Caspian Sea, with its more or less pronounced sea character, the largest truly freshwater lake of the area is Lake Ladoga (area 18,766 km^2, maximum depth more than 200 m—similar to Lake Ontario). This is an example of the large Baltic glacial lakes bearing several species of so-called relict fauna identical with species living in North America. These species seem to be sensitive indicators of the changes in water quality. The lake was used during the nineteenth century as the drinking-water supply of St. Petersburg (now Leningrad). In connection with this use, detailed botanical and microbiological investigations were made (Bolokhontsev, 1909; Bolokhontsev and Grimm, 1911). The complicated morphometry of the lake results in a differentiation of water masses. In the deep northern end of the lake, the surface temperature never exceeds 12–13° C. In the shallow southern part, the surface temperature may reach 20° C. The deep part seems to be still untouched by eutrophication, having an oxygen content of 90 to 100 percent saturation. In recent decades, several marginal areas have shown increased results of man's activity (Pravdin, 1956). This is due mainly to increased industrialization of the area, paper manufacturing being dominant, and a corresponding increase in urban drainage. For example, the Volkhov Bay, influenced by paper mills on the inflowing Syaz River, showed an increased pH and a decreased oxygen content. Changes in fish fauna here are remarkably similar to changes in the other large lakes of the region, for example, Lake Onega.

The smaller Baltic lakes also are becoming increasingly eutrophic, as reported, for example, from the Mazurian lake district in Poland. In the group of Mazurian lakes surveyed limnologically by Stangenberg (1936), Wiszniewski (1953), Gieysztor and Odechowska (1958), Olszewski and Paschalski (1959), Patalas (1960 a, b, and c), and Olszewski (1965), the most profound effect of effluents was observed on the shallow Lake Niegocin. According to

Olszewski and Laskowska (1959), the urban effluents of a small town, including effluents of the local industry (small gasworks, dairy, canning of fish), are deposited here after treatment in an insufficient Immoff's tank. First published observations from 1901 (Cohn, 1903) showed a transparency of 3.0 to 6.5 m, more than in the surrounding lakes. In 1938, Brandt (1944) observed during calm summer weather an oxygen depletion up to 6 m from the surface. In 1951, a decrease of transparency (2.3 to 3.7 m) was noted. A dominant feature of the phytoplankton became a bloom of *Melosira*. An indication of the forthcoming eutrophication of the lake is seen in the disappearance of the whitefish (*Coregonus lavaretus maraena*) at the end of the last century.

Changes in the deepest of Polish lakes—Lake Hancza (area about 300 ha, depth 108.5 m)—become evident when observations by Sczepański (1961) are compared with those by Kozminski (1932). The oxygen content in the hypolimnion decreased slightly (about 1 mg per liter), and the pH at the surface increased. These changes might indicate a beginning eutrophication.

Hazardous effects of pollution by urban effluents and industry were observed on several other lakes of the area. Lake Maly Sasek (area 319 ha, maximum depth 3.7 m) is influenced by the town of Sczytno, with 13,000 inhabitants plus slaughterhouses, a dairy, a brewery, and a linen factory (Granops, 1965). Lake Beldany (area 928 ha, maximum depth 46 m) is polluted by a hardboard plant (Lossów, 1965).

When studying physical limnology of mountain lakes, Olszewski (1953) found morphometry, exposure to wind action, and the level of productivity to be the causes of meromixis in addition to the total area recognized previously by Findenegg. Investigations of the effect of exposure on thermal conditions in both mountain and Mazurian lakes (Olszewski, 1959a) led to distinguishing three types of exposure. Patalas (1960c, 1961) expressed the effect of wind on the depth of thermocline determined for 69 lakes as $E = 4.4 \sqrt{D}$ (where E is the depth of epilimnion in m, D = mean effective axis of the lake in km), finding a difference of only ± 20 percent among the lakes investigated. Czeczuga (1959, 1962) and Czeczuga and Baszyński (1963) studied reasons for the metalimnetic oxygen minima (and H_2S production) but failed to find correlations with both increased biomass of bacteria and concentration of planktonic Entomostraca assumed by previous authors. Paschalski (1964) presented an elaborate system of physical typology of lakes based mainly on papers on Mazurian lakes.

After stating his ideas on the decreased fishery value of lakes that have an oxygen deficit at the bottom, Olszewski (1959b, 1961) organized an experiment in which the hypolimnetic water was siphoned into the outflow (without any requirement for energy). During the first year, the oxygen content increased, but so did the temperature of the hypolimnion. Correlated phytoplankton maxima were observed; zooplankton numbers increased and the zone rich in animals extended much deeper than it previously had. Effects that the experiment had in later years have been published very briefly by Olszewski (1967).

Turning to the more southern parts of the area, we may select Lake Balaton (area 600 km^2, volume 2 km^3, maximum depth 11 m) as an example of the very shallow, highly alkaline lakes of this territory (Sebestyén, 1962). Entz (1949/50) showed there is no thermal stratification in the lake; the warming of the whole water column proceeds very rapidly in spring with a rise from 2.7 to 20° C in 1 month. The phosphorus content of the open lake (Entz, 1959) ranges from 80 μg per liter to several hundred μg per liter, and the nitrate nitrogen content ranges from 100 to 500 μg per liter. The algal production of the lake is not very high because of increased turbidity caused by precipitated carbonates (Entz and Fillinger, 1961). Felföldy (1959) experimented with *Chlorella* cultures on natural and enriched Balaton water. The addition of phosphorus and nitrates did not provoke a marked increase in the photosynthetic capacity of the cultures. On the other hand, the addition of CO_2 had a marked effect. In later experiments with a culture of *Coelastrum microphorum* isolated from Balaton (Felföldy, 1960), the effect of CO_2 was not proved.

In spite of the concentrated recreation around the lake, the water, according to Papp et al. (1960), is still usable for water supply. At present, the polluting effect of municipal sewage discharged into the lake does not extend to distances greater than 200 m from the inlets. The authors explained the rapid self-purification processes as an effect of macrophytes densely overgrowing the inlet areas. No direct data were presented to support this explanation.

We have the pleasure to report here on large lakes of the area for which existing data show that eutrophication is not yet evident. Among these lakes is the zoogeographically well-known Lake Ochrida (area 350 km^2, maximum depth 286 m, mean depth 145 m, 693 m above sea level), on the Balkan Peninsula, which bears many endemic animal genera and species. The life of the lake was excellently described by Stankovič (1960). No indication of increasing eutrophy of this highly oligotrophic lake of subtrophic type is to be found in the published data on the oxygen conditions and phytoplankton (Kozarov, 1958) or on the zooplankton. A comparison of the zooplankton populations of 1926 and 1954 by Serafimova-Hadžišče (1957) indicates only very insignificant changes.

In closing this part, we should like to stress that there has been no broad review, similar to the one prepared by the Food and Agriculture Organization of the United Nations, of all the lakes in the territory that have recently become eutrophic. The total number of such lakes is not known, and we are not able to present any figures. However, we do not think this is the main task of the present symposium. Having mentioned recent changes in some of the widely known large lakes, we should like to turn to some particular questions of eutrophication processes, contributed, we think, mostly by limnologists of Eastern Europe.

RESERVOIRS

Man-made lakes have been studied intensively in Eastern Europe. Although the main concern has been water quality, the economic importance of fishery has also been considered. The limnology of reservoirs, differing in several respects from that of lakes, is a relatively new field of study that has not yet been summarized.

The number of publications devoted to reservoirs is enormous. A world bibliography of papers on reservoirs published up to 1956 (Zhadin and Ivanova, 1959) lists 600 Soviet papers. With the number increasing at a rate of about 75 a year, there are now about 2,000 Soviet papers on the subject, and about half of them are devoted to fish. Intensive reservoir construction and studies on water quality in reservoirs are in progress in Poland, Czechoslovakia, and East Germany. Studies of the more southern areas of Eastern Europe have appeared. Hence, it is impossible to present here a picture that is anywhere near complete. To present a picture that is at least somewhat adequate, we have selected several well-studied reservoirs representing different main types to illustrate the results of studies pertinent to eutrophication.

Rybinskoe Reservoir   For some years this Russian reservoir, which was filled during 1941, was the largest reservoir in the world (capacity, 25 km^3). With an area of 4,550 km^2, it is extremely shallow, the mean depth reaching just 5.5 m. At the inflow to the reservoir the Volga River has a mean throughflow of 1,142 m^3/sec, which results in a mean theoretical renewal time of about 7 months (Rutkovskii and Kurdina, 1959). For these reasons, this is an example of a large, lake-like but shallow reservoir with a eutrophic inflow.

The study of the reservoir was initiated long before its existence, as early as 1934 (Mosevich and Mosevich, 1954). A list of literature devoted to it includes 613 papers (Smetanich, 1963).

The quality of water in the reservoir is due mainly to the way inflowing water passes through the reservoir. The throughflow, as studied by Butorin and Litvinov (1963), is given in Figure 1. Arrows indicate the direction of flow; numbers give the speed expressed in cm/sec. The main inflow of the Volga River is seen in the bottom of the figure. The highest flow, 30 to 40 cm/sec, decreases in the widened part to 5 to 7 cm and increases again in the vicinity of the dam (in the middle right) up to 10 to 15 cm. Two other areas of the increased currents, following the old beds of the smaller inflows, are to

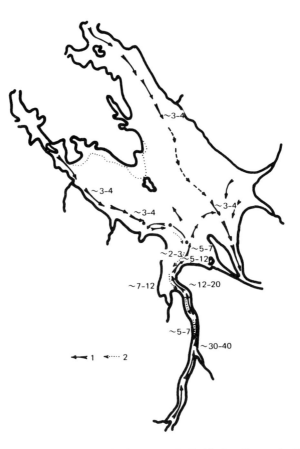

FIGURE 1 Direction and speed of currents in Rybinskoe Reservoir, USSR. 1, main currents; 2, back-currents. Speed is expressed in centimeters per second. (Butorin and Litvinov, 1963.)

be distinguished, with a flow velocity of 3 to 4 cm/sec. In the remaining central part of the reservoir, wind-directed irregular flows have velocities usually below 1 cm/sec.

Short-term (daily) fluctuations of the water level are not large (several centimeters), because of the enormous capacity. However, an important feature of the reservoir limnology was observed in connection with the intermittent operation of the power-generating station. Back-currents, indicated in Figure 1 by dotted arrows, attained speeds as high as 12 cm/sec. These back-currents will contribute greatly to the mixing of the inflowing water with the "old" reservoir water.

Temperature stratification is slight except in periods of calm weather. In summer the temperature reaches a maximum of 26° C and is usually above 22° C. At the same time, the temperature of the bottom strata is about 20° C.

Edelshtein (1963) studied the relation of the formation of the thermocline to meteorological phenomena (wind speed and direction) and hydrological phenomena (air and water temperature changes, conductivity, and speed and direction of currents). The absence of temperature stratification seems to be due also to the intake of water on turbines. Zevin et al. (1961) showed that during maximum intake (more than 1,000 m^3/sec) there is an identical speed of currents in all layers of the reservoir near the dam, which indicates that all layers are sucked into the outflow.

The mixing phenomena of the inflow and reservoir water are evident from the differences in the total mineral content. During periods of drought, the total mineral content of the reservoir water is about 60 percent lower than that of the inflowing water. Phosphate phosphorus is very high, 10 to 80 μg per liter, with a considerable drop during summer season (Mosevich and Mosevich, 1954). A total elimination of phosphate phosphorus was not observed even during heaviest blooms of the blue-green algae, which are typical of the large Soviet reservoirs. Guseva (1952) listed 30 large Soviet reservoirs that had water-blooms (out of a total of 50) and stated that the blooms interfered with water treatment in some of them.

According to Guseva (1955) and Priimachenko (1960), 70 to 90 percent of the Rybinskoe Reservoir phytoplankton during summer are *Aphanizomenon flos aquae* and *Microcystis aeruginosa* (expressed in numbers of cells). The remainder and the dominant phytoplankton in other periods are Diatomaceae.

Long-term water-level fluctuations have a marked effect on the development of phytoplankton in the shallow littoral areas. During years with high water levels, the quantity of phytoplankton increases because of the amount of biogenic elements diluted from soil and vegetation while the level is rising (Guseva, 1958). The decomposition of macrophytic plants was studied from this point of view (Krashennikova, 1958; Korelyakova, 1958, 1959). The influence of the elements leached from soil was recognized as an important factor influencing production during the first years of the reservoir's existence. Usually, for the first 3 to 7 years, the bacterial numbers (Kuznetsov, 1958), primary production, and zooplankton amount are much higher than later. The same is true of fish crops.

Primary production of phytoplankton is very high, reaching 50 g of C per m^2 (Sorokin, 1958). A comparison of the dark and light bottle oxygen method, which is widely used in the area (Vinberg, 1960), and the ^{14}C method showed a general similarity. The allochthonous organic matter and phytoplankton brought by the river have a marked influence on phytoplankton of the Rybinskoe Reservoir (Sorokin, 1961). Grazing by zooplankton was found to be an important factor in phytoplankton biology (Sushchenya, 1961).

Slapy Reservoir As an example of eutrophication in a riverine "canyon" reservoir constructed mainly for generating power, we include here some results of the study of Slapy Reservoir in Czechoslovakia as found by the laboratory we came from. Morphological features of the reservoir and their reflection in limnological conditions are illustrated in Figure 2.

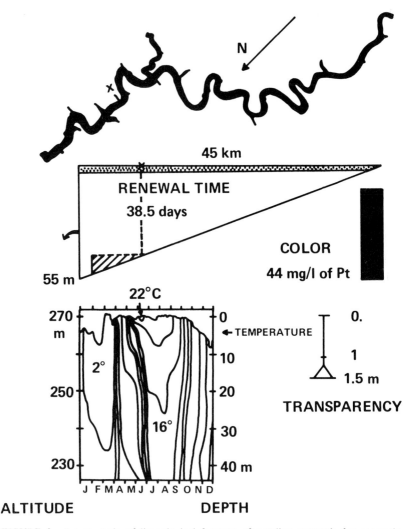

FIGURE 2   An example of limnological features of a valley reservoir for generating power: Slapy Reservoir, Czechoslovakia. (Adapted from Straškraba, 1966.)

Table 2 summarizes the findings of different zooplankton species in the inflowing river and in the reservoir during 1959. It is evident that zooplankton in the reservoir is almost completely independent of that in the river, which has an inflow of about 50 m³/sec (Straškraba and Hrbáček,

TABLE 2 Comparison of the Frequent Zooplankton Species of Slapy Reservoir and the Inflowing Vltava River, 1958–1960[a]

Species	Vltava River near Kamýk	Slapy Reservoir Point 9
Asplanchna priodonta	+	+
Bdelloidea g. sp.	+	−
Brachionus angularis	+	+
B. calyciflorus	+	+
B. quadridentatus	+	+
B. urceus	+	+
Conochilus spp.	+	+
Euchlanis deflexa	+	−
Filinia longiseta	−	+
Kellicotia longispina	−	+
Keratella quadrata	+	+
K. cochlearis	+	−
Lecane luna	+	−
Mytilina spp.	+	−
Polyarthra spp.	+	+
Testudinella patina	+	−
Synchaeta sp.	+	+
Alona quadrangularis	+	−
A. guttata	+	−
Bosmina coregoni	−	+
B. longirostris	+	+
Ceriodaphnia quadrangula	−	+
Chydorus sphaericus	+	−
Daphnia cucullata	−	+
D. hyalina	−	+
Diaphanosoma brachyurum	−	+
Leptodora kindtii	−	+
Limnosida frontosa	−	+
Simocephalus vetulus	−	+
Eudiaptomus gracilis	−	+
Acanthocyclops vernalis	−	+
Cyclops vicinus	−	+
Diacyclops bicuspidatus	−	+
Megacyclops viridus	−	+
Mesocyclops leuckarti	−	+
Thermocyclops crassus	−	+

[a]Adapted from Straškraba and Hrbáček (1966).

1966). This is not true to the same degree for phytoplankton; the similarity between river and reservoir phytoplankton is much greater, for lake conditions eliminate some species of river plankton and support others (Javornický *et al.*, 1962). Conditions inside the reservoir seem to be limiting for the formation of the reservoir plankton. It appears unreasonable, therefore, to give much attention to the populations of the inflowing river.

A high degree of eutrophication of the reservoir is evident from the presence of heavy blooms of Cyanophyceaea (*Aphanizomenon flos aquae* and *Microcystis aeruginosa*). The amount of these corresponded to one third of the zooplankton, if expressed in the same units (mg Kjeldahl nitrogen per dm^2). Cyanophycean blooms formed a green soup very unpleasant for recreation and treatment of waters. When nutrient content (nitrogen compounds and both total and phosphate phosphorus) in Slapy Reservoir was compared with the nutrient contents of other reservoirs and with ponds and backwaters with and without Cyanophycean blooms (Hrbáček, 1964), no clear-cut correlation was revealed between presence of blooms and N and P compounds.

Other factors, such as depth conditions and fish stock, seem to cooperate here. Blooms were observed in deeper waters at much lower concentrations of phosphorus. Extraordinarily high numbers of fish decreased the *Aphanizomenon* blooms. However, at least 50 μg per liter of total phosphorus and 10 μg per liter of phosphate phosphorus seem to be necessary for the appearance of *Aphanizomenon* and *Microcystis* in considerable numbers. A large-scale study of the appearance of blooms in Czechoslovak reservoirs and ponds was organized by the Institute of Hygiene, Prague. Because of methods difficulties in local hygienic laboratories, the nutrient content of the bodies of water was not estimated. According to studies of 23 ponds and 44 reservoirs, it was determined that Cyanophycean blooms do not occur at altitudes greater than 500 m above sea level, at annual average surface temperatures below $11°$ C, at average Secchi disc readings above 270 cm and average 1 percent illuminations below 250 cm, at average alkalinity above 5 mEq per liter, and at average pH below 6.5 (Štěpánek *et al.*, 1963). These values, however, are often exceeded without blooms necessarily occurring.

Hrbáček *et al.* (1966) showed a decrease of the average concentrations of N and P compounds in the outflow of the reservoir as compared with the inflow. The estimated budgets for BOD_5, COD, and total phosphorus are shown in Table 3. A comparison with the scarce similar data on other reservoirs and lakes does not show Slapy Reservoir to be unique with respect to the amounts retained. Phosphate phosphorus is always lower in the surface layers of the reservoir; a definite U-shaped trend due to a decrease in summer months may be explained by (1) uptake of phosphorus by algae, (2) sedimentation of seston during the summer, and (3) the entering of inflowing water, rich in phosphates, to surface layers during the spring.

TABLE 3 Estimated Budget for BOD$_5$, COD, and Total Phosphorus, Slapy Reservoir[a]

| | Amounts in Inflowing and Outflowing Water (metric tons) | | Retained in Reservoir | |
Analyzed	Inflow	Outflow	Amount (metric tons)	Percent
April 1–August 31, 1959				
BOD$_5$	4,830	2,630	2,200	46
COD	51,500	36,900	14,600	28
Phosphorus	198	62	136	69
September 1, 1959– March 31, 1960				
BOD$_5$	4,790	3,000	1,790	37
COD	50,200	37,700	12,500	25
Phosphorus	118	81	37	31
April 1–August 31, 1960				
BOD$_5$	5,720	3,660	2,060	36
COD	60,900	45,100	15,800	26
Phosphorus	201	81	120	60
September 1, 1959– August 31, 1960[b]				
BOD$_5$	10,510	6,660	3,850	37
COD	111,100	82,800	28,300	25
Phosphorus	319	162	157	49

[a]Source: Hrbáček et al. (1966).
[b]Representative year.

Procházková (1966), analyzing the seasonal changes of nitrogenous compounds, found a correlation of nitrate nitrogen with variations in the amount of inflowing water and surface-level fluctuations. Algae of the reservoir mostly seem to prefer ammonia nitrogen to nitrates, even during a high surplus of the nitrates.

A large-scale experiment concerned with eutrophy in reservoirs was made possible by the construction of another reservoir of similar character immediately above Slapy Reservoir. All the processes of nutrient retention observed previously in Slapy Reservoir occurred now in Orlík Reservoir. Heavy blooms of Aphanizomenon and Microcystis occurred there. Water from lower strata, much richer in Cyanophyceae and zooplankton, but somewhat poorer in nutrients than previously, was released to Slapy. The long-term changes in stratification of Slapy Reservoir after the upperlying Orlík was established are illustrated by temperature changes evident in Figure 3, which shows a gradual sharpening of the thermocline in recent years. This sharpening is a result of the influence of changed temperature of the inflow.

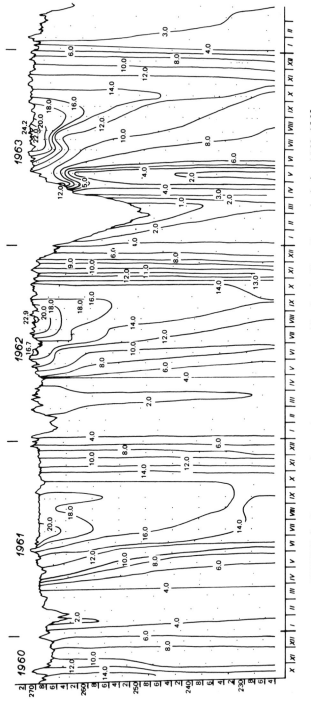

FIGURE 3 Long-term changes in temperature stratification in Slapy Reservoir during 1960–1963.

78

It is also connected with the changes in the renewal time of reservoir water due to low precipitation during the years studied. Schräder (1958), in studying thermal conditions in several reservoirs in East Germany, recognized higher summer temperatures in deeper strata of reservoirs with a shorter mean renewal time. As evident from the example of Slapy Reservoir, we also find occurrence of higher summer temperatures when comparing years rich and poor in precipitation in one and the same reservoir. The inflowing water seeks depths where the reservoir water has a corresponding density, which is due largely to its temperature. A seasonal cycle can be recognized in which inflowing water, as its temperature changes throughout the year, seeks varying layers in the reservoir, substantially influencing eutrophication processes.

After establishment of Orlík Reservoir above Slapy and the indicated changes in throughflow conditions, the extent of Cyanophycean blooms decreased considerably, as shown in Figure 4. Javornický (1966) found that, in the period before Orlík was constructed, diatoms grew fairly well in samples of Slapy water exposed to surface radiation. At that time, however, stratification was slight, Diatomaceae were transported to lower strata into unfavorable light conditions, and Cyanophyceae were more successful. The primary production of the total phytoplankton declined markedly, the decreased Cyanophyceae being replaced by other phytoplanktonic groups, mainly diatoms. The zooplankton standing crop does not decrease much in spite of the decreased total primary production, indicating that phytoplankton other than Cyanophyceae blooms is much more valuable for zooplankton nutrition. Cyanophyceae seem to be mostly available to zooplankton after decaying, and the corresponding energy losses for bacterial respiration cause the different efficiency. This is also one of the reasons for the unfavorableness of Cyanophyceae with respect to water quality.

Long-term changes in the total bacterial numbers (yearly means based on 3-week sampling intervals) at the surface of Slapy Reservoir are shown in Figure 5. During 1961 and 1962 the freshly flooded areas of the reservoir constructed upstream influenced the number of bacteria in the lower reservoir. A similar phenomenon has been regularly observed on flat and shallow Soviet reservoirs during spring floods when a huge area of dry land is flooded (Kusnetsov, 1958).

Looking at this experiment, we may see an example of remarkable "diseutrophication" connected with a change in the reservoir arrangement. This change shows the uniqueness of problems of reservoir eutrophication and a need for basic studies, causal rather than merely descriptive, to understand it.

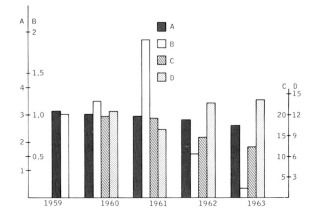

FIGURE 4 Plankton changes in Slapy Reservoir during 1959–1961 (before the upperlying reservoir was built) and during 1962–1963 (after it was built).
A Zooplankton standing crop in mg/dm^2 Kjeldahl N.
B Cyanophycean standing crop in mg/dm^2 Kjeldahl N.
C Gross primary production of phytoplankton in mg/dm^2 O$_2$ per day.
D Abundance of phytoplankton except Cyanophyceae in cells ×10^6 per liter.

Note the different scales: A and B, left; C and D, right. (Straškraba, 1966.)

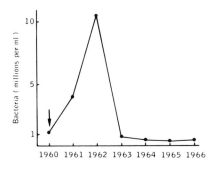

FIGURE 5 Changes in the yearly means of total bacterial numbers at the surface of Slapy Reservoir.

Saidenbach Reservoir If possible, reservoirs for drinking-water supply should be situated in untouched areas. This possibility, however, does not exist in many areas, and eutrophication problems arise. The Saidenbach Reservoir in East Germany, extensively studied by a group of limnologists from Leipzig, will be used as an example of reservoirs in which eutrophication problems have arisen. Characteristic features of the reservoir, built in 1933, are as follows: area, 146 ha; volume 22 million m^3; maximum depth, 30 m; mean renewal time, about 1 year (Klapper, 1957/58, 1961; Beuschold, 1959, 1961; and Rudolph, 1966).

Figure 6 shows characteristic features of a reservoir of this type. One feature is the relatively deep water, which is kept cold because of the slow throughflow of the reservoir by a small volume of water (temperatures at the bottom 4 to 5° C). Although a seasonal drop in water level occurs in summer, short-term fluctuations in water level are negligible. In connection with the pronounced stratification, there is in summer a depletion of oxygen at the bottom, the degree of depletion depending on organic production.

The quality of the water in the reservoir is affected to a large extent by the character of the watershed, for this will determine the kind and amount of nutrients that will enter the reservoir (Wetzel, 1958, 1962).

There are two other much smaller reservoirs in the vicinity, with a watershed area free of inhabitants and with water of good quality. Saidenbach Reservoir, with about 10,000 inhabitants living in small villages of the 76 km^2 watershed area, often has water-treatment troubles caused by abundant algae. In untouched highland streams of the area, Mädler (1961) found a very low phosphorus content (below 5 μg per liter of PO_4 and below 20 μg per liter of total phosphorus). Streams flowing through fields and meadows of this area, which is not very highly cultivated, contained 20 to 75 μg per liter of total phosphorus, whereas the water in creeks passing villages contained more than 2,000 μg per liter. The high phosphorus content came from both dunghills and sculleries (detergents).

The importance of drainage of soil particles from cultivated areas was evident when phosphorus contents of streams during dry weather and during melting of snow were compared. In phosphorus-poor streams, the original concentration increased as much as 20-fold, and the amount carried increased as much as 200-fold. A rough comparison of the data indicated that a strong 18-day rain will bring to a reservoir via a small brook about the same amount of phosphorus that will be brought in 1 year by the effluents of villages having about 10,000 inhabitants.

Hedlich (1961, 1966) made direct measurements of phosphorus entering surface and subsurface drainage from wooded slopes. He used special collectors and lysimeters installed on a recently cleared area and inside a thin older growth and a more dense younger growth. In both growths, 90 percent of the trees were spruce. The observed concentrations of phosphate phosphorus in the water drained from the surface were about 100 times higher than those in the reservoir and about 10 times higher than those drained from 20 cm below the surface. The old thin spruce forest showed the lowest retention capacity for phosphorus.

Much care is given in some areas to preventing depletion of sources of phosphorus, including botanical and forestry measures (Klapper, 1964).

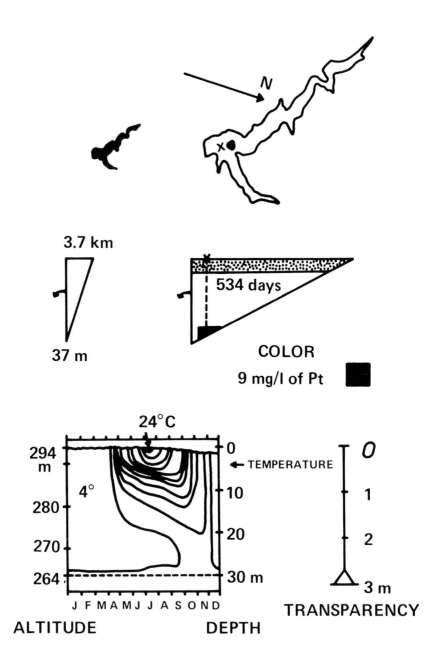

FIGURE 6   An example of limnological features of a reservoir for drinking-water supply: Klíčava Reservoir, Czechoslovakia. (Adapted from Straškraba, 1966.)

Sedlice Reservoir A series of very extensive publications by a group of limnologists of the Hygienic Institute of Prague is devoted to the Sedlice Reservoir in Czechoslovakia. This is an example of a reservoir with a great rate of flow and a mean theoretical renewal time of about 8 days. It is a small reservoir (area, 36 ha; volume, 2 million m^3; maximum depth, 13 m) situated in a hilly area (450 m above sea level) on the confluence of two brooks with a combined throughflow of 2.75 m^3/sec. The small capacity is responsible for considerable fluctuations in both water level and renewal time. During maximum daily flows, the water is renewed twice a day; during minimum flow, about once in 30 days. A small hydroelectric power station, built in 1927, is installed on the reservoir. For power generation, bottom water is taken by a side outlet pipe situated about 450 m upstream from the dam.

It was felt the excentric position of the outlet pipe would allow the throughflowing upper part and an undisturbed part immediately above the dam to be kept separate and thus make possible the storing of water in a reservoir for drinking-water purposes. This proved to be difficult because of large and irregular fluctuations in water level and because of quick renewal of the reservoir water and the small extent of the lower part. However, data concerning water-quality changes in a reservoir with a great rate of flow were accumulated. Temperature stratification is slight, the maximum summer bottom temperatures reaching 15 to 16° C; chemical stratification often is disrupted by droughts. Oxygen depletion occurred at bottom during lower throughflows occurring continuously during winter months. The high oxygen consumption and production of gases by bottom deposits dropped markedly from the inflow to the dam (Cǔta *et al.*, 1962). At distances of 800, 900, 1,000, and 1,400 m from the inflow, the total production of gases for a period of about 9 months was 2,290, 1,290, 1,118, and 252 liters, representing a ratio of 100:55:51:12. The sedimented matter freed large amounts of CO_2 for phytoplankton reassimilation.

Štěpánek (1960) observed the influence of meteorological factors on the quantitative development of phytoplankton in the reservoir. No relation of the atmospheric pressure and nannoplankton development was observed. In accordance with the observations on Hungarian ponds (Kiss, 1952), an effect of the pressure changes in initiating a rise of water-bloom organisms from the bottom was observed (Štěpánek and Zelinka, 1962). Štěpánek (1960) found highly significant statistical correlations between the duration of sunshine, total radiation, height of sun above the horizon, and other parameters of the radiation and the quantitative development of phytoplankton. The statistical procedure used does not allow distinguishing between causal and successive effects, and no consideration was given to the specific reservoir conditions.

Chalupa (1959) studied phosphorus and iron in the reservoir. Analyses of phosphate phosphorus in rain (Chalupa, 1960b) showed 0 to 160 μg per liter of P_2O_5 in 24 determinations distributed over half a year. Maxima coincided with the phosphate manuring of fields in the area. A rough recalculation of the contribution of phosphorus from precipitations to the trophogenic layer (assuming stratification prevents phosphorus from other sources from entering) showed that this source is very low in comparison with the inflow (0.2 percent); however, the increase in surface layers may occasionally influence phytoplankton development.

Experiments were made (Chalupa, 1960a) to prove the possibility of eliminating dissolved inorganic phosphates directly in the body of water by using their ability to be adsorbed on flocculated ferrihydroxyd at a ratio of 5:1 $(Fe:P_2O_5)$. If excessive amounts of ferrisulfate were added, the elimination of phosphates was neither quantitative nor permanent if usual phosphate concentrations (tens of micrograms per liter) were present. A decrease of about one order was observed only during presence of hundreds of micrograms per liter.

Previous experiences by Guseva (1940, 1952) in eradication of Cyanophycean blooms by algicides in Soviet reservoirs were used in selecting several preparations based on Cu salts. A solution of $AgNO_3$ and $CuSO_4$ (1:14 to 1:15 w/w), diluted to achieve the final concentrations of 50 μg of Ag per liter and 300 μg of Cu per liter, proved to be the most successful agent. *Aphanizomenon* in the experimental bags was eradicated in a few hours, and within a few days it was eradicated from the reservoir surface by spraying from a boat or an airplane. No harm to zooplankton and fish was observed; however, there was an increase in nannoplankton organisms several weeks after the decomposition of algae. A slight decrease in the oxygen content was noted.

Sládečková-Vinníková (1957, 1958) Sládecková (1960, 1962) studied the quantitative distribution of organisms attached to slides at various depths. Sládeček and Sládečková (1963) tried to determine the periphyton production on these artificial substrates. Sládečková (1966) proposed using the character of periphyton on slides, as recognized by the naked eye, as a simple means for distinguishing trophogenic, hypolimnetic, and oxygenless zones in the reservoir.

Summarizing the part on reservoirs, we should like to stress the primary importance of the retention time and hydrology of the inflow and outflow for the physical, chemical, and biological processes in reservoirs. This basic differential factor is manifested in the range of water-level fluctuations, temperature, and stratification conditions decisive for the degree of the utilization of nutrients by the phytoplankton population and feeding of zooplankton. An understanding of these interrelations and their manifestation in the individual conditions of the given reservoir seems to give us a key for understanding the uniqueness of the eutrophication processes in reservoirs.

BACTERIA IN BODIES OF WATER OF DIFFERENT EUTROPHICATION

Bacteriological research on lakes and reservoirs in Eastern Europe provided comparative data on numbers of bacteria, biomass, and production in bodies of water of different eutrophication and pollution.

Numbers of bacteria were determined by direct counting on membrane filters. It is believed that the disadvantage of this method, mainly the inclusion of dead and inactive bacteria, does not lessen the value of its greatest advantage—no selectivity. By use of differential stains (Giemsa and light green), it was found that the percentage of dead bacteria in surface water does not exceed 10 percent of the total counts (Kusnetsov, 1959). However, the percentage of living, but inactive, bacteria is not found by this method.

Numbers of bacteria determined by incubation on beef-peptone agar, on the other hand, are negligible in comparison with direct counts.* Nevertheless, exceptions were found in particular instances, indicating some special situations; for example, in freshly flooded areas in newly built reservoirs, the total numbers of bacteria reach high values without a corresponding increase in the numbers of colonies on agar plates.

Although it is known that numbers of bacteria tend to increase with increasing pollution, the general relation has not been proved and calculated. Analysis of the 200 samples showed a linear relation between the biochemical oxygen demand in 5 days, chosen as a criterion of pollution, and numbers of bacteria on a log–log scale. The relation is different for numbers estimated by different methods.

When direct counts were made (Figure 7), the correlation coefficient was 0.917 for 195 samples, and the regression equation was log numbers of bacteria = 1.165 log BOD_5 + 5.262. From the figure it is evident that all the data, from both stagnant and running water, fit into the same equation. The colonies on beef-peptone agar after 2 days of incubation, on the other hand, provide more-scattered data for stagnant waters (Figure 8). The data from running water and sewage plotted against BOD_5 in log–log scale fit into a regression line with the equation log numbers of bacteria = 1.784 log BOD_5 + 2.364 and correlation coefficient 0.910 for 96 samples. The correlation coefficient for bacterial data from stagnant water and BOD_5 was low, r = 0.203 for 99 samples. It is noteworthy that the data on BOD_5 and bacteria on agar plates from a Swedish and a New Zealand river fit rather well into the equation derived from data from Czechoslovak rivers.

*With increasing pollution and increasing total numbers of bacteria, the percentage of bacteria growing on agar plates increases. In waters with bacterial numbers (determined by direct counts) above 10 million per ml, the difference between the two methods is negligible (Straškrabová-Prokešová, in preparation).

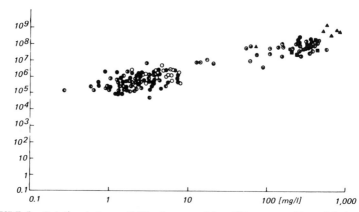

FIGURE 7   Relation between BOD$_5$ (expressed in milligrams per liter of O$_2$) and the total number of bacteria (expressed in number per milliliter) in bodies of water polluted to different degrees. Symbols represent different types of water bodies, including reservoirs, reservoir effluents, brooks, rivers, treated sewage, sewage-sedimentation tanks, sewage, and activation tanks. (Adapted from Straškrabová, 1968.)

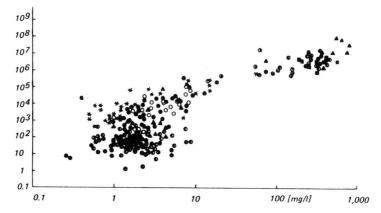

FIGURE 8   Relation between BOD$_5$ (expressed in milligrams per liter of O$_2$) and the number of bacteria on beef-peptone agar plates after 2 days of incubation (expressed in number per milliliter) in bodies of water polluted to different degrees. In addition to the symbols shown in Figure 7, this figure shows asterisks (*), which refer to data given by Stone (1965) for a New Zealand river. (Adapted from Straškrabová, 1968.)

Methods for estimating production of bacteria in natural water were developed in the area. The reproduction rate of bacteria was calculated on the basis of changing numbers of bacteria in bottles incubated in the lake (Ivanov, 1955; Czeczuga, 1961). The generation time, g, is the time required for the number of bacteria to double. Values of g were based on direct counting. If only bacteria growing on beef-peptone agar were followed, generation times were 50 to 80 percent shorter. The maximum reproduction rate quoted in a fertilized fishpond was g = 4 hours for direct counts and g = 1.5 hours for bacteria on agar plates. If very short incubation times were used, changes in number of bacteria on membrane filters incubated on natural water were quite similar to those occurring in the water itself. This allows use of incubation on membrane filters as another method for estimating the reproduction rate of bacteria (Straškrabová-Prokešová, 1967).

The assimilation of ^{14}C-labeled carbonates by bacteria was also used for studying bacterial production. Although this method is the most convenient for autotrophic bacteria, it was also proposed for measuring the reproduction rate of heterotrophic bacteria. Romanenko (1964) and Sorokin (1964, 1965) believe the percentage of inorganic carbon incorporated into the newly formed biomass of heterotrophic bacteria is rather constant and could be used for calculating natural conditions.

Table 4 includes results of the determinations of total numbers of bacteria

TABLE 4 Bacteriological Data on Lakes and Reservoirs in Eastern Europe, by Trophic Degree[a]

Reference	Number of Bacteria[b]	Bacterial Biomass[c]	Generation Time[d]	Daily Production/Biomass Ratio
EUTROPHIC				
Belyazkaya (1958)	1.8–8	2.8–10.5	–	–
Czeczuga (1961)	2.2	–	25–41	0.64–0.91
Ivanov (1955)	0.9–1.8	–	14.5–35	0.63–1.35
Kusnetsov (1958)	2.0–3.7	1.8	–	–
Romanenko (1965)	4–10	–	–	–
MESOTROPHIC				
Belyazkaya (1958)	0.8–3	0.5–1.5	–	–
Czeczuga (1961)	0.6–1.4	–	8.3–59	0.58–2.28
Ivanov (1955)	0.3–0.5	–	78–220	0.31
Kusnetsov (1958)	0.5–1.6	–	–	–
Romanenko (1965)	1–3	–	–	–
Straškrabová-Prokešová (1967)	0.4–1.1	0.3–1.2	15.2–75.9	–
OLIGOTROPHIC				
Czeczuga (1961)	0.8–1	–	81–121	0.14–0.41
Ivanov (1955)	0.2	–	218	0.15
Kusnetsov (1958)	0.05–02	0.13	–	–
Romanenko (1965)	0.2–0.9	–	–	–

[a]Data limited to those in which bodies of water of different types are compared.
[b]Millions per milliliter of surface water.
[c]Milligrams per liter.
[d]Hours.

(direct counts), biomass of bacteria calculated from microscopic counting and sizing, determination of the generation time of the natural bacterial populations, and calculated daily-production–biomass ratios from bodies of water of different trophy, based on the above methods. Results of a few authors comparing different bodies of water were selected. The original designation of the trophic degree is used.

The present estimates of bacterial production should be considered as preliminary, since the methods are in development and regular seasonal observations have been few.

The top panel of Figure 9 compares the biomass of bacteria with the biomass of both phytoplankton and zooplankton calculated from summer averages for three lakes in Bielorussia (after data by Petrovich, 1961). As is evident, the biomass of bacteria is lower than that of the other groups. The bottom panel of Figure 9 illustrates the ratio of daily production for phytoplankton and bacteria (after data by Ivanov, 1955). The reproduction of bacteria was found to increase much quicker than the primary production when bodies of water with increasing eutrophication were compared. This might indicate an allochthonous source of organic substances in eutrophic bodies of water or a relatively lower degree of the utilization of primary production by zooplankton and its consequent availability for bacterial decay.

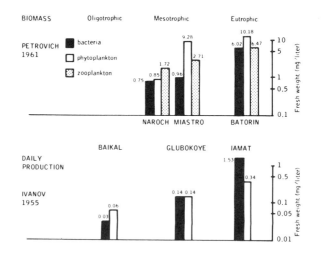

FIGURE 9   Comparison of bacterial biomass and production with those of phytoplankton and zooplankton in bodies of water of different eutrophication.

BIOCENOTICAL RELATIONS IN A BODY OF WATER AS A WATER-QUALITY FACTOR

At higher trophic levels, the direct relation between the nutrient content of the body of water, its primary production, and amount of zooplankton seems no longer to be proportional. Figure 10 gives some selected examples of the summer means of primary production and zooplankton standing crop in different bodies of water. This figure indicates other factors were working here. Recent results in this area and elsewhere indicate that recognition of the biocenotical interrelations of different trophic links is one of the important contributions to the understanding of the question of water quality. Reciprocal interactions must be taken into account. The influences of nonliving factors of the environment (e.g., nutrient supply) on primary producers, of primary producers on herbivores, and so on, are generally recognized. Equally important but often overlooked are the actions of predators on herbivores, herbivores on primary producers, and primary producers on nutrient supply.

A considerable body of information was accumulated concerning the influence of fish, the highest trophic link of the trophic chain in freshwaters, on the lower links. Experimental observations showed that when fish populations are high, the zooplankton is composed of small Cladocerans and Rotifers, nannoplankton algae are fairly developed, and Secchi disc readings are low. When fish populations are low, the zooplankton is composed mainly of larger Cladocerans; Rotifers become scarce. Phytoplankton is now formed by a smaller number of larger species, and Secchi disc readings are increased (Hrbáček *et al.*, 1961; Hrbáček, 1962; Novotná and Kořínek, 1966; Ertl, 1966).

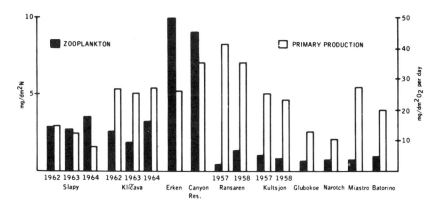

FIGURE 10   Summer means of zooplankton standing crop and primary production in different lakes. (Straškraba, 1966.)

Changes in zooplankton populations seem to be due mainly to the selective feeding of fishes on larger species; however, Hrbáček et al. (1961) demonstrated that because of higher numbers of smaller zooplankton organisms having intensive metabolism, the metabolism of the plankton association speeds up without a change in the nutrient supply. The phytoplankton is affected by the increased needs of the zooplankton. Grygierek (1962, 1965) demonstrated increased reproduction rates in *Daphnia longispina* during periods when fish populations are higher. Figure

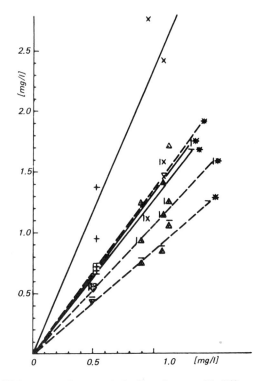

FIGURE 11   Night oxygen decreases in bodies of water with different fish stock related to the amount of Kjeldahl nitrogen in the water. Yearly means of oxygen decrease are expressed in milligrams per liter of O_2 per night. Kjeldahl nitrogen is expressed in milligrams per liter of N.

+ and ⨉   water bodies with high fish stock
△ and ▽   water bodies with low fish stock
| values weighted by volume, for both high fish stock (solid lines) and low fish stock (broken lines)
—   values corrected for diffusion, for both high fish stock (solid lines) and low fish stock (broken lines)

(Hrbáček, 1968.)

11 shows the decrease of the oxygen content during the night as a measure of the oxygen consumption plotted against the Kjeldahl nitrogen content in ponds with a different fish stock. A greater decrease in ponds with more fish is evident, indicating different metabolic rates of the association at the same nutrient level.

Although we are far from understanding all the interrelations, it is obvious that these questions are directly related to water-quality problems both as a means of gaining a better understanding of the changes in the composition of zooplankton and phytoplankton and as a possible means of regulating water quality by regulating fish populations (Hrbáček, 1965).

REFERENCES

Belyazkaya, Ya. S. 1958. Seasonal changes of the total number and biomass of bacteria in the water of three lakes of different types [in Russian]. Mikrobiologiya 27:113–119.

Beuschold, E. 1959. Chemische und biochemische Verhältnisse in den Talsperren des Bodenwerkes und ihren Inflüssen sowie Schutzmassnahmen gegen Verunreinigung. Wasserwirtschaft-Wassertechnik 9:475–479.

Beuschold, E. 1961. Limnologische Untersuchungen am Hauptbecken der Saidenbach-Talsperre. Int. Rev. Ges. Hydrobiol. 46:18–42.

Birshtein, Ya. A. 1953. Probable changes of the hydrobiological regime of the Caspian Sea [in Russian]. Tr. Vsesoyuz. Gidrobiol. Obsch. 5:3–12.

Bolokhontsev, E. N. 1909. Botanical-biological observations of Lake Ladoga [in Russian]. St. Petersbourg.

Bolokhontsev, E. N., and M. D. Grimm. 1911. Bacteriological observations of Lake Ladoga water [in Russian]. St. Petersbourg.

Brandt, A. 1944. Ueber Zelluloseabbau in Seen. Arch. f. Hydrobiol. 40:778–819.

Butorin, N. V., and A. S. Litvinov. 1963. On currents of the Rybinskoe reservoir [in Russian]. Tr. Inst. Biol. Vnutr. Vod 6:270–302.

Chalupa, J. 1958. Limnological study of the Reservoir Sedlice near Želiv. Physical-chemical and chemical part I. Sborník Vys. Školy Chem. Techn., Techn. Vody, Praha 2(2):151–300.

Chalupa, J. 1959. Limnological study of the Reservoir Sedlice near Želiv. III. Chemical part I. Iron and dissolved inorganic phosphates. Sborník Vys. Školy Chem. Techn., Techn. Vody, Praha 3(2):21–166.

Chalupa, J. 1960a. Pokusné řízení jakosti vody eliminací rozpuštěných anorganiských fosforečňanů. Experimental water quality control by the elimination of dissolved inorganic phosphates [in Czech, English summary]. Čsl. Hygiena 5:606–613.

Chalupa, J. 1966b. Eutrofisace nádrží srážkovým fosforem. Eutrophication of reservoirs by atmospheric phosphorus [in Czech, English summary]. Sborník Vys. Školy Chem. Techn., Techn. Vody, Praha 4(1):295–308.

Cohn, L. 1903. Untersuchungen über das Plankton des Löwentin und einiger anderer Seen Masuriens. Z. f. Fischerei 10(4):227–260.

Čuta, J., M. Vorderwinklerová-Čulíková, and V. Bernátová. 1962. Produkce plynu a rozklad sedimentu v údolní nádrži Sedlice na Zělvice. Gasbildung und Sediment-zersetzung in der Talsperre Sedlice [in Czech, German summary]. Sborník Vys. Školy Chem. Techn., Techn. Vody, Praha 6(1):157–176.

Czeczuga, B. 1959. O minimum i maksimum tlenowym v metalimnionie Jezior Rajgrodzkich. On oxygen minimum and maximum in the metalimnion of Rajgrod Lakes [in Polish, English summary]. Acta Hydrobiol. 1:190–122.

Czeczuga, B. 1961. Intesywnośc rozmnozania sie i produkcja bakterii Jezior Rajgrodzkich v okresie letnim. Intensity of multiplication and production of bacteria in Rajgrodzkie lakes in summer [in Polish, English summary]. Polsk. Arch. Hydrobiol 9:349–360.

Czeczuga, B. 1962. An attempt at establishing the production and numerical relations of bacterioplankton biomass. Acta Hydrobiol. 4:1–20.

Czeczuga, B., and T. Baszyński. 1963. Niektore danie hydrochemiczne wod Jeziora Rajgridzkiego. Some hydrochemical data on Lake Rajgrodzkie [in Polish, English summary]. Polsk. Arch. Hydrobiol. 11:267–274.

Datsko, V. G. 1961. Importance of phytoplankton primary production in the organic matter budget, as exemplified at Azow, Black and Caspian Seas [in Russian]. Primary production in seas and inland waters, Minsk:47–51.

Edelshtein, K. K. 1963. On thermocline layer and its dynamics in the Rybinsk Reservoir [in Russian]. Tr. Inst. Biol. Vnutr. Vod 6:250–257.

Entz, B. 1949/50. Some physical and chemical conditions of the water of Lake Balaton, investigated from September 1948 to April 1949. Ann. Biol. Tihany 19:69–81.

Entz, B. 1959. Chemische Charakterisierung der Gewässer in der Umgebung des Balatonsees (Plattensee) und chemische Verhältnisse des Balatonwassers. Ann. Biol. Tihany 26:131–202.

Entz, B., and M. Fillinger. 1961. Adatok a Balaton fékykklímájának ismeretéhez. Angaben zur Kenntnis des Lichtklimas des Balatons [in Hungarian, German summary]. Ann. Biol. Tihany 28:49–89.

Ertl, M. 1966. Zooplankton and chemistry of two backwaters of the Danube River. Hydrobiol. Stud. 1:267–295.

Felföldy, L. 1959. Experiments with algal cultures for determining some properties of Balaton Lake. Ann. Biol. Tihany 26:211–222.

Felföldy, L. 1960. Photosynthetic experiments with unicellular algae of different photosynthetic type. Ann. Biol. Tihany 27:193–220.

Gieysztor, M., and Z. Odechowska. 1958. Observations on the thermal and chemical properties of Mazurian Lakes in the Gizycko Region. Polsk. Arch. Hydrobiol. 4:123–152.

Granops, M. 1965. Stan zanieczyszczenia wod rzeky Sawicy i jeziora Maly Sasek na Mazurach. Degree of the pollution of Sawica River and Sasek Maly Lake in Mazurian district [in Polish, English summary]. Zeszyty Naukowe WSR Olsztyn 20:115–126.

Grygierek, E. 1962. The influence of increasing carp fry population on crustacean plankton. Roczn. Nauk Rol. B, 81:189–210.

Grygierek, E. 1965. Wplyw ryb na faune skorupiakow planktonowych. The effect of fish on Crustacean plankton [in Polish, English summary]. Roczn. Nauk Rol. B, 86:147–168.

Guseva, K. A. 1940. Influence of copper on algae. Mikrobiologiya 9:5.

Guseva, K. A. 1952. Water-blooms, their causes, prognoses and eradication [in Russian]. Tr. Vsesoyuz. Gidrobiol. Obsch. 4:13–92.

Guseva, K. A. 1955. Phytoplankton of the Rybinskoe Reservoir [in Russian]. Tr. Biol. Sta. Borok 2:2–23.

Guseva, K. A. 1958. Influence of water-level fluctuations of Rybinskoe Reservoir on phytoplankton development [in Russian]. Tr. Biol. Sta. Borok 3:112–124.

Hedlich, R. 1961. Die Wirkung der Nährstoffauswaschung auf Kahlschlägen aus eine

Trinkwassertalsperre (Neunzehnhain). Wiss. Z. Karl-Marx-Univ. Leipzig, Math.-Naturwiss. Reihe 10(1):89–92.

Hedlich, R. 1966. Über die Auswaschung von Pflanzennährstoffen aus Waldböden und ihr Einfluss auf die Wasserqualität von Talsperren. Wiss. Z. Karl-Marx-Univ. Leipzig, Math.-Naturwiss. Reihe 15(1):217–227.

Hrbáček, J. 1962. Species composition and the amount of the zooplankton in relation to the fish stock. Rozpravy ČSAV, Řada MPV 72(10):1–116.

Hrbáček, J. 1964. Contribution to the ecology of water-bloom-forming blue-green algae. Aphanizomenon flos aquae and Microcystis aeruginosa. Verh. Internat. Verein. Limnol. 15:837–846.

Hrbáček, J. 1965. Beziehungen zwischen Nährstoffgehalt, Organismenproduktion und Wasserqualität in Talsperren. Wiss. Z. Karl-Marx-Univ. Leipzig, Math.-Naturwiss. Reihe 14(2):265–273.

Hrbáček, J. 1968. Relation of productivity phenomena to the water quality criteria in ponds and reservoirs. Fourth Intern. Conf. Water Pollut. Res., Proc. Sec. 3, Paper 4:1–8.

Hrbácek, J., M. Dvoráková, V. Kořínek, and L. Procházková. 1961. Demonstration of the effect of the fishstock on the species composition of zooplankton and the intensity of metabolism of the whole plankton association. Verh. Internat. Verein. Limnol. 14:192–195.

Hrbáček, J., L. Procházková, and V. Straškrabová-Prokešová. 1966. The relationship between the chemical characteristics of Vltava River and Slapy Reservoir with an appendix by C. O. Yonge: Chemical budget for Slapy Reservoir. Hydrobiol. Stud. 1:41–84.

Ivanov, M. V. 1955. Methods of the determination of bacterial production in a water-body [in Russian]. Mikrobiologiya 24:79–89.

Javornický, P. 1966. Light as the main factor limiting the development of diatoms in Slapy Reservoir. Verh. Internat. Verein. Limnol. 16:701–712.

Javornický, P., J. Komárek, and J. Růžička. 1962. Fytoplankton Slapské údolní nádrže v letech 1958-1960. Phytoplankton of the Slapy Reservoir during 1958-1960 [in Czech, English summary]. Sborník Vys. Školy Chem. Techn., Techn. Vody, Praha 6(1):349–387.

Kiss, J. 1952. Meteorobiological investigation of micro-organisms causing water-blooms [in Russian]. Acta Biol. Hung. 3(2):159–220.

Klapper, H. 1957/58. Biologische Untersuchungen an den Einläufen und Vorbecken der Saidenbach-Talsperre (Erzgebirge). Wiss. Z. Karl-Marx-Univ. Leipzig, Math.-Naturwiss. Reihe 7(1):11–47.

Klapper, H. 1961. Die Eutrophierung von Trinkwassertalsperren und ihre möglichen Auswirkungen, dargestellt am Beispiel der Saidenbach-Talsperre. Wiss. Z. Karl-Marx-Univ. Leipzig, Math.-Naturwiss. Reihe 10(1):81–87.

Klapper, H. 1964. Die "dritte Reinigungsstufe" und das Problem der Eutrophierung von Seen und Trinkwasserspeichern. Wiss. Z. Karl-Marx-Univ. Leipzig, Math.-Naturwiss. Reihe 13(1):53–60.

Korelyakova, I. L. 1958. Some observations on decay of overwintering littoral vegetation of Rybinskoe Reservoir [in Russian]. Biul. Inst. Biol. Vodokhran. 1:22–25.

Korelyakova, I. L. 1959. Decay of mowed littoral vegetation [in Russian]. Biul. Inst. Biol. Vodokhran. 3:13–16.

Kozarov, G. 1958. Zasitenost na gornata cheterotepmna termicka zona na Ochridskoto ozero so O_2 i odnosot i kon gustinata na populacijata na fitoplanktonot i temperaturata na vodata. Saturation of the upper heterothermic zone with oxygen in Ohrid

Lake and its relation to the density of phytoplankton and water temperature. Zbornik Na Rabotita 6(14):1–12.

Kozminski, Z. 1932. O stosunkach tlenowych jeziora Hancza na Suwalczyznie. Ueber die Sauerstoff-Verhältnisse in den Hancza-See, Suwalki-Seengebirge, Polen [in Polish, German summary]. Arch. Hydrobiol. Ryb. 6:66–85.

Krashennikova, S. A. 1958. Microbiological processes of decay of aquatic vegetation in littoral region of Rybinskoe Reservoir [in Russian]. Biul. Inst. Biol. Vodokhran. 2:3–6.

Kuznetsov, S. I. 1958. Bacterial numbers of Rybinskoe Reservoir [in Russian]. Biul. Inst. Biol. Vodokhran. 14:8–10.

Kuznetsov, S. I. 1959. Die Rolle der Mikroorganismen im Stoffkreislauf der Seen. VEB Deutcher Verlag ol. Wissenschaften. Berlin.

Liatti, C. I. 1932. Hydrochemical observations on Lake Sevan [in Russian]. Materialy Po Issledov. Ozera Sevan 4(2):23–40.

Lossów, K. 1965. Obserwacjestanu zanieczyszczenia jeziora Beldany. Observations of state of the pollution from Beldany Lake [in Polish, English summary]. Zeszyty Naukowe WSR Olsztyn 20:99–114.

Mädler, K. 1961. Untersuchungen über den Phosphorgehalt van Bächen. Int. Rev. Hydrobiol. 46:75–83.

Mechkova, T. M. 1947. The animal plankton of Lake Sevan [in Russian]. Tr. Sevanskoi Gidrobiol. Sta. 10:1–67.

Mechkova, T. M. 1953. Zooplankton of Lake Sevan—biology and productivity [in Russian]. Tr. Sevanskoi Gidrobiol. Sta. 13:5–170.

Mechkova, T. M. 1961. Die Veränderungen desbiologischen Regimes des Sewansees im Zusammenhang mit dem Sinken seines Wasserstandes. Verh. Internat. Verein. Limnol. 14:204–207.

Mechkova, T. M. 1962. Present plankton conditions of Lake Sevan [in Russian]. Tr. Sevanskoi Gidrobiol. Sta. 16:15–88.

Mechkova, T. M. 1966. On the effect of a monomolecular film of fatty alcohols used as an evaporation depressor on the biological regime of basins [in Russian, English summary]. Gidrobiologicheskii Zhurnal 22:3–8.

Mordukhai-Boltovskoi, F. 1960. Caspian fauna in the Azow-Black Sea Basin [in Russian]. Edit. Acad. Sci., URSS, Moskva-Leningrad, 286 pp.

Mosevich, N. A., and M. V. Mosevich. 1954. Principal characters of the hydrochemical regime and microbiological processes in Rybinskoe Reservoir [in Russian]. Tr. Problem. i Temat. Sov. 2:11–21.

Novotná, M., and V. Kořínek. 1966. Effect of the fishstock on the quantity and species composition of the plankton of two backwaters. Hydrobiol. Stud. 1:297–322.

Olszewski, P. 1953. The thermal conditions of Mountain Lakes. Bull. Acad. Polon. Sci., Cl. Sci. Math. Nat., (A)1951:239–290.

Olszewski, P. 1959a. Stopnie nasilenia wplywu wiatru na jeziora. Graduation in the intensity of the wind effects on lakes [in Polish, English summary]. Zeszyty naukowe WSR Olsztyn 9(4):111–132.

Olszewski, P. 1959b. Usuwanie hypolimnionu jezior. Removing of hypolimnion of lakes [in Polish]. Zeszyty naukowe WSR Olsztyn 9(4):331–339.

Olszewski, P. 1961. Versuch einer Ableitung des hypolimnischen Wassers aus einem See. Verh. Internat. Verein. Limnol. 14:855–861.

Olszewski, P. 1965. Some physical and chemical properties of water of Mazurian Lakes. Mazury, Hydrobiol. Commity Polish Acad. Sci., 29–37.

Olszewski, P. 1967. Die Ableitung des hypolimnischen Wassers aus einem See. Inform. Bull. EFPW 14:87–89.

Olszewski, P., and M. Laskowska. 1959. Uwagi o stanie zanieczyszczenia Wielkich jezior Mazurskich. On pollution of Mazurian lakes [in Polish]. Zeszyty Naukowe WSR Olsztyn 9(4):196–200.

Olszewski, P., and J. Paschalski. 1959. Wstepna charakterystyka limnologiczna niektorych jezior Pojezierza Mazurskiego. Preliminary limnological characterization of some lakes in the Mazurian Lake District [in Polish, English summary]. Zeszyty Naukowe WSR Olsztyn 9(4):1–109.

Papp, Sz, B. Karoly, G. Margit, H. Laszlo, and S. Kalman. 1960. A Balaton vizének komplex égészségügyi vizsgálata. Comprehensive sanitary investigation of water from Lake Balaton [in Hungarian, English summary]. Hidr. Közl. 40:304–315.

Paschalski, J. 1964. Circulation types of lakes. Polsk. Arch. Hydrobiol. 12:383–408.

Patalas, K. 1960a. Charakterystyka skladu chemicznego wody 48 jezior okolic Wegorzewa. Characteristics of chemical composition of water in 48 lakes of Wegorzewo District [in Polish, English summary]. Roczniki Nauk Roln. B, 77(1):243–297.

Patalas, K. 1960b. Stosunki termiczne i tlenowe oraz przezroczystość wody w 44 jeziorach okolic Wegorzewa. Thermal and oxygen conditions and transparency of water in 44 lakes of Wegorzewo District [in Polish, English summary]. Roczniki Nauk Roln. B, 77(1):105–222.

Patalas, K. 1960c. Mieszanie wody jako czynnik określajacy intensywność krazenia materii w roznych morfologicznie jeziorach okolic Wegorzewa. Mixing of water as the factor defining intensity of food materials circulation in morphologically different lakes of Wegorzewo District [in Polish, English summary]. Roczniki Nauk Roln. B, 77(1):223–242.

Patalas, K. 1961. Wind- und morphologiebediente Wasserbewegungstypen als bestimmender Faktor für die Intensität des Stoffkreislaufes in nordpolnischen Seen. Verh. Internat. Verein. Limnol. 14:59–64.

Petrovich, P. 1961. The relation between phytoplankton, bacteria, zooplankton and large aquatic plant standing crops and production in Naroch, Miastro and Batorin Lakes [in Russian]. Primary production in seas and inland waters, Minsk, 381–385.

Pravdin, I. F. 1956. Fishery resources of Lake Ladoga and their utilisation [in Russian]. Izv. Vsesoyuzn. Nauchn. Issledovat. Instituta Ozern. i Rechn. Rybnovo Chozyaistva 38:3–11.

Priimachenko, A. D. 1960. Main regularities of composition and biomass distribution of phytoplankton in water reservoirs of some rivers of the USSR [in Russian]. Tr. Inst. Biol. Vodokhran. 3:59–86.

Procházková, L. 1966. Seasonal changes of nitrogen compounds in two reservoirs. Verh. Internat. Verein. Limnol. 16:693–700.

Romanenko, V. I. 1964. Heterotrophic assimilation of CO_2 by the aquatic microflora [in Russian]. Mikrobiologiya 33:679–683.

Romanenko, V. I. 1965. Comparative characteristic of microbiological processes in reservoirs of different types [in Russian]. Tr. Inst. Biol. Vnutr. Vod 9:233–246.

Rudolf, K. 1966. Untersuchungen über die Ernährungsbeziehungen in der Freiwasserregion der Saidenbach-Talsperre. Wiss. Z. Karl-Marx-Univ. Leipzig, Math.-Naturwiss. Reihe 15(1):229–245.

Rutkovskii, V. I., and T. N. Kurdina. 1959. Hydrobiological budget of Rybinskoe Reservoir for the period 1947–1955 [in Russian]. Tr. Inst. Biol. Vodokhran. 1:5–24.

Schräder, Th. 1958. Thermische Verhältnisse in Talsperren. Gewässer u. Abwässer 21:68–88.

Sczepański, A. 1961. Limnological characteristics of Lake Hancza [in Polish, English summary]. Polsk. Arch. Hydrobiol. 9:9–18.

96 STRAŠKRABA AND STRAŠKRABOVÁ

Sebestyén, O. 1962. Ergenbnisse der Balaton-Forschung der letzten fünfzehn Jahren 1946–1960. Ann. Biol. Tihany 29:217–266.

Serafimova-Hadžišče, J. 1957. Zooplankton na Ochridskoto ozero vo tekot na 1952, 1953 i 1954 godina. Le zooplancton du lac d'Ohrid au cours des années 1952, 1953 et 1954 [in Serbocroation, French summary]. Posebni izdanija, Sta. hydrobiol. Ohrid 1:3–65.

Sládeček, V., and A. Sládečková. 1963. Limnological study of the reservoir Sedlice near Želiv. XXIII Periphyton production. Sborník Vys. Školy Chem. Techn., Techn. Vody, Praha 7(2):77–134.

Sládečková-Vinníková, A. 1957. Studie nárostů v Sedlické nádrži na Želivce I. Study of growths in the Sedlice Reservoir on the Želivka River I [in Czech, English summary]. Vodní Hospodářství 7:183–184.

Sládecková-Vinníková, A. 1958. Studie nárostů v Sedlické údolní nádrzi II. Study of growth in the Sedlice Reservoir on the Želivka River II [in Czech, English summary]. Vodní Hospodářství 8:156–157.

Sládečková, A. 1960. Limnologická studie o nádrži Sedlice u Želiva. XI. Vertikální zonace nárostů v průběhu prvého celoročního sledování (od června 1957 do července 1958). Limnological study of the Reservoir Sedlice near Želiv. XI. Periphyton stratification during the first year-long period (June 1957-July 1958) [in Czech, English summary]. Sborník Vys. Školy Chem. Techn., Techn. Vody, Praha 4(2):143–262.

Sládečková, A. 1962. Limnologická studie o nádrži Sedlice u Želiva. XX. Vertikální zonace nárostů y průbehu celorocního sledování–od srpna 1958 do cervence 1959. Limnological study of the Reservoir Sedlice near Želiv. XX. Periphyton stratification during the second year-long period–August 1958-June 1959 [in Czech, English summary]. Sborník Vys. Školy Chem. Techn., Techn. Vody, Praha 6(1):221–292.

Sládečková, A. 1966. The significance of the periphyton in reservoirs for theoretical and applied limnology. Verh. Internat. Verein. Limnol. 16:753–758.

Slobodchikov, B. Ya. 1951. Hydrochemical regime of Lake Sevan, data of 1947-1948 [in Russian]. Tr. Sevanskoi Gidrobiol. Sta. 12:5–28.

Slobodchikov, B. Ya. 1955. Oxygen conditions of Lake Sevan, data from 1947-1948 [in Russian]. Tr. Sevanskoi Gidrobiol. Sta. 14:165–181.

Smetanich, V. S. 1963. Recent literature on Rybinskoe Reservoir [in Russian]. Tr. Inst. Biol. Vnutr. Vod 6:309–317.

Sorokin, Yu. I. 1958. Primary production of organic matter in the water column of Rybinskoe Reservoir [in Russian]. Tr. Biol. Sta. Borok 3:66–88.

Sorokin, Yu. I. 1961. Photosynthetic production in Volga-reservoirs during late June 1959 [in Russian]. Biul. Inst. Biol. Vodokhran. 11:3–6.

Sorokin, Yu. I. 1964. On the trophic role of chemosynthesis in water bodies. Int. Rev. Ges. Hydrobiol. 49:307–324.

Sorokin, Yu. I. 1965. On the trophic role of chemosynthesis and bacterial biosynthesis in water bodies. Mem. Inst. Ital. Idrobiol. 18, suppl.:187–205.

Stangenberg, M. 1936. Szkic limnologiczny na tle stosunkow hydrochemicznych Pojezierza Suwalskiego. Limnologische Charakteristik der Seen des Suwalki-Gebietes auf Grund der hydrochemischen Untersuchungen [in Polish, German summary]. Rozpr. i Spraw. Inst. Bad. Lasów Panstw Warszawa A 19.

Stankovič, S. 1960. The Balkan lake Ohrid and its living world. Den Haag, Uitgeverej Dr. W. Junk, 356 pp.

Štěpánek, M. 1960. Limnological study of the Reservoir Sedlice near Želiv. X. Hydrobioclimatological part: The relation of the sun radiation to the primary production of nannoplankton. Sborník Vys. Školy Chem. Techn., Techn. Vody, Praha 4(2):21–142.

Štěpánek, M., J. Chalupa, and L. Mašínová. 1960. Application of algicide Ca 350 in reservoirs. Sborník Vys. Školy Chem. Techn., Techn. Vody, Praha 4(2):375–402.

Štěpánek, M., and M. Zelinka. 1962. Limnological study of the reservoir Sedlice near Želiv XIX. The influence of atmospheric pressure upon the phytoplankton development [in Russian, English summary]. Sborník Vys. Školy Chem. Techn., Techn. Vody, Praha 6(1):193–220.

Štěpánek, M., J. Binovec, and J. Chalupa. 1963. Vodní květy v ČSSR. Water blooms in Czechoslovakia [in Czech, English summary]. Sborník Vys. Školy Chem. Techn., Techn. Vody, Praha 7(2):175–264.

Stone, H. M. 1965. A bacteriological and chemical survey of the Hutt River, New Zealand. New Zealand J. Sci. 8:54–65.

Straškraba, M. 1966. Interrelations between zooplankton and phytoplankton in the Reservoirs Slapy and Klíčava. Verh. Internat. Verein. Limnol. 16:719–726.

Straškraba, M., and J. Hrbáček. 1966. Net-plankton cycle in Slapy Reservoir during 1958-1960. Hydrobiol. Stud. 1:113–153.

Straškrabová, V. 1968. Bacteriologische Indikation der Wasserverunreinigung mit abbaubaren Stoffen. Limnologica Berlin 6:29–36.

Straškrabová-Prokešová, V. 1967. Seasonal changes in reproduction rate of water bacteria in two reservoirs. Verh. Internat. Verein. Limnol. 16:1527–1533.

Stroikina, V. T. 1952. Phytoplankton of Lake Sevan [in Russian]. Tr. Sevanskoi Gidrobiol. Sta. 13:171–212.

Sushchenya, L. 1961. Utilization of primary production in the subsequent links of a food chain [in Russian]. Primary production in seas and inland waters, Minsk, 386–396.

Vinberg, G. G. 1960. Primary production of water bodies [in Russian]. Minsk Edit. Acad. Sci. USSR, 439 p.

Vineckaya, I. I. 1961. Primary production in the Northern Caspian Sea [in Russian]. Primary production in seas and inland waters, Minsk, 52–66.

Vladimirova, K. S. 1947. Phytoplankton of Lake Sevan [in Russian]. Tr. Sevanskoi Gidrobiol. Sta. 9:69–144.

Wetzel, A. 1958. Bemerkungen über die Biologie von Trinkwassertalsperren. Desinf. u. Gesundh.-Wesen 50(11):97–101.

Wetzel, A. 1962. Biologische Beschaffenheit u. Gütezustand von Trinkwasser-Talsperren. Wasserwirtschaft-Wassertechnik 12:55–61.

Wiszniewski, J. 1953. Uwagi w sprawie typologii jezior polskich. Remarques sur la classification typologique des lacs de Pologne [in Polish, French summary]. Polsk. Arch. Hydrobiol. 1:11–23.

Zevin, A. A., V. I. Rogozhkin, and N. G. Fesenko. 1961. On the character of water movements in the dam areas of Cimlyanskoe, Gorkovskoe, Kuibyshevskoe and Stalingradskoe Reservoirs [in Russian]. Gidrochim. Materialy 32:113–121.

Zhadin, V. N., and M. B. Ivanova. 1959. Survey of world literature on water reservoirs [in Russian]. Tr. VI. Sov. Po Probl. Biol. Vnutr. Vod, 615–622.

SHOJI HORIE
Otsu Hydrobiological Station, Kyoto University, Shiga-Ken, Japan

Asian Lakes

BASIC CONCEPTS

Since modern limnological work began in Europe most investigations have been carried out by biologists, particularly zoologists, on the biota of lakes and swamps. On the basis of comparative hydrobiology, these biological investigators separated harmonic lakes into two groups: oligotrophic and eutrophic. The investigators, including such distinguished limnologists as E. Naumann and A. Thienemann, did a tremendous amount of work in classifying lakes as to their trophy.

Their basic idea seems to be that lake eutrophication is unilateral. Early in their development, most lakes are oligotrophic, but as time passes, their basins gradually fill with both autochthonous and allochthonous material; the hypolimnion and the tropholytic layer become smaller and smaller; organic production increases rapidly as chemically reduced nutrient salts appear in the water; dissolved oxygen is consumed by the rapid metabolism stemming from the increased organic production. Finally, each basin consists only of the epilimnion and trophogenic layer, and these are moving toward swamp and grassland.

Lake succession from oligotrophy through mesotrophy to eutrophy, or change in type from tanytarsus through chironomus to corethra, has been the limnologists' model of lake transition. For the lakes that the limnologists have studied—lakes in Northern Europe or in young volcanic regions—this model is valid. But lakes in Northern Europe have appeared only during the past 10,000 years, since the deglaciation of the Würm (Wisconsin) continental ice sheet. Investigations have dealt with hydrobiological features of recent or comparatively recent times.

When lake succession is considered from the geological point of view, problems arise. Paleolimnological data obtained in ancient lakes that passed through Pleistocene and Holocene climatic changes indicate that the climatic changes have complicated the history of lake trophy. Crustal deformation joins in this complication.

Generally speaking, changes in the atmosphere or lithosphere lead to changes in the hydrosphere and biosphere. For instance, fluctuations of solar radiation, precipitation, or evaporation, and movements of the earth's crust (as through volcanic activity or upheaval or subsidence of the lake's basin) change the degree of penetration of the sun's rays or the depth of the lake water. Inasmuch as oligotrophy and eutrophy are roughly determined by the relation between two volumes of water (that is, the trophogenic layer and the tropholytic layer), the gradual increase in lake trophy must have been interrupted by changes in the atmosphere and lithosphere that have occurred many times. Tropical and polar lakes, however, are insensitive to these changes.

Even in lakes that have been formed during the last 10,000 years, oligotrophy and eutrophy might alternate to some degree. It is well known, for example, that the air temperature was higher during the postglacial Hypsithermal Interval than it is now. Insofar as the present circulation pattern is controlled by temperature, even such a minor oscillation of air temperature as this might affect lakes, particularly those located on the boundary between dimictic and warm monomictic lakes. Accordingly, lakes might also alternate between aerobic and anaerobic, conditions that control metabolism in the lake.

If we could live for several thousand years, we would be able to witness these changes. Fluctuations of temperature between glacial and interglacial ages must have shifted zones of lacustrine climate back and forth. This shift must have occurred horizontally, from pole to Equator, as well as vertically on the earth's surface, and on a hemispheric or global scale (Horie, unpublished data). From this point of view, I hope to show that *a lake is an unstable microcosm.*

Let us direct our attention to lake eutrophication as it now exists. Because the process of eutrophication can be detected only through observations made over a period of many years, data from one observation are not sufficient for discerning the process. Unfortunately the period of study is inadequate, even for lakes in Japan itself (though we have numerous limnological data on each lake), to say nothing of lakes in other Asiatic countries and former territories of Japan. On the other hand, even though we obtain data over a period of many years, we still must consider daily or monthly fluctuations in data. In my most recent limnological study (for 24 hours at the center of Lake Biwa-ko), I found that every sample differed slightly from all preceding samples. Of course, a lake has its unique character. But if we are to understand it more than superficially we must observe it continually and sample and analyze it over a long period. Eutrophication has come to be noticed by the general public as well as by limnologists. In order to deal with eutrophication, I have classified the process into local eutrophication and global eutrophication. Included in local eutrophication are local movements of the earth's crust, cultivation of the shoreline, and artificial oscillation of lake level, as described by Rodhe (1964). Global eutrophication includes climatic changes. At the moment, we probably see the results of a combination of local and global eutrophication. In addition, the influence of industrial pollution and sewage, which are local phenomena, is becoming a serious problem. These agents differ from other agents of eutrophication in their high speed and, frequently, in their sapropellization of lake water.

According to my study on the developmental history of Japanese ancient lakes, all of which seem to be harmonic, oligotrophy and eutrophy apparently have alternated. This alternation means that the mechanism is reversible. But insofar as the influence of the polluted water and sewage is eliminated by the buffer action of the lake, influx of pollution might bring about the lake's eutrophication. A close relation has already been verified between the amount of sewage and the appearance of algal water bloom. However, when the amounts of sewage exceed a lake's buffer action or dilution power, the eutrophic lake changes to a worse state, defined as *Eusaprobität* by Sládeček (1961). Then only limited kinds of organisms tend to develop extensively, just as in the case of disharmonic lake types: acidotrophic, dystrophic, siderotrophic, and so on. After the critical point is reached, the lake continues to move in only one direction; that is, it is irreversible. Sládeček (1961) has tried to classify lake-water pollution. Similar work was carried out by Thomas (1944), Cyrus and Cyrus (1947), Šrámek-Hušek (1950, 1956), and Tsuda (1964). Following the classification of Sládeček, increased pollution causes the transition from *Limnosaprobität* to *Isosaprobität*, in which case the water contains no dissolved oxygen but a small amount of hydrogen sulfide. Indicative organisms are *Paramoecium putrinum, Colpidium colpoda, Glaucoma scintillans, Tetrahymena pyriformis, Uronema bütschlii, Dexiotrichides centralis, Tillina magna, Vorticella microstoma, Polytoma uvella, Trepomonas agilis, Sarcina, Streptococcus, Spirillum, Zoogloea,* and *Amoeba limax.* Biological oxygen demand is usually 40 to 400 mg per liter of O_2. Worse than *Isosaprobität* is *Metasaprobität*, in which there is no dissolved oxygen but in which there are 50 to 60 mg per liter of hydrogen sulfide. Indicative organisms are *Cercobodo longicauda, Bodo putrinus, Oicomonas mutabilis, Trigonomonas compresia, Tetramitus pyriformis, Hexamitus, Peloploca, Spirillum,* and *Lamprocystis.* Biological oxygen demand is usually

200 to 700 mg per liter of O_2. Worse than *Metasaprobität* is *Hypersaprobität*. In this case, there is no dissolved oxygen, but several mg per liter of hydrogen sulfide are found. Indicative organisms are bacteria. Biological oxygen demand is usually 500 to 1,500 mg per liter of O_2. A typical example of *Hypersaprobität* is sapropel. This the primary step caused by sewage and industrial pollution is eutrophication. But when pollution accumulates past the lake's buffering capacity, the influence of the pollution causes a change to *Eusaprobität,* accelerating formation of sapropel step by step toward *Hypersaprobität.* This is what I mean by "sapropellization." Its speed depends on various factors, including size of lake, volume of lake water, physicochemical and hydrobiological character of the lake itself, growth rate of population around the lake, degree of artificial purification, and amount of agricultural runoff. During the last two decades, however, sapropellization has become noticeable in many parts of the world because of the evolution of industry; such artificial deformation of lake water to sapropellization is much faster than natural movement toward a higher degree of eutrophy of lake water. *Hypersaprobität* also moves toward *Ultrasaprobität,* causing more serious problems among people.

As an example of lake eutrophication moving toward sapropellization, I should like to mention Lake Suwa-ko, situated in central Japan (Figures 1 and 2). Its area is 14.5 km^2, and its volume is 0.06 km^3 (Horie, 1962).

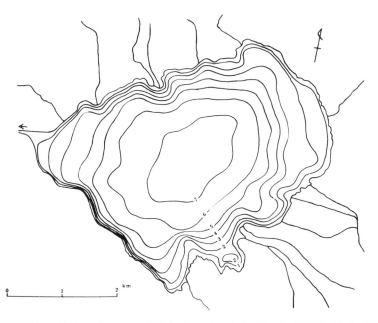

FIGURE 1   Bathymetric map of Lake Suwa-ko. After Tanaka (1918), Horie (1962).

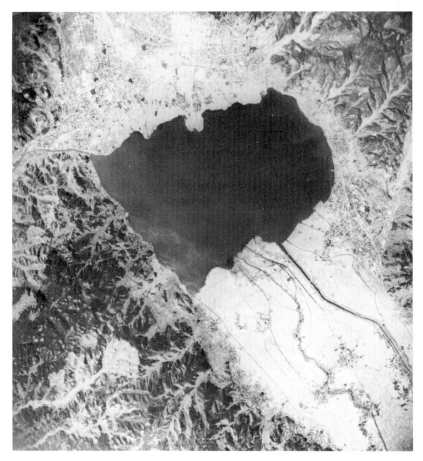

FIGURE 2   Aerial photograph of Lake Suwa-ko. (U.S. Air Force photo.)

Around the lake are big factories. But the most important activity is the hotel business, since Suwa, with its hot springs, is a famous health resort. Accordingly, this area is densely populated, residents numbering about 180,000. (In addition, the resort attracts many visitors throughout the year.) It has been said that Lake Suwa-ko is a typical eutrophic lake, but during the last 20 years, lake-water pollution, mainly by sewage from hotels, is much more noticeable than it was previously. According to Koidsumi and others (Koidsumi *et al.*, 1967; Koidsumi *et al.*, 1968), people living near the lake have noticed that the lake has become badly polluted. The waters in the littoral region of the urban areas have become heavily stained; the Secchi disc transparency is below 50 cm. The transparency in the pelagic zone, however, decreased only slightly between 1904 and 1966.

Dominant rooted aquatic plants in 1966 were as follows (figures denote estimated total standing crop of the plant, expressed as tons of wet weight):

Emerged	Floating	Submerged
Zizania latifolia 234	*Trapa natans* 630	*Hydrilla verticillata* 597
Nuphar japonicum 142		*Potamogeton malaianus* 276
Phragmites communis 63		*Vallisneria asiatica* 16

The total area occupied by the plants was about 0.92 km^2, which constitutes 6.3 percent of the area of the lake (14.5 km^2). In the late summer, when the growth reached the maximum, the total standing yield of the plants in the lake amounted to about 1.958 tons in wet weight and to about 220 tons in dry weight. The wet quantity is just twice as large as that reported in 1949 (970 tons). Ecological succession in the hydrophytes was quite noticeable. Generally, the submerged plants have decreased and the floating and emerged plants have become luxuriant.

Suspended solids are composed mainly of organic constituents, which account for 60 to 80 percent of the total weight. Suspended solids are presently 5 to 15 times as plentiful as in the years before 1950, both in total amount and in amount of organic ingredients. Filtering the lake water through any filter material yields water that is almost colorless and transparent. Accordingly, the recent striking pollution is apparently due to the marked increase of substances in suspension. The most important suspended solids making water opaque are plankton organisms, which are abundant. Analysis shows that the amounts of nitrogen and phosphorus in lake water, essential nutrients for plants, have recently increased greatly.

In addition, suspension of reducing substances and upward movement of anaerobic water frequently cause fish kills amounting to 50 to 60 tons.

Rapid change in the trophy of Lake Suwa-ko as a result of human activity is thus obvious. A sewage countermeasure is planned. Unless vigorous action is taken, early sapropellization of the lake is considered inevitable.

DOCUMENTATION OF EUTROPHICATION

In the Asian part of the continent of Eurasia, the earth's crust is more unstable than anywhere else in the world. Here the crust zone shows the Himalayan Orogeny, derived from the Tethys Geosyncline, and many active and inactive volcanoes, which extend from Kamchatka to Indonesia through Japan and the Philippines. There are also many large rivers flowing on the plains, for example, the Yangtze, Huang, Mekong, and Irrawaddy.

A characteristic feature of continental Asia is rather scanty distribution of the glacial morphology formed by the continental ice sheet. Such a morphological character must be kept in mind when we compare the limnology of the Asian continent with that of Europe and North America.

Because this area extends from a polar climate to a tropical climate and also from a humid, oceanic climate to an arid, continental climate, variability of the limnological character is distinct. In this area, lakes of ancient origin are found—lakes that began their limnetic history in the Tertiary period. Examples are Lakes Baikal and Khanka in Siberia and Lake Biwa-ko in Japan. Possible examples are Lake Lanao in the Philippines and Lakes Matana, Towuti, and Mahalona in Celebes.

Accordingly, we might say that the Asian continent has almost all types of lakes. It is, however, unfortunate that the data on eutrophication are meager when we compare them with data from North America and Europe. Although Japanese limnologists studied many lakes of Asia before and during the Second World War, a considerable amount of their data is not yet published. It is desirable that these data be published soon in occidental language; when limnological investigations in these lakes are resumed, changes of trophy during these years will be clarified.

JAPAN

There are more than 600 lakes in Japan (Horie, 1962). These lakes are of many kinds and have various origins. Limnologically, Japan is the best known country in Asia.

In 1927, kokanee salmon were transplanted from Lake Towada-ko to Lake Akan-ko, which has an area of 11.8 km^2 and a maximum depth of 36.6 m. This was done because fishermen had caught most of the native kokanee salmon in Lake Akan-ko. At that time an increase of phytoplankton was observed, but even in 1935 zooplankton such as *Daphnia* and *Acanthodiaptomus* was found as the main contents of the stomach of kokanee salmon, though the sign of *Vegetationstrübung* was noticed. However, in the summer of 1944, water bloom of *Anabaena flos-aquae* appeared; in the autumn of 1966, *Ceratium hirundinella* and *Melosira varians* were dominant, and the Crustacea that I have mentioned were scanty. Transparency in 1955 was about half what it was in 1917. Although the influence of polluted water and sewage has been particularly noticeable since 1955, symptoms of eutrophication appeared around 1935 (Ueno, 1966).

Lake Haruna-ko is a volcanic lake in central Japan. Its area is 1.23 km^2, and its maximum depth is 13 m. Yoshimura (1933a, b, and 1937) compiled data on the eutrophication of this lake.

As for plankton succession, in September 1927 Ueno collected:

Daphnia longispina longispina (O. F. Müller)   most abundant
Bosmina longirostris (O. F. Müller)   abundant
Diaptomus pacificus Burckhardt   common
Volvox globator Leeuwenhoek   abundant

Volvox, Daphnia, and *Diaptomus,* which are common in the mountainous mesotrophic lake, were dominant, and none of the blue-green algae was noticed.

In 1930, at the end of September, the Gunma Fishery Experimental Station collected: *Anabaena flos-aquae, A. spiroides, Navicula* sp., *Amphora* sp., *Melosira* sp., *Asterionella* sp., *Ulothrix zonata, Volvox globator, Volvox* sp., *Asplanchna priodonta, Cyclops* sp., and *Diaptomus* sp.

Since Miyadi found *Anabaena* in 1929, it may have begun to appear around 1928–1929. In 1930, no *Daphnia longispina,* which was abundant before, was collected. In the summer of 1931, the existence of *Anabaena* also was noticed. On June 26, 1932, Yoshimura observed that *Anabaena* had formed thick water bloom in parts shallower than 5 m and that the water was yellow-green, like eutrophic lakes in the plains. As for the chemical elements (Table 1), the total residue in the water was rather moderate for Japan.

TABLE 1 Chemical Composition of Water in Lake Haruna-ko, June 26, 1932 (milligrams per liter)[a]

Components	Depth (m)			
	0	5	10	12
Na^+	9.4	8.9	–	6.9
Ca^{2+}	12.9	–	–	13.1
Mg^{2+}	2.4	–	–	2.0
Cl^-	6.0	–	–	6.0
SO_4^{2-}	–	10.5	8.0	–
SiO_2	24.2	24.1	25.8	27.8
HCO_3^-	19.3	21.0	27.2	29.9
Fe	–	–	0.3	3.0
Mn	–	–	0.1	2.0
$KMnO_4$ consumption	2.7	2.7	2.9	9.5
Nitrogen as N				
Free NH_3	0.09	0.08	0.05	0.19
N_2O_5	0.02	0.015	0.01	0.02
Albuminoid NH_3	0.10	0.06	0.05	0.10
Phosphate as P_2O_5				
Free	0.005	0.01	0.015	0.07
Total	–	0.06	–	–
Total Residue	–	73.3	79.8	90.8

[a]Source: Yoshimura (1933a).

Silicate is the most important constituent of this lake. On the whole, the water showed no characters worth mentioning. Slight stratifications of all the salts were seen. Except in the bottom layer, where large amounts of iron and manganese had accumulated, the water was very poor in humic substance, as evidenced by the small amount of potassium permanganate that was required to oxidize the organic matter in it. The total amount of free ammonia, nitrate, and albuminoid nitrogen was about 0.15 mg per liter—the critical value separating the oligotrophic from the eutrophic type. The stratification of the soluble phosphate was very clear, the surface water being nearly devoid of it. The total amount of phosphate, which probably is the limiting factor for the production of phytoplankton in this lake, was rather large. On June 26, 1932, no *Volvox* or *Diaptomus* was collected; instead, *Eudorina elegans* was taken.

As for the benthic fauna, at the time of Miyadi's visit in 1929, the mud even at the greatest depth was said to be gray or a typical gyttja, whereas Yoshimura found that mud below a depth of 10 m became blackish or exhibited the character of sapropel. Grayish mud or diatom gyttja developed only between 5 and 10 m; it contained large amounts of both planktonic and benthic diatoms, such as *Melosira, Surirella, Navicula,* and *Pleurosigma.* The shallower zone was composed of sand, mingled with the detritus of water and land plants. Miyadi collected a good deal of *Chironomus plumosus* and *Tubifex* from the profundal region. *Plumosus* was found most abundantly between 9 and 12 m, *Tubifex* between 9 and 11 m. They appeared to avoid the deepest bottom, where the dissolved oxygen was depleted. Yoshimura made many vain attempts to collect the benthic fauna from bottoms deeper than 10 m. For every five attempts, Yoshimura obtained only one *Tubifex,* and he did not find *Chironomus plumosus* at all. There were a few full-grown larvae and pupae of *Plumosus, Chironomus connectens,* and Tanypinae in the zone between 5 and 9 m. The amounts in that zone were nearly equal to those of the previous catch by Miyadi in August 1929.

The fact that the collections were made in different seasons accounts, of course, for the decrease of *Plumosus* in the profundal region; some eggs laid at the beginning of summer had hatched in August. But the number of full-grown larvae of *Plumosus* (hatched the previous year) was much less at the time of Yoshimura's visit. The decrease is undoubtedly due to the tendency of the mud to change to sapropel as a result of a decrease in dissolved oxygen (Table 2, Figure 3). It may be said that eutrophication has proceeded from time to time.

Increase of dissolved nutrients contributes to the formation of water bloom of a blue-green alga, *Anabaena.* Increase of organic matter reduces the quantity of dissolved oxygen in the bottom layer and changes the gyttja into sapropel, the poisonous property of which prevents the growth of *Chironomus plumosus* and *Tubifex* in the profundal region. Hydrogen sulfide

TABLE 2   Physicochemical Changes in Lake Haruna-ko, by Date[a]

Depth (m)	Temperature June 26, 1932 (^{o}C)	pH August 2, 1929	pH July 29, 1931	pH June 26, 1932	pH August 17, 1936	Carbon Dioxide as CO_2, June 26, 1932 Free (mg/liter)	Carbon Dioxide as CO_2, June 26, 1932 Carbonate (mg/liter)
0	18.0	>7.6	9.5	8.5	7.8	−1.1	13.9
1	17.5	−	−	−	7.8	−	−
2	16.7	−	−	8.5	7.8	−	−
3	16.5	>7.6	9.2	−	7.8	−	−
4	16.1	−	−	8.3	7.8	−	−
5	15.8	7.6	−	8.1	7.6	0.4	15.1
6	13.9	−	6.5	7.1	6.5	−	−
7	13.1	7.4	−	6.8	6.4	−	−
8	11.9	−	−	6.6	6.4	−	−
9	10.6	−	6.2	6.5	6.3	−	−
10	9.9	6.8	−	6.4	6.4	1.5	19.6
11	9.3	−	−	6.4	6.4	−	−
12	9.2	6.8	6.5	6.4	6.4	6.0	21.5
12.5	9.2	−	−	6.4	−	−	−

Dissolved Oxygen

Depth	August 6, 1929 (cc/liter)	(%)	June 26, 1932 (cc/liter)	(%)	August 17, 1936 (cc/liter)	(%)
0	5.67	110	5.93	104	5.41	−
1	−	−	−	−	−	−
2	−	−	6.43	110	5.26	−
3	7.53	139	−	−	−	−
4	−	−	6.39	107	5.39	−
5	5.85	101	6.21	104	6.12	−
6	−	−	6.03	97	−	−
7	5.08	81	−	−	2.30	−
8	−	−	5.03	78	1.28	−
9	2.89	44	−	−	0.16	−
10	0.37	5	0.58	7	0.06	−
11	0.08	−	0.15	2	−	−
12	0.00	−	0.00	0	0.00	−
12.5	−	−	−	−	−	−

[a]Source: Yoshimura (1933a, 1937).

kills a number of gibels during the autumnal partial circulation. A more interesting fact is that in the middle of August 1936, Yoshimura found that *Anabaena* had decreased remarkably, and he inferred that the decrease occurred in 1934. In addition, *Volvox, Daphnia,* and *Diaptomus,* together with *Dinobryon divergens,* reappeared. Thus, during a few years, oligotrophication progressed in this lake. Continual records of transparency and pH are shown in Tables 2 and 3.

Because pollution and sewage are rare in this mountainous lake, I believe the change is due to climatic events, not to local conditions induced by human beings. Onuki (1950) found that between May and October of 1948, dominant limnoplankton of this lake were *Asterionella formosa* (May 15), *Dinobryon divergens* (June 30), *Dinobryon divergens* and *Asterionella formosa* (July 10), *Anabaena planctonica* (August 31), and *Bosmina longirostris* and *Eudorina elegans* (October 3).

From these data, I conclude that Lake Haruna-ko is unstable. If climatic elements such as temperature or precipitation are favorable to eutrophication, eutrophic tendency takes place, and vice versa. Accordingly, it seems likely

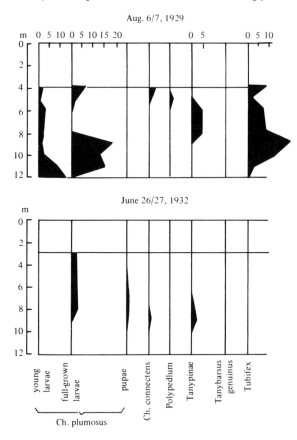

FIGURE 3   Number of benthic fauna in Lake Haruna-ko in summer of 1929 compared with number in summer of 1932 (number per 0.02 m^2 of surface). After Yoshimura (1933a).

TABLE 3   Changes in Transparency in Lake Haruna-ko, 1906 to 1936[a]

Date	Transparency (m)	Color
July 4, 1906	9.5	V
July 20, 1908	6.0	–
May 13, 1924	5.7	–
Aug. 6, 1929	3.2	–
Sept. 25, 1930	4.0	–
June 26, 1932	2.3	X–XI
June 29, 1932	1.3	IX
Aug. 30, 1932	1.2	XVI–XVII
Oct. 27, 1932	2.0	X
Aug. 17, 1936	4.9	VI–VII

[a]Source: Yoshimura (1933a, 1937).

that *temporary fluctuation of lake trophy* (Horie, 1966) occurs in this lake, which has a mean depth of 8.1 m. Future work on the paleolimnology of this lake must be of interest from the standpoint of lake eutrophication. It is unfortunate that since 1950 we have had no data on Lake Haruna-ko comparable with our data on Lakes Yogo-ko and Biwa-ko.

Lake Yogo-ko, a tiny lake situated in the northernmost part of the Lake Biwa basin, is the first lake in which paleolimnological work in Japan has been carried out (Horie, 1966).

In recent years, changes in natural water quality have become noticeable in connection with the increased activity of people. In addition, changes of climate affect the physics of the lake, causing changes in its chemistry and biota. With respect to this point, the nature of the stratification of water of Lake Yogo-ko must be noted. Having cited A. Tanaka's opinion, Yoshimura (1936) stressed that Lake Yogo-ko is located on the boundary between the temperate and tropical lakes in Japan with regard to water temperatures. This was also suggested by Miyadi and Hazama (1932), who wrote as follows: "The ice cover seldom forms and the reversal stratification, if it occurs at all, is temporary and soon disturbed by the wind. The lake is homothermous in January, March, October and December." My observation in February 1962 was also made soon after the melting of thin ice that had covered most of the lake. Accordingly, if air temperature drops slightly, the lake will show indirect stratification. Therefore, in limnological investigations in Lake Yogo-ko, which is located on the dimictic–warm monomictic boundary and is sensitive to minor oscillations of climate, attention must be paid to the climatological facts.

One proof of change in lake trophy is change in transparency, for which I have compiled data covering the last 40 years. Obviously, the value drops, but it is difficult to discern whether the drop occurs gradually or rapidly. Data from the summer of 1951, however, suggest that the present low transparency already existed in 1951 (Figure 4). It should be kept in mind that the northern shore of the lake is a paddy field.

The following discussion of chemical changes in the water of Lake Yogo-ko is based on sporadic data from past observations. Such data, of course, are not adequate for a detailed study; only a general trend can be described.

Yoshimura (1931a) analyzed surface water collected September 27, 1930, as follows:

pH	6.9
SiO_2	3.3 mg/liter
Cl	6.0 mg/liter
Ca	4.7 mg/liter
$KMnO_4$ consumption	10.5 mg/liter
N	0.175 mg/liter

He supplemented the data in Table 4 with the data given above. A year later he published the remaining data on water that were collected September 27, 1930 (Yoshimura, 1932a), as follows:

NH_3	80 mg/m^3
N_2O_5	15 mg/m^3
Albuminoid NH_3	80 mg/m^3
Sum	175 mg/m^3
P_2O_5 (soluble)	60 mg/m^3

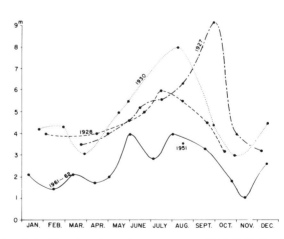

FIGURE 4   Lake Yogo-ko, change in transparency, 1927-1962 (Horie, 1967b).

In addition, Yoshimura (1932b) reported 1.0 mg/liter of CaO and the data on chemical composition contained in Table 5.

According to later data published by the same author (Yoshimura, 1935), the nitrogen content of the surface water of Lake Yogo-ko on December 15, 1930, was 0.07 mg/liter.

M. Tanaka's papers on the chemistry of the lake water (Tanaka, 1953, 1954 a and b) showed that the amount of total Fe, Fe^{2+}, Mn, P, and Si increased remarkably as the amount of dissolved oxygen in the profundal part of the lake decreased in mid-August. The increases were as follows (mg/liter): Fe^{2+}, 6.85; Mn, 2.25; P dissous, 0.66; Si dissous, 6.1.

Another important study was carried out by T. Koyama (unpublished data), who analyzed water collected from Lake Yogo-ko on August 21, 1951 (Table 6).

Further chemical data are given in Table 7, which is from Horie (1968). Comparing the data of Yoshimura, Tanaka, Koyama, and Horie, one clearly sees increases in nitrogen, silicate, and chemical oxygen demand. These data and the data on transparency variation support the conclusion that the productivity of this lake has been increasing during the last 40 years.

TABLE 4   Chemical Analysis, Lake Yogo-ko (Sample Collected September 27, 1930)[a]

Depth (m)	Constituent (mg/liter)			
	SiO_2	Fe_2O_3	Mn	CaO
0	2.3	1.0	0.04	1.2
9	3.8	1.6	0.8	–
12	10.1	20.0	1.8	–

[a]Source: Yoshimura (1931 b and c).

TABLE 5   Chemical Analysis, Lake Yogo-ko (Sample Collected September 27, 1930)[a]

Depth (m)	Constituent (mg/liter)		
	CO_2		
	Free	Fixed	Ca
0	8.4	7	4.6
9	5.2	12	4.2
12	10.8	24	3.9

[a]Source: Yoshimura (1932b).

TABLE 6   Chemical Composition of Water at Station 1, Lake Yogo-ko, August 21, 1951 (mg/liter)[a]

Depth (m)	Ca	Mg	Cl	SO_4	Fe^{2+}	Fe (Unfiltered)		Mn
						Fe Total	Fe Sol-uble	
0	2.94	1.28	6.66	1.09	<0.01	0.215	0.042	<0.02
2.5	–	–	–	–	<0.01	0.244	0.047	<0.02
5	2.90	1.33	7.00	1.09	<0.01	0.265	0.070	0.12
7.5	–	–	–	–	<0.01	0.320	0.094	0.17
10	3.40	1.44	6.58	1.87	2.66	3.58	3.28	2.05

	Al	SiO_2	P	NH_4-N	NO_2-N	NO_3-N	$KMnO_4$ consumption
0	0.223	3.86	<0.01	<0.02	<0.001	<0.02	12.2
2.5	0.353	4.78	<0.01	<0.02	<0.001	<0.02	10.0
5	0.243	5.27	<0.01	<0.02	<0.0015	<0.02	11.2
7.5	0.302	5.78	<0.01	<0.02	<0.001	<0.02	10.3
10	0.288	7.07	0.392	<0.02	<0.001	<0.02	11.0

[a]Source: T. Koyama (unpublished data).

TABLE 7   Chemical Composition of Water, Lake Yogo-ko, September 3, 1962 (mg/liter)[a]

Depth (m)	Fe^{2+}	Mn	SiO_2	Ca	Cl	H_2S	Total Sulfide	NH_4-N
0	–	Trace	6.00	3.00	5.61	–	–	0.11
8	–	0.20	6.70	5.48	5.69	–	–	0.20
12	4.000	1.10	12.00	7.16	6.36	0.01	0.055	2.00

	NO_2-N	NO_3-N	Total -N	PO_4-P	Total -P	COD	Total Residue	Ignition Loss
0	0.0004	0.020	0.375	0.0036	0.0107	13.40	54.0	11.1
8	0.0018	0.036	0.510	0.0124	0.0204	13.33	64.5	14.5
12	0.0042	0.760	2.050	0.810	4.800	13.82	86.5	16.8

[a]Source: Horie (1968).

As for the paleolimnology of this lake, I have obtained cores totaling 9.8 m. According to a previous study (Horie, 1966), symptoms of eutrophication occurred many years ago. That event at the 6-m level might have occurred because of the drop of lake level in this closed lake (Figure 5). Another point

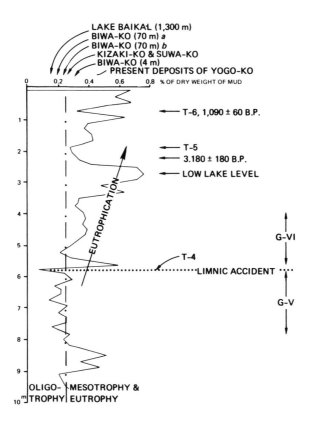

FIGURE 5   Developmental process of Lake Yogo-ko as inferred from amount of nitrogen. After Horie (1966).

of interest is the beautiful parallelism of nitrogen and carbon contents in the core sample (Figure 6). It strongly suggests that eutrophication caused by the drop in lake level is accompanied by the supply of allochthonous organic material. Eutrophication in this tiny lake was brought about by the appearance of a warm, dry climate, which caused an increase in both nitrogen and carbon. More organic material was produced as the climate warmed, supplying more carbon of allochthonous origin to the lake. At the same time, the drier climate brought about a drop in the level of the closed lake, causing a reduction in redox potential and therefore increasing metabolic activity and increasing the nitrogen content of the water. Also, K. Negoro's preliminary data on diatom fossils are interesting (Table 8). Correlation between chemical data and microfossil data indicates that lake trophy tends to alternate between oligotrophy and eutrophy.

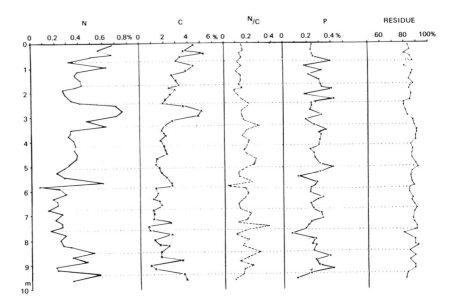

FIGURE 6   Fluctuation of chemical elements in core sample of Lake Yogo-ko. After
Horie (1967a).

TABLE 8   Diatom Fossils in the Core Sample of Lake Yogo-ko[a]

Depth (cm)	Amount	Dominant Species
7	*	*Melosira italica, Melosira granulata*
27	*	
47	**	
55	****	
75	****	
95	***	
115	****	
127	****	*Stephanodiscus carconensis*
147	***	
167	****	
177	**	
186	*	
206	*	
226	****	
236	**	
254	***	*Melosira italica, Melosira granulata*
274	***	*Melosira italica, Melosira granulata*
294	***	*Melosira italica, Melosira granulata*
311	*	*Cyclotella comta*
331	**	*Melosira italica, Melosira granulata*

TABLE 8–Continued

Depth	Amount	Dominant Species
351	*	*Cyclotella comta*
361	**	*Stephanodiscus carconensis*
373	**	*Cyclotella comta*
393	**	*Melosira italica, Melosira granulata*
413	**	*Cyclotella comta*
433	*	
447	*	
467	*	
487	*	
501	*	*Cyclotella comta*
521	*	
541	0	
561	*	*Stephanodiscus carconensis*
576	0	
588	0	
608	0	
628	0	
638	**	*Stephanodiscus carconensis*
651	*	*Stephanodiscus carconensis*
671	0	
691	*	*Stephanodiscus carconensis*
711	*	
725	**	*Stephanodiscus carconensis*
736	*	
756	*	*Cyclotella comta*
776	*	
796	*	
816	**	*Melosira islandica?*
825	**	
845	**	
865	**	
885	**	*Cyclotella comta*
905	*	*Stephanodiscus carconensis*
918	*	
938	*	*Melosira islandica?*
958	**	

[a]Source: Negoro (1968).

Thus, climatic change probably controls trophy in closed lake basins in which everything of both autochthonous and allochthonous origin must accumulate for many years, but it is difficult to find similar evidence in the core sample of an open lake, even in a shallow lake like Lake Yogo-ko; organic detritus and the nutrient salts frequently are lost by discharge. Moreover, the phosphorus content in this core is considered to be of inorganic origin.

Lake Biwa-ko is the largest lake in Japan and is a typical ancient lake (Horie, 1961). It is oligotrophic. Its mean depth is 41.2 m. Its content of dissolved oxygen is 60 to 70 percent in the hypolimnion during the summer stagnation period. In a spot far from shore, the pH of the surface water is 8.6 during the summer. Fortunately, data on transparency were collected during the last four decades by the Shiga Fishery Experimental Station. Actually, the first observation began in 1908, but at that time data collection was sporadic. Systematic observation started in 1922 and has continued to the present, except for the periods 1931–1935 and 1943–1946. I plotted the monthly observations from the center of that lake (Figure 7). The transparency decreased gradually between 1922 and 1942, but a dramatic drop occurred in 1950. Although the data of recent years are not yet published, it seems that transparency continues to be low. Low figures in the summer of 1938 may have been due to heavy rainfall, whereas low transparency (less than 2.5 m) in 1950 and 1959 was probably due to typhoons that hit Lake Biwa-ko. In addition, a typhoon is considered to have been the cause of the low transparency in the summer of 1953.

Finding the cause of the drop in lake transparency is difficult. Local influences, such as shore cultivation, construction, and influx of river water from devastated forestland must supply high amounts of inorganic particles. But the fact that considerable cultivated land surrounds this lake should not be overlooked. The Japanese economy, which was destroyed by the Second World War, has been recovering since the end of the war, and farmers have used more and more fertilizer and other agricultural chemicals. The cultivated land is irrigated with river water; seepage must carry nutrients into the lake. I cannot say whether the drop in transparency occurred rapidly or an accumulation of small influences caused the sudden drop in 1950. (A similar situation may exist in Lake Yogo-ko.)

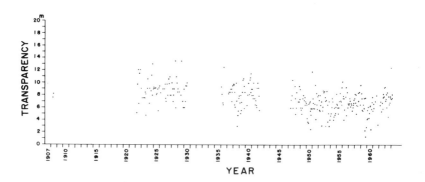

FIGURE 7   Transparency variation in Lake Biwa-ko, 1908-1963. (Compiled by Horie from data of the Shiga Fisheries Experimental Station.)

Another cause of eutrophication of this lake is the change of climate on a worldwide scale. It is well known that the temperature has been rising since the end of the Second World War. Higher temperature must cause higher productivity in every lake. The influence of global eutrophication on the drop of transparency in Lake Biwa-ko will be determined by comparing data from the temperate region of the Northern Hemisphere.

In 1965 I obtained a core sample more than 6 m long at the center of this lake. I tried several kinds of analyses together with radiocarbon dating on two horizons of the core (Figures 8 and 9). Although the relation between ignition loss and residue is controlled by several factors, the sedimentation rate may be inferred from the relation. Heavy rainfall and turbulence tend to carry more inorganic matter to the center of the lake (several kilometers from the nearest shore). Since this is so, strata between 1 and 2 m and between 3 and 4 m might have been formed in a shorter time than other strata. On the

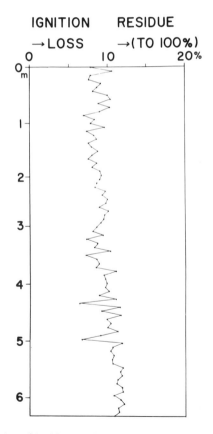

FIGURE 8   Oscillation of ignition loss in core sample, Lake Biwa-ko (Horie, in press).

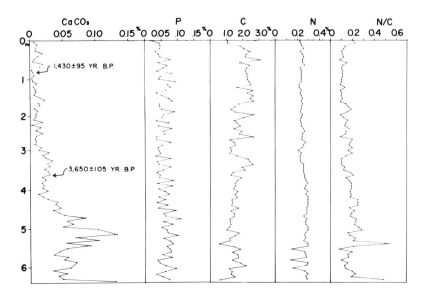

FIGURE 9   Fluctuation of chemical elements in core sample, Lake Biwa-ko (Horie, in press).

other hand, higher productivity accompanied by higher temperature in the lower half of the core is suggested. Ignition loss in the top layer, which suggests the present-day situation, is 10 percent. The amount of calcium carbonate is regarded as an indicator for past oscillation of water temperature. The calcium carbonate content of the top layer of sediment—as far as I analyzed it—was 0.03 percent; therefore, it is apparent that the temperature below a depth of 4 m was much higher than it is today. The stratum below 4 m might have been formed during the Hypsithermal Interval. Phosphorus in the core is regarded as inorganic and as having no connection with lake productivity (Toyoda *et al.*, 1968). Complicated fluctuation is shown in carbon content, which is regarded as of allochthonous origin. The fluctuation has been interpreted in connection with the change of circulation pattern caused by temperature oscillation (Horie, in press). In any event, carbon oscillation in Lake Biwa-ko is different from that in Lake Yogo-ko, which has a tiny closed basin. Another interesting point is the nitrogen curve. Between the top of the sample and approximately the 5.5-m level, no significant fluctuation in nitrogen has been recognized. This lack of fluctuation suggests the maintenance of a similar lake type, that is, the oligotrophic type. Probably, it is due to the size of the lake basin, which has a mean depth of 41.2 m, compared with 7.4 m in Lake Yogo-ko. Minor changes of climate that easily change the proportion between the trophogenic layer and the tropholytic layer in a lake as small as Yogo have probably been ineffective in a lake as big and deep as Biwa. However, below the 5.5-m level, three layers of a more oligotrophic lake type show alternate oligotrophy and eutrophy.

Lake Suwa-ko is also an ancient lake, having lacustrine sediments more than 400 m thick. A core sample that long still has not been treated from the limnological point of view. I have an 8-m core from the center of the lake. The most interesting fact that it reveals is the existence of a layer of peat at a depth of 6 m (Figure 10). The figure indicates that Lake Suwa-ko experienced dystrophy in the near past and that the lake basin was again occupied by water. In other words, there are two limnetic cycles in this diagram, though their mechanism is still unknown (Horie, 1966 and in press). It is possible to recognize the instability of lake trophy and its multidirectional trend during approximately the last 3,000 years.

Another important lake is Kizaki-ko, which is situated in a big trough running from Lake Suwa-ko. From the paleoclimatic point of view, both Kizaki and Suwa are noteworthy. They are close to high mountains in central Japan in which extensive Pleistocene glaciation took place. Lake Kizaki-ko is mesotrophic and an open lake at present. But when we consider the geomorphological evidence of the surrounding area and the small size of the inflowing river, we can see that if active evaporation had occurred before, this lake might easily have deformed to a closed-lake morphology.

Paleolimnological data of one kind are shown in Figure 11. Alternation of

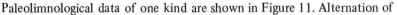

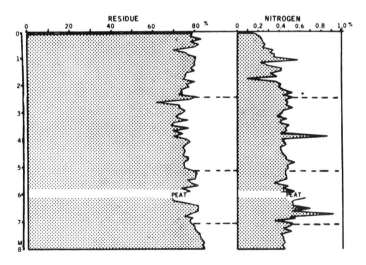

FIGURE 10   Fluctuation of lake trophy discovered in core sample, Lake Suwa-ko (Horie, 1966).

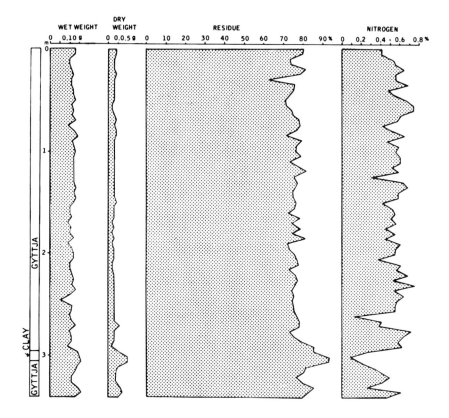

FIGURE 11   Fluctuation of lake trophy in core sample, Lake Kizaki-ko (Horie, 1966).

lake types is more noticeable here than in Lakes Biwa-ko and Suwa-ko. Because of the size of the lake basin and because the lake is in the middle of a huge trough close to a former glacial area, Lake Kizaki-ko could easily change from today's mesotrophy to oligotrophy or eutrophy.

From the geographical point of view, this lake is regarded as one that did not change its circulation pattern in response to the fluctuation of air temperature.

Although we have only begun to investigate the cause of alternation of lake types, I hope to stress that the present type (mesotrophic-eutrophic) and the morphology of the closed lake basin are essential conditions for that alternation. These examples are found in Yogo and Kizaki in contrast to completely oligotrophic Biwa-ko and to completely eutrophic Suwa-ko, which are rather unresponsive to change of environment.

CHINA

Outside of Japan, eutrophication has been little studied in Asia. But some work was carried out by Japanese limnologists before 1945. Most of the data that follow were obtained by Yamasaki (1945 a, b, c, d, and e).

In Manchuria, the *Brachionus urceus-Daphnia magna* community appears in polluted waters. Where pollution is less severe, *Daphnia carinata* and *Moina macrocopa* are found. *Closterium* sp., *Scenedesmus* sp., and *Pediastrum* sp. also appear. *Pediastrum boryanum* is a noticeable organism where pollution occurs.

In the northern part of the mainland of China are many lakes, but limnologists have studied them only to a small extent. Actually, they are astatic lakes scattered in the floodplains of the loess region along the River Huang. Every summer heavy rainfall raises the level of these lakes, but during the winter and spring, the level drops as the result of active evaporation and scanty precipitation. Range of fluctuation in the lake level is less than several meters, since the slope of the earth's surface in this region is very slight. Basin-making crustal movement may exist along the Great Canal in the easternmost part of China. Of course, such movement is favorable for the formation of lakes and swamps.

The average inclination of the River Huang in the plain is 1/6200. Accordingly, drainage is poor, and people dig holes to obtain soil so they can protect their houses from river floods. Small ponds are formed in the holes by the rainfall, which is heavy during the summer. Furthermore, this widespread plain consists of loess without any vegetation. Because loess is easily eroded by the river, it forms natural levees along the river. The soil has a pH of about 8. Water temperature of the small ponds is more than 40° C during the summer. Consequently, lake water tends to be alkaline, having a pH greater than 9.8. Even under the ice of the lake situated inside the Peking Castle, the water has a pH of 8.0 to >8.8. Because of tiny particles of suspended loess, the transparency of these lakes in northern China is extremely low, and the water is easily stirred by the wind. Total residue is more than 10,000 ppm. In addition, much polluted water from the towns flows into these ponds. As a result, lakes become more and more eutrophic and finally become disharmonic. This is a condition that restricts organisms. *Brachionus plicatilis,* which is found in highly saline lakes, brackish water, and water discharged from factories, is most common, being prevalent among the summer plankton in northern China. This species coexists with *Spirulina* sp., which has not been found in the waters of Japan but is common in the polluted waters in China. *Brachionus calyciflorus* and *Brachionus angularis* are found in less-polluted waters, together with *Microcystis* sp. The other noticeable phenomena are the *Daphnia magna* and *Moina* sp. *Daphnia magna* is found in large amounts in astatic polluted water. It increases in April and May; after that, when the water temperature goes above 30° C, it disappears. It appears again in autumn, together with *Brachionus urceus* and *Euglena. Moina* sp. is also common in shallow ponds during the warmest time of summer.

The author wishes to express his thanks to Professors Koidsumi and Koyama for their permission to cite unpublished data.

This paper is a contribution from the Otsu Hydrobiological Station, University of Kyoto, Number 186.

REFERENCES

Cyrus, B., and Z. Cyrus. 1947. A map of the purity of flows in the catchment areas of the Elbe, Danube, and Oder. Práce a studie Stát. úst. hydrol., Praha, Nr. 64.

Horie, S. 1961. Paleolimnological problems of Lake Biwa-ko. Mem. Coll. Sci., Univ. Kyoto, B. 28:53–71.

Horie, S. 1962. Morphometric features and the classification of all the lakes in Japan. Mem. Coll. Sci., Univ. Kyoto, B. 29:191–262.

Horie, S. 1966. Paleolimnological study on ancient lake sediments in Japan. Verh. Int. Verein. Limnol., Bd. 16, 274–281.

Horie, S. 1967a. On the problem of the crustal deformation in lake basin [in Japanese, English summary]. Disaster Prev. Res. Inst. Annuals, 10 A. 599–606.

Horie, S. 1967b. Limnological studies of Lake Yogo-ko (II). Bull. Disaster Prev. Res. Inst., Univ. Kyoto. 17:31–46.

Horie, S. 1968. Limnological studies of Lake Yogo-ko (III). Bull. Disaster Prev. Res. Inst., Univ. Kyoto. 17:21–28.

Horie, S. In press. Late Pleistocene limnetic history of Japanese ancient lakes Biwa, Yogo, Suwa, and Kizaki. Mitt. Int. Verein. Limnol.

Koidsumi, K., Y. Sakurai, and S. Kawashima. 1967. Standing crop of higher aquatic plants in Lake Suwa [in Japanese, English summary]. Jap. J. Limnol. 28:57–63.

Koidsumi, K., Y. Sakurai, S. Kawashima, and T. Nagasawa. 1968. Present status and recent trend of apparent eutrophication and pollution of water in Lake Suwa with special reference to their relations to the amounts of suspended and dissolved substances. Jap. J. Ecol. 18:167–171.

Miyadi, D., and N. Hazama. 1932. Quantitative investigation of the bottom fauna of Lake Yogo. Jap. J. Zool. 4:151–196.

Negoro, K. 1968. An analytical study of diatom shells in the bottom deposits of Lake Yogo-ko. Mem. Fac. Sci., Kyoto Univ., Ser. Biol. 1:121–124.

Onuki, Y. 1950. On the phytoplankton of Lake Haruna [in Japanese, English summary]. Jap. J. Limnol. 15:51–55.

Rodhe, W. 1964. Effects of impoundment on water chemistry and plankton in Lake Ransaren (Swedish Lapland). Verh. Int. Verein. Limnol. Bd. 15:437–443.

Sládeček, V. 1961. Zur biologischen Gliederung der höheren Saprobitätsstufen. Arch. f. Hydrobiol., Bd. 58, 103–121.

Srámek-Husek, R. 1950. Biologická kontrola odpadnich vod. (Tschechisch; Biologische Kontrolle der Abwässer.) Péč o čistotu vod. Bd. 1:95–109.

Šrámek-Hušek, R. 1956. Zur biologischen Charakteristik der höheren Saprobitätsstufen. Arch. f. Hydrobiol., Bd. 51:376–390.

Tanaka, A. 1918. Limnological study of Lake Suwa (Koshō-gaku jō yori mi ta ru Suwa-ko no Kenkyū). Vol. 1, Tokyo, Iwanami Shoten; Kamisuwa, Miyazaka Nisshin-dō. 936 p.

Tanaka, M. 1953. Etude chimiquè sur le metabolisme minéral dans les lacs. J. Earth Sci., Nagoya Univ. 1:119–134.

Tanaka, M. 1954a. Etat du fer dans les eaux des lacs. Bull. Chem. Soc. Jap. 27:89–93.

Tanaka, M. 1954b. Aluminium dans les eaux des lacs. Bull. Chem. Soc. Jap. 27:98–102.

Thomas, E. A. 1944. Versuche über die Selbstreinigung fliessenden Wassers. Mitt. a.d. Geb. d. Lebensmitteluntersuchung u. Hygiene. Bd. 35:199–218.

Toyoda, Y., S. Horie, and Y. Saijo. 1968. Studies on the sedimentation in Lake Biwa from the viewpoint of lake metabolism. Mitt. Int. Verein. Limnol. 14:243–255.

Tsuda, M. 1964. Biology of polluted waters (Osui Seibutsu Gaku). Tokyo, Hokuryu-kan. 258 p.

Ueno, M. 1966. Eutrophication of lakes and its related problems [in Japanese]. Kōnan Women's College Researches 3:113–131.

Yamasaki, M. 1945a. Limnology and limnoplankton in Peking [in Japanese]. Contribution, Physiol. Ecol., Fac. Sci., Univ. Kyoto; No. 40. 9 p.

Yamasaki, M. 1945b. Limnology and limnoplankton in the northern part of Honan [in Japanese]. Contribution, Physiol. Ecol., Fac. Sci., Univ. Kyoto; No. 41. 20 p.

Yamasaki, M. 1945c. Limnology and limnoplankton at T'aiyüan [in Japanese]. Contribution, Physiol. Ecol., Fac. Sci., Univ. Kyoto; No. 42. 9 p.

Yamasaki, M. 1945d. Disharmonic lakes in the alluvial plain of northern China [in Japanese]. Contribution, Physiol. Ecol., Fac. Sci., Univ. Kyoto; No. 43. 17 p.

Yamasaki, M. 1945e. Study on the regional limnology of Manchuria [in Japanese]. Contribution, Physiol. Ecol., Fac. Sci., Univ. Kyoto; No. 51. 29 p.

Yoshimura, S. 1931a. Chemical components of Japanese lakes [in Japanese]. Jap. J. Limnol. 1:25–31.

Yoshimura, S. 1931b. Contribution to the knowledge of the stratification of iron and manganese in lake water of Japan. Jap. J. Geol. Geogr. 9:61–69.

Yoshimura, S. 1931c. Contribution to the knowledge of hydrogen ion concentration in the lake water of Japan [in Japanese]. Geogr. Rev. Jap. 7:848–876, 943–969.

Yoshimura, S. 1932a. Contributions to the knowledge of nitrogenous compounds and phosphate in the lake waters of Japan. Imp. Acad., Proc. 8:94–97.

Yoshimura, S. 1932b. Calcium in solution in the lake waters of Japan. Jap. J. Geol. Geogr. 10:33–60.

Yoshimura, S. 1933a. Rapid eutrophication within recent years of Lake Haruna, Gunma, Japan, Jap. J. Geol. Geogr. 11:31–41.

Yoshimura, S. 1933b. Secular changes in the transparency of lake waters. Imp. Acad., Proc. 9:502–505.

Yoshimura, S. 1935. Amounts of nitrogen compounds dissolved in water of Japanese harmonic lakes [in Japanese]. Bull. Jap. Soc. Sci. Fish. 4:183–189.

Yoshimura, S. 1936. A contribution to the knowledge of deep-water temperatures of Japanese lakes. Jap. J. Astr. Geophys. 13:61–120, 195–200; 14:57–83.

Yoshimura, S. 1937. Annual succession of plankton of Lake Haruna-ko in recent years [in Japanese, English summary]. Ecol. Rev. 3:1–9.

W. T. EDMONDSON
University of Washington, Seattle

Eutrophication in North America

When I accepted the assignment to prepare a paper on the eutrophication of
lakes in North America, other than the Great Lakes, I had in mind giving a
picture of the spread of human population across the continent—how the
people cleared the land, developed the cities, and made various uses of the
natural resources. Many lakes were used as receptacles for sewage or they
received drainage from farms and became productive. The lakes that people
encountered varied greatly in different localities, and in each area, lakes were
initially in different stages of succession, and therefore in different productive
conditions. The effect of human activities had a different impact on the
different kinds of lakes with their different amounts of water supply and
different amounts of dissolved solids.

All of this must have happened, but to get much good documentation for
lakes across the continent is impracticable. We have many good examples of
lakes obviously made productive artificially, but often we do not know much
of what they were like before they became objectionable. Also, the picture is
clouded by regional differences in basic productivity. Some lakes are
productive for natural reasons—edaphically, morphometrically, or both—
without evident human influence, as Klamath Lake (Phinney and Peek,
1961). Lake Oneida, in New York, is shallow and produces very dense
myxophycean blooms. It produced nuisances even before European settlement; in the eighteenth century Indians called it "stinking green" (D. F.
Jackson, personal communication). Even in a center of limnological research
it is difficult to form an idea of what the lakes originally were like and to

identify quantitatively the part that domestic enrichment has had in their change.

Probably in the future, when paleolimnology is even more developed than it is now, we will learn more about the past of these lakes. For the present, I have prepared a much simpler paper than I first had in mind and am emphasizing certain selected aspects of eutrophication.

In preparing this paper, I have been influenced by the knowledge that the symposium on eutrophication came about largely because of public concern for our natural water resources. I am, therefore, concentrating on the problem of artificial enrichment. This paper emphasizes individual lakes that have been enriched in nutrients by human activity with consequent changes in productivity and abundance of organisms. I am not surveying the general status of natural aquatic community succession or the principles of natural changes in productivity. These concepts have developed progressively in several parts of the world. However, it seems obvious that any changes must be evaluated against a background of the basic principles of productivity and succession. Any prediction about the effects of particular degrees of enrichment or relief from enrichment will be limited by knowledge of the basic system.

PRODUCTIVITY AND ABUNDANCE

In organizing this material, I have regarded the productivity of a lake as a rate that is set by the simultaneous operation of a number of controlling factors, one of which is the nutrient supply. A given nutrient supply to two lakes would result in two rates of productivity if there are striking differences in other features, such as shape, isolation, or climate. Further, the way a given basic primary productivity is translated into amounts of organisms of different kinds will be affected by morphological features of the lake, character of dissolved material, and other properties.

In natural succession it would seem that noticeable changes in abundance of such organisms as algae or rooted plants may be caused as much by morphological changes as by actual increases in nutrient supply. Thus, the increase in the relative amount of bottom area in the littoral as a lake ages permits increase in rooted vegetation and masses of benthic algae. I want to suggest that the progressive increase in amount of bottom exposed to light increases productivity by way of direct supply of nutrients to algae in contact with the bottom.

Because multiple causation establishes a particular rate of production at any moment (and therefore establishes the total annual production in a lake), we can expect lakes to respond to various changes in their circumstances. In many cases we cannot explain very well why a given lake does what it does. But we can predict (qualitatively at least) the change that will be induced by a particular change in conditions. Thus, I did not have much worry in 1957 about making two predictions about Lake Washington. One was that if the discharge of sewage into Lake Washington continued as it was going, the lake would continue to deteriorate and would become unacceptable to the public. The other was that prompt, complete removal of effluent soon would be followed by prompt regression of the lake to a lower productivity and a much more acceptable condition. I did not try very seriously to predict what conditions would result from various degrees of partial diversion.

The properties that generate nuisance conditions, especially the "pea soup" type of blue-green alga bloom that bothers people, is not productivity or high rate of turnover as such, but the abundance of particular types of organisms. Usually, production of dense crops for prolonged periods requires an elevated basic productivity level, so there will naturally be a correlation between nutrition and nuisance.

One scientific advantage in studying artificial eutrophication resulting from addition of rich nutrient supplies is that it permits a quasiexperimental approach to studying variation in one factor (admittedly, a complex factor) while others (basin morphology, climate) are held constant. Thus, good limnological studies of a variety of enriched lakes can contribute to our basic understanding of the control of productivity and species composition of the biota in freshwater communities.

NUISANCE CONDITIONS

Much of the literature on, and reputation of, eutrophication puts emphasis on the production of algal scums and odor nuisances (Hasler, 1947; Prescott, 1960). But there are other aspects of high production that make trouble. One of these is in lakes used for supply of drinking water; unsatisfactory tastes can be generated by biological conditions that are otherwise acceptable. For example, Lake Skaneateles, in New York, used as the water supply of Syracuse, is beautifully clear (Secchi transparency 11.2 m) and attractive, but populations of *Synedra* develop, producing a very unsatisfactory taste without notably affecting transparency, odor, or recreational use of the lake. Further, the *Synedra* problem is becoming more prevalent in reservoirs in New York State (D. F. Jackson, personal communication; Jackson and Meier, 1966). Algae are not the only organisms that use inorganic nutrients; massive growths of rooted or floating vascular plants offer serious problems in many places.

Another feature of productive lakes is the great abundance of insects. During periods of emergence, the adults may occur in such numbers as to interfere with various sorts of recreational or other outdoor activities, and masses of floating exuviae may be objectionable to swimmers. Chironomids are probably the most frequent offenders, but the Clear Lake gnat (*Chaoborus*) is a well-known example (Lindquist and Deonier, 1942). Other insects, such as mosquitoes and Ephemeroptera, also are offenders. Richer lakes produce more insects, but how extensively insect production may be increased in artificially enriched lakes seems not to be known (Provost, 1958).

Of course, eutrophication is not necessarily an irreversible process, at least on a small scale. Some years ago, artificial lakes were made at the headwaters of the Colorado River. Because the land that was being flooded was rich in nutrients, the lakes were initially productive. But with the passage of time the nutrients were largely leached out, and productivity decreased (R. W. Pennak, personal communication). Similar changes may have taken place in lakes in natural succession after glaciation (Mackereth, 1966). In another case, a reservoir flooded a rather rich forest soil, and the initial bottom fauna was the kind that is characteristic of productive lakes. Later, after the soil had leached and had been covered with silt, the fauna changed to one that is characteristic of less productive lakes.

The most direct approach to a review of the state of eutrophication in North America is to study lakes that have been strongly affected by human activities and to compare their early condition with the later one. I have had a hard time finding good examples. My difficulty does not mean that eutrophication is rare on our continent; quite the contrary. It means simply that limnological studies have rarely been made of unaffected lakes that later became affected. Most limnological studies have been either on lakes that were affected at the time of study or on lakes that have not yet been affected. But few studies have continued long enough to follow changes.

A number of large, unproductive lakes are showing subtle signs of increased productivity. Between 1928 and 1965 the mean Secchi disc transparency of Lake Cayuga changed from 17.3 to 7.9 m, and that of Lake Seneca changed from 29.6 to 10.3 (J. P. Barlow, personal communication). At present, some lakes are being studied in a way that may well provide valuable information in the future if these lakes become changed. Examples are Lake Tahoe (Goldman and Carter, 1965), Bear Lake, in Utah (Sigler and Nyquist, personal communication), and Lake Champlain (Henson and Potash, 1966; Henson *et al.*, 1966). Public concern about these situations is reaching the popular literature (Douglas, 1968).

If we are to regard artificial eutrophication as a limnological experiment, there is the awkward problem of control; very few situations are provided with the experimental elegance of the upper and lower basins of Zürich (Thomas, 1957; Hasler, 1947). In some cases, however, we can see what happens after amelioration of enrichment. This is analogous to the off—on light experiments in studies of photosynthesis.

One might suppose that strong sewage enrichment would give a good opportunity for testing current hypotheses about relations between productivity and community structure (species diversity). However, if the waste is complex and contains both toxic and nutritional substances that are effective in different dilutions, one may not be able to test the hypothesis with the data (Maloney, 1966).

REGIONAL STUDIES

In addition to the specific cases discussed in a later section, some investigations are being carried out on groups of lakes in regions where human influence varies. These are special cases of the kind of study in which correlations are established among related variables (Moyle, 1945; Vollenweider, 1965; Pennak, 1958; Rawson, 1960; Larkin and Northcote, 1958). There is, therefore, some chance of finding the effect of artificial eutrophication on lake populations when the sample of lakes is uniform enough and the differences in human effect are large enough. The restriction on uniformity is a stringent one. A good example of this on a small scale is provided by Nygaard (1955), although, ideally, one needs a much larger group of lakes to provide the necessary statistical basis. In Nygaard's study the most productive lakes were those with sewage or agricultural influence.

Deevey (1940) showed on a small scale how lake regions based on different geological formations could be quite different. R. J. Benoit (personal communication) recently revisited many of Deevey's lakes and found significant increases in phosphorus concentrations. Some of the lakes produce more algal blooms than they did during Deevey's study. Other studies that permit this kind of evaluation are those of Wetzel (1966) and Dean (1964).

The comparative approach needs to be carried out against a background of information on geology, climatology, and other factors influencing productivity. The basic character of water supply will be important; the supply of carbon must vary greatly with the alkalinity of water, and the susceptibility to response by enrichment with effluents rich in phosphorus and nitrogen or other elements must vary. Regional differences in the chemistry of the water will presumably cause artificial enrichment to manifest itself in various ways.

The distribution of the human population and the waste disposal and agricultural practices of the population also influence eutrophication. Richard

Vollenweider (personal communication) has pointed out interesting correlations between production of livestock and agriculture and lake problems in the United States. Useful compilations have been presented by Ackerman and Löf (1959), American Water Works Association (1966, 1967), Livingstone (1963), Missingham (1967), Nichols (1965), Sawyer (1965), and the U.S. Geological Survey (1954 a and b).

LITERATURE

It is not practicable to present here a full review of the background limnological literature. Fortunately, several major sources of information exist. The most comprehensive is *Limnology in North America* (Frey, 1963). I have had occasion to review some of the general problems in detail and suggest reference to that source for additional comments about general background (Edmondson, 1968).

A great deal of relevant information about the conditions in bodies of water throughout the continent is found in reports of various state and federal agencies that have been given rather limited distribution in the form of mimeographed reports. Some of this information is presented in this paper, but a complete survey has not been attempted.

Additional sources of useful references are Mackenthun (1965), Mackenthun *et al.* (1964), Stewart *et al.* (1966), Ingram *et al.* (1966), and Sawyer (1954).

CASE HISTORIES

As mentioned earlier, relatively few lakes have been studied in detail before and after strong modification by artificial enrichment except small lakes that have been deliberately fertilized; they are out of the scope of this paper. However, nuisances now exist in regions that were formerly known for their clear lakes. In some cases, these lakes have been, or are being, thoroughly studied after deterioration. While the initial condition is not certainly known from study, something can be said about it on the basis of general, comparative limnological knowledge coupled with intelligent evaluation of informal reports of property owners about conditions in earlier years. Lakes in large parts of New England and the Northwest appear to be unproductive, whereas lakes in large areas in the Midwest are productive. Some of the clearest examples of the effect of enrichment come from New England and the Northwest, but some well-known nuisance lakes are in the Midwest. In these three regions and in others, there are many reports of high nutrient concentrations (Mackenthun, 1965).

LAKE SEBASTICOOK, MAINE

Until some time between 1940 and 1950, Lake Sebasticook had apparently been in acceptable condition. According to local residents, however, it began to produce algal nuisances during the 1950's and by 1966 was obnoxious. A detailed study of the lake has been made, and the study included some measurement of the nutrient income (U.S. Department of Health, Education, and Welfare, 1966). In addition to sewage, the lake was receiving through its tributaries a remarkable variety of industrial wastes high in organic nutrients and organic material. For example, one industry released sodium pyrophosphate as well as waste whole potatoes and potato slices. Not all the unsatisfactory waste-disposal practices concerned eutrophication; in one inlet, waste from a woolen factory had formed a mat of floating wool about 6 inches thick and 400 feet long. It seems clear that the algal problems stem very largely from the several concentrated sources of nutrients that contributed 75 percent of the income of phosphorus to the lake. Agricultural drainage contributed only about 2 percent of the phosphorus.

The phytoplankton was dominated by the blue-green *Gomphosphaeria* (*Coelosphaerium*), *Anacystis,* and *Anabaena* at various times. The total quantity was large enough to be noticeable at all times, and during the summer, typical "green paint scums" formed. The maximum cell mass found was 40 μg/ml, except on one occasion, when an unusual value of 560 was found, presumably resulting from concentration by flotation and wind. The transparency of the water is largely governed by phytoplankton in this lake, and the extreme values recorded were 0.9 and 2.6 m.

In accordance with the highly concentrated nature of the artificial enrichment, relatively high concentrations of phosphorus and nitrogen were found in the lake, with surface wintertime maxima of 1.2 mg per liter of inorganic nitrogen, 0.011 of dissolved phosphorus, and 0.05 of total phosphorus.

Two kinds of microfossils in sediments were examined: diatoms and cladocera. Diatoms and Chydorus were much more abundant in the top inch of sediment than in the deeper sediments.

No information is given about changes in the sources of nutrients from the time of establishment, but it appears from the paleolimnological work that pronounced changes took place around 1957.

LAKE WINNISQUAM, NEW HAMPSHIRE

Lake Winnisquam, formerly in satisfactory condition (Hoover, 1938), now produces myxophycean blooms to the extent that copper sulfate treatment has been given. The two Secchi disc records for summer in 1936 and 1938 were 4.6 and 7.1 m. At present, oxygen is reduced to low concentrations during summer in much more of the lake than was true in 1938 (P. J. Sawyer, personal communication).

The nutrient income has been studied. The lake receives about 40 percent of its inorganic nitrogen and 68 percent of its inorganic phosphorus from sewage (T. P. Frost, personal communication, and unpublished report by Metcalf and Eddy, engineers). As usual, the sewage concentrations were very much higher than those in the natural water supply to the lake. Experiments showed that sewage-plant effluent stimulated the growth of algae in lake water, and the algae removed nutrients from the effluent.

The rates of photosynthetic carbon fixation are relatively high, averaging about 0.32 g/m^2 per day for the summer.

LAKES AT MADISON, WISCONSIN

The lakes at Madison, especially Lake Monona, are well known, and the details have been published extensively (Hasler, 1947; Sawyer, 1947), although little is available on recent conditions (see Edmondson, 1968 for review).

The situation at Madison was confused for a long time. Concepts of the nature and control of biological productivity had not developed far enough at the time of public trouble with the lake to permit real understanding. There was a long period of dispute and legal actions (J. J. Flannery, unpublished M.S. thesis, University of Wisconsin, Madison, 1949). One of the problems was that the lakes are productive; in the early days Lake Mendota was capable of developing conspicuous algal scums even without much obvious enrichment. But the condition of Mendota in the period 1938–1942 did not match that of Lake Monona during its maximum enrichment with sewage at the same time.

Early in this century, Lake Monona began to produce odor nuisances described by Flannery as being "like odors from a foul and neglected pig sty." Starting in 1912, copper sulfate was applied to control the algae, the amount used being regulated according to conditions and thus serving as a rough index of the degree of nuisance. Eventually it was resolved to stop disposing of sewage effluent into Lake Monona. After complete sewage diversion in 1936, the lake began slowly to improve as measured by the quantity of copper sulfate required to control the algae. During the Second World War, there was a slight temporary reversal of improvement when the lake received some sewage again, but the lake has not required copper sulfate since 1962 (Figure 1). It still produces large populations of algae as measured by cell count, but these are not predominately scum-forming species (G. Fred Lee, personal communication).

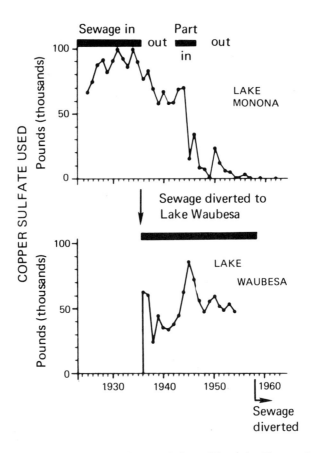

FIGURE 1   Algal nuisances in lakes at Madison, Wisconsin. The graphs show the amount of copper sulfate required to control the algae, which serves as a rough index of the nuisance condition of the lake. Above, Lake Monona; below, Lake Waubesa, downstream from Monona. The black bars show the periods during which the lakes received sewage effluent. In 1959 sewage was diverted from the entire chain of lakes. For more detail see Edmondson (1967); Mackenthun *et al.* (1964).

The whole situation was not cured by the diversion of sewage from Lake Monona because the effluent was merely moved downstream to Lake Waubesa, which, although evidently a naturally eutrophic lake, had not been producing the worst kind of nuisances. It immediately started doing so, and copper sulfate was required in large doses. In 1958, effluent was diverted from the entire chain of lakes and emptied into the Yahara River, outlet of the lower lake. The condition of the stream has deteriorated considerably, but the lakes appear improved (Mackenthun *et al.,* 1960).

Thus, we have here what amounts to an experimental study of the effect of sewage effluent on lakes, with on–off conditions. While qualitatively there seems to be no doubt about the character of the effect, quantitatively, little information is available in print about the relevant limnological conditions, even in recent years.

The situation stimulated a detailed study of sources of nutrients to the entire chain of lakes at Madison. This study led to the first development, as far as I know, of a detailed nutrient budget of a lake (Sawyer, 1947). The study, made in 1942–1944, showed that Lake Waubesa received at least 75 percent of its inorganic nitrogen and 88 percent of its inorganic phosphorus from sewage effluent.

Lake Mendota itself appears to be deteriorating and is the subject of increasing study. Nutrient income is difficult to calculate because nutrients in large amounts appear to be carried in marsh drainage that cannot be measured. Even without these nutrients, the annual phosphorus income is 21,400 kg (Lake Mendota Problems Committee, 1966), much larger than the income calculated earlier from less complete data (Edmondson, 1961).

RESERVOIRS

Some awkward problems are provided by newly created reservoirs that turn out to be productive and generate nuisances, interfering with projected water supply, recreation, and other uses. It seems only natural that such a lake, made with nutrient-rich water, should be productive, especially if much of it is relatively shallow, and if the shore and bottom are highly developed (Damann, 1951; Neel et al., 1963; Neel, 1963).

It seems to be fairly common experience for newly created reservoirs to be more productive in their first years and to settle down later to be less productive. However, productive conditions may persist, and in many of these cases, little can be done other than to accept the fact that eutrophic water will support high production. For example, some reservoirs on the Missouri River develop relatively high nutrient content, over 1 mg per liter of total phosphorus (Damann, 1951).

LAKE WASHINGTON

Lake Washington (Figure 2) will receive detailed consideration here because it was the subject of two limnological studies before it became heavily enriched (Scheffer and Robinson, 1939; Comita and Anderson, 1959). Thus, changes that took place in the lake during artificial eutrophication can be specified in some detail. Further, the sewage has now been diverted from the lake, and if all goes well, the lake can be studied as it regresses toward a less productive condition.

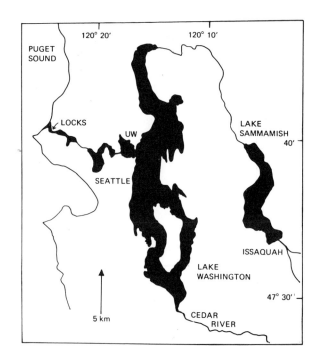

FIGURE 2   Location map for Lake Washington and Lake Sammamish. Lake Washington is connected to Puget Sound, but the level is held higher by locks. The University of Washington is shown by UW. Charts 6449 and 6447, prepared by the U.S. Coast and Geodetic Survey, show details of Lake Washington. Gould and Budinger (1958) have published a bathymetric map of Lake Washington. Walcott (1961) has published a bathymetric map of Lake Sammamish.

Considerable information has been published on Lake Washington in a number of preliminary reports; hence, only the main features will be outlined here. For literature, see Edmondson *et al.* (1956) and Edmondson (1961, 1963, 1966, and, especially, 1968).

Lake Washington has steep sides on most of its perimeter, and it has relatively little littoral development (Gould and Budinger, 1958). Soft sediments lie mostly below the euphotic zone, and only a little of the illuminated part of the bottom is suitable for growth of rooted plants. The major water supply to the lake, the Cedar River, has low concentrations of nutrients, usually less than 10 μg per liter of total phosphorus and less than 100 μg per liter of total nitrogen (Table 1). The volume of water entering through the inlets in 1957 was 31 percent of the volume of the lake, but flow is somewhat higher in most years (U.S. Geological Survey, unpublished data).

TABLE 1   Total Phosphorus in Cedar River and Sammamish River Inlets of Lake Washington, 1957[a]

Phosphorus Concentration[b]	Occurrence[c]	
	Cedar River	Sammamish River
0–5	61	1
5.1–10	37	8
10.1–15	2	13
15.1–20	–	17
20.1–25	–	24
25.1–30	–	12
30.1–35	–	4
35.1–40	–	14
40.0–45	–	3
45.1–50	–	–
50.1–55	–	–
55.1–60	–	–
60.1–65	–	1

[a]Based on twice-weekly measurements. Source: Hollis M. Phillips, Seattle Department of Engineering.
[b]Micrograms per liter.
[c]Percent of observations.

The city of Seattle lies between Puget Sound and the west side of Lake Washington. Early in this century, the lake was used for disposal of raw sewage, and unsatisfactory conditions developed. In the early 1930's, the sewage was diverted to Puget Sound, and for a while, the pollution of the lake was reduced. However, Seattle was expanding north and south, and smaller towns around the lake were growing. In 1940 a floating bridge was built across the lake, which permitted considerable suburban development on the east side. In 1941 a two-stage sewage-treatment plant was established, and by 1952, 10 plants had been built with direct entry into the lake. An additional one was built on the Sammamish River, an inlet to the lake, in 1959. In addition, some of the smaller streams were heavily contaminated with drainage from septic tanks (Edmondson, 1961, Figure 4).

The study in 1933 by Scheffer and Robinson (1939) took place at a time when the lake was less polluted than it had been a few years before. In 1950, when Comita and Anderson studied the lake, it was receiving effluent from six plants giving secondary treatment, the capacity of which was approximately 50 percent of the maximum later development (Comita and Anderson, 1959). The largest of the plants, at Lake City, opened late in 1952. During the period of development of new plants, each established plant was serving a growing population.

Both of the early studies of the lake were general limnological studies and were not made with special reference to pollution problems. Nevertheless, it became obvious during the 1950 work that the condition of the lake was different from its 1933 condition in several features, including the winter concentrations of phosphate and the oxygen deficit. It seemed likely that the lake was responding to an increase in nutrition. In 1952 the Washington State Pollution Control Commission made many measurements of phosphate, nitrate, and chlorophyll at 21 places in the lake (Peterson, 1955). This study followed an earlier pollution study (Peterson et al., 1952). In June 1955, Dr. George Anderson observed by chance a very noticeable bloom of *Oscillatoria rubescens* in the lake, and new limnological studies were immediately initiated (Edmondson et al., 1956; Edmondson, 1961).

Public concern had been growing about the sewerage situation in the entire Seattle metropolitan area, and the obvious deterioration of the lake gave focus to the concern. After a great deal of discussion and a certain amount of difficulty, the Municipality of Metropolitan Seattle (Metro) was organized by public vote and was charged with developing an effective sewage-disposal system for the entire area at a total cost of about $121 million (League of Women Voters, 1966; Clark, 1967). Not all this money can be assigned to cleaning up the lake, since part of the operation is to stop raw sewage from entering Puget Sound and part is to carry out other improvements, but much of the money can be assigned to improving the condition of the lake (Edmondson, 1967). The limnological knowledge of the lake and straight-forward predictions of future possibilities were influential in causing a majority of citizens in the area to vote this large expenditure. It was clearly a case of an informed decision being made to protect water resources.

Before the formation of Metro, a detailed engineering study had been made. It included some analyses of nutrients in the sewage effluent and streams entering the lake (Brown and Caldwell, 1958). Additional nutrient analyses were made by the City of Seattle Engineers Office. This work permits calculation of the nutrient income for 1957, and estimates can be made for earlier years. Additional analyses were made by Metro (Glenn Faris, personal communication).

Construction of the diversion works extended over a period of 5 years, starting with the three plants at the south end of the lake on February 20, 1963, and ending at the north end on March 30, 1967, except for one small plant finally removed in February 1968.

Limnological work has consisted mainly of repeated measurements in water and sediments of features related to productivity and to chemical conditions that control or are affected by biological activity and can serve as measures of the rates of activity. Thus, abundance of phytoplankton and zooplankton are frequently determined. Primary productivity is measured by oxygen production and carbon-14 uptake. The reproductive rate of zooplankton is expressed as the rate of egg production. Concentrations of phosphorus and nitrogen in various combinations (inorganic and organic), oxygen, carbon dioxide, and other substances are measured. All these properties have changed in a systematic way over the years.

The most easily noticeable change in the lake is in its transparency as measured by Secchi disc. Because inflow is at a minimum in summer, changes in transparency stem largely from changes in phytoplankton. The seasonal minimum value of 1950 was 2.1 m. The lowest value ever seen was 0.7 m in 1966, and values less than 1.0 were common in the period 1962–1965. Unfortunately, the Secchi disc was not used by Scheffer and Robinson. The quantity of phytoplankton as measured by total cell volume, by chlorophyll content of water samples, and by abundance of particular species increased greatly following enrichment (Figure 3, and Edmondson, 1966, Figure 1).

Photosynthetic productivity by phytoplankton is at a high level, comparable with that seen in other enriched or deliberately fertilized lakes. The greatest oxygen production yet observed was 6.95 g/m^2 per day in June 1966. Measurements were first made in 1957. In September 1957, oxygen production was 4.48. At Bare Lake, Alaska, after fertilization, the maximum was 6.0 (Nelson and Edmondson, 1955). Carbon fixation directly measured is also high. The maximum, recorded in September 1966, is 6.09 g/m^2 per day. Some species of zooplankton show an increased abundance and rate of egg production; others do not (Edmondson, 1966).

It is difficult to relate the plankton conditions of recent years to those of 1933 because of the lack of truly quantitative data for the earlier year. Scheffer and Robinson concluded that the lake is "distinctly oligotrophic," but they recorded several blue-green algae and stated that *Anabaena lemmermanni* could be "abundant in early summer and very common again in fall." However, the maximum dry weight of net plankton was only 0.55 mg/l.

The nutrient concentrations in the water greatly increased during the period of enrichment. Because of the oligotrophic character of the water supply to the lake (in Naumann's sense), mostly from the Cedar River, and because of the volume of water entering the lake, dilution of the existing stores of nutrients should be relatively rapid. Indeed, the concentrations have started to decrease since diversion of effluent started. Lake Washington usually develops maximum concentrations of phosphate and nitrate during winter, and the concentrations decrease rapidly during the spring growth of phytoplankton. The winter concentrations of phosphate have decreased progressively since the first diversion of sewage. The maximum phosphate concentration occurred in 1963, but the mean for the first 3 months of the year was highest in 1964. Nitrate has not changed so regularly, but it is less concentrated now (Figures 3 and 4).

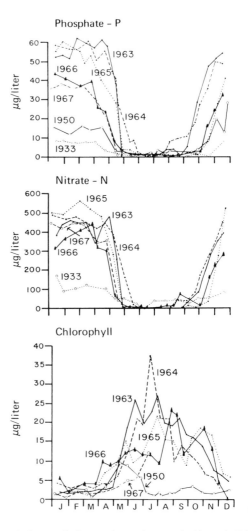

FIGURE 3   Seasonal changes in inorganic nutrients and chlorophyll in surface water of Lake Washington in various years.

The N:P ratio in the water has changed in a way suggesting that this ratio, and perhaps certain others, can be used diagnostically (Figure 5). In 1933, during the spring phytoplankton pulse, the concentrations of nitrate and phosphate decreased. By the end of the pulse, phosphate became undetectable, but nitrate was still present. Just the reverse was true in 1964, with a distinct excess of phosphate.

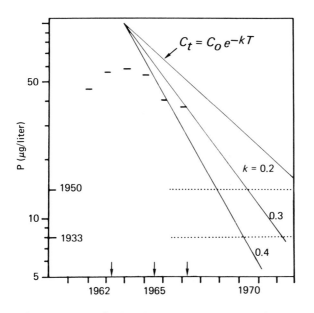

FIGURE 4   Concentration of phosphate phosphorus in surface water of Lake Washington. Short horizontal lines show the mean concentration for the first 3 months of each year. Oblique lines show how concentrations would decrease according to a simple model involving continuous and complete mixing. C_o is the initial concentration; C_t is the concentration at the time t after the interval T; k is the dilution coefficient, and is related to the ratio of annual volume of inflow to volume of lake (see also Rainey, 1967). For more detailed and complete treatment, see Piontelli and Tonolli (1964). Arrows on the base line show progress of sewage diversion. Construction started in 1963 and ended in 1967. By the middle of 1965, about half the effluent had been diverted.

The sewage entering Lake Washington was much richer in phosphorus than was the natural water supply (Tables 2 and 3). Presumably, the change in the ratio results from the heavy enrichment with phosphorus. The ratio should return toward its original values as the lake becomes diluted.

Considerable emphasis has been placed in this discussion on concentrations of nutrients, but, of course, they must not be regarded as static entities. The changes recorded in Figure 3 represent changes in the balance between input and output. The rates at which the concentrations change are significant. Concentration does have significance; the rate of uptake may be proportional to concentration in certain ranges. Further, the concentration of dissolved nutrients present at any one moment represents a stock that can be converted to algae. Thus, the concentration of phosphate phosphorus present in March in Lake Washington is about matched by that present in the plankton 3 months later (Edmondson, 1968, Figure 1).

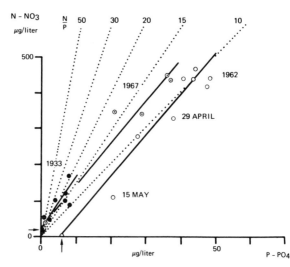

FIGURE 5   Correlation between surface-water values for phosphate and nitrate in Lake Washington. Values are plotted for the times of spring increase of phytoplankton in three years when the concentrations of nutrients were decreasing rapidly (two points are labeled with date to illustrate). Dotted lines shown are for equal N:P ratios. Note that in 1933, nitrate was in excess at the time phosphate approached zero. In 1962, a considerable excess of phosphate existed as nitrate was becoming exhausted.

TABLE 2   Total Phosphorus in Effluent, Lake City and Renton Sewage-Treatment Plants, 1957[a]

Phosphorus Concentration[b]	Occurrence[c]	
	Lake City Plant	Renton Plant
0–1	6	
1.1–2	16	
2.1–3	23	4
3.1–4	23	8
4.1–5	15	18
5.1–6	14	20
6.1–7	2	30
7.1–8	1	18
8.1–9	1	–
9.1–10	–	–
10.1–11	–	–
11.1–12	–	2

[a]Based on twice-weekly measurements. Source: Hollis M. Phillips, Seattle Department of Engineering.
[b]Milligrams per liter. Compare with Table 1.
[c]Percent of observations.

TABLE 3   Nitrogen and Phosphorus in Streams and in Sewage Effluent Entering Lake Washington in 1957: Amounts and Ratios[a]

Nitrogen and Phosphorus	In Streams		In Sewage Effluent	
	Amount (kg per year)	Ratio	Amount (kg per year)	Ratio
Inorganic N (NO_3 + NO_2)	226,500	} 19.9	19,600	} 0.52
Inorganic phosphate P	11,400		37,700	
Dissolved Kjeldahl N	1,077,200	} 32.3	133,300	} 2.67
Total dissolved P	33,300		49,900	
Total Kjeldahl N	1,168,800	} 29.5	163,100	} 3.27
Total P	39,600		49,900	
Total N (sum of inorganic N and total Kjeldahl N)	1,395,300	} 35.2	182,700	} 3.67
Total P	39,600		49,900	

[a]Source: Hollis M. Phillips, Seattle Department of Engineering.

The nuisance-forming Myxophyceae have a reputation for occurring in lakes with high concentrations of organic nitrogen, a point that has sometimes been given a causal interpretation, although this interpretation has been disputed (Edmondson, 1967). In Lake Washington the most rapid growth of the *Oscillatoria* species takes place at a time when inorganic nitrogen is near its annual maximum concentration. No particular change took place in dissolved organic nitrogen during the period of rapid decrease of nitrate (Figure 6).

Phosphate accumulation in the hypolimnion during summer stratification has been very large, but this accumulation cannot be used directly as a measure of the net rate of regeneration from the bottom because the effluent from three treatment plants entered the hypolimnion directly. The distribution of material from the Lake City treatment plant has been studied with rhodamine B by the Department of Oceanography, University of Washington. The material spread in a layer less than 2 m thick for several kilometers from the outfall (C. A. Barnes and E. E. Collias, personal communication).

LAKE SAMMAMISH

The outlet of Lake Sammamish forms the inlet to the north end of Lake Washington (Figure 2). The lake itself resembles a small-scale Lake Washington, being elongate and steep-sided. It has an area 0.226 and a volume 0.122 that of Lake Washington. The land around Lake Sammamish has not been nearly as heavily populated as that around Lake Washington, but the lake

TABLE 4 Highest Values Ever Observed for Certain Variables in Lake Washington and Lake Sammamish[a]

	Phytoplankton		Seston[d]	Total P[e]	Nitrate N[e]	Oxygen Production[f]
	Volume[b]	Composition[c]				
Lake Washington	10.8	98	9.8	226	690	7.0
Lake Sammamish	4.0	72	9.5	190	545	3.7

[a]Data for Lake Sammamish are for July 1964–December 1965. Source: Isaac et al. (1966)
[b]Millions of cubic microns per milliliter.
[c]Percent of colonial Myxophyceae, including filaments.
[d]Milligrams per liter, dry weight, uncombusted.
[e]Micrograms per liter.
[f]Grams per square meter per day.

TABLE 5 Comparison of Nutrient Incomes, Lake Washington and Lake Sammamish

	Area[a]	Volume[b]	Mean Depth[c]	Annual Increment[d]		Total P Income	
				Total P	Inorganic N (NO$_3$–NO$_2$)	Areal[e]	Volumetric[f]
Lake Washington	87.615	2,884	32.9	92,600	246,100	1.057	0.032
Lake Sammamish	19.8	350	17.7	22,800	54,500	1.152	0.065

[a]Square kilometers.
[b]Millions of cubic meters.
[c]Meters.
[d]Kilograms.
[e]Kilograms per square kilometer.
[f]Kilograms per million cubic meters.

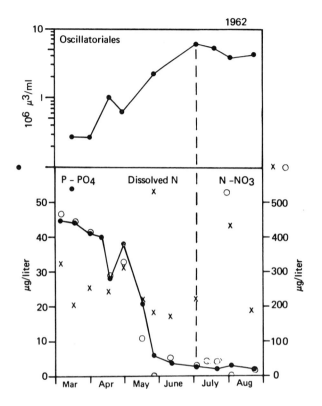

FIGURE 6   Growth of population of Oscillatoriales (mostly *Oscillatoria agardhii,* volumetric units, log scale), nitrogen, and phosphorus during spring and summer 1962. Note that the most rapid growth of algae was in the presence of high concentrations of inorganic nutrients. Inorganic phosphate and nitrate decline rapidly during April and May, and the increase of the algal population slowed down as the inorganic nutrients reached a low concentration. Dissolved nitrogen is Kjeldahl nitrogen passing through a Millipore filter, including ammonia. It did not show such a large regular change.

receives effluent from a small treatment plant, and houses are being built in increasing numbers near the lake in places where septic-tank overflow can enter the lake.

Not much is known about the original condition of the lake; only infrequent and sketchy limnological studies were made before 1964. Apparently for morphometric reasons, low oxygen concentrations were developed in the hypolimnion as early as 1913 (3.9 mg per liter), but they were lower in 1957 (0.8 mg per liter). (Data for 1913 from Kemmerer *et al.,* 1924; data for 1957 from J. Shapiro, personal communication.)

In anticipation of the possible development of difficulties, limnological studies have been carried out by the Municipality of Metropolitan Seattle

(Isaac *et al.*, 1966; Dalseg *et al.*, 1966). A further evaluation of the water budget has been attempted by L. W. Kirkpatrick (unpublished M.S. thesis, University of Washington, 1967).

These studies indicate that Lake Sammamish, although certainly not producing nuisance conditions, appears to be showing a response to the enrichment it is receiving, and in some properties it approaches Lake Washington (Table 4). It seems likely that possible increases of population in the area could cause difficulties in the lake unless proper sewerage is installed. Secchi-disc transparency ranged between 2.0 and 4.0 m in 1964–1965, but J. Shapiro (personal communication) found 1.9 and 2.8 on two occasions in 1957. The spring bloom in April 1965 was composed almost entirely of diatoms. In October, when about 72 percent of the phytoplankton was composed of colonial Myxophyceae (including filamentous species), the total phytoplankton volume was only about a quarter of the greatest volume ever observed.

These maximum values for Lake Sammamish are relatively high, but the annual means are much lower; for example, the annual mean total phosphorus is less than 50 μg per liter.

A comparison of the nutrient income with that of Lake Washington is instructive (Table 5). The annual nutrient income is 54,500 kg of inorganic nitrogen (nitrate and nitrite) and 22,800 kg of total phosphorus. On areal and volumetric bases, these amounts are well above the corresponding income of Lake Washington, and the question naturally arises: Why does Lake Sammamish not produce larger populations of algae than Lake Washington? It may be significant that the concentrations of the materials in the water entering the two lakes are very different because a larger proportion of the income entered directly into Lake Washington as very concentrated effluent. That is, sewage effluent delivered directly into Lake Washington had mostly concentrations of phosphate phosphorus in the range of 5 to 10 mg per liter, and about half the phosphate was income of the lake delivered in this form. Lake Sammamish does not receive effluent directly but through Issaquah Creek. The most concentrated effluent entering Lake Sammamish had 0.23 mg per liter of total phosphorus, and the largest stream had 0.16 mg. Thus, even though on a relative basis the income of Sammamish is higher, the concentration is lower. This kind of information must be used in evaluating nutrient budgets, but exactly how is not yet clear.

CONCLUSION

The problem of eutrophication in North America has been approached in a variety of ways with regard to scientific investigation and practical control measures. While there appears to be no particular unity of approach, perhaps one of the outstanding features of the study of artificially enriched lakes in North America has been the development of nutrient budgets, that is, measurement of the rate of income of various substances from different sources, the outgo, and storage. However, this approach needs further development. It is common to express the "loading" as kilograms per hectare, as in agriculture, but this makes comparison among lakes difficult (Sawyer, 1947; Edmondson, 1961). Allowing for depth and calculating the loading on a volumetric basis is an improvement, but it is still necessary to take account of the concentration at which nutrients are delivered to lakes and to consider interaction of nutrients.

Continued development of relevant bioassay techniques can be expected to help with this problem and to increase the accuracy of predictions about the effect of various management programs.

A good many artificially eutrophic lakes in North America are situated in such a way that some degree of amelioration is possible with present techniques, although the techniques are expensive. Moreover, public awareness seems to be developing. This, in addition to improvements in sewage techniques in the forseeable future, may lead to considerable improvement in deteriorated lakes. It is to be hoped that any major amelioration programs will be accompanied by useful limnological studies beyond "monitoring" programs.

The work on Lake Washington reported here has been supported mostly by the National Science Foundation (Grants G 6167 and G 24949, 1958 to date), the National Institutes of Health (Grant F6 4623, 1956–1958), and the State of Washington Fund for Research in Biology and Medicine, Initiative 171.

It is impracticable to list all the persons who have participated in the work on Lake Washington, but special mention should be made of David E. Allison, Donald J. Hall, and Joseph Shapiro.

I have acknowledged, in the text, my indebtedness to some of the persons who have furnished information in correspondence. In addition, I should mention the following: R. Benoit, R. A. Crossman, D. W. Cummins, E. B. Henson, G. F. Lee, D. Nyquist, Ruth Patrick, H. D. Putnam, W. F. Sigler, and J. L. Yount.

REFERENCES

Ackerman, E. A., and G. O. G. Löf. 1959. Technology in American water development. The Johns Hopkins Press, Baltimore, Maryland. 710 p.

American Water Works Association Task Group 2610P. 1966. Nutrient associated problems in water quality and treatment. 58:1337–1355.

American Water Works Association Task Group 2610P. 1967. Sources of nitrogen and phosphorus in water supplies. 59:344–366.

Brown and Caldwell (civil and chemical engineers). 1958. Metropolitan Seattle sewerage and drainage survey: A report for the city of Seattle, King County, and the State of Washington. 558 p.

Clark, E. 1967. How Seattle is beating water pollution. Harper's Magazine 234:91–95.

Comita, G. W., and G. C. Anderson. 1959. The seasonal development of a population of *Diaptomus ashlandi* Marsh, and related phytoplankton cycles in Lake Washington. Limnol. Oceanogr. 4:37–52.

Dalseg, R. D., G. W. Isaac, and R. I. Matsuda. 1966. A survey of stream conditions in Issaquah Creek. Municipality of Metropolitan Seattle Water Quality Series No. 3. 24 p.

Damann, K. E. 1951. Missouri River basin plankton study, 1950; Report by Environmental Health Center, U.S. Public Health Service, Cincinnati, Ohio. Limited distribution.

Dean, J. W. 1964. The effect of sewage on a chain of lakes in Indiana. Hydrobiologia 24:435–440.

Deevey, E. S. 1940. Limnological studies in Connecticut. V. A contribution to regional limnology. Amer. J. Sci. 238:717–741.

Douglas, W. O. 1968. An inquest on our lakes and rivers. Playboy Magazine 15:96–98, 177–181.

Edmondson, W. T. 1961. Changes in Lake Washington following an increase in the nutrient income. Verh. Intern. Ver. Limnol. 14:167–175.

Edmondson, W. T. 1963. Pacific Coast and Great Basin, p. 371–392 *In* D. G. Frey [ed], Limnology in North America. University of Wisconsin Press, Madison, Wisconsin.

Edmondson, W. T. 1966. Changes in the oxygen deficit of Lake Washington. Int. Soc. Theor. and Appl. Limnology, Proc. 16:153–158.

Edmondson, W. T. 1967. Why study blue-green algae? p. 1–6 *In* Environmental requirements of blue-green algae. U.S. Dep. Interior, Pacific Northwest Water Laboratory.

Edmondson, W. T. 1968. Lake eutrophication and water quality management: The Lake Washington case, p. 139–178 *In* Water quality control. University of Washington Press, Seattle, Washington.

Edmondson, W. T., G. C. Anderson, and D. R. Peterson. 1956. Artificial eutrophication of Lake Washington. Limnol. Oceanogr. 1:47–53.

Frey, D. G. [ed], Limnology in North America. University of Wisconsin Press, Madison, Wisconsin. 734 p.

Goldman, C. R. 1963. Primary productivity measurements in Lake Tahoe (Appendix I), p. 154–163 *In* Comprehensive study on protection of Lake Tahoe basin. Engineering-sciences, Inc., for the Lake Tahoe Area Council.

Goldman, C. R., and R. C. Carter. 1965. An investigation of rapid carbon-14 bioassay of factors affecting the cultural eutrophication of Lake Tahoe, California. J. Water Pollut. Control Fed. 37:1044–1059.

Gould, H. R., and R. F. Budinger. 1958. Control of sedimentation and bottom configuration by convection currents, Lake Washington, Washington. J. Marine Res. 17:183–198.

Hasler, A. D. 1947. Eutrophication of lakes by domestic sewage. Ecology 28:383–395.

Henson, E. B., and M. Potash. 1966. A synoptic survey of Lake Champlain, summer 1965. Univ. Michigan, Great Lakes Res. Div. Pub. No. 15:38–43.

Henson, E. B., M. Potash, and S. E. Sundberg. 1966. Some limnological characteristics of

Lake Champlain, U.S.A. and Canada. Int. Soc. Theor. and Appl. Limnol., Proc. 16:74–82.

Hoover, E. E. 1938. Biological survey of the Merrimack watershed. New Hampshire Fish and Game Department Survey Report No. 3. 238 p.

Ingram, W. M., K. M. Mackenthun, and A. F. Bartsch. 1966. Biological field investigative data for water pollution surveys. U.S. Dep. Interior, Fed. Water Pollut. Control Admin. 139 p.

Isaac, G. W., R. I. Matsuda, and J. R. Welker. 1966. A limnological investigation of water quality conditions in Lake Sammamish. Municipality of Metropolitan Seattle Water Quality Series No. 2. 47 p.

Jackson, D. F., and H. F. A. Meier. 1966. Variations in summer phytoplankton populations of Skaneateles Lake, New York. Int. Ass. Theor. Appl. Limnol., Proc. 16:173–183.

Kemmerer, G., J. F. Bovard, and W. R. Boorman. 1924. Northwestern lakes of the United States: Biological and chemical studies with reference to possibilities in production of fish. Bull. U.S. Bur. Fish. 39:51–140.

Lackey, J. B. 1945. Plankton productivity of certain southeastern Wisconsin lakes as related to fertilization. II. Productivity. Sewage Works J. 17:795–802.

Lackey, J. B., and C. N. Sawyer. 1945. Plankton productivity of certain southeastern Wisconsin lakes as related to fertilization. I. Surveys. Sewage Works J. 17:573–585.

Larkin, P. A., and T. G. Northcote. 1958. Factors in lake typology in British Columbia, Canada. Int. Ass. Theor. Appl. Limnol., Proc. 13:252–263.

Lake Mendota Problems Committee. 1966. Report on the nutrient sources of Lake Mendota. University of Wisconsin, Madison, Wisconsin. 41 p.

League of Women Voters. 1966. The big water fight. Stephen Greene Press, Brattleboro, Vermont.

Lindquist, A. W., and C. C. Deonier. 1942. Emergence habits of the Clear Lake gnat. J. Kansas Entomol. Soc. 15:109–120.

Livingstone, D. A. 1963. Chemical composition of rivers and lakes *In* Data of geochemistry. U.S. Geol. Survey professional paper 440-G.

Mackenthun, K. M. 1965. Nitrogen and phosphorus in water, an annotated selected bibliography of their biological effects. Robert A. Taft Sanitary Engineering Center, Cincinnati, Ohio. U.S. Public Health Serv. Pub. No. 1305. 111 p.

Mackenthun, K. M., W. M. Ingram, and R. Porges. 1964. Limnological aspects of recreational lakes. U.S. Dep. Health, Education, and Welfare. U.S. Government Printing Office, Washington, D.C.

Mackenthun, K. M., L. A. Leuschow, and C. D. McNabb. 1960. A study of the effects of diverting the effluent from sewage treatment upon the receiving stream. Wisconsin Acad. Sci. Arts Lett., Trans. 49:51–72.

Mackereth, F. J. H. 1966. Some chemical observations on post-glacial lake sediments. Roy. Soc. (Edinburgh), Trans. 250(B):165–213.

Maloney, T. F. 1966. Detergent phosphorus effect on algae. J. Water Pollut. Control Fed. 38:38–45.

Missingham, C. A. 1967. Occurrence of phosphates in surface waters and some related problems. J. Amer. Water Works Ass. 59:183–211.

Moyle, J. 1945. Some chemical factors influencing the distribution of aquatic plants in Minnesota. Amer. Midland Natur. 34:402–420.

Neel, J. K. 1963. The impact of reservoirs, p. 575–593 *In* D. G. Frey [ed], Limnology of North America. University of Wisconsin Press, Madison, Wisconsin.

Neel, J. K., H. P. Nicholson, and A. Hirsch. 1963. Main stem reservoir effects on water quality in the central Missouri River. U.S. Public Health Service. Region VI. Water Supply and Pollution Control.

Nelson, P. R., and W. T. Edmondson. 1955. Limnological effects of fertilizing Bare Lake, Alaska. U.S. Fish and Wildlife Service Fishery Bull. 102:414–436.

Nichols, M. S. 1965. Nitrates in the environment. J. Amer. Water Works Ass. 57:1319–1327.

Nygaard, G. 1955. On the productivity of five Danish waters. Int. Soc. Theor. Appl. Limnol., Proc. 12:123–133.

Pennak, R. W. 1958. Regional lake typology in northern Colorado, U.S.A. Int. Soc. Theor. Appl. Limnol., Proc. 13:263–283.

Peterson, D. R. 1955. An investigation of pollutional effects in Lake Washington. Washington Pollut. Control Comm. Tech. Bull. 18. 18 p.

Peterson, D. R., K. R. Jones, and G. T. Orlob. 1952. An investigation of pollution in Lake Washington. Washington Pollut. Control Comm. Tech. Bull. 14. 29 p.

Phinney, H. K., and C. A. Peek. 1961. Klamath Lake, an instance of natural enrichment, p. 22–27 *In* Algae and metropolitan wastes. Transactions of the 1960 seminar, Robert A. Taft Sanitary Engineering Center, Cincinnati, Ohio. U.S. Public Health Service Pub. 61–3.

Piontelli, R., and V. Tonolli. 1964. Il tempo di residenza delle acque lacustri in relazione ai fenomeni di arricchimento in sostanze immesse, con particolare reguardo al Lago Maggiore. Mem. Ital. di Idrobiol. 17:247–266.

Prescott, G. W. 1960. Biological disturbances resulting from algal populations in standing waters, p. 22–37 *In* The ecology of algae. Pymatuning special Pub. No. 2. University of Pittsburgh, Pittsburgh, Pennsylvania.

Provost, M. W. 1958. Chironomids and lake nutrients in Florida. Sewage Indust. Wastes 30:1417–1419.

Rainey, R. H. 1967. Natural displacement of pollution from the Great Lakes. Science 155:1242–1243.

Rawson, D. S. 1960. A limnological comparison of twelve large lakes in Northern Saskatchewan. Limnol. Oceanogr. 5:195–211.

Sawyer, C. N. 1947. Fertilization of lakes by agricultural and urban drainage. J. New England Water Works Ass. 61:109–127.

Sawyer, C. N. 1952. Some new aspects of phosphates in relation to lake fertilization. Sewage Indust. Wastes 24:768–776.

Sawyer, C. N. 1954. Factors involved in disposal of sewage effluents to lakes. Sewage Indust. Wastes 26:317–325.

Sawyer, C. N. 1965. Problem of phosphorus in water supplies. J. Amer. Water Works Ass. 57:1431–1439.

Scheffer, V. B., and R. J. Robinson. 1939. A limnological study of Lake Washington. Ecol. Monogr. 9:95–143.

Stewart, R. K., W. M. Ingram, and D. M. Mackenthun. 1966. Water pollution control: Waste treatment and water treatment. Selected biological references on fresh and marine waters. U.S. Dep. Interior, Fed. Water Pollut. Control Admin.

Thomas, E. A. 1957. Der Zürichsee, sein Wasser und sein Boden. Jahrb. vom Zürichsee 17:173–208.

U.S. Department of Health, Education, and Welfare. 1966. Fertilization and algae in Lake Sebasticook, Maine. Federal Water Pollution Control Administration, Technical Services Program, Robert A. Taft Sanitary Engineering Center, Cincinnati, Ohio. 124 p.

U.S. Geological Survey. 1954a. The industrial utility of public water supplies in the

United States, 1952. Part 1: States east of the Mississippi River. Water Supply Paper 1299. 639 p.

U.S. Geological Survey. 1954b. The industrial utility of public water supplies in the United States, 1952. Part 2: States west of the Mississippi River. Water Supply Paper 1300. 462 p.

Vollenweider, R. A. 1965. Materiali ed idee per una idrochimica delle acque insubriche. Mem. Ist. Ital. Idrobiol. 19:213–286.

Wetzel, R. G. 1966. Variations in productivity of Goose and hypereutrophic Sylvan Lakes, Indiana. Invest. Indiana Lakes and Streams 7:147–184.

Wolcott, E. E. 1961. Lakes of Washington. Vol. 1. Western Washington. Water Supply Bull. No. 14. Washington Dep. Conservation. 619 p.

A. M. BEETON
University of Wisconsin, Madison

Changes in the Environment and Biota
of the Great Lakes

Changes in the Great Lakes and concern over these changes are not new. Evidence of decreased fish populations was present as early as 1871 (Milner, 1874). Numerous references to declines in the commercial catch are found in early volumes of *Transactions of the American Fisheries Society.* Smiley (1882), in his paper on changes in the fisheries of the Great Lakes during 1870–1880, mentions that there was a decrease in the average size of whitefish (*Coregonis clupeaformis*), and lake trout (*Salvelinus namaycush*) and that some areas were seriously or totally depleted, persumably by overfishing.

Pollution was also of concern even this early. Around 1845, whitefish were migrating 20 miles up the Oconto River from Green Bay to spawn. Spawning runs had ceased by 1880, after sawdust covered the river bottom and an area extending 2 miles into the bay. A 90-percent decrease in whitefish catch in this part of Green Bay was attributed to sawdust pollution. Similar situations were known to exist elsewhere.

Pollution of harbors, bays, tributaries, and some inshore areas of the lakes has been recognized for many years, but the idea that the lakes may be undergoing accelerated eutrophication is recent (Beeton, 1961). The earlier concept, as expressed in the report of the International Board of Inquiry for the Great Lakes Fisheries (1943), was that the decline of fish in Lake Erie was due to direct activities of man, including pollution, and not to a change in environment. Consequently, it appears that pollution was considered a manageable problem that if controlled, would permit return to the prepollution condition and would not lead to long-term changes in the environment.

Present information shows that the environment and biota of various areas

150

of the Great Lakes have changed in varying degrees. A number of these changes can be considered indirect evidence or indices of eutrophication (Beeton, 1966). Very little information is available, however, on actual increases in nutrients, that is, eutrophication.

Relatively little information is available on past conditions in most areas of the lakes, since extensive limnological studies have been undertaken only within the past 15 years. The main information on past conditions comes from commercial fisheries statistics, water-quality data from water intakes, and the few early limnological studies. Each of these data sources presents certain problems of interpretation. Commercial fish-production records used in this report are from Baldwin and Saalfeld (1962 and supplement) unless otherwise noted. These data show major declines for almost all the species. A number of factors, however, limit the usefulness of these data. Among the factors obscuring the real trend in abundance are problems of species identification, changes in fishing effort, different distribution of fish, changes in fishing regulations, market conditions, changes in fishing gear and methods, and lack of information on the sports fishery. Data from water intakes may cover many years, but frequently the records are discontinuous. Analytical methods may have changed so much that early and recent data are not comparable, or the data may not be representative of the lake proper.

Changes in the environment and biota of the lakes, in most instances, can be related to man's activities. Environmental changes can be considered in three main categories: (1) pollution of inshore areas, including harbors and tributaries; (2) long-term changes in the open waters of the lakes; and (3) long-term changes in the sediments. The first category presents a problem in all the lakes and represents conditions that are maintained by continued pollution of these waters. With proper pollution abatement, substantial improvement in water quality could occur in these areas in a relatively short time because of the tremendous influx of open-lake waters of high quality. Long-term changes in the open waters of Lakes Erie, Michigan, and Ontario have several characteristics of eutrophication associated with small lakes. However, major changes in Lake Erie appear to be more closely related to substantial changes in the sediments. Biotic changes may be due directly to activities of man—overfishing, for example, or introduction of new species such as carp or smelt. Or the changes may be indirectly due to activities of man—discharge of wastes, for example, or construction of the Welland Canal, which permitted migration of the sea lamprey (*Petromyzon marinus*) and alewife (*Alosa pseudoharengus*) from Lake Ontario into the upper lakes.

LAKE SUPERIOR

Lake Superior has the largest surface area of any freshwater lake in the world—31,820 mi^2—and a mean depth of 487 ft. Uppermost in the chain of

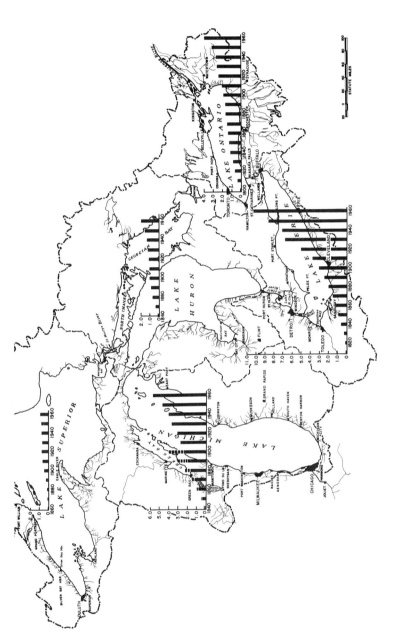

FIGURE 1   Population growth in the basins of the Great Lakes. Bar graph shows population in millions for every 10 years. Broken bars represent Chicago population growth prior to completion of the Chicago Sanitary Canal in 1900. Broken line designates watershed. Census data from Dominion Bureau of Statistics (1961) and U.S. Bureau of the Census (1951, 1961).

the Great Lakes, it discharges into Lake Huron via the St. Marys River (Figure 1). Most of the Lake Superior basin consists of pre-Cambrian rock of the Canadian Shield. These rocks weather slowly, and the chemical content of Lake Superior water is substantially lower than that of other lakes and is similar in some aspects to rainwater (Table 1).

Lake Superior and its watershed have been economically important for minerals, lumbering, commercial fishing, and recreation. Nevertheless, the population has grown slowly, as shown by an increase from about 400,000 in 1900 to almost 800,000 in 1960 (Figure 1). Hence, Lake Superior has been free of many of the consequences usually associated with proximity to a large, fast-growing population. Of first importance among the consequences from which it has been relatively free is pollution.

Physicochemical characteristics, such as high transparency, low specific conductance, and high dissolved oxygen content found at all depths throughout the year, all indicate oligotrophy (Beeton, 1965). This is further substantiated by the nature of the benthos, dominance of salmonids in the fish population, and the small amount of plankton. Diatoms, which make up most of the phytoplankton, ranged from 68 to 455 per ml in the open lake, according to Holland (1965). The primary productivity, which averages 163 mg of C per m^2 per day, is as high as that observed in some eutrophic lakes, but in terms of carbon fixed per cubic meter per day it is low (Putnam and Olson, 1966). The high productivity per m^2 of surface is to be expected with the deep photic zone and extensive mixing of a large clear lake. The crustaceans *Mysis relicta* and *Pontoporeia affinis* are the dominant deepwater

TABLE 1 Chemical Characteristics of Lake Superior Water and Precipitation

Chemical	Lake Superior Water[a] (ppm)	Precipitation (ppm)
Calcium	12.4	–
Magnesium	2.8	–
Potassium	0.6	0.3–0.6[b]
Sodium	1.1	0.26[c]
Chloride	1.9	0.6[c]
Sulfate	3.2	3.3[c]
Nitrate	0.28–0.52[d]	0.52[e]
Phosphorus	0.005	0.01[b]

[a] Beeton and Chandler (1963).
[b] Voigt (1960).
[c] Larson and Hetlech (1956).
[d] Putnam and Olson (1960).
[e] Carroll (1962).

organisms. *Heterotrissocladius,* which is associated with oligotrophic conditions (Brundin, 1958), is the dominant midge genus in the deep areas. The oligochaete fauna consists of *Euilyodrilus vejdovskiji, Limnodrilus hoffmeisterii, Peloscolex variegatus, Rhyacodrilus* spp., *Stylodrilus heringianus,* and *Tubifex kessleri americanus* (Brinkhurst, personal communication). The sphaeriid clams *Pisidium* and *Sphaerium* are also important.

ENVIRONMENTAL CHANGES

No changes have been detected in Lake Superior, except for some localized pollution of harbors, bays, and tributaries. Concentrations of calcium, chloride, sodium-plus-potassium, and sulfate have remained the same since at least 1886 (Figure 2). A slight downward trend in total dissolved solids is not significant, and concentrations have probably been somewhere between 55

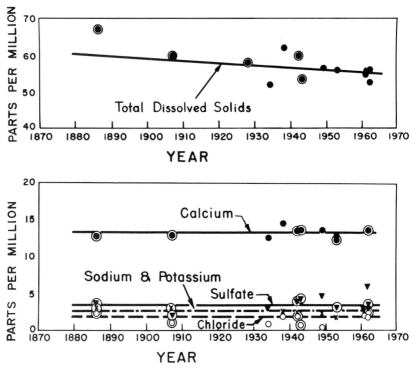

FIGURE 2   Chemical characteristics of Lake Superior. Circled points represent open-lake data (from sources listed by Beeton, 1965); chloride, sulfate, and total dissolved solids for 1962 are from Duluth, Minnesota, and Sault Ste. Marie, Michigan, intakes (U.S. Public Health Service, 1962).

and 60 ppm during the past 81 years. Phosphorus concentrations are somewhat higher in the waters of the littoral current, which flows from west to east along the south shore and then into the open lake, and may indicate some phosphorus increase in these waters (Beeton *et al.*, 1959). This littoral current receives the outflow from various tributaries and urbanized areas and carries it to Whitefish Bay and the outflow of Lake Superior. Consequently, the current may divert some pollution from the open lake.

BIOLOGICAL CHANGES

Some major changes in the fish population of Lake Superior have resulted from man's activities. The early fisheries, about 1900, consisted of lake trout, whitefish, lake herring (*Coregonus artedii*), chubs (*Coregonus* spp.), and walleye (*Stizostedion vitreum vitreum*). The exact importance of lake herring relative to chubs is not known, since catch figures for this group of fish were combined for the early years; however, most of the catch was probably lake herring. Lake herring became the major species in the commercial catch commencing about 1900. Chubs did not attain major importance until recently. By 1940, the catch, in order of importance, consisted of lake herring, lake trout, whitefish, chubs, and walleye. The subsequent decline in the lake trout and whitefish populations (Figure 3) and the successful establishment of the smelt (Van Oosten, 1937) resulted in a fishery of lake herring, chubs, smelt (*Osmerus mordax*), whitefish, walleye, and lake trout. The decline in the lake trout and the whitefish coincides with the establishment of the sea lamprey. Commercial catch records probably do not show the actual beginning of the decline in lake trout, since Hile *et al.* (1951) presented evidence suggesting that the continued high lake trout production during the late 1940's was due to a substantial increase in fishing effort. The successful control of the sea lamprey (Baldwin, 1964) has led to rapid recovery of the lake trout (Smith, 1968). The re-establishment of this major predator and the recent introduction of salmon will undoubtedly result in additional changes in the fish population.

LAKE MICHIGAN

Lake Michigan ranks sixth in size among the world's lakes, with a surface area of 22,420 mi^2 and a mean depth of 276 ft. The lake proper has two major basins, one in the south and the other in the north. The physicochemical characteristics of Green Bay and the limited exchange of bay waters with the lake make the bay almost a separate lake. Green Bay is 118 mi long and

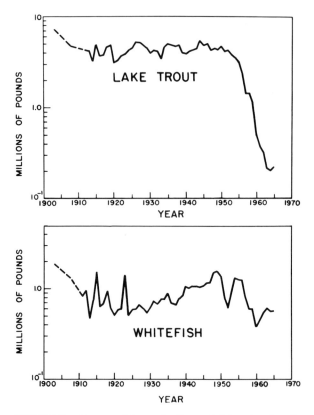

FIGURE 3   Commercial production of lake trout and
whitefish in Lake Superior, 1900–1965. Broken lines represent production during periods when annual data were not
available.

averages 23 mi in breadth. Lake Michigan waters flow into Lake Huron via
the Straits of Mackinaw. Extensive glacial deposits cover most of the bedrock
in the basin. The northern part of the watershed is forested, the central part is
primarily farmland, and the southern part is highly urbanized. The basin has
experienced significant population growth (Figure 1). Rapidly expanding
urban populations have used the lake increasingly for water supply and waste
disposal. Population pressure on the lake had increased to about 4 million by
1900, but some of the effect of this population increase was alleviated by
completion of the Chicago sanitary canal in 1900 to divert sewage from the
lake. The population of Chicago is not included in the total for the basin after
1900, since the Chicago area is not considered in the Lake Michigan
watershed after this date. Nevertheless, the population increased from 2.7
million to 5.7 million from 1910 to 1960.

Some of the physicochemical characteristics of Lake Michigan, such as the high dissolved oxygen content at all depths throughout the year and the relatively high transparency of open lake waters (Beeton, 1965), indicate oligotrophy. Concentrations of total dissolved solids, calcium, magnesium, total alkalinity, and sulfate approximate those of Lake Erie. The phosphorus content (0.013 ppm), however, is low (Beeton and Chandler, 1963).

The biological characteristics of Lake Michigan are those associated with oligotrophy (Beeton, 1965). The species of the benthic community are similar to those of Lake Superior, although they are more abundant. The only study on primary productivity in the lake indicates that the rate of fixation of carbon (193 mg of C per m^2 per day) is not much higher than that for Lake Superior (Saunders, 1964). Salmonids were the dominant fishes until the invasion and population explosion of the alewife during the 1950's.

ENVIRONMENTAL CHANGES

Chloride, sulfate, and total dissolved solids have increased substantially during the past 90 years (Figure 4). Calcium and magnesium have remained constant. Magnesium data are not presented, but concentrations have been about 10 ppm since 1877. Additional data, primarily from the Milwaukee water plant for the period 1928 to 1966, substantiate previous conclusions as to the extent of increase in total dissolved solids, sulfate, and chloride (Beeton, 1965) and make it possible to better estimate the degree of change. Total dissolved solids have increased by 30 ppm in 90 years at a rate of 0.33 ppm per year. Sulfate increased 13 ppm, and chloride about 6 ppm. The chloride data from Milwaukee show the same trends as those presented by Owenby and Willeke (1965) for Chicago (Figure 4). A change in the buildup of chloride may have occured around 1900. This change may be related to the completion of the Chicago sanitary canal in 1900, which diverted substantial sewage from the lake. This must remain a speculation, however, because of the small amount of data for the early years. Insufficient phosphorus data are available to show any change, and the published data do not agree. Data published by Beeton and Moffett (1965) give an average value of 0.013 ppm total phosphorus for the lake exclusive of the extreme southern end. More recent results for a comparable area are 0.02 ppm total phosphate (Risley and Fuller, 1965). Phosphate concentrations are higher in the extreme southern end of the lake (Risley and Fuller, 1965).

Data from the Milwaukee water plant show that organic nitrogen (albuminoid ammonia) has increased and that inorganic nitrogen (nitrate) has decreased during the past 38 years (Figure 5). Apparently, a greater amount of the inorganic nitrogen has become tied up in the organic fraction. A similar relationship is found when different areas are compared on the basis

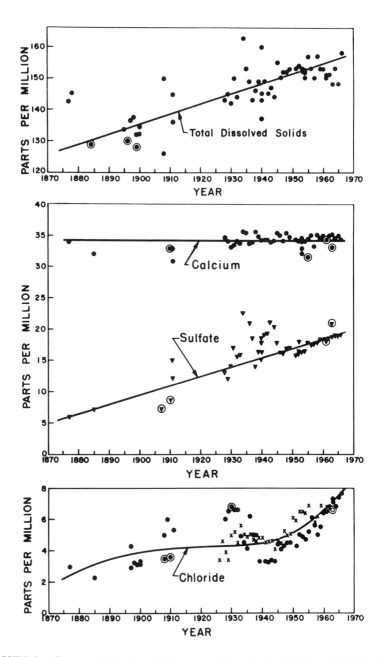

FIGURE 4   Changes in the chemical characteristics of Lake Michigan. Circled points represent open-lake data (from Beeton, 1965); Milwaukee, Wisconsin, water intake 1928–1966. Chloride data from Chicago, Illinois, intake are represented by crosses.

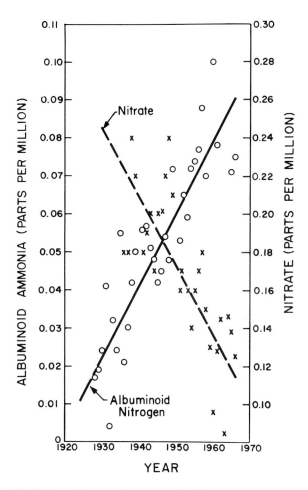

FIGURE 5 Changes in nitrate-N (x) and albuminoid ammonia (O) at the Milwaukee, Wisconsin, intake in Lake Michigan.

of surveys by the U.S. Public Health Service in 1962 and 1963. Nitrate concentrations were 0.12 ppm in the southern part of the lake and 0.19 ppm in the central part. Risley and Fuller (1965, p. 169) attributed the lower nitrate values in the southern part to "uptake by the plankton, which was more numerous in the southern tip of the lake." Allen (1966) found that nitrates were much lower and that total and soluble phosphorus was much higher in the highly productive waters of Green Bay than in Lake Michigan. Nitrate decreased to the extent that it was not measurable in September 1965.

The extent of depletion of dissolved oxygen in the southern part of Green Bay is evidently more severe now than in the past. The lowest dissolved oxygen concentrations were between 2 and 3 ppm in 1938–1939 (Wisconsin State Committee on Water Pollution, 1939) but between 0.0 and 1.0 ppm in 1955–1956 (Balch et al., 1956).

BIOLOGICAL CHANGES

Several changes have been observed in Lake Michigan plankton. It appears that the cladoceran *Bosmina coregoni* has been replaced by *B. longirostris*, although the exact taxonomic status of the Lake Michigan bosminids remains in doubt. *Diaptomus oregonensis* has become an important copepod species (Beeton, 1965). A brackish water copepod, *Eurytemora affinis*, is a new addition to the zooplankton (Robertson, 1966). Data from the Chicago water intake show that the plankton abundance has increased at an average of 13 organisms per ml per year, 1926 to 1958 (Damann, 1960). This may represent conditions only in the extreme southern part of the lake, however, since the amount of plankton in the open lake has been about one third that reported for the Chicago intake. The plankton has had two peaks of maximum abundance (spring and fall) at the Chicago intake but only one (early summer) at the Milwaukee intake (Damann, 1966) and in the open lake (Holland, in press), indicating environmental differences in these areas. Two periods of maximum abundance may be typical of productive waters of the Great Lakes, since Holland also reported two maxima for Green Bay. Two diatoms not previously reported from Lake Michigan, *Stephanodiscus hantzschii* and *S. binderanus* (probably *Melosira binderana*, Kutz.), have become abundant enough at the Chicago water intake to cause filtration problems (Vaughn, 1961). Blooms of the blue-green alga *Aphanizomenon* were observed in Green Bay in 1938–1939; *Schizothrix* was reported in 1952 (Surber and Cooley, 1952) and in 1963–1965. Evidently growths of *Cladophora* have increased substantially, since floating mats of detached *Cladophora* have become a serious problem on the beaches in recent years.

Some organic enrichment of the bottom deposits may have taken place during the past 34 years. Oligochaetes and the amphipod *Pontoporeia affinis* were more abundant in 1964 than in 1931–1932 (Robertson and Alley, 1966). Cook and Risley (1963) also reported a greater number of organisms per square meter in 1962 than in 1931–1932 and found that amphipods (probably *P. affinis*) made up 48 percent and oligochaetes 39 percent of the fauna, whereas they had made up 65 percent and 24 percent, respectively, of the earlier fauna (Eggleton, 1937). The oligochaetes *Aulodrilus* spp. and *Potamothrix* spp., which R. O. Brinkhurst (personal communication) associates with eutrophic conditions, are important in the benthos of the southern tip of the lake, but they have not been found elsewhere in the open lake.

Major changes have occurred in the species composition and abundance of benthic organisms in southern Green Bay. Nymphs of the mayfly *Hexagenia*, which were an important component of the benthic community, occurred in 31 percent of the samples from a survey in 1938–1939, but they were found in only one area in 1952 and had disappeared by 1955 (Balch *et al.*, 1956). Midge larvae increased substantially between 1939 and 1952. The distribution and abundance of oligochaetes has changed during the past 27 years. A greater number of oligochaetes are found today, but the zone of maximum abundance is found farther out in the bay from the mouth of the Fox River (Figure 6). Evidently, conditions near the river mouth have become unsuitable for even the more pollution-tolerant organisms. The dominant organism near the river mouth was the oligochaete *Limnodrilus hoffmeisterii*. At some places *L. hoffmeisterii* and *L. claparedeanus* were the only organisms. *Peloscolex multisetosus* and *L. cervix* were important species farther out in the bay. An abundance of *L. hoffmeisterii* and the presence of few other organisms are associated with gross organic pollution (Brinkhurst,

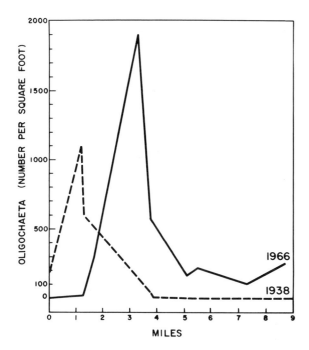

FIGURE 6   Change in abundance and distribution of oligo-chaetes in southern Green Bay between 1938 and 1966. "Miles" indicates distance into the bay from the mouth of Fox River. Data from Wisconsin State Committee on Water Pollution (1939) and Howmiller and Beeton (1967).

1966). Abundance of the other species would indicate advanced eutrophication.

The first change in the fish population of Lake Michigan was the decimation of the sturgeon (*Acipenser fulvescens*). The commercial catch records show that 3.8 million lb were taken in 1879, but many more were caught and destroyed to eliminate them from the fishing grounds, since they damaged gear used for more valuable species (Smith, 1968). Further commercial fishing for this species was prohibited in 1929 when the catch had declined to a few thousand pounds, but the species has not increased. The environment may no longer be suitable for sturgeon, since it was usually more abundant in the areas that have suffered the severest pollution.

Lake herring, lake trout, yellow perch (*Perca flavescens*), whitefish, chubs, and suckers (*Catostomus* spp.), in this order, were important in the fishery about 1900. By the late 1930's the commercial catch, in order of importance, consisted of lake trout, chubs, lake herring, yellow perch, smelt, suckers, carp (*Cyprinus carpio*), and whitefish. The carp was introduced in the 1880's (International Board of Inquiry for the Great Lakes Fisheries, 1943). Smelt were introduced into the drainage in 1912 and subsequently migrated to the other lakes (Van Oosten, 1937). Fish populations have changed substantially in the last 20 years. The most apparent changes in the commercial catch records are in lake trout, whitefish, and lake herring. The alewife has become the dominant species in the lake, and it is assuming prominence in the commercial fishery (Figure 7). The 1965 fishery consisted of alewife, chubs, carp, yellow perch, whitefish, and smelt.

The catch of lake trout dropped from 5.4 million lb in 1945 to less than 500 lb in 1953. Before 1945 the lake trout had provided a stable fishery without major fluctuations. The decline of the lake trout in Lake Michigan was very sudden and started when the sea lamprey population was relatively small. If these fish were being exploited at the optimum rate, the additional mortality caused by lamprey may have been sufficient to bring about the collapse in the lake trout population (Smith, 1968). Other changes were taking place in the environment that probably placed an additional stress on the lake trout. Nevertheless, all evidence to date indicates that sea lamprey predation triggered the decline of the lake trout.

The whitefish catch declined from 3.5 million lb in 1949 to a low of 25,000 lb in 1957 (Figure 7). This decline was undoubtedly caused primarily by lamprey predation. The present upward trend in the commercial catch may be among the first indications of successful control of the lamprey in Lake Michigan.

The decrease in catch of lake herring, from 7.7 million lb in 1954 to 116,000 lb in 1962, took place when the alewife was becoming abundant in the lake. The lamprey, as well as the alewife, has probably contributed to the decline of the lake herring, but it should be noted that the major lake herring fishery was in Green Bay, where accelerated eutrophication probably contributed to collapse of the lake herring population.

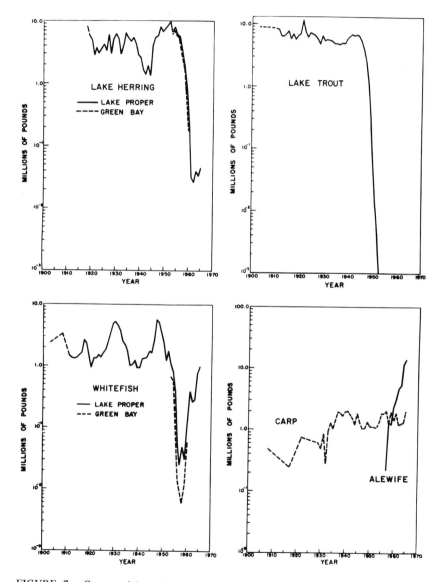

FIGURE 7 Commercial production of alewife, carp, lake herring, lake trout, and whitefish in Lake Michigan.

Chub populations have changed significantly, although this is not obvious from the commercial catch records. The chub fishery became important after the collapse of the Lake Erie cisco fishery in 1925. By 1932 the larger chubs (*Coregonus johannae* and *C. nigripinnis*) were becoming scarce, and the mesh size of the nets was gradually reduced to capture the smaller species (Hile and Buettner, 1955). Increased commercial fishing and sea lamprey predation resulted in substantial changes in the chub population to the extent that chubs of intermediate size (*C. alpenae, kiyi, reighardi,* and *zenithicus*), which made up about 66 percent of the population in the 1930's, had declined to about 24 percent in 1955 and to about 6 percent in 1960 (Smith, 1964). The smallest species, *C. hoyi,* an important food of lake trout and too small for lamprey predation, made up less than a third of the population in 1930 and about 94 percent in 1960. This species has responded to its changed environment as shown by an increase in mean length of 2 in. and an increased growth rate (Smith, 1968).

Further changes are to be expected in the fish population, since the major effort to control the sea lamprey will be completed this year (1967). A program is under way to re-establish the lake trout, the coho salmon (*Oncorhynchus kisutch*) has been introduced, and the chinook salmon (*O. tshawytscha)* will be introduced this year (1967). Re-establishment of the lake trout probably will be successful. Introductions of the salmon may not be successful in terms of natural reproduction, although the coho are growing rapidly and evidently are feeding heavily on alewife.

LAKE HURON

Lake Huron is slightly larger than Lake Michigan, with a surface area of 23,010 mi^2 and a mean depth of 195 ft (Beeton and Chandler, 1963). Major inflow to the lake is from Lakes Michigan (55,000 ft^3/sec) and Superior (73,300 ft^3/sec). The lake consists of four distinctly different major areas: the lake proper, Georgian Bay, North Channel, and Saginaw Bay. Georgian Bay and the North Channel lie in pre-Cambrian rock. The rest of the bedrock formation is of various Paleozoic periods, the Mississippian and Pennsylvanian being most important in the western part of the basin. Glacial deposits cover most of the bedrock except for the pre-Cambrian. The northern and northeastern parts of the watershed are forested; the remainder is mainly farmland. Population growth has been relatively small (Figure 1), especially in the Canadian part of the basin, where the population increased from about 414,000 in 1900 to 679,000 in 1960. The major urban centers, which are in the Saginaw River drainage area, account for most of the almost 1 million increase in population since 1900.

The physicochemical and biological characteristics of Lake Huron proper are those usually associated with oligotrophy (Beeton, 1965). Total dissolved solids are low, transparency is high, and dissolved oxygen is near saturation at all depths throughout the year. Concentrations of major ions and total dissolved solids fall between those of Lakes Michigan and Superior. Calcium, magnesium, and sodium concentrations in Lake Huron are within 1 ppm of calculated concentrations, judging from rate of inflow and content of these ions in water from Lakes Michigan and Superior. Concentrations of chloride, sulfate, and total dissolved solids in Lake Huron are 2 to 10 ppm greater than those calculated from rate of inflow and chemical content of waters from Lakes Michigan and Superior. These greater concentrations suggest some additional contributions of these chemicals from within the Lake Huron basin.

Salmonids are the dominant fishes, plankton concentrations are low, and the benthic community is dominated by *M. relicta* and *P. affinis*. The midge fauna is dominated by the genus *Heterotrissocladius.*

The outer half of Saginaw Bay is similar to the lake proper, but the inner part is shallow and very productive. The bay constitutes only 5 percent of the total lake area, but it has produced about 40 percent of the total commercial fish catch (Beeton *et al.,* 1967). In the inner bay, which is 10 to 30 percent Saginaw River water, concentrations of the major ions are 2 to 7 times greater than in the lake proper. The river is considered heavily polluted (Surber, 1957), but sufficient exchange takes place between lake and bay waters to modify the influence of the river. The more pollution-tolerant oligochaete *L. hoffmeisterii* dominates the benthos in areas of the bay where river water is most prevalent (Brinkhurst, 1967). Other species of oligochaetes, *Peloscolex ferox, Potamothrix* spp., *Aulodrilus* spp., and *Paranais litoralis,* which are associated with eutrophy, are common in the inner bay. A number of "clean water" forms, including *Hexagenia limbata,* have constituted part of the fauna in areas not directly affected by the Saginaw River (Surber, 1957).

ENVIRONMENTAL CHANGES

Chloride, sulfate, and total dissolved solids have increased slightly in the past 30 to 40 years (Figure 8). The source for these increased concentrations must be mainly in the Lake Huron basin, as indicated above. Even the present concentrations of chloride, sulfate, and total dissolved solids in Lake Michigan water would not account for the concentrations in Lake Huron, as suggested by Beeton (1965), after dilution by Lake Superior water. Much of the increased chemical content can be attributed to urbanization and industrial growth in the Saginaw Valley. The Saginaw River usually has a high chloride (60 to 600 ppm) and sulfate (70 to 115 ppm) content (Beeton *et al.,* 1967).

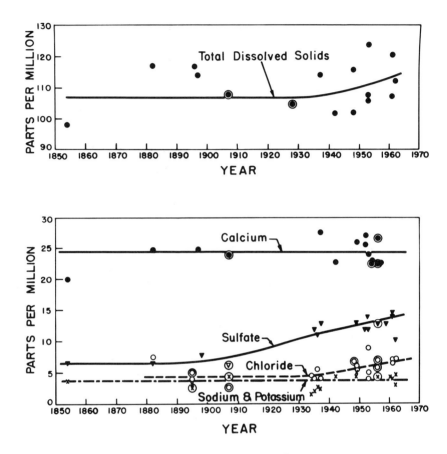

FIGURE 8   Changes in the chemical characteristics of Lake Huron. Circled points represent open-lake data (From Beeton, 1965); 1854, 1882, and 1897 data are for the Detroit River (Beeton, 1961, Table 3); chloride, sulfate, and total dissolved solids for 1962 are from Port Huron, Michigan, intake (U.S. Public Health Service, 1962).

BIOLOGICAL CHANGES

Little information is available on past biological conditions, except for data from the commercial fishery. Consequently, information is not available on biological changes, although the importance of oligochaetes, especially those associated with eutrophy, in the benthos of Saginaw Bay may indicate changes there. Comparison of data presented by Surber (1957) for 1954–1955 and by Schuytema and Powers (1966) for 1965 suggests a significant decline in the abundance of the mayfly *Hexagenia*.

Several significant changes have taken place in the fish population, as shown by the commercial catch. Lake trout, lake herring, yellow perch, walleye, whitefish, and suckers were important in the fishery at the turn of the century. The relative importance of species had not changed much by 1940. Lake herring, lake trout, walleye, whitefish, suckers, yellow perch, and carp dominated the catch, but the catch of yellow perch had declined, and carp had become an important commercial species. The statistics from the commercial fishery do not, however, necessarily indicate the abundance of these species of lower value. The next 25 years was a period of dramatic change; chubs, carp, yellow perch, whitefish, suckers, and walleye, in that order, were important in the catch of 1965. Total production for the lake was only 8.2 million lb compared with 14.6 million lb in 1940 and 21.6 million lb in 1900.

Decreased production was caused primarily by collapse of the lake herring and lake trout populations. Each contributed 4 to 6 million lb to the earlier fisheries (Figure 9). The increased production of chubs from less than 0.5 million lb in earlier years to 2.8 million lb and the carp catch of 1.5 million lb in 1965 did not compensate for the loss of the lake herring and lake trout.

Lamprey predation is considered the major cause of the decline of the lake trout. The sea lamprey became established earlier in Lake Huron than in Lakes Michigan and Superior (Applegate and Moffett, 1955). Before 1940, the fishery had produced 4 to 6 million lb annually without major fluctuations (Figure 9). The catch decreased first in the lake proper, then in Saginaw and Georgian bays. The slight change in rate of decline, 1950–1952, was due to a temporarily increased production in Georgian Bay. The lake trout fishery was not especially important or stable in Saginaw Bay, where production fluctuated between 9,000 and 325,000 lb between 1900 and the early 1950's. Decline of the whitefish in the 1930's is attributed to excessive exploitation by fishermen using deep trap nets (Van Oosten et al., 1946). The whitefish continued to decline, however, after the use of trap nets was restricted to water less than 80 ft deep. One exceptionally abundant year class, 1943, contributed to increased production, expecially in Saginaw Bay, in 1946–1948. This year class disappeared rapidly there, production being less than 1,000 lb in the bay in 1955, 1956, and 1958 (Figure 9). The greater catch in 1950–1954 was due to increased production in Georgian Bay, but this population also declined rapidly. Production since 1955 has been mainly from Lake Huron proper. The continued decline of the whitefish was probably caused by heavy sea lamprey predation. Changes in production for lake herring reflect changes in catch in Saginaw Bay, where lake herring production was greatest (Figure 9). Decreased production between 1942 and 1954 probably was due to abandonment of the pound-net fishery (Hile and Buettner, 1959). Production has declined significantly in all areas of the lake

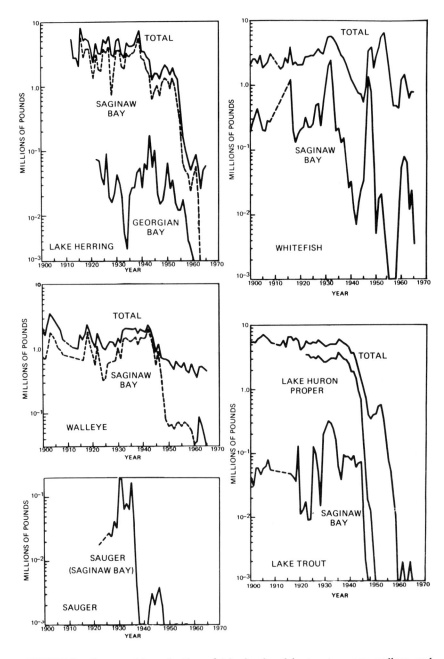

FIGURE 9   Commercial production of lake herring, lake trout, sauger, walleye, and whitefish in Lake Huron.

since 1954, the greatest decline being in Saginaw Bay. Walleye production has declined, primarily because of the collapse of the fishery in Saginaw Bay (Figure 9). Overfishing apparently was not a significant factor, but pollution has been considered important (Hile, 1954). The sauger (*Stizostedion canadense*) has not been an important commercial species in Lake Huron, although production was 210,000 lb, mostly from Saginaw Bay, in 1930 (Figure 9). Production data are included here because of the significant decline in catch after 1935. The catch has been less than 1,000 lb since 1948.

Several factors have been significant in the changing fish populations of Lake Huron. Sea lamprey predation, overfishing, and changes in the environment must all have contributed to collapse of the major fisheries. Changes in the fish populations of Saginaw Bay cannot be attributed only to sea lamprey predation. The decline of the lake herring has occurred in the past 10 years, well after the sea lamprey reached its peak. The changing environment of Saginaw Bay may be of importance, but the lake herring has also declined in Georgian Bay. The other important change during this period has been the buildup in the alewife population. The dramatic decline in the sauger, walleye, and whitefish populations in Saginaw Bay indicates development of an environment not suitable for these species.

LAKE ERIE

Lake Erie is the shallowest of the Great Lakes, with a mean depth of 58 ft (Beeton and Chandler, 1963). It is substantially smaller than the upper lakes, with a surface area of 9,930 mi^3, although it ranks twelfth in size among the world's lakes. It is divided into three areas by Point Pelee and Long Point (Figure 1). The shallow western basin (mean depth about 24 ft) lies west of a line connecting Point Pelee and Sandusky, Ohio. The central basin, between Long Point and Point Pelee, makes up the greater part of the lake, with depths between 60 and 80 ft. The eastern basin, east of a line connecting the base of Long Point and Erie, Pennsylvania, is the deepest part, with a maximum depth of 210 ft. The major inflow is into the western basin from Lake Huron via the St. Clair and Detroit rivers (mean flow 178,000 ft^3/sec). The chemical content of the Detroit River is different from that of Lake Huron, however, except for water in the main shipping channel. The river receives enough industrial and municipal wastes to increase the chloride content 2½ times (Ownbey and Kee, 1967).

The Lake Erie basin shows the greatest population growth among the Great Lakes (Figure 1). The population increased from about 3 million in 1900 to about 10.1 million in 1960. The Canadian part of the watershed is used primarily for agriculture, but the United States part is heavily populated, urbanization being especially advanced in the western half.

The high total dissolved solids, low transparency, and low dissolved oxygen of the hypolimnion of Lake Erie are characteristic of eutrophy (Beeton, 1965). Biological characteristics of the western and central basins also indicate eutrophy. These characteristics are high plankton abundance, blooms of blue-green algae, warmwater fish, and large populations of oligochaetes and midges in the benthos. Primary productivity of the western basin (98 to 7,340 mg of C per m^2 per day) is greater than in many small lakes considered highly productive (Saunders, 1964). The eastern basin has components of a fauna associated with oligotrophy—*Mysis relicta, Pontoporeia affinis,* and some salmonid fishes—and deep, cold water with sufficient dissolved oxygen to maintain this fauna. Rawson (1960) suggested that lakes with similar characteristics were morphometrically oligotrophic.

ENVIRONMENTAL CHANGES

Total dissolved solids, calcium, chloride, sodium-plus-potassium, and sulfate have increased significantly during the period of record (Figure 10). Total dissolved solids were about 56 ppm higher in 1965 than in 1910. Calcium, chloride, sodium-plus-potassium, and sulfate have increased by 8, 16, 5, and 12 ppm, respectively, during the same period. Magnesium has not changed. Nitrogen and phosphorus data are available, but much of this information is not usable because of uncertainty as to the units in which the data were recorded (Beeton, 1961). Data from the few open-lake studies of the western basin indicate that ammonia-N increased fivefold and total nitrogen increased about threefold between 1930 and 1958. Total phosphorus concentrations appear to have doubled between 1942 and 1958.

Studies of seasonal and local changes in dissolved oxygen indicate a much greater oxygen demand in the lake today than in the past. Synoptic surveys conducted jointly by a number of organizations in 1959 and 1960 demonstrated the extent and severity of oxygen depletion when the lake is stratified (Figure 11). Dissolved-oxygen concentrations of less than 3 ppm were found in bottom waters in about 75 percent of the total area of the central basin. Subsequent studies by the U.S. Public Health Service (1965) substantiated the conclusion that low dissolved-oxygen concentrations probably affect a similar area each year after the lake is thermally stratified. A few other observations have been made of low dissolved oxygen since 1930, but the degree of depletion and the area affected have become greater (Carr, 1962). Only 5 days of calm weather and subsequent stratification were necessary for dissolved oxygen to drop below 3 ppm in the western basin in 1963, whereas 28 days were required in 1953 (Carr et al., 1965). Oxygen depletion in the hypolimnetic waters is probably due to the high oxygen demand of the sediments. The 5-minute uptake of dissolved oxygen by

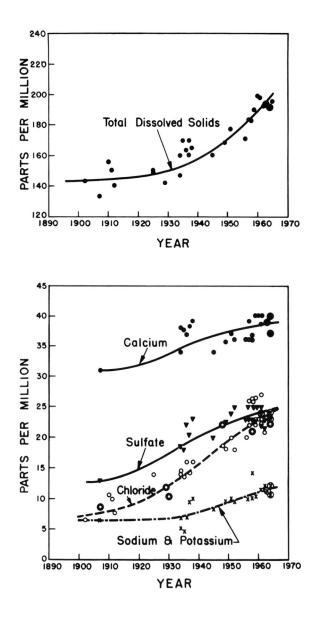

FIGURE 10   Changes in the chemical characteristics of Lake Erie. Circled points represent open-lake data (from Beeton, 1965); data for 1963 and 1964 are from the Lake Erie office of the Federal Water Pollution Control Administration; chloride, sulfate, and total dissolved solids for 1961 and 1962 are from Buffalo, New York, intake (U.S. Public Health Service, 1962).

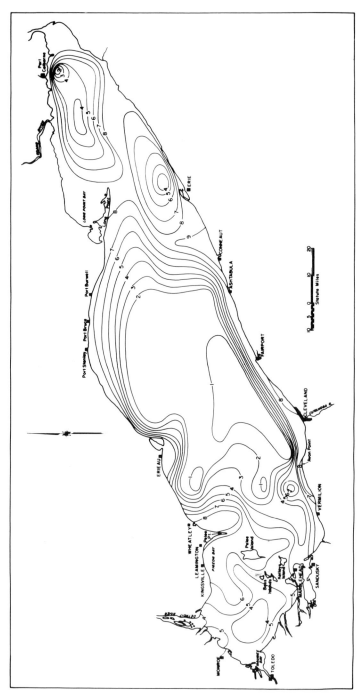

FIGURE 11   Distribution of dissolved oxygen (ppm) in the bottom waters of Lake Erie, 1960 (from Beeton, 1963).

sediments in the western basin was 400 to 600 mg/g of sediments, 200 to 600 mg/g in the western part of the central basin, 25 to 400 mg/g in the eastern part of the central basin, and 50 to 400 mg/g in the eastern basin (J. F. Carr, personal communication). All the surveys mentioned previously showed that the severest oxygen depletion developed in the western part of the central basin, where oxygen demand of the sediments was greatest, and conditions improved toward the east (Figure 11). The western basin is usually homothermous. Consequently, oxygen depletion is not detected as frequently as in the deeper central basin, but the high oxygen demand of the sediments removes the oxygen rapidly when stratification occurs.

Present information on past conditions does not offer any evidence of change in transparency of Lake Erie waters. Wright (1955) presented only a brief summary of Secchi disc measurements made in the western basin in 1929–1930. Re-examination of the original field notes, which include 228 Secchi disc records, shows that transparency was about the same in 1929–1930 as in 1958. The average disc depths at a number of stations, which were revisited frequently in 1929–1930, ranged from 0.8 to 1.5 m. The average disc depths ranged from 1.1 to 1.7 m in 1939–1941 and were 1.5 m in 1958 (Beeton, 1961). Some decrease in transparency would be expected in view of the increased abundance of plankton, but transparency probably has been determined by the amount of nonplanktonic material kept in suspension by turbulence in the shallow western end. The increase in plankton, while significant, evidently has little effect on the transparency relative to the quantities of other materials in suspension.

An increase of 2° F in the mean annual temperature of Lake Erie since 1918 is a consequence of the general climatic warming (Beeton, 1961).

BIOLOGICAL CHANGES

Major changes in the benthos of the western basin first directed our attention to accelerated eutrophication. Although declines in fish populations had been observed, these declines usually were not considered to be a result of major changes in the environment. Since 1930, nymphs of the mayfly *Hexagenia,* which formerly dominated the benthic community of western Lake Erie, have almost disappeared, and oligochaetes have become the dominant organisms (Figure 12). Pollution-tolerant species of fingernail clams, midges, and snails also have become considerably more abundant (Carr and Hiltunen 1965). Brown (1953) used the same criterion as Wright (1955) for indicating the degree of pollution, that is, the number of oligochaetes per square meter, and concluded that the zones of heavy and moderate pollution had moved 5.5 and 8 mi, respectively, lakeward from Maumee Bay between 1930 and 1951. The benthos of most of the central basin consists of animals considered

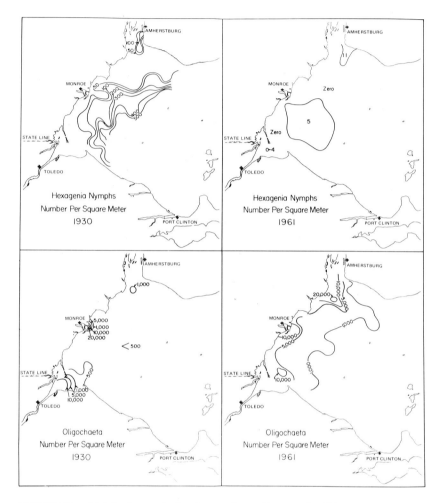

FIGURE 12   Distribution and abundance of mayfly nymphs (*Hexagenia* spp.) and oligochaetes in western Lake Erie, 1930 and 1961 (from Beeton, 1965).

to be pollution-tolerant, although organisms such as *Hexagenia* may have been abundant in the past (Davis, 1966). Some oligotrophic forms, such as *Mysis relicta* and *Pontoporeia affinis,* live in the eastern part of the central basin and in the eastern basin. The distribution of oligochaetes offers further evidence of differing environmental conditions from west to east (Brinkhurst *et al.,* 1968). Most of the species common in the upper lakes are found in the eastern basin. Species that are especially tolerant of bad organic pollution (*Limnodrilus hoffmeisterii, L. cervix,* and *Peloscolex multisetosus*) are abundant in the mouth of the Detroit River. Species associated with eutrophic conditions (*Aulodrilus* spp., *Branchiura sowerbyi, L. hoffmeisterii, Peloscolex ferox,* and *Potamothrix* spp.) are important in the western and central basins.

Severe oxygen depletion probably has been a major factor for change in the benthos, especially for *Hexagenia.* Britt (1955) found that after a period of low dissolved oxygen in 1953, the *Hexagenia* population was reduced in the western basin to about 10 percent of its former abundance.

Blooms of blue-green algae are starting to appear in Lake Erie (Casper, 1965). Between 1919 and 1963, the abundance of phytoplankton increased threefold, spring and fall maxima were greater and lasted longer, and different diatom genera became dominant (Davis, 1964). *Melosira binderana,* which was not reported from the United States until 1961, has become a major species, making up as much as 99 percent of the total phytoplankton at times around the islands (Hohn, 1966). Copepods and cladocerans increased in abundance between 1939 and 1958 (Bradshaw, 1964). *Diaptomus siciloides,* which was found occasionally in 1929 and 1930 and is usually found in eutrophic waters, has become a dominant zooplankter (Davis, 1966). A brackish water copepod, *Eurytemora affinis,* has become established in the lake (Engel, 1962).

Lake Erie has produced, and continues to produce, about 50 million lb of fish per year, about 50 percent of the total Great Lakes production. The species composition of the catch has changed markedly. The major species, in order of importance in the 1899 catch, were lake herring (cisco), blue pike, carp, yellow perch, sauger, whitefish, and walleye. The lake herring fishery collapsed after 1925, the sauger started to decline about 1920, and the walleye was becoming more abundant (Figure 13). By 1940, the commercial fishery had changed; it was dominated by blue pike, whitefish, yellow perch, walleye, sheepshead (*Aplodinotus grunniens*), carp, and suckers. The past 25 years has been the period of greatest change in the fish populations. Blue pike, lake herring, sauger, and whitefish have almost disappeared from the lake (Figure 13). The dramatic collapse of the blue pike population resembles that of the lake trout in Lake Michigan. Apparently conditions were not suitable for reproduction, and no recruitment to the blue pike population occurred after the mid-1950's. The decline in the whitefish population started about the same time, and it was almost as rapid as that of the blue pike. The lake herring continued to decline after collapse of the fishery in the 1920's. Good year classes contributed substantially to the commercial fishery in 1938–1939 and 1945–1946 (Figure 13), indicating that the population had the ability to recover rapidly when conditions were suitable. Sauger production continued the downward trend, which started around 1920, until 1950, when the population collapsed. The commercial fishery turned to other

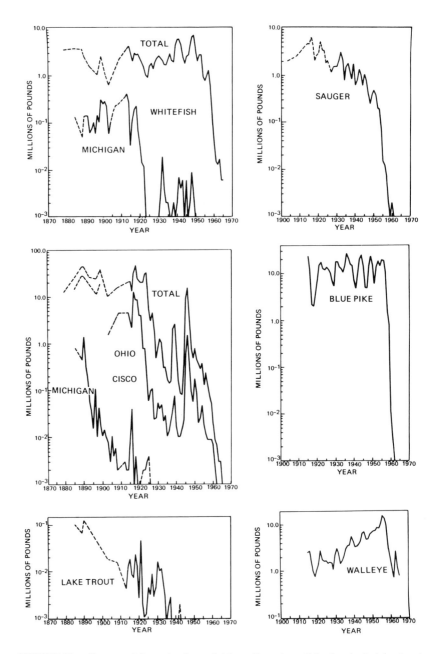

FIGURE 13   Commercial production of blue pike, cisco (lake herring), lake trout, sauger, walleye, and whitefish in Lake Erie. Broken lines represent production during periods when annual data were not available.

abundant species, especially yellow perch and smelt. The smelt started to become important in the fishery in 1952. The fishery, in order of importance, consisted of yellow perch, smelt, sheepshead, whitebass (*Roccus chrysops*), carp, catfish (*Ictalurus punctatus*), and walleye in 1965.

No single factor has been responsible for bringing about the changes in the fish populations of Lake Erie. The sea lamprey has never been as important in Lake Erie as in the upper lakes, since only a few streams are suitable for lamprey spawning. Fishing pressure on some species has obviously been intense. The effect of fishing is difficult to determine, however; the combined commercial and sports fishing effort is not known, although the sports fishing pressure on some species must be significant. The marked changes in the environment must have affected most species. The lake trout fishery was never important, but the long-term decline of this species and its eventual disappearance (Figure 13) indicate development of an unsuitable environment. It also appears that the areas of the lake suitable for lake herring and whitefish became smaller. These species are now found primarily in the eastern part of the lake. These changes are shown by their disappearance first from the Detroit River and then from western Lake Erie. The early production in the Detroit River was as much as 1 million lb, mostly whitefish and lake herring, in the 1870's and 1880's (International Board of Inquiry for the Great Lakes Fisheries, 1943). The catch was declining by 1890. Lake herring was the first to disappear: only a few hundred lb per year were being caught around 1900, and the last recorded catch was 100 lb in 1913. Whitefish production was less than 100,000 lb after 1888, except for 1912, and had declined to 300 lb in 1925, the last recorded catch. The lake herring and whitefish production declined in the Michigan waters of the extreme western end of the lake long before it did elsewhere in the lake (Figure 13). Collapse of the blue pike and sauger populations occurred during the period when extensive oxygen depletion and changes in the benthos were first reported.

LAKE ONTARIO

Lake Ontario, with a surface area of 7,520 mi^2, is the smallest of the Great Lakes and ranks fourteenth among the world's lakes (Beeton and Chandler, 1963). The western and central areas are deep, and most of the shallow waters are in the eastern end. The mean depth is 283 ft. The major inflow (83 percent) is from Lake Erie via the Niagara River (195,800 ft^3/sec). The St. Lawrence River carries the outflow (233,900 ft^3/sec) to the Atlantic Ocean. Much of the watershed use is agricultural. The large metropolitan areas of Toronto and Hamilton are in the western part of the basin, and Rochester is on the southern shore. The Lake Ontario area was settled earlier than the rest of the Great Lakes basin, and the population was about 1.4 million by 1860 (Figure 1). Population increase has not been as great as that in the basins of Lakes Erie and Michigan. The population did not increase significantly between 1920 and 1940, but it did increase from 2.6 million in 1940 to 3.8 million in 1960. Some of the population increase for the Buffalo metropolitan area probably should be included in the Lake Ontario figures, since much of the industrial and municipal wastes from Buffalo enter the Niagara River and consequently have more effect on Lake Ontario than on Lake Erie.

Lake Ontario has some characteristics associated with eutrophic conditions and others indicating oligotrophy. The inflow of nutrient-rich waters from Lake Erie certainly is sufficient to stimulate high organic production, but the depth of the lake probably does not facilitate full utilization of the nutrients. Most of the lake is deeper than 120 ft. Consequently, it appears that Lake Ontario fits well into what Rawson (1960) called morphometrically oligotrophic. The relatively low transparency, high total dissolved solids, and high specific conductance indicate eutrophic conditions (Beeton, 1965). Measurements of dissolved oxygen in 1966 did not confirm the low values reported previously that were cited by Beeton as evidence of eutrophication. Concentrations in the deep waters were usually 90 to 100 percent of saturation, although occasionally 70 percent saturation was found in one area in the shallow eastern part in 1966 (Dobson, 1967). Concentrations of the major ions are only a few ppm greater in Lake Ontario than in Lake Erie. Phosphorus concentrations in Lake Ontario, reported as about 0.01 to 0.028 ppm by Sutherland et al. (1966), evidently are somewhat lower than in Lake Erie, although we do not have sufficient published data to support that conclusion.

The biota has many characteristics of oligotrophy. *Mysis relicta, Pontoporeia affinis,* and an oligochaete fauna similar to that of Lake Superior make up most of the deep-water benthos. Salmonids are important in the fish population. Phytoplankton is abundant at times, but it consists primarily of diatoms; green and blue-green algae are of minor importance (Nalewajko, 1966). The preponderance of the diatoms *Melosira islandica* and *Asterionella formosa* in the open lake suggests oligotrophy, whereas the dominance of *Stephanodiscus tenuis* in inshore waters may indicate higher nutrient concentrations along shore (Nalewajko, 1966).

ENVIRONMENTAL CHANGES

Increases in total dissolved solids, calcium, chloride, sodium-plus-potassium, and sulfate in Lake Ontario (Figure 14) are the same as in Lake Erie, as would be expected, since the main inflow to Lake Ontario is from Lake Erie. The

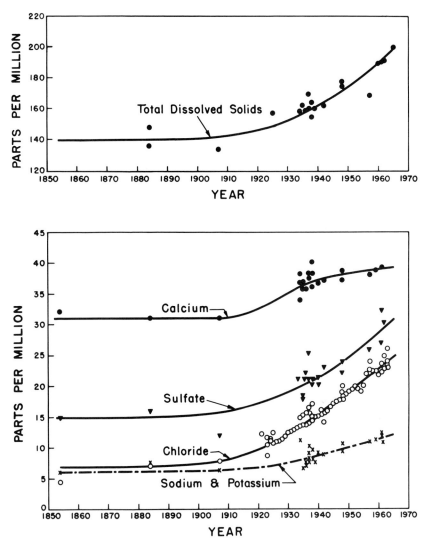

FIGURE 14   Changes in the chemical characteristics of Lake Ontario. Most chloride data for 1923–1963 are from Schenk and Thompson (1965); chloride, sulfate, and total dissolved solids for 1961 and 1962 are from St. Lawrence River, Massena, New York (U.S. Public Health Service, 1962). For sources of other data, see Beeton (1965).

somewhat higher concentrations of salts in Lake Ontario, compared with those in Lake Erie, are probably related to the growth of the Hamilton, Rochester, and Toronto metropolitan areas. Agreement among the few early analyses (1854, 1884, and 1907) is close and indicates that increases in the chemical content of Lake Ontario and Lake Erie started around 1910. Additional data presented by Dobson (1967) on the rate of change of calcium, chloride, sodium, sulfate, total salt content, and specific conductance agree closely with the trends in Figure 14. He concluded that sodium was increasing 21 percent per decade, chloride 19 percent, sulfate 6 percent, and calcium 3 percent.

Schenk and Thompson (1965) found that free ammonia, chloride, and turbidity increased at the Toronto water intake during the period 1923–1963. Free ammonia concentrations doubled, with the greatest increases occurring in 1961 and 1962. Much of this increase may indicate increased local pollution alone, since ammonia values are lower elsewhere in the lake. Turbidity has almost tripled, but, as in the case of ammonia, this reflects changing conditions in the vicinity of Toronto. Chloride concentrations at the intake were similar to those in the open lake, however, and these data have been included in Figure 14.

BIOLOGICAL CHANGES

The mean annual amount of plankton almost doubled at the Toronto water intake between 1923 and 1954, and a shift in dominant genera similar to that observed in Lake Erie occurred (Schenk and Thompson, 1965). *Asterionella* has beeen replaced or has shared dominance with *Cyclotella* and *Melosira* since 1938. This observation may apply to the entire lake, since the dominant genera are the same ones that were important in the open lake (Nalewajko, 1966).

The commercial fishery was well developed in Lake Ontario before the first statistics were collected; hence, records of early change are not available. It is known, however, that some changes took place in the fish population before 1900. The Atlantic salmon (*Salmo salar salar*), which had ascended various streams from Lake Ontario in the pioneer days, declined in abundance and had almost disappeared by 1880 (International Board of Inquiry for the Great Lakes Fisheries, 1943). Undoubtedly the main cause of decline was changes in the streams, with development of settlements, that made the streams unsuitable for spawning.

The commercial fishery has never been as important as in the other lakes, and there has been great fluctuation in the catch of the major species (Figure 15). Consequently, the relative importance of a species in the commercial catch has changed frequently during the period of record, although certain long-term trends are apparent. Lake herring and chubs, catfish and bullheads (*Ictalurus* spp.), yellow perch, whitefish, northern pike (*Esox lucius*), suckers, and sturgeon, in order of importance, dominated the catch in 1899. The combined lake herring and chub catch was 1.4 million lb, and catfish and

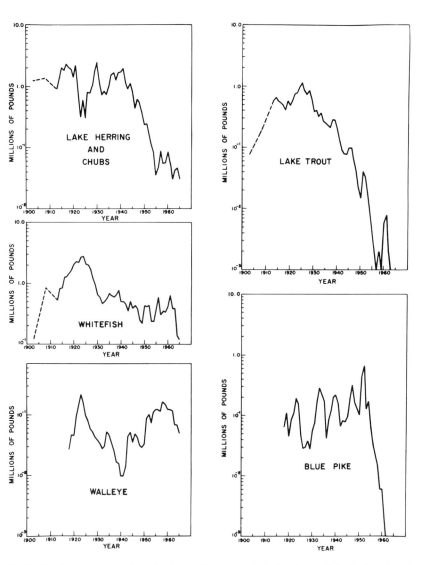

FIGURE 15   Commercial production of blue pike, lake herring and chubs, lake trout, walleye, and whitefish in Lake Ontario.

bullheads (mostly bullheads) and yellow perch yielded over 0.5 million lb. The 1940 production consisted of lake herring and chubs, carp, whitefish, blue pike, lake trout, catfish and bullheads, and yellow perch.

The catch in 1965 reflected the major declines in abundance of lake herring, lake trout, whitefish, and blue pike (Figure 15); it consisted of yellow perch, carp, eel (*Anguilla rostrata*), smelt, catfish and bullheads, sunfish (*Lepomis* spp.), and whitefish. Total production was about 2.9 million lb, a substantial decrease from the 4.3-million-lb catch of 1940 and the 7.5-million-lb catch of 1890. Whitefish production was around 0.5 million lb before and after the peak production years of 1914–1929, with a previous low of 122,000 lb in 1903 (Figure 15). Present walleye production is about the same as it was before the 1921–1925 and 1951–1962 peaks. The catch of eel has not changed significantly over the years; it has assumed a more prominent place in the catch because of the decline in other species. Carp production was not of any importance until 1914 and reached a catch of 1 million lb only once (in 1939) after that. The white perch (*Roccus americanus*) has made its way into Lake Ontario and is becoming important in the commercial fishery.

The major change is the collapse of the lake herring, lake trout, and blue pike fisheries (Figure 15), which indicates the development of an environment unsuitable for these species. The collapse of the lake herring population is similar to what occurred in Lake Erie, although the major decline started 15 years after that of the Lake Erie population. Lake trout production has been 1 million lb at least twice since 1867—in 1879 and 1925. The decline of the lake trout population started around 1930 and was more gradual than in the other lakes. The sea lamprey has been in Lake Ontario for many years (Baldwin, 1964) and does not appear to be the cause of the lake trout decline. The collapse of the blue pike population in Lake Ontario occurred during the same period as the collapse in Lake Erie, and it is remarkably similar to it (Figures 13 and 15).

CONCLUSION

Some change in the biota or the environment has taken place in all of the Great Lakes, and almost all changes have been brought about either directly or indirectly by man's activities. The effects of man's activities are obvious in the changing fish populations. Carp and smelt were purposely introduced, and construction of the Welland Canal permitted migration of the alewife and sea lamprey around Niagara Falls. The effect of the alewife, carp, and smelt on the populations of native fishes is not known, although it appears that decline of the lake herring populations in Lakes Huron and Michigan coincided with establishment of the alewife in these lakes. Rapid decline of the lake trout in each of the upper lakes is undoubtedly the result of predation by the sea lamprey. The near-extinction of the sturgeon in all the lakes was caused by overfishing (Smith, 1968).

Other changes, such as increases in chemical content, increased abundance of plankton, and changes in the benthos, have been more subtle and were not recognized until conditions were substantially altered. It is not happenstance, however, that the major changes have taken place in the lakes having large metropolitan areas in their basins. The greatest population growth has been in the Lake Erie basin, and this lake has shown the greatest changes in the environment and biota. Increases in the chemical content and abundance of plankton in Lake Ontario closely parallel changes in Lake Erie and do not indicate changes occurring just within the Lake Ontario basin. The effect of a rapidly increasing population is beginning to show up in Lake Michigan; changes have been more gradual than in Lake Erie and probably will continue to be more gradual because the volume of Lake Michigan is much greater than that of Lake Erie. The extent of change in Lake Michigan undoubtedly would have been much greater if the Chicago sanitary canal had not been constructed to divert wastes from Chicago away from Lake Michigan. The long-term outlook for Lake Michigan is not encouraging, since the net addition and flowthrough of water is only 55,000 ft^3/sec and most of the major tributaries are seriously polluted. The possibility of improving conditions in Lake Erie is somewhat better, since high-quality Lake Huron water enters the St. Clair and Detroit rivers. Abatement of pollution of the rivers should lead to eventual improvement of conditions in Lake Erie, since it is theoretically possible to exchange the entire volume of the lake in about 3 years because of the magnitude of the flowthrough and the shallowness of the lake.

Many of the changes that have taken place in Lakes Erie, Michigan, and Ontario indicate accelerated eutrophication. Increases in nitrogen and phosphorus and decreases in dissolved-oxygen content are accepted indices of eutrophication (Hasler, 1947). Increases in the major ions may indicate environmental change not necessarily associated with eutrophication, but these increases probably reflect what is happening to the nutrients. Most of the changes in the biota have considerable significance as indices of eutrophication. Changes in the species composition and increased abundance of plankton and decline and disappearance of salmonid fishes have occurred in a number of small lakes undergoing eutrophication (Hasler, 1947).

It appears, however, that the important changes in the Great Lakes are those taking place in the sediments because of the tremendous amounts of allochthonous materials entering the lakes. About 1.4 million lb of suspended solids are discharged daily to the Detroit River in municipal and industrial wastes (Vaughn and Harlow, 1965). The major changes in the benthos and the extensive depletion of dissolved oxygen offer evidence of change in the sediments. Present information suggests that depletion of dissolved oxygen in the hypolimnetic waters of western and central Lake Erie is greater than could be accounted for by organic synthesis in the epilimnion. The oxygen demand of Lake Erie sediments is about 3 times that of Lake Michigan sediments and at least 10 times that of Lake Huron sediments. Changes in the fish population of Lake Erie may also be closely related to changes in the sediments, since all Great Lakes fishes, except sheepshead, have demersal eggs. Consequently, most fish are part of the benthos during a critical period in their life histories. Elsewhere in the Great Lakes there is evidence of change in the sediments. Significant changes have occurred in the benthos of southern Green Bay. Increased abundance of benthic organisms in Lake Michigan, especially in the southern end, indicates organic enrichment of the sediments. Oligochaete species, which are associated with eutrophic conditions, are important in the benthic communities of southern Green Bay, southern Lake Michigan, Saginaw Bay, Lake Erie, and in the littoral zone of Lake Ontario.

Cooperation of many persons who provided water-quality data is gratefully acknowledged. Special thanks are due to Mr. F. A. Underwood and Mr. T. E. Dolan of the Milwaukee Water Works. Miss Marian Pierce compiled and tabulated the census data used in Figure 1. Mr. Ratko Ristic prepared the figures.

This paper is Contribution Number 7 from the Center for Great Lakes Studies, University of Wisconsin, Milwaukee.

REFERENCES

Allen, H. E. 1964. Chemical characteristics of south-central Lake Huron. Univ. Mich., Great Lakes Res. Div., Pub. 11:45–53.

Allen, H. E. 1966. Variations in phosphorus and nitrate in Lake Michigan and Green Bay, 1965. Ninth Conf. Great Lakes Res., Ill. Inst. Tech. (abstract).

Applegate, V. C., and J. W. Moffett, 1955. The sea lamprey. Sci. Amer., April, 1955.

Balch, R. F., K. M. Mackenthun, W. M. Van Horn, and T. F. Wisniewski. 1956. Biological studies of the Fox River and Green Bay. Wis. Comm. Water Pollution, Bull. WP102. 74 p.

Baldwin, N. S. 1964. Sea lamprey in the Great Lakes. Canadian Audubon Magazine, Nov.–Dec., 1964. 7 p.

Baldwin, N. S., and R. W. Saalfeld. 1962. Commercial fish production in the Great Lakes 1867–1960. Great Lakes Fish. Comm., Tech. Rep. No. 3, 1966 and supplement (1967).

Beeton, A. M. 1961. Environmental changes in Lake Erie. Trans. Amer. Fish. Soc. 90:153–159.

Beeton, A. M. 1963. Limnological survey of Lake Erie 1959 and 1960. Great Lakes Fish. Comm., Tech. Rep. No. 6. 32 p.

Beeton, A. M. 1965. Eutrophication of the St. Lawrence Great Lakes. Limnol. Oceanogr. 10:240–254.

Beeton, A. M. 1966. Indices of Great Lakes eutrophication. Univ. Mich., Great Lakes Res. Div., Pub 15:1–8.

Beeton, A. M., and D. C. Chandler. 1963. The St. Lawrence Great Lakes. *In* Limnology in North America, University of Wisconsin Press, Madison, Chapter 19, 535–558.

Beeton, A. M., J. H. Johnson, and S. H. Smith. 1959. Lake Superior limnological data. U.S. Fish and Wildlife Service, Spec. Sci. Rept. Fish. No. 297. 177 p.

Beeton, A. M., and J. W. Moffett. 1965. Lake Michigan chemical data, 1954–55, 1960–61. U.S. Fish and Wildlife Service, Data Report No. 6. 102 p.

Beeton, A. M., S. H. Smith, and F. F. Hooper. 1967. The physical limnology of Saginaw Bay, Lake Huron. Great Lakes Fish. Comm., Tech. Rept. No. 12. 56 p.

Bradshaw, A. S. 1964. The crustacean zooplankton picture: Lake Erie 1939–49–59; Cayuga 1910–51–61. Verh. Int. Verein. Limnol. 15:700–708.

Brinkhurst, R. O. 1966. The Tubificidae (Oligochaeta) of polluted waters. Verh. Int. Verein. Limnol. 16:854–859.

Brinkhurst, R. O. 1967. The distribution of aquatic oligochaetes in Saginaw Bay, Lake Huron. Limnol. Oceanogr. 12:137–143.

Brinkhurst, R. O., A. L. Hamilton, and H. B. Herrington. 1968. Components of the bottom fauna of the St. Lawrence Great Lakes. Univ. Toronto, Great Lakes Inst., No. PR33. 50 p.

Britt, N. W. 1955. Stratification in western Lake Erie in summer of 1953; effects on the *Hexagenia* (*Ephemeroptera*) population. Ecology 36:239–244.

Brown, E. H., Jr. 1953. Survey of the bottom fauna at the mouths of ten Lake Erie, south shore rivers. *In* Lake Erie Pollution Survey. Ohio Dept. Nat. Res., Div. Water, Final Rep. p. 156–170.

Brundin, L. 1958. The bottom faunistic lake-type system. Verh. Int. Verein. Limnol. 13:288–297.

Carr, J. F. 1962. Dissolved oxygen in Lake Erie, past and present. Univ. Michigan, Great Lakes Res. Div., Pub. 9:1–14.

Carr, J. F., V. C. Applegate, and M. Keller. 1965. A recent occurence of thermal stratification and low dissolved oxygen in western Lake Erie. Ohio J. Sci. 65:319–327.

Carr, J. F., and J. K. Hiltunen. 1965. Changes in the bottom fauna of western Lake Erie from 1930–1961. Limnol. Oceanogr. 10:551–569.

Carroll, D. 1962. Rain water as a chemical agent of geologic processes–a review. U.S. Geological Survey, Water-Supply Paper 1535G. U.S. Government Printing Office, Washington. D.C.

Casper, V. L. 1965. A phytoplankton bloom in western Lake Erie. Univ. Michigan, Great Lakes Res. Div., Pub. 13:29–35.

Cook, G. E., and C. Risley, Jr. 1963. The distribution of phytoplankton and benthic fauna of Lake Michigan: a preliminary report. Paper presented at Amer. Chem. Soc., N.Y., Sept. 19, 1963. 15 p.

Damann, K. E. 1960. Plankton studies of Lake Michigan. II. Thirty-three years of continuous plankton and coliform bacteria data collected from Lake Michigan at Chicago, Illinois. Trans. Amer. Microscop. Soc. 79:397–404.

Damann, K. E. 1966. Plankton studies of Lake Michigan. III. Seasonal periodicity of total plankton. Univ. Mich., Great Lakes Res. Div., Publ. 15:9–17.

Davis, C. C. 1964. Evidence for the eutrophication of Lake Erie from phytoplankton records. Limnol. Oceanogr. 9:275–283.

Davis, C. C. 1966. Biological research in the central basin of Lake Erie. Univ. Mich., Great Lakes Res. Div., Pub. 15:18–26.

Dobson, H. H. 1967. Principal ions and dissolved oxygen in Lake Ontario. Proc. 10th Conf. Great Lakes Res., 337–356.

Dominion Bureau of Statistics. Census of Canada, 1961. Queen's Printer, Ottawa.

Eggleton, F. E., 1937. Productivity of the profundal benthic zone in Lake Michigan. Papers Mich. Acad. Sci. Arts Lett. 22:593–611.

Engel, R. A., 1962. *Eurytemora affinis,* a Calanoid copepod new to Lake Erie. Ohio J. Sci. 62:252.

Hasler, A. D. 1947. Eutrophication of lakes by domestic drainage. Ecology 28:383–395.

Hile, R. 1954. Fluctuations in growth and year-class strength of the walleye in Saginaw Bay. U.S. Fish and Wildlife Service, Fish. Bull. 56:7–59.

Hile, R., and H. J. Buettner. 1955. Commercial fishery for chubs (ciscoes) in Lake Michigan through 1953. U.S. Fish and Wildlife Service, Special Sci. Rep. No. 163. 49 p.

Hile, R., and H. J. Buettner. 1959. Fluctuations in the commercial fisheries of Saginaw Bay 1885–1956. U.S. Fish and Wildlife Service, Res. Rep. No. 51. 38 p.

Hile, R., and P. H. Eschmeyer, and G. F. Lunger. 1951. Status of the lake trout fishery in Lake Superior. Trans. Amer. Fish. Soc. 80:278–312.

Hohn, M. 1966. Analysis of plankton ingested by *Stizostedium vitreum vitreum* (Mitchell) fry and concurrent vertical plankton tows from southwestern Lake Erie, May, 1961 and May, 1962. Ohio J. Sci. 66:193–197.

Holland, R. E. 1965. The distribution and abundance of planktonic diatoms in Lake Superior. Univ. Mich., Great Lakes Res. Div. Pub. 13:96–105.

Holland, R. E. in press. Seasonal fluctuations of Lake Michigan diatoms. Limnol. Oceanogr.

Howmiller, R. P., and A. M. Beeton. 1967. Bottom fauna investigation in lower Green Bay. Paper presented at the Midwest Benthological Soc.. Carbondale, Ill., 1967.

International Board of Inquiry for the Great Lakes Fisheries. 1943. Report and Supplement. U.S. Government Printing Office, Washington, D.C. 213 p.

Larson, G., and C. Hetlech. 1956. Mineral composition of rainwater. Tellus 8:191.

Milner, J. W. 1874. Report on the fisheries of the Great Lakes. U.S. Comm. Fish. Rep. (1872–73), Part 2:1–78.

Nalewajko, C. 1966. Composition of phytoplankton in surface waters of Lake Ontario. J. Fish. Res. Bd. Canada 23:1715–1725.

Ownbey, C. R., and D. A. Kee. 1967. Chlorides in Lake Erie. Proc. 10th Conf. Great Lakes Res., 382–389.

Ownbey, C. R., and G. E. Willeke. 1965. Long-term solids buildup in Lake Michigan water. Univ. Mich., Great Lakes Res. Div., Publ. 13:141–152.

Putnam, H. D., and T. A. Olson. 1960. An investigation of nutrients in western Lake Superior. Univ. Minn., School of Public Health. 46 p.

Putnam, H. D., and T. A. Olson. 1966. Primary productivity at a fixed station in western Lake Superior. Univ. Mich., Great Lakes Res. Div., Pub. 15:119–128.

Rawson, D. S. 1960. A limnological comparison of twelve large lakes in northern Saskatchewan. Limnol. Oceanogr. 5:195–211.

Risley, C., Jr., and F. D. Fuller. 1965. Chemical characteristics of Lake Michigan. Univ. Mich., Great Lakes Res. Div., Pub. 13:168–174.

Robertson, A. 1966. The distribution of calanoid copepods in the Great Lakes. Univ. Mich., Great Lakes Res. Div., Pub. 15:129–139.

Robertson, A., and W. P. Alley. 1966. A comparative study of Lake Michigan macrobenthos. Limnol. Oceanogr. 11:576–583.

Saunders, G. W. 1964. Studies of primary productivity in the Great Lakes. Univ. Mich., Great Lakes Res. Div., Pub. 11:122–129.

Schenk, C. F., and R. E. Thompson. 1965. Long-term changes in water chemistry and abundance of plankton at a single sampling location in Lake Ontario. Univ. Mich., Great Lakes Res. Div., Pub. 13:197–208.

Schuytema, G. S., and R. E. Powers. 1966. The distribution of benthic fauna in Lake Huron. Univ. Mich., Great Lakes Res. Div., Pub. 15:152–163.

Smiley, C. W. 1882. Changes in the fisheries of the Great Lakes during the decade 1870–1880. Trans. Amer. Fish.-Cultural Assoc. (Trans. Amer. Fish. Soc.) 11:28–37.

Smith, S. H. 1964. Status of the deepwater cisco population of Lake Michigan. Trans. Amer. Fish. Soc. 93:209–230.

Smith, S. H. 1968. Species succession and fishery exploitation in the Great Lakes. J. Fish. Res. Board Canada, 25(4):667–693.

Surber, E. W. 1957. Biological criteria for the determination of lake pollution. In Biological problems in water pollution. U.S. Public Health Service, Washington, D.C. p. 164–174.

Surber, E. W., and H. L. Cooley. 1952. Bottom fauna studies of Green Bay, Wisconsin, in relation to pollution. U.S. Public Health Service and Wis. Comm. Water Pollution. 7 p.

Sutherland, J. C., J. R. Kramer, L. Nichols, and T. D. Kurtz. 1966. Mineral-water equilibria, Great Lakes: silica and phosphorus. Univ. Mich., Great Lakes Res. Div., Pub. 15:439–445.

U.S. Bureau of the Census. 1951. U.S. census of population: 1950 (population of counties from earliest census to 1950). U.S. Government Printing Office, Washington, D.C. Vol. 1.

U.S. Bureau of the Census. 1961. U.S. census of population: 1960. U.S. Government Printing Office, Washington, D.C. Vol. I.

U.S. Public Health Service. 1962. National water quality network. Annual compilation of data October 1, 1961–September 30, 1962. U.S. Public Health Service Publ. 663. 909 p.

U.S. Public Health Service. 1965. Report on pollution of Lake Erie and its tributaries. I. Lake Erie. 50 p.

Van Oosten, J. 1937. The dispersal of smelt, Osmerus mordax (Mitchell), in the Great Lakes region. Trans. Amer. Fish Soc. 66:160–171.

Van Oosten, J., R. Hile, and F. W. Jobes. 1946. The whitefish fishery of Lakes Huron and Michigan with special reference to the deep-trap-net fishery. U.S. Fish and Wildlife Service, Fish. Bull. 50:297–394.

Vaughn, J. C. 1961. Coagulation difficulties at the South District filtration plant. Pure Water 13:45–49.

Vaughn, R. D., and G. L. Harlow. 1965. Report on pollution of the Detroit River, Michigan waters of Lake Erie, and their tributaries. Summary, conclusions, and recommendations. U.S. Public Health Service, Div. Water Supply and Pollution Control. 59 p.

Voigt, G. K. 1960. Alteration of the composition of rainwater by trees. Amer. Midland Natur. 63:321–326.

Wisconsin State Committee on Water Pollution, State Board of Health, and the Green Bay Metropolitan Sewerage District. 1939. Investigation of the pollution of the Fox and East Rivers and of Green Bay in the vicinity of the City of Green Bay. 242 p.

Wright, S. 1955. Limnological survey of western Lake Erie. U.S. Fish and Wildlife Service, Spec. Sci. Rep. Fish. 1939. 341 p.

H. B. N. HYNES
University of Waterloo, Toronto, Ontario, Canada

The Enrichment of Streams

The term "eutrophic," which was coined for application to lakes, has acquired so many connotations of aging and evolution of the environment that it cannot properly be applied to running water. Unlike a lake, a stream has no allotted life-span; it is an ongoing phenomenon, and if it ages, it does so only in the sense of erosion toward base level. It does not steadily become more and more productive and finally choke to death like a body of still water. Indeed, only marked geological changes, especially as they affect climate, can extinguish streams; rather, this was true until modern man set about, almost systematically, destroying watercourses by interfering with patterns of runoff.

Streams, however, change along their length in a number of well-documented ways. Usually, the further they flow, the more water they carry, and this increase in discharge is correlated in a fairly complex manner with alterations in width, depth, slope of the bed, and transport of materials, all of which, except slope, tend to increase (Leopold *et al.*, 1964). The total load of suspended matter and, usually, the concentration of dissolved solids also tend to increase. Since dissolved solids are the basis of plant growth, what happens to them in streams is important to us in considerations of enrichment.

Another important downstream change occurs in the temperature. Whether the source of the water is spring or runoff, the annual temperature amplitude tends to increase in a downstream direction until a river in its lower reaches is more or less in equilibrium with the mean monthly air temperature (Schmitz, 1961). Therefore, besides being richer in dissolved nutrients than the headwaters, the lower reaches are warmer than the headwaters in summer. In spring-fed rivers, the lower reaches are colder in winter than the headwaters, but this is not a point of great significance to us here because the summer is the growing season for most plants. We should further note that running water, except when it is near springs, becomes very cold in winter and does not possess the $4°C$ refugium that is so important in lakes. Although rivers sometimes stratify thermally for short periods, especially under ice (Shadin, 1956), stratification is an evanescent phenomenon, so nutrients are not locked up as happens in lakes.

We should also note that as the load of suspended solids increases downstream, the depth to which light penetrates declines; so, despite the warmer, richer water, the potential for plant growth may not increase. This may not have applied to all rivers in the distant past, although it clearly did apply to some. Why else did the Indians call the Missouri the "big muddy," and why did the Egyptian Pharaohs distinguish between the White Nile and the Blue Nile? We must, however, appreciate that the streams and rivers about whose enrichment we worry are almost as man-made as our managed forests and ranched grasslands. Often they are as tidily controlled, fenced, confined, and regulated as our plowed fields. I think that because they still break loose from control at times, because they are more like mustangs than children's ponies, we often tend to forget that they are controlled, and we regard them as still being part of the wilderness.

The pristine stream flowed, in most parts of the world, through a wooded valley, and often through swamps where the channel was broken into distributaries. It was thus fairly densely shaded during the summer, or all year round in the tropics. Even in the lower, wider reaches, the forest hung over the banks and shaded the shallows, so only the deep water was exposed to sunlight during the main growing season. It is not surprising, therefore, that the plants that have evolved to cope most successfully with the rigors of the running-water habitat are shade plants, such as mosses and Rhodophyceae, and cold-water species, particularly diatoms, that can flourish early and late in the season when there are no leaves on deciduous trees. Only in special places like waterfalls, stony shoals, and rocky outcrops in midstream is a stable substratum well exposed to the light, where it can provide a habitat for light-loving algae and mosses or the highly specialized Podostemaceae, the only group of angiosperms fully adapted to running water (Gessner, 1955).

We can assume, then, that apart from the plankton, which is significant only in large rivers, the unaltered stream was a poor producer of plant material, so the fertility of the water was of little consequence to primary production. This is indeed emphasized by the discovery (Ruttner, 1926) that running water is more fertile than still water simply because its turbulent flow prevents the formation of zones of nutrient depletion around plants. Many studies have since shown that the metabolism of plants is enhanced by water movement (Whitford, 1960; Whitford and Schumacher, 1961, 1964; Gessner and Pannier, 1958; McIntyre, 1966 a and b). Except in extreme instances, therefore, low fertility of the water can hardly have been a limiting factor.

On the other hand, many types of aquatic plants are opportunists, and they invade suitable sites with great rapidity. Exposed shallows are thus quickly colonized by a wide variety of species, even though their hold on such habitats is tenuous and often temporary. It is well established that both rooted plants and algae are much affected by instability and washout and that, while they may grow densely at times, they rarely achieve a permanent vegetative cover (Butcher, 1933, 1947; Gessner, 1955; Douglas, 1958). As a result, even though streams are potentially fertile, production is patchy.

Odum (1956) devised a method, based on changes of oxygen concentration of the water, for measuring primary production in streams. He concluded, from the study of published data, that streams are very productive and that those "recovering from pollution" are probably the areas of highest primary production on the planet. However, his sample was biased toward enriched streams. Most studies made since have revealed that community respiration is greater than photosynthesis (Hoskin, 1959; Nelson and Scott, 1962, Duffer and Dorris, 1966).

We know less about plankton production than about the benthos. But we do know, from many studies, that phytoplankton occurs in significant amounts only in rivers, in the *potamon*— to use the terminology of Illies and Botosaneanu (1963)— where relatively deep water is open to the sky. We also know that phytoplankton is seasonal and much reduced by floods and turbulence and that significant growth and production occur only at temperatures greater than about 12°C (Knöpp, 1960). So, because it is transient by its very nature, phytoplankton does not store biomass for later consumption. Its contribution to production must be confined to the warm months. It produces no "hay" for the rest of the year and is an unreliable and erratic crop like the opportunist flora of the benthos.

We must conclude, then, that normal rivers and streams are net consumers of organic matter, most of which is allochthonous and arrives in temperate climates mostly as an annual allocation of leaf-fall. I have stressed elsewhere that the whole biocenosis in a normal stream is geared to this arrangement with respect to diets of animals, life cycles, and longitudinal distribution (Hynes, 1963). What happens, then, when we enrich a normal stream?

Before that can be determined, we must distinguish between two kinds of enrichment. When we add organic matter, such as sewage, we increase the amount of allochthonous organic matter. Within limits, either directly or indirectly via the medium of some organism, such as *Sphaerotilus*, that thrives on dissolved organic matter and is probably always present in small amounts, we merely enhance the amount of animal production. This has been demonstrated very nicely by Warren *et al.* (1964) in their experimental addition of sugar to parts of a small stream in Oregon. Beyond those limits, we have organic pollution. The limits can be defined by the amount by which the oxygen content of the water is lowered under extreme conditions (not, may it be noted, the oxygen *demand* of the effluent, because the effect varies with turbulence and temperature) and by the rate of clogging of the stream bed with particulate matter and induced bacterial growth. But organic pollution is another topic because it involves a change of the whole biocenosis and is different in kind from enrichment (Hynes, 1960). We can, therefore, ignore it in the context of this discussion.

Downstream, however, of reaches that are definitely organically polluted, or where effluents have been well oxidized before discharge, we have the second type of enrichment in the form of plant-nutrient salts. These are primarily nitrate and phosphate; we need not consider potassium because it is so common a constituent of minerals that it is rarely, if ever, in short supply in running water. Similarly, it would seem unlikely that the various micronutrients that are known to be needed by algae (Eyster, 1964) are limiting factors in running-water environments.

We also increase the nitrate and phosphate supply to streams by other, more insidious, means. Agricultural land in Wisconsin has been shown to lose about 12 lb of nitrate N and 0.6 lb of phosphate P per square mile per day by leaching (Sawyer, 1947), and 0.1 lb of phosphorus has been given as a corresponding figure for agricultural land in Illinois (Engelbrecht and Morgan, 1961). Runoff from urban land contributes considerably more than this (Weibel *et al.,* 1964). And almost everywhere we are increasing our rate of application of fertilizers to cultivated land and our areas of urban sprawl. Moreover, there is increasing evidence that soil erosion and washout of banks contribute large amounts of phosphate to streams (Owen and Johnson, 1966; Webber and Elrick, 1967). Almost every agricultural or constructional activity in a watershed, therefore, must increase the fertility of the water.

During the past few decades, we have steadily increased our use of synthetic detergents, which are usually packed with phosphatic fillers, and this has resulted in a threefold to fourfold increase in the phosphate content of sewage over its prewar value (Stumm and Morgan, 1962). Detergents are also widely used in rural areas, by farmers, campers, and holiday cottagers; hence, a great deal of this material is entering streams far from sewage works. Indeed, it is a common observation of stream biologists who have been active in the field during the past three decades that streams and rivers now foam more readily than they formerly did.

When one considers that each square mile of landscape supports 1.4 miles of channel or less, it becomes clear that the amounts of nutrients being added per unit area of stream by leaching and runoff are enormous. The average relationship is such that

$$L = 1.4 \ A^{0.6},$$

where L = length of channel and A = area drained (Leopold et al., 1964). The amounts of nutrients are probably at least equal to the normal rates of addition of fertilizer to land used for growing high-priced crops.

If one adds to this the fact that we have cleared the forest and bush from the banks of millions of miles of channel, and so removed the summertime limiting factor of low light intensity, it becomes clear that, even without adding effluents, we have made our streams into potentially better primary producers than they used to be. Our negative contributions to this process are that we have straightened channels, drainage swamps, and skinned the beavers, so streams are more liable to spate and drought than they were; and by bad land use we have increased the turbidity of the water. These negative factors are, however, in process of reduction, possibly to levels lower even than their pristine ones, because of the widespread construction of impoundments. Even the Missouri River now flows as fairly clear water from dam to dam.

I contend, therefore, that rivers and streams are now enriched almost everywhere and that they are better places for plant growth than they were a century ago.

Whenever streams are enriched—particularly through the more spectacular enrichment by sewage, which contains 15 to 35 mg per liter of saline nitrogen and 2 to 4 mg per liter of phosphorous (Sawyer, 1952)—certain types of plants tend to occur in great abundance. These include some species of *Potamogeton, Stigeoclonium, Cladophora, Ulothrix, Rhizoclonium, Oscillatoria, Phormidium, Gomphonema, Nitzschia, Navicula,* and *Surirella* (Hynes, 1960), all of which may, of course, be found in unenriched streams, but far less abundantly. We do not know why these particular genera are encouraged while others are not, but we do know that the same general phenomenon occurs, without human intervention, when a river flows through an area of nutrient-rich rocks. It occurs, for example, where the Sundays River crosses the Ecca series of rocks in South Africa—rocks that are rich in nitrate and phosphate (Oliff et al., 1965). It has also been shown experimentally that on the shores of the Great Lakes *Cladophora* growths are stimulated by the addition of phosphate to the water (Neil and Owen, 1964).

It would seem that the combination of enrichment and the opening up of the stream bed to the sky, which go more or less together wherever man interferes with a watershed, have the effect of displacing to upstream situations many of the ecological conditions originally found only in the downstream reaches of a large river. These conditions are high levels of illumination and nutrients and high summer temperatures. But there is one considerable difference: in a large river, most of the bed lies deep below the surface and much of it is unstable anyway; hence, it is unsuitable for plant growth. Only on the occasional stable shoals is much growth possible, and there we often find opportunist species, many of which (*Potamogeton, Myriophyllum, Cladophora, Stigeoclonium, Oscillatoria, Phormidium*) are characteristic of enriched and cleared small streams. In streams, on the other hand, the water is mostly shallow, and much of the bed is hard and provides at least a temporary home for the opportunist species.

Enrichment, therefore, seems to move elements of the flora of the *potamon* into the *rhithron*. And this, together with the direct effect of greater temperature changes and the occasional nocturnal reduction of oxygen caused by respiration of the great masses of plants, encourages a change of fauna from that typical of cool, stony streams to that characteristic of river shoals. The biocenosis is shifted upstream, and much of the enormous potential fertility of river water is realized in a way that is rare in large rivers. Perhaps we have here the explanation of the great production of streams recovering from pollution.

A considerable difference from large rivers arises, however, from the fact that the volume of water per unit area in a stream is much less than in a river, so the biomass per unit volume is very high. This high concentration of biomass results in very large chemical changes caused by biotic activity and sometimes in very low oxygen tensions at night in warm weather (Edwards, 1962). This low oxygen tension can result in catastrophic fish kills that would never occur in the larger volume of water in a river. Similarly, smaller streams, simply because they are small and they drain a smaller area, are more liable to spates than are rivers. A storm that floods a small valley hardly affects the river into which it drains. Thus, the great biomass of plants in an enriched stream is especially subject to sudden washout, and there are many instances on record of masses of plants being piled against bridges or being washed into small lakes where they putrefy.

Enrichment and clearing of streams, therefore, affects them deleteriously from many points of view. They no longer support the fish species that they did, and they may even cause trouble because of plant growth. Unlike lakes, however, they probably do not retain excess nutrients for long. One would expect enrichment to pass almost as quickly as the water, but we have little information about this. It does seem, however, that some enrichment is retained for awhile in finely divided bottom deposits, as was inferred by Huntsman (1948), who piled fertilizer beside streams in barren parts of Nova Scotia to observe the effect on production of trout.

The effect of enrichment on larger rivers is far less well documented; the reason may be that there is very little effect. From our earlier considerations it will be clear that unenriched rivers are fertile, warm, and open to the sky, and that their production is limited by lack of suitable sites for plant growth. Addition of extra nutrients would, therefore, not be expected to greatly alter the situation.

Normally, most of the growth of plants actually in the water is planktonic. River plankton is usually dominated by rotifers and diatoms. Although some plankton occurs even in very muddy rivers—for example, the Missouri in its original state (Berner, 1951)—turbidity seems usually to be the major limiting factor (Williams, 1964, 1966). Turbidity, as we have seen, tends to be increased more or less in parallel with nutrient status by human activity, so the net result of enrichment by man is probably quite small. There is, however, a marked tendency of Chlorophyceae and Myxophyceae to appear in the plankton in warm weather; and these groups include many of the nuisance algae of waterworks. High temperatures are often correlated with low discharges and low turbidities, so that all factors are most favorable for plankton growth. We must assume, therefore, that enrichment is likely to increase the probability of unwanted blooms in warm weather, especially if it is caused by effluents rather than by erosion and runoff, which are, of course, at a minimum during periods of low discharge. Blooms have often been observed in rivers, and sometimes the numbers of algae have been found to be correlated with the nutrient status of the water, for example, nitrate in the River Oka in Russia (Mokeeva, 1964). On the other hand, Claus (1961) found no correlation between numbers of algae in the Danube near Vienna and any supposed chemical factor that he measured. Undoubtedly, the situation is far from simple, and I do not think it is an overstatement to say that we do not know what effect enrichment has on the *potamon*. We merely suspect that it probably increases algal production.

If this is, in fact, the situation, the increasing construction of dams may be expected to enhance the effect. Impoundments reduce the size and number of spates, and they decrease the turbidity. They also raise the water temperature in summer. All these changes lead to better conditions for the growth of phytoplankton. As the dam lakes provide a continual source of planktonic algae, the result can be only increased primary production in the rivers and fuller development of their potential fertility.

In summary, then, we can say that enrichment produces fairly obvious effects on small watercourses and that in moderation these may be beneficial from our point of view, since they increase production of such things as game fish. Greater amounts produce definitely deleterious effects, although up to a point this may not be true in countries where the value of fish is reckoned in weight of protein rather than quality. In larger rivers the effect is uncertain, but probably bad, and we may make it worse by constructing impoundments. It should, however, be emphasized here that in contrast to a lake, which can be made eutrophic and then probably remains in that state, a stream or river has to be continuously enriched. It can, therefore, be rescued and restored.

REFERENCES

Berner, L. M. 1951. Limnology of the lower Missouri River. Ecology 31:1–12.

Butcher, R. W. 1933. Studies on the ecology of rivers. I. On the distribution of macrophytic vegetation in the rivers of Britain. J. Ecol. 21:58–91.

Butcher, R. W. 1947. Studies on the ecology of rivers. VII. The algae of organically enriched waters. J. Ecol. 35:186–191.

Claus, G. 1961. Monthly ecological studies on the flora of the Danube at Vienna in 1957–58. Verh. Int. Ver. Limnol. 14:459–565.

Douglas, B. 1958. The ecology of the attached diatoms and other algae in a small stony stream. J. Ecol. 46:295–322.

Duffer, W. R., and T. C. Dorris. 1966. Primary productivity in a southern Great Plains stream. Limnol. Oceanogr. 11:143–151.

Edwards, R. W. 1962. Some effects of plants and animals on the conditions in fresh-water streams with particular reference to their oxygen balance. Int. J. Air Water Pollut. 6:505–520.

Engelbrecht, R. S., and J. J. Morgan. 1961. Land drainage as a source of phosphorus in Illinois surface waters, p. 74–79 In Algae and metropolitan wastes. U.S. Public Health Service SEC TR W 61–63.

Eyster, C. 1964. Micronutrient requirements for green plants, expecially algae, p. 77–85 In D. F. Jackson [ed] Algae and man. Plenum Press, New York.

Gessner, F. 1955. Hydrobotanik I Energiehaushalt. Deutscher Verlag der Wissenshaft, Berlin. 517 p.

Gessner, F., and F. Pannier. 1958. Der Sauerstoffverbrauch der Wasserpflanzen bei verschiedenen Sauerstoffspannungen. Hydrobiologia 10:323–351.

Hoskin, C. M. 1959. Studies of oxygen metabolism of streams in North Carolina, p. 186–192 In Publication 6, Institute of Marine Science, University of Texas.

Huntsman, A. G. 1948. Fertility and fertilisation of streams. J. Fish. Res. Bd. Canada 7:248–253.

Hynes, H. B. N. 1960. The biology of polluted waters. University of Liverpool Press, Liverpool. 202 p.

Hynes, H. B. N. 1963. Imported organic matter and secondary productivity in streams. XVI Int. Congr. Zool., Proc. 4:324–329.

Illies, J., and L. Botosaneanu. 1963. Problèmes et méthodes de la classification et de la zonation écologique des eaux courantes, considerées surtout du point de vue faunistique. Mitt. int. Ver. Limnol. 12. 57 p.

Knöpp, H. 1960. Untersuchungen über das Sauerstoff-Produktions-Potential von Flussplankton. Schweiz. Z. Hydrobiol. 22:152–166.

Leopold, L. B., M. G. Wolman, and J. P. Miller. 1964. Fluvial processes in geomorphology. Freeman, San Francisco. 522 p.

McIntyre, C. D. 1966a. Some effects of current velocity on periphyton communities in laboratory streams. Hydrobiologia 27:559–570.

McIntyre, C. D. 1966b. Some factors affecting respiration of periphyton communities in lotic environments. Ecology 47:918–930.

Mokeeva, N. P. 1964. The algoflora of the Oka River [in Russian]. Trud. Zool. Inst. 32:92–105.

Neil, J. H., and G. E. Owen. 1964. Distribution, environmental requirements and significance of *Cladophora* in the Great Lakes, p. 113–121 *In* Publication 11, Great Lakes Research Division, University of Michigan.

Nelson, D. J., and D. C. Scott. 1962. Role of detritus in the productivity of a rock outcrop community in a Piedmont stream. Limnol. Oceanogr. 7:396–413.

Odum, H. T. 1956. Primary production in flowing waters. Limnol. Oceanogr. 1:102–117.

Oliff, W. D., P. H. Kemp, and J. L. King. 1965. Hydrobiological studies on the Tugela River system. V. The Sundays River. Hydrobiologia 26:189–202.

Owen, G. E., and M. G. Johnson. 1966. Significance of some factors affecting yields of phosphorus from several Lake Ontario watersheds, p. 400–410 *In* Publication 15, Great Lakes Research Division, University of Michigan.

Ruttner, F. 1926. Bemerkungen über den Sauerstoffgehalt der Gewässer and dessen respiratorischen Wert. Naturw. 14:1237–1239.

Sawyer, C. N. 1947. Fertilization of lakes by agricultural and urban drainage. J. New England Water Works Assn. 61:109–127.

Sawyer, C. N. 1952. Some new aspects of phosphorus in relation to lake fertilization. Sewage Ind. Wastes 24:768–776.

Schmitz, W. 1961. Fliesswasserforschung–Hydrographie und Botanik. Verh. int. Ver. Limnol. 14:541–586.

Shadin, V. I. 1956. Life in rivers [in Russian]. Jizni Presnih Vod S.S.S.R., Moscow 3:113–256.

Stumm, W., and J. J. Morgan. 1962. Stream pollution by algal nutrients. 12th Ann. Conf. Sanit. Eng., Univ. Kansas Proc. 16–26.

Warren, C. E., J. H. Wales, G. E. Davis, and P. Doudoroff. 1964. Trout production in an experimental stream enriched with sucrose. J. Wildlife Management 28:617–660.

Webber, L. R., and D. E. Elrick. 1967. The soil and lake eutrophication. Tenth Conference on Great Lakes Research, Proc. 404–412.

Weibel, S. R., R. J. Anderson, and R. L. Woodward. 1964. Urban land runoff as a factor in stream pollution. J. Water Pollut. Control Fed. 36:914–924.

Whitford, L. A. 1960. The current effect and growth of fresh-water algae. Amer. Microscop. Soc., Trans. 79:302–309.

Whitford, L. A., and G. J. Schumacher. 1961. Effect of current on mineral uptake and respiration by a fresh-water alga. Limnol. Oceanogr. 6:423–425.

Whitford, L. A., and G. J. Schumacher. 1964. Effect of a current on respiration and mineral uptake in *Spirogyra* and *Oedogonium.* Ecology 45:168–170.

Williams, L. G. 1964. Possible relations between plankton–diatom species numbers and water quality estimates. Ecology 45:809–823.

Williams, L. G. 1966. Dominant planktonic rotifers of major waterways in the United States. Limnol. Oceanogr. 11:83–91.

BOSTWICK H. KETCHUM
Woods Hole Oceanographic Institution, Massachusetts

Eutrophication of Estuaries

Many attempts to define estuaries on a scientific basis have been criticized because each definition seems either to include too much or too little of the coastal waters to suit the scientist who is evaluating the definition (Pritchard, 1967; Caspers, 1967). Perhaps the best approach is the one presented in the *CF Letter*, published by The Conservation Foundation, Washington, D.C. An article in the May 22, 1967, issue states:

An estuary, where the fresh water of a river meets the tide of the sea, is:
    Wasteland? Rich resource? Marine nursery? Recreation area? Source of raw materials for industry? None of these? All?
    The answer depends on who's talking.
    To the scientist, an estuary is "very fertile and productive of plant and animal life—more productive, in general, than either land or sea"—and can produce "more harvestable food per acre than the best midwestern farmland."
    To some in crowded urban centers, an estuary is "unused" space to be developed for housing projects, industrial sites, roads, marinas, golf courses, amusement parks, sewage treatment plants.
    To commercial and sport fishermen, estuaries are vital: Over 90 percent of the total harvest of seafood taken by American fishermen comes from the continental shelf and about two thirds of that volume are species whose existence depends on the estuarine zone or which must pass through the zone enroute to spawning grounds.
    To a builder seeking a permit to dredge and fill an estuary, it was "not a veritable paradise," for "while wildlife present on the marsh may include shorebirds, ducks and other marine life, it also includes such vermin as rats, mice and mosquitoes."

To another developer seeking to fill and build upon an estuary, only concern about "the love life of fiddler crabs" stood in his way.

To the sand and gravel industry, estuaries are "a principal source" of those raw materials for construction.

To some others, estuaries are to be left untouched for enjoyment and relaxation, for recreation without development, for closeness to nature.

Although a precise definition of an estuary may be difficult, we all realize that most of the major harbors of the world are in estuaries or in rivers that flow into the sea there. Many dense urban populations have developed on the shores of estuaries or the banks of their rivers. Traditionally, maintaining adequate navigation and using the waterways for disposing of our civilization's wastes have been two of the major demands on estuarine waters. For decades it has been obvious that pollution of rivers and estuaries has profoundly affected the food resources that are dependent on these bodies of water. Shad and salmon no longer run in many of our eastern Atlantic streams, and areas suitable for growing and harvesting shellfish are constantly becoming more drastically curtailed. In recent years, the burgeoning demand for recreational facilities has led to development of a multimillion-dollar boating and sport fishing industry, which requires clean waters in our rivers and estuaries.

It is obvious, therefore, that there are conflicting demands on the waters of the estuaries and of the rivers that ultimately wend their way into estuaries. We need these waters for our industrial and urban development, but we also need to preserve them both for the fisheries they sustain and for the recreational value they offer. We are learning how to produce industrial wastes and to crowd people into more compact urban developments faster than we are learning how to dispose of the resultant wastes in ways that will not infringe on our enjoyment of our natural resources. We must learn to protect these resources so that our descendants can look with pride at what has been accomplished in the twentieth century and not with dismay at the destruction of our natural resources.

CHARACTERISTICS OF ESTUARINE CIRCULATION

I will mention a few of the essential characteristics of estuarine circulation as they relate to the distribution of pollutants. I will not go into detail because this is covered by Carpenter, Pritchard, and Whaley in this volume (page 210). The estuary offers advantages not offered by the river in its ability to dilute and disperse added contaminants.

In the river itself, the volume of water available to dilute a pollutant is furnished simply by the river flow, which carries the contaminant down-stream at a rate determined solely by the river flow and the geometry of the river bed. In the estuary, the circulation is more complex, although the net seaward flow is also determined by the rate of river flow. If no mixing were involved, this fresh river water would merely flow seaward as a layer on top of undiluted seawater. Mixing is involved, however, and salinity gradually increases down the estuary as river water mixes with more and more seawater. Seawater must flow into the estuary to provide the salt needed to balance the system. In a steady-state condition, the volume of seawater entering the estuary in a given unit of time equals the volume flowing out; there is no augmentation of the net seaward flow. The seawater thus entrained with the freshwater does, however, increase the diluting capacity of the mixed water that is escaping from the estuary. This effect can be evaluated by using the distribution of salt water and freshwater in the estuary.

The amount of freshwater contained in any given sample of brackish water can be calculated from the salinity, since

$$F = \left(1 - \frac{S}{\sigma}\right),$$

in which F is the fraction of freshwater in the sample, S is the salinity of the sample, and σ is the salinity of the "source" seawater. If the average freshwater content of a complete cross section is known, the volume available for the dilution of the pollutant at that location can be approximated. To obtain the fraction of freshwater in a complete cross section of the estuary, it is necessary to integrate the values from top to bottom and from bank to bank. The volume available for the dilution of the pollutant in a given period is determined approximately by dividing the rate of river flow by the fraction of freshwater in the cross section. If the section is 50 percent freshwater, two volumes must move seaward to move one volume of river water seaward. Closer to the mouth of the estuary, where the amount of freshwater has been reduced to 10 percent, ten volumes must move seaward to remove the river water. A more precise determination of the diluting volume requires detailed knowledge of the circulation. But this simple calculation shows that the total volume available for dilution increases in the seaward direction.

Estuaries have a unique mechanism for transporting pollutants upstream from the point at which they are introduced. The pollutant will be carried upstream as well as downstream by tidal action and mixing, and the oscilla-tory character of tidal currents complicates the estuarine circulation. On the flooding tide the entire water mass may move landward, reversing the normal current provided by river flow. On the ebbing tide the entire water mass may move seaward. In stratified estuaries the net seaward flow, which carries the

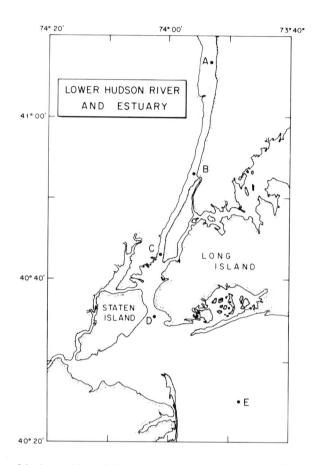

FIGURE 1   Location of hydrographic and chemical observations in the lower Hudson River and estuary and in the approaches to New York harbor.

freshwater out of the estuary, may be limited to the surface layers, whereas
the flow in the deeper waters may be landward in order to maintain the salt
balance. Thus, even if the pollutant were introduced at the mouth of the
estuary, it would be carried upstream to the same extent that salt is carried
upstream, and by the same mechanism.

   A conservative pollutant is not changed by biological processes or
exchanges with the atmosphere; its distribution is determined solely by the
circulation and mixing of the waters. If such a pollutant is introduced into
the river, its distribution within the estuary will be proportional to the
distribution of freshwater, because the same mechanisms will mix and dilute
pollutant and river water alike (Ketchum, 1955). If a pollutant is introduced

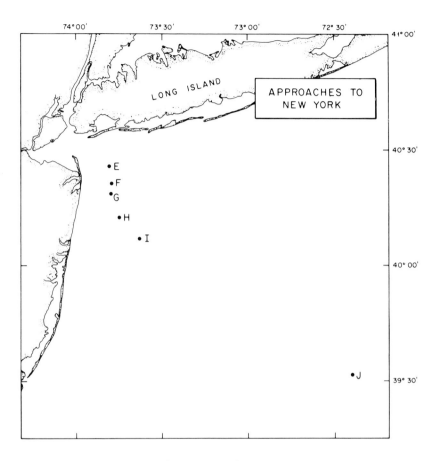

into the estuary, greater dilution will be provided, but the downstream distribution will still change in proportion to the further changes of freshwater, and the upstream distribution will be proportional to the distribution of salt above the point of introduction.

DISTRIBUTIONS IN THE HUDSON ESTUARY

Some observations we have made on the estuary of the Hudson River will serve to illustrate the principles that have been mentioned above. The location of stations where samples were taken is shown in Figure 1. Our observations extended into the offshore coastal water well beyond the mouth of the Hudson River, which many people would define as the outer limit of the estuary. The first of these offshore stations is needed to identify the characteristics of the source sea water that is mixing with the Hudson River water. Data from the other offshore stations are included only to show the consistency of these offshore waters.

The distributions of salinity, of oxygen deficit, and of total phosphorus are presented in Figure 2 for samples taken at the surface and at 10 m. From the oxygen and phosphorus data it is clear that additional pollution is being added to the estuary at Station C, just off the southern tip of Manhattan Island.

The salinity varied from a minimum value of 7.56 ppt in the surface waters at Station A to values of 31.2 to 31.4 ppt at Station E. The estuary was stratified with the water at 10 m, having salinities from 2 to 7 ppt higher than the surface waters. A given salinity of the surface water was found 5 to 10 mi farther upstream in the deep water. This salinity distribution illustrates the

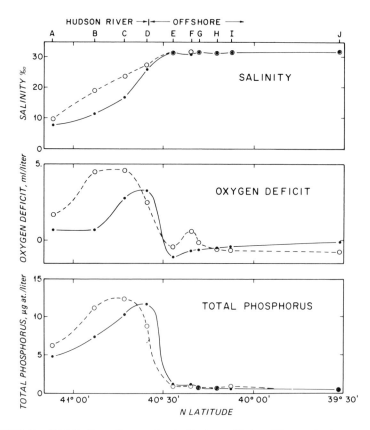

FIGURE 2   Distribution of properties in the lower Hudson River and estuary and in the waters off New York City.

upstream transport of salt in the deep water, or conversely, the downstream transport of river water at the surface.

The oxygen deficit is the difference between the saturation value and the observed oxygen content. This deficit is greatest at the 10-m depth at Station C, where the oxygen content was decreased to 17 percent of saturation. The offshore coastal waters were generally supersaturated, with a maximum value of 121 percent for the surface water at Station E. It should be mentioned, however, that this supersaturation in the offshore surface waters does not improve the oxygen balance of the estuary; the source seawater is drawn from greater depths, where the oxygen concentration varied from 59 to 85 percent of saturation.

The maximum concentration of total phosphorus, 12.3 μg at. per liter, was also observed at a depth of 10 m at Station C. In this sample, 70 percent of the phosphorus was present as inorganic phosphate, 16 percent was particulate, and 14 percent was dissolved organic phosphorus. Though the concentrations of total phosphorus decreased by dilution both landward and seaward from this location, the fraction found in particulate organic matter in the surface waters was 25 percent at Station A and 36 percent at Station E. These results show that the natural phytoplankton population was growing and assimilating the phosphorus that was added to the estuary in the inorganic form.

The relationship between the distribution of total phosphorus and the distributions of salt water and freshwater is shown in Figure 3. The water at a depth of 10 m at Station C, where the maximum total phosphorus was observed, had a salinity of 23.77 ppt (So). The water at a depth of 30 m at Station E was taken as the source of seawater that mixes within the estuary. The salinity of this water was 32.11 ppt, and the total phosphorus content was 1.77 μg at. per liter. From these salinities we know that the freshwater made up 32 percent of the mixture present at 10 m at Station C. Upstream from this station the phosphorus data are linearly related to distribution of salt, given by the ratio Sx/So, and downstream to the distribution of freshwater, given by the ratio Fx/Fo (Ketchum, 1955). It appears, however, that we did not observe the maximum concentration of phosphorus in the estuary; extrapolation of two linear parts of the relationship to their intersection suggests that the maximum value would be found somewhat north of Station C and would be about 10 percent greater than the highest concentration we observed. Extrapolation of the linear upstream relationship suggests that water containing no seawater would have a phosphorus content of somewhat less than 1 μg at. per liter, a value lower than that actually observed in the source seawater that mixes within the estuary. The four open circles at the seaward end of the plot (Figure 3) are samples of surface seawater from outside the estuary; this water clearly is not involved in mixing within the estuary.

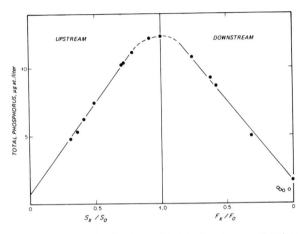

FIGURE 3   Distribution of total phosphorus related to distribution of salt water upstream and distribution of fresh-water downstream.

SECONDARY POLLUTION

Much effort has been expended in treating domestic sewage waste in an effort to remineralize the sewage and thus decrease the organic material and the biological-oxygen demand that is introduced into natural flowing waters. If, however, the nutrients essential for plant growth are still in the effluent, the normal phytoplankton populations will bloom, utilizing these fertilizing elements, and will resynthesize organic material. The amount of organic matter that can be produced is determined by the amount of fertilizer added. If all the fertilizing elements are released to natural waters, the organic matter that is formed could equal the amount that was removed with considerable labor and expense in the treatment plant.

Seawater that has not been depleted contains a remarkable balance of fertilizing elements that are essential for plant growth (Redfield, 1958; Redfield et al., 1963). The elements that most commonly limit the phytoplankton production in the sea are phosphorus and nitrogen. There is a nearly constant ratio between the changes in the amounts of these elements and the carbon fixed in photosynthesis or the oxygen utilized in the decomposition of organic material. Redfield et al. discuss the ratios of utilization of various nutrient elements in seawater and give the following as the usual atomic ratios of change that result from biological activity:

$$\Delta O: \Delta C: \Delta N: \Delta P = -276:\ 106:\ 16:1$$

These ratios are not constant in the sense that applies to the stoichiometric combining ratios in chemistry, but they are useful in evaluating the changes that might be expected in a normal habitat. It should be emphasized, however, that these are the ratios of the biological requirements for the elements; they are not necessarily the ratios of concentration of these elements in the environment. In inshore waters, for example, nitrogen is apt to be a limiting element and to be completely utilized before the phosphorus is exhausted (Riley, 1967; Ketchum *et al.*, 1958). These ratios cannot be used dependably for inferring changes in all elements from changes that have been measured in only one element. The dependability of the inference is affected by differences in the composition of living matter and by differences in the rates of the biological processes of use and regeneration.

Measurement of phosphorus is useful, however, for understanding excess pollution. Routine analysis is comparatively easy, and with three analyses it is possible to separate the inorganic, the particulate organic, and (by difference) the dissolved organic fraction of phosphorus in the water.

Complete analysis of the nitrogen cycle is considerably more difficult. Nitrogen appears in three available inorganic forms: nitrate, nitrite, and ammonia nitrogen. Furthermore, the total amount of nitrogen in the water may be modified by nitrogen fixation or denitrification. Total nitrogen, therefore, cannot be considered a conservative property.

C. S. Yentsch (unpublished data) has used a relationship between chlorophyll and inorganic phosphate in describing the eutrophication of estuarine waters. His results are presented in Figure 4. Since there is a time lag between the introduction of a fertilizing element and the development of the phytoplankton population utilizing the element, there is a wide range of chorophyll values for any given phosphate content. Both chlorophyll and phosphate, however, show a continuous progression from the natural unpolluted seawater to the heavily polluted or eutrophic waters. Yentsch postulates a phosphate value of 2.8 μg at. per liter as the approximate upper limit of unpolluted water. Redfield *et al.* (1963) indicate that the amount of oxygen dissolved in deep ocean water is adequate to decompose the amount of organic material that could be produced by the utilization of this amount of phosphorus in the growth of phytoplankton populations.

By employing the ratios given above, we may compute the changes in concentration of the products of decomposition as a function of the increasing phosphate concentration. Figure 5 shows the results of such an analysis for winter and summer conditions in northern temperate latitudes. The calculation is based on an estuarine water with a salinity of 30 ppt, corresponding to a freshwater content of 14 percent in average seawater. A winter temperature of 2°C and a summer temperature of 20°C are assumed. These temperatures give oxygen-saturation values of 706 and 460 μg at. per

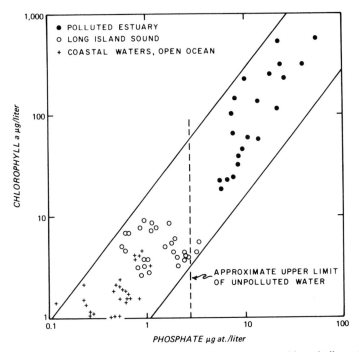

FIGURE 4   Relationship between inorganic phosphate and chlorophyll content of waters ranging from unpolluted seawater to polluted estuaries (C. S. Yentsch, unpublished data).

liter for these two conditions. Two assumptions are inherent in the calculations.

1. Phosphorus is the limiting factor and all other nutrients are present either in excess or in quantities that will meet the requirements of the plant.

2. Photosynthesis has gone to completion, leaving the oxygen content of the water in equilibrium with the atmosphere; that is, the oxygen produced in photosynthesis has escaped from the water.

The decomposition of the organic material that is formed decreases the oxygen content to zero. This process is followed by denitrification and, finally, in an overenriched environment, by sulfate reduction and the production of hydrogen sulfide. The oxygen demand of the organic matter that is produced by photosynthesis equals the available supply of oxygen when the phosphorus enrichment reaches 2.55 μg at. per liter in winter or 1.7 μg at. per liter in summer. These limits of concentration may be accepted as danger signals in evaluating the eutrophication of an estuary.

This is a conservative definition of the problem because it is assumed that the only oxygen available is that which can be dissolved at saturation in the water. Oxygen will be added to the system both by exchanges with the atmosphere and by photosynthesis. If all of the oxygen produced by photosynthesis were trapped within the system, the phosphate values could be double those given before anoxic conditions would be produced. Because of supersaturation, however, some of the oxygen produced in photosynthesis will be lost to the atmosphere. The correct danger limits for phosphate enrichment, therefore, will lie somewhere between the values given and double those values.

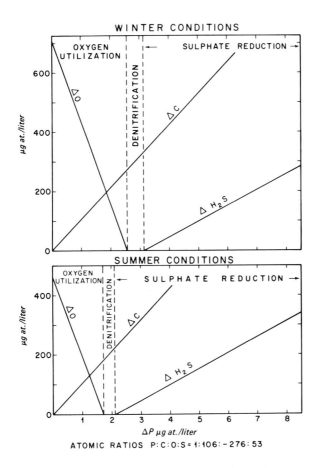

FIGURE 5   Products of decomposition of organic matter in seawater as related to change in phosphate concentration (modified from Redfield *et al.*, 1963).

CONCLUSION

Eutrophication of estuaries, in moderation, could be of benefit to mankind. In excess, it leads to the development of obnoxious species, to the elimination of desirable species, and, ultimately, to the development of anoxic conditions and the elimination of all higher forms of life. Even if pollutants are added only in mineral form, the natural populations will synthesize organic matter, which exhausts the oxygen content of the water as it decomposes.

Even while we are fertilizing natural waters to excess and creating undesirable and obnoxious conditions in our rivers and estuaries, we are concerned about available food supplies for the rapidly expanding populations of the world. The sea is a valuable source of animal protein, but the total productivity is limited by lack of nutrient elements. Small marine areas, such as lagoons and salt ponds, are effectively fertilized to increase productivity. But the deliberate purchase of fertilizers for any major area may be too costly for the probable returns.

When these two problems are considered separately, a solution seems almost hopeless. However, the characteristics of pollution that create problems are the same as the characteristics needed to increase the productivity of the sea and to augment the food supply. A drastic change in viewpoint and attitude may be needed. We might solve the two problems much more easily by combining them than by continuing our efforts to solve them separately.

This solution might be considered merely an extension of the oriental use of "night soil." The ecology of the sea is, however, much more complex than the garden plot by the back door. The marine ecologist believes that he understands fairly well the relationships between plant production of organic matter and the essential nutrients in the water. Conditions that lead to efficient conversion of phytoplankton to zooplankton and of these to fish are poorly understood. Clearly, more research is needed on these basic ecological problems. Engineers and ecologists must join forces in an attack on these problems if we are to leave descendants who are well fed and who have a clean environment that they can enjoy.

This paper is Contribution Number 1960, Woods Hole Oceanographic Institution, Woods Hole, Massachusetts. This study was supported in part by U.S. Atomic Energy Commission Contract Reference NYO 1918-154.

REFERENCES

Caspers, H. 1967. Estuaries: Analysis of definitions and biological considerations, p. 6–8 *In* G. H. Lauff [ed] Estuaries. Pub. 83. American Association for the Advancement of Science, Washington, D.C.

Ketchum, B. H. 1955. Distribution of coliform bacteria and other pollutants in tidal estuaries. Sewage Indust. Wastes 27:1288–1296.

Ketchum, B. H., R. F. Vaccaro, and N. Corwin. 1958. The annual cycle of phosphorus and nitrogen in New England coastal waters. Sears Foundation: J. Mar. Res. 17:282–301.

Pritchard, D. W. 1967. What is an estuary: physical viewpoint, p. 3–5 *In* G. H. Lauff [ed] Estuaries. Pub. 83. American Association for the Advancement of Science, Washington, D.C.

Redfield, A. C. 1958. The biological control of chemical factors in the environment. Amer. Sci. 46:205–221.

Redfield, A. C., B. H. Ketchum, and F. A. Richards. 1963. The influence of organisms on the composition of sea water, p. 26–77 *In* M. N. Hill [ed] The sea. Vol. 2. Interscience, New York and London.

Riley, G. A. 1967. Mathematical model of nutrient conditions in coastal waters. Bull. Bingham Oceanogr. Coll. 19(2):72–80.

J. H. CARPENTER

D. W. PRITCHARD

R. C. WHALEY

The Johns Hopkins University, Baltimore, Maryland

Observations of Eutrophication and Nutrient Cycles in Some Coastal Plain Estuaries

Distributions of nitrogen and phosphorus compounds in the northern half of the Chesapeake Bay and in the Potomac River downstream from Washington, D.C., have been observed for the past several years during monthly cruises conducted by the Chesapeake Bay Institute. Several smaller tributaries (the South, Severn, Magothy, Back, Chester, and Miles Rivers and the Eastern Bay) were also surveyed during these cruises. Measured at each station (Figure 1) were temperature, salinity, dissolved oxygen, pH, alkalinity, chlorophyll, transparency, inorganic phosphate, total filtrate phosphate, total phosphate, nitrate, nitrite, ammonia, total nitrogen, productivity, and counts of net phytoplankton and zooplankton. The purpose of these surveys was to determine existing conditions and nutrient levels in order to provide a basis for considering permissible increases in nutrient discharges or, in some areas, desirable reductions. This paper is a review of the salient features of the observed distributions.

UPPER CHESAPEAKE BAY

The upper Chesapeake Bay is primarily the estuary of the Susquehanna River, which drains central Pennsylvania and a part of New York State. The mean annual discharge from the Susquehanna of 43,000 ft^3/sec represents about 86 percent of the freshwater flow into the bay above the mouth of the Potomac River. Weekly samples of Susquehanna River water were collected during 1965–1966 to monitor the input into the upper bay.

210

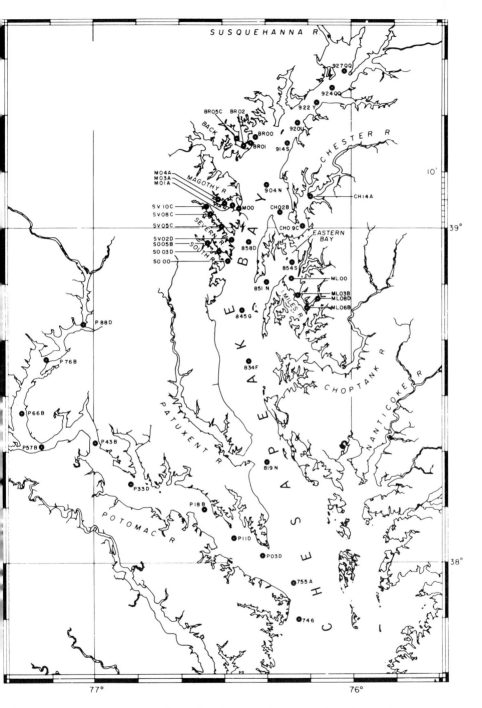

FIGURE 1   Locations of sampling points in the northern part of the Cheaspeake Bay and the Potomac estuary.

The watershed of the Susquehanna River has extensive agricultural areas and a population in excess of 1 million; considerable quantities of phosphorus and nitrogen are added to the river from these sources. The effects of these inputs are considerably modified by the passage of the river water through reaches where the cross section is broad and shallow and through a series of reservoirs where biological removal occurs. As a result, the river water that reaches the Chesapeake Bay has a total phosphorus content of about 1.5 μg at. per liter during winter and spring and about 1.0 μg at. per liter during summer and fall. Most (85 percent) of the nitrogen in this water is in the form of nitrate; in spring, concentrations are 80 to 105 μg at. per liter, whereas in other seasons they are 40 to 60 μg at. per liter.

A significant part of the suspended sediments carried by the Susquehanna is trapped in the reservoirs in the lower reaches of the river. Consequently, the river water is relatively clear by the time it reaches the bay, with suspended-solid contents of 10 to 20 mg per liter, except for brief periods during early stages of high discharge.

The city and metropolitan area of Baltimore, Maryland, located 35 miles south of the entrance of the Susquehanna River into the bay, has a population of 1,400,000. The Back River treatment plant of Baltimore produces about 150 million gal/day of treated effluent. About 100 million gal/day of this effluent is used as industrial water by the Bethlehem Steel Corporation and is discharged into Baltimore Harbor; the remainder is discharged into the Back River, a tributary estuary of the Chesapeake Bay located just north of Baltimore Harbor.

The effects of these treated sewage discharges are not observable in the Chesapeake Bay below Baltimore. During the prolonged drought of 1965, discharge of the Susquehanna River was 4,000 ft^3/sec during July, August, and September. The admixture of this inflowing freshwater with seawater produced a density-driven circulation in the bay off Baltimore with a flow in the upper layer of about three times the freshwater discharge, or 12,000 ft^3/sec. This flow would provide a dilution for the sewage discharge of 1 to 50, which corresponds to a possible increase of 6 μg at. per liter of phosphorus and 36 μg at. per liter of nitrogen in the mixture. Such increases are not observed in the bay.

Algal growth in the Back River is profuse; chlorophyll concentrations exceeding 60 μg per liter were observed throughout the period March through November 1965, with 400 μg per liter observed in October. Decreases in chlorophyll and total phosphate at the mouth of the Back River of fivefold to sevenfold are greater than can be expected to result from local dilution; deposition of the algal cells in the sediment appears to be the most probable process. Eutrophication in the Back River is intense, but the effects are limited to this local area.

For the two thirds of the effluent that is used as industrial water, an added constituent, in the form of waste ferrous sulfate, appears to have further effects in reducing the phosphorus additions to the bay. The hydrolysis and oxidation of the added ferrous iron (100 tons/day) should be more than adequate to precipitate the phosphate (4 tons/day) in the effluent. The observed distribution of total iron in Baltimore Harbor suggests that more than 90 percent of the iron is deposited within a few miles of the discharge area. Analyses of the sediments in the region show large iron enrichment, confirming this suggestion.

The distribution of nitrate in the upper Chesapeake Bay is easily contoured, as shown in Figure 2 for March 1965, and the pattern of the contours suggests that the inflow of the Susquehanna River, which enters the bay north of Turkey Point, is the major source of nitrate. Similar patterns were observed during the springs of 1964 and 1966. By the middle of April the upper bay has a rather uniform nitrate distribution, with concentrations of about 45 μg at. per liter. From this time through September, a general decrease in nitrate is observed throughout the bay. By September 1965, the nitrate concentrations were less than 1 μg at. per liter (Figure 3).

The distributions of phosphorus contrast strongly with the distributions of nitrate. The dissolved inorganic phosphate rarely exceeds 0.2 μg at. per liter and does not show any seasonal or spatial pattern; that is, there is no pronounced longitudinal gradient. Total phosphate values are in the range of 1 to 2 μg at. per liter. The disappearance of some 45 μg at. per liter of nitrate during the period May through August is not accompanied by changes in phosphate. With "normal" composition of algal cells, the nitrogen-to-phosphorus-atom ratio would be 15 to 1, so utilization of 45 μg at. per liter of nitrate would require 3 μg at. per liter of phosphorus, or more than the total observed at any time. The phosphorus concentrations are frequently 1.5 μg at. per liter; so we must conclude, disregarding nitrogen regeneration, that phosphorus is cycled at least twice during the period May through August. Throughout this period, more than one half of the total phosphorus is present as dissolved organic phosphate, indicative of continual production of this fraction.

Nitrogen regeneration on the bottom is not a major process in the upper Chesapeake Bay during late spring and summer. During July 1965, for example, stations in the area off Annapolis, Maryland, showed nitrate concentrations of 4 to 6 μg at. per liter, without significant variation with depth, and ammonia concentrations of 3 to 5 μg at. per liter in the upper layer and 10 to 12 μg at. per liter in the lower layer. River flow was 4,000 ft^3/sec and had been uniform for the previous month. The observed salinities in the region showed a 3 ppt difference between the upper layer and the lower layer. The longitudinal salinity gradient was 0.1 ppt per mile. For this

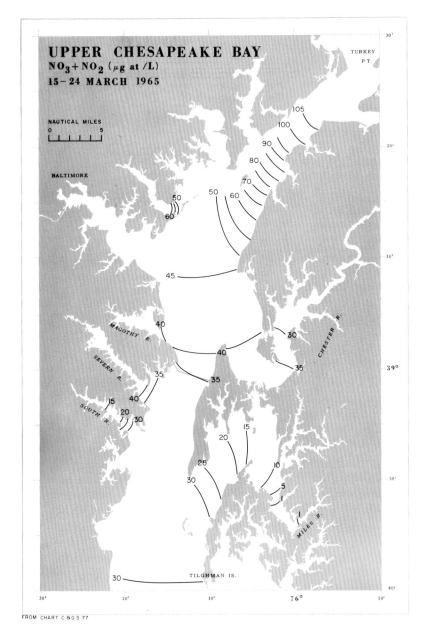

FIGURE 2   Surface nitrate concentrations (NO_3 plus NO_2) observed in the upper Chesapeake Bay, March 15 to 24, 1965. Data expressed in μg at. per liter.

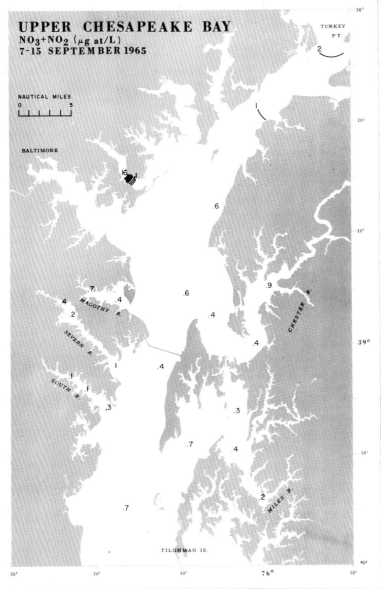

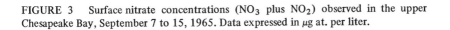

FIGURE 3   Surface nitrate concentrations (NO_3 plus NO_2) observed in the upper Chesapeake Bay, September 7 to 15, 1965. Data expressed in μg at. per liter.

region the salinity distribution indicated a flow of 16,000 ft³/sec in the upper layer, which is a drift of 0.5 mi per day. This amounts to an increase in salinity of 0.5 ppt per day in the moving water of the upper layer, primarily as a result of mixing (exchange) with the lower layer. Since the difference in salinity between the upper layer and lower layer is 3 ppt, the time change in salinity corresponds to a vertical exchange rate of 1.7 percent per day. With the 7 μg at. of NH_3 per liter difference between the layers, an addition of only 0.12 μg at. of NH_3 per liter per day to the upper layer is indicated.

Measurements of radiocarbon productivity show that roughly 0.2 mg per liter per day of organic carbon is being produced in the upper layer during the summer. This value is uncertain by perhaps a factor of 2, either high or low. However, this rate of carbon fixation corresponds to an uptake of 4 μg at. of nitrogen per liter and 0.5 μg at. of phosphorus per liter per day. At this time, concentrations of inorganic nitrogen, inorganic phosphate, and dissolved organic phosphate are roughly as follows (μg at. per liter):

Inorganic nitrogen (nitrate, nitrite, and ammonia)	8
Inorganic phosphate	0.1
Dissolved organic phosphate	0.8

Therefore, the observed carbon fixation would result in utilization of all the available nitrogen and phosphorus in 2 days or, taking into consideration the uncertainties, between 1 and 4 days. Since the chlorophyll, nitrogen, and phosphorus concentrations do not change rapidly, and since regeneration on the bottom or in the lower layer is not adequate to supply nitrogen and phosphorus at this rate, as shown above, rapid regeneration of phosphorus and nitrogen in the upper layer, coupled with removal of chlorophyll, is indicated. Grazing by zooplankton seems the most likely manner in which these two processes could occur. The standing crops of zooplankton appear more than adequate to consume the production.

While nutrient cycling between the upper layer and the lower layer and bottom does not seem to be important, deposition of nitrogen may be an important part of the nitrogen balance. As noted above, some 45 μg at. of nitrate nitrogen per liter disappears from the water column during the late spring and summer. Estimates of the rates of sediment accumulation average around 1 mm per year. With 50 percent water content and 5 percent carbon, this amounts to 25 g of carbon per m² or 2 g at. per m². The corresponding nitrogen deposition would be 500 mg at. per m². The observed nitrate decrease amounts to 450 mg at. per m² for the mean water depth of 10 m. Uncertainty regarding the sedimentation rate will not permit a precise calculation of nitrogen deposition, but the estimates are not discordant, and the loss of nitrate appears to be adequately accounted for by deposition.

POTOMAC RIVER

The Potomac River is tidal to just above Washington, D.C.; however, the salt-water intrusion extends only to 20 to 30 miles below Washington. The freshwater region below Washington will be called the tidal river to distinguish it from the remainder of the estuary where the presence of salt water causes vertical density stratification and the characteristic estuarine pattern of two-layer circulation. The sewerage systems of the Washington metropolitan area discharge 6 tons of phosphorus and 10 tons of nitrogen per day directly to the Potomac in the form of treated effluent. A series of cruises from March 1965 through April 1966 was conducted to observe the distributions of these materials in the tidal river and estuary.

The mean freshwater flow of the Potomac is 11,000 ft^3/sec. As shown in Figure 4, the study period was one of extreme summer drought, with periods of 1,000 ft^3/sec discharge and brief periods of 80,000 to 90,000 ft^3/sec discharge. The Potomac watershed has extensive agriculture, but a smaller population than the Susquehanna watershed. There are no significant reservoirs or broad reaches, so upstream phosphorus and nitrogen removal is less than that observed in the Susquehanna discharge. Nitrate levels of 100 to 150 μg at. per liter and phosphorus levels of 5 μg at. per liter are present in the river just above Washington during periods of high discharge. During

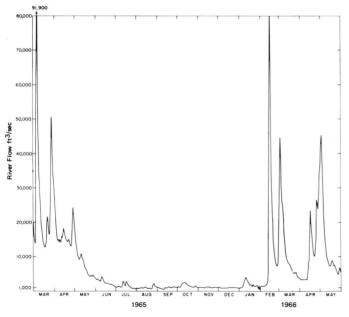

FIGURE 4   Discharge of Potomac River during study period.

periods of lower flow, nitrate was found to be 50 to 70 μg at. per liter and phosphate 3 to 4 μg at. per liter.

The additions of phosphorus and nitrogen below Washington produce very large concentrations locally. For example, at 3,000 ft³/sec discharge, the phosphorus increase would be 180 μg at. per liter.

Such increases are not observed very far downstream. The total nitrogen distribution shows a seasonal pattern, with higher concentrations during January, February, and March (Figure 5). Nitrate increases are the principal contributions to this pattern, with high nitrate concentrations occurring downstream in the estuary during March. Nearly complete removal of nitrate in the estuary occurs by June; nitrate levels are very low throughout the summer and fall (Figure 6).

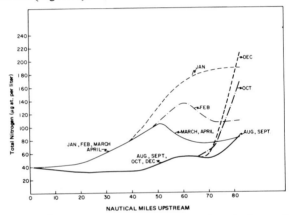

FIGURE 5   Longitudinal distribution of total nitrogen in the Potomac River.

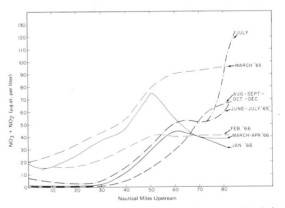

FIGURE 6   Longitudinal distribution of nitrate plus nitrite in the Potomac River.

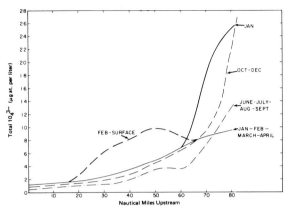

FIGURE 7   Longitudinal distribution of total phosphate in the Potomac River.

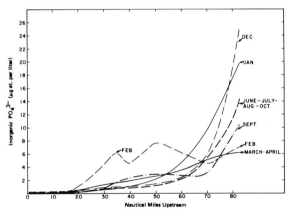

FIGURE 8   Longitudinal distribution of inorganic phosphate in the Potomac River.

The total phosphate distributions (Figure 7) show increases in the late fall and winter in the tidal river, movement downstream into the estuary during the high flow, then relatively moderate and constant concentrations throughout the summer and fall. Inorganic phosphate concentrations (Figure 8) are high in the tidal river and are an appreciable part of the total phosphatic concentration in the tidal river. In the estuarine part of the Potomac, however, inorganic phosphate concentrations greater than 0.5 μg at. per liter occur only after high river flow, and the relative concentrations of total and inorganic phosphate are similar to those found in the upper Chesapeake Bay.

The large decreases in total phosphate and inorganic phosphate concentrations downstream from Washington suggest loss of phosphate from the water to the sediment. However, quantitative interpretation is complicated by the variable rate of river flow, which by itself would produce a longitudinal gradient downstream from the Washington discharge. The travel time from Washington to the mouth is so long that water that passed Washington at quite different rates is present in different parts of the estuary. Since the concentration of added phosphorus and nitrogen depends on the rate of river flow past the discharge point, this time variation in flow rate produces a spatial variation in concentration downstream.

The high concentrations of chlorophyll (Figure 9) in the tidal river below Washington are produced primarily by *Microcystis aeruginosa,* which floats to form highly visual discolorations and collects on the shore line in unattractive mats. Lack of effective grazing on this alga may contribute to the large standing crop and to ultimate deposition of the algal colonies in the sediments.

Phosphorus and nitrogen concentrations in the estuarine part of the Potomac River during 1965 were not greatly different from those observed in the upper Chesapeake Bay. Correspondingly, the chlorophyll concentrations were not much higher in the estuarine Potomac than in the upper Chesapeake Bay. However, the zooplankton populations in the Potomac were frequently threefold to fivefold greater than those found in the upper Chesapeake Bay. The zooplankton are not adequately sampled in the small (100 ml) volumes used for chemical analyses. The phosphorus and nitrogen content of the zooplankton may represent an appreciable part of the total water-column content. For example, Acartia ranged from 5 to 85 individuals per liter, with 30 being a frequent value. This density would amount to about 1 μg at. of

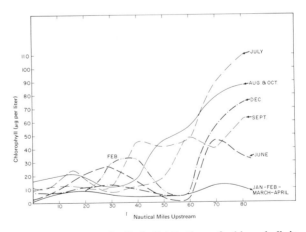

FIGURE 9   Longitudinal distribution of chlorophyll in the Potomac River.

phosphorus per liter in the region of the estuary where the "total" phosphorus was 1 to 1.5 μg at. per liter. An adequate inventory will require analyses on zooplankton collected from large volumes and an extensive sampling program to estimate the zooplankton populations with the required precision.

The decrease in dissolved oxygen in the lower layer shows that eutrophication may be more pronounced in the estuarine Potomac than is indicated by phytoplankton counts and chlorophyll concentrations. During 1965, oxygen depletion occurred to levels less than 1 ml per liter in areas of the Potomac; comparable (salinity and vertical stratification) areas in the upper Chesapeake Bay did not show depletion to less than 1 ml per liter. Review of the dissolved-oxygen data taken by the Chesapeake Bay Institute during the period 1950–1960 shows that the dissolved-oxygen concentrations decreased more rapidly in the late spring and that low (near zero) concentrations persisted for longer periods in the estuarine Potomac than in the upper Chesapeake Bay. These very low oxygen values are confined to the water layers below the halocline, with near-saturation and supersaturation values occurring at the surface in both the Potomac estuary and the upper Chesapeake Bay.

This paper is Contribution Number 108, Chesapeake Bay Institute and Department of Oceanography, The Johns Hopkins University. This work was supported by contracts with the Bureau of Environmental Hygiene, Maryland State Department of Health; the U.S. Department of Health, Education, and Welfare; and the Federal Water Pollution Control Administration, U.S. Department of the Interior.

IV.

DETECTION

AND MEASUREMENT

OF EUTROPHICATION

FRANK F. HOOPER
The University of Michigan, Ann Arbor

Eutrophication Indices and Their Relation to Other Indices of Ecosystem Change

Speakers preceding me have documented eutrophication in many geographic areas and, in so doing, have pointed out meaningful indices. Those who follow me will discuss indices in dealing with specific events and phenomena that are useful in detecting and measuring eutrophication. Perhaps the most useful things that can be done at this stage are to consider the purposes that indices might serve, outline desirable characteristics of indices, examine some of the indices that have been used or suggested, and consider how these indices measure up to requirements.

One important need for good indices of this process is to provide adequate documentation of past changes. By this I mean data or phenomena that biologists, ecologists, sanitary engineers, and representatives of government can agree upon as being meaningful in assessing the rate of change of our aquatic communities. In part, then, good indices will provide a common language and will assist in securing agreement on the significance of known events. Lack of criteria and indices acceptable to those interested in environmental quality has effects that are illustrated when an enrichment problem arrives at government's doorstep. Engineers discuss the problem in relation to pounds of BOD. Biologists talk about the presence or absence of key species of fish, bottom fauna, or plankton. Ecologists discuss density, diversity, or homeostasis. All would agree on the desirability of having indices that all professional workers accept as important, but I do not think that anyone here is so naive as to believe that we can accomplish this feat. A somewhat more modest objective might be to tell the diverse group of people concerned with environmental quality why various specialists think that the indices and criteria they propose are useful and significant.

A second and compelling reason for searching for proper indices is that they might enable us to scale the rate of change and to anticipate events detrimental to man's interest before they occur. There is not much value in having indicators that tell us "the barn has just burned down." Up to the present, we have not been very successful in predicting the status of our aquatic communities. Also, there is little value in having indices that document the rather obvious and sudden changes resulting from gross insults to our environment, such as the discharge of untreated effluents into small lakes. The uses I propose for eutrophication indices are to characterize the natural process and to assist in measuring instances of relatively slow and creeping changes in the ecosystem arising from relatively mild eutrophication.

Scaling the rate of change might do more than enable us to take preventive and corrective action. It might help us to understand the causes of accelerations and decelerations and their impact on resources. I think it is tremendously important for us to know whether a given level of cultural enrichment is likely to bring about a sudden and dramatic collapse of an existing community or a slow, gradual change extending over several decades.

CRITERIA OF USEFUL INDICES

I have suggested two reasons for providing indices. If we accept them, what are the criteria of a useful index? I would say that there are four.

1. A good index should discriminate between changes associated with nutrient level and those associated with other categories of environmental change. Obviously, most indices will not be conservative in regard to all environmental changes except fluctuations in enrichment level. It will be difficult, if not impossible, to find an index that will discriminate between normal changes, such as base leveling by runoff, and erosion from the slow changes in nutrient supply associated with changing land use. An index should differentiate between changes in enrichment level and seasonal effects as well as changes associated with short-term climatic cycles.

2. A good index should have considerable sensitivity to levels of enrichment, because we want it to help us anticipate changes that are contrary to man's interest. It does not seem likely that the index will have sufficient sensitivity unless, in some way, minor and temporary increases in nutrients are integrated over a long period. The great value of indicator organisms is that the presence or absence of certain species, in effect, carries out such an integration and relieves the biologist from continuous monitoring and surveillance.

3. A good index should have properties similar to those that geologists specify for a good index fossil: properties that are widespread geographically

and short-lived geologically. An index of the eutrophication process, perhaps, should be a property or a biological characteristic widespread among aquatic environments but short-lived and sensitive to changes in enrichment levels, especially levels arising from cultural events.

The requirement that indices be widespread ecologically and geographically is obviously one of convenience. Ubiquitous indices are eagerly sought for setting standards and requirements. It is becoming clear, however, that the search for ubiquity in the setting of water standards is fraught with inconsistency and compromise.

4. A good index should have other practical properties. It should be suitable for long-term surveillance and monitoring by many generations of scientists. Nationwide, or even worldwide, surveillance and monitoring networks may be required to properly assess the impact of man on his environment. Eutrophication indices are certainly among the most important environmental parameters to be considered. Simplicity in technology, interpretation, and collection of data are further practical attributes of a good index.

Few, if any, indices will satisfy these exacting requirements. Nevertheless, it may be helpful to examine, from the standpoint of these criteria, indices that have been used or proposed.

EUTROPHICATION INDICES UTILIZED OR PROPOSED

OXYGEN BUDGETS

The indirect effects of enrichment on dissolved oxygen content and oxygen budgets of the deeper lakes have been discussed at this meeting and in the literature (Edmondson *et al.*, 1956; Sylvester *et al.*, 1956). It seems clear that changes in the shape of the oxygen curve as well as changes in the hypolimnetic oxygen deficit are meaningful indices.

Oxygen budget data and oxygen curves appear to be relatively conservative with respect to changes other than enrichment. The relation between hypolimnetic oxygen deficits and phosphate concentrations, which was pointed out by Edmondson *et al.* (1956), suggests a degree of specificity for this index. Much of the usefulness of hypolimnetic oxygen budgets comes from the integration of periodic changes in nutrient level over an entire season—a desirable time span. Collection and interpretation of data present no serious problems, but a reasonable amount of effort is required to produce data that are sensitive to minor changes. The chief shortcoming of the oxygen budget index appears to be lack of ubiquity. It is not applicable to environments other than the deeper stratified lakes. Other indices must be used for shallower lakes that are approaching senescence.

TRANSPARENCY

Changes in transparency of the water column are indicative of the abundance of plankton organisms; transparency changes have been used to assess the rate of eutrophication. Transparency data are relatively easy to collect, but they may be somewhat difficult to interpret because they do not discriminate between various types of suspended matter. Many uncertainties arise when one attempts to correlate transparency changes with changes in nutrient level (Edmondson, 1956). Transparency measurements are apt to be of relatively low sensitivity unless they are recorded continuously over an extended period. Although of limited usefulness by themselves, they may be of considerable value when taken in conjunction with other data.

MORPHOMETRIC INDICES

As eutrophication proceeds, basins become shallower and shorelines become more regular. Various morphometric indices—depth and shore development for example—can be used to characterize existing conditions and to compare lakes in various stages of aging. Taken together with sedimentation rates, morphometric data can be used with an analysis of past rates of change. Morphometric indices appear to have limited use in the detection and measurement of recent man-induced changes such as those I am considering in this presentation.

NUTRIENTS AND ASSOCIATED IONS

Changes in concentrations of plant nutrients and other ions that are indirectly associated with eutrophication by their appearance in domestic drainage have been used as indices. The advantages of directly measuring critical materials in this process are obvious. However, in few, if any, instances do we have long-term measurements of nitrogen and phosphorus compounds and trace elements that can be compared with such other indices as oxygen budgets, indicator organisms, and transparency changes. In most cases, ions associated with urbanization and domestic use of water, rather than the nutrients themselves, have been used as indicators.

Sulfate, chloride, sodium, potassium, calcium, and total dissolved solids were used by Beeton (1965) as indices of the eutrophication of the St. Lawrence Great Lakes. Most of these ions are conservative water properties and in the larger lakes, certainly, they fluctuate less seasonally than various forms of inorganic nitrogen and phosphorus; thus, they constitute more stable indices. Although there is considerable uncertainty about the role that many of these associated ions play in production of organic matter, changes in concentrations of these ions as well as of total dissolved solids seem to be well correlated with other indices—oxygen levels and changes in indicator organisms, for example (Beeton, 1965).

The sensitivity of direct measurements of nitrogen and phosphorus compounds might be too limited unless measurements are monitored almost continuously. Minor temporal changes in nutrient input will certainly be overlooked. Furthermore, certain basic objections have been raised to orthophosphate measurement as an indication of the availability of nutrients to aquatic plants. A paper by Rigler (1966) indicates discrepancies between concentrations of inorganic phosphorus determined chemically and concentrations determined radiobiologically. An abundance of data indicating rapid exchange of inorganic phosphorus within lake basins (Hutchinson and Bowen, 1950); Hayes *et al.,* 1952) casts doubt on the significance of values of transient or steady-state levels of orthophosphate as commonly measured in standing waters. These matters must be clarified further before orthophosphate can be considered a precise index. Total phosphorus is perhaps more indicative of nutrient level than is orthophosphate. Since much of the phosphorus incorporated into phytoplankton and other suspended debris can be regenerated by bacteria, total phosphorus is a better index for assessing allochthonous phosphorus inputs.

A more desirable index would assess the pool of biologically active nutrients of a given body of water. This pool should include not only the available phosphorus of the water itself but the phosphorus associated with plankton, macrophytes, biological films, encrusting algae, and that available in bottom muds. Turnover time or residence time of phosphorus as determined by Hayes *et al.* (1952) with ^{32}P seems to be a measure of all phosphorus participating in biological processes. These data suggest that bodies of water with long turnover times have large pools of participating phosphorus and that short turnover times are characteristic of oligotrophic environments. Measurements of turnover times are probably conservative regarding short-term fluctuations and thus may be a stable index. However, the subject has not been investigated fully, and further work is needed. The usefulness of this index is limited by the practicalities of treating large bodies of water with ^{32}P. However, results from a number of small-scale, *in situ* experiments in a lake might be added together to yield data characteristic of the lake as a whole.

Nitrogen budgets of freshwater lakes are complicated by nitrogen fixation, denitrification, and unknown exchange rate of nitrates with the surrounding groundwaters. Hence, there have been few attempts to derive indices based on inorganic nitrogen.

BIOLOGICAL INDICATORS

The appearance of new species of organisms and the disappearance of existing species have perhaps been the most useful indices of natural and cultural eutrophication. The various kinds of organisms used as indicators are discussed in detail in the following papers in this section. Previous presentations, as well as many published records, provide an abundance of evidence associating the disappearance, with cultural eutrophication, of planktonic, benthic, and fish species. Sequences of microfossils and inorganic materials within the lake sediments have been used as indicators of the past changes in lakes (Hutchinson and Wollack, 1940; Deevey, 1942).

Because good indicator organisms seem to respond to a summation of relatively minor changes occurring sporadically in nutrient level, they may provide considerable sensitivity to change (Patrick, 1957). Many of the problems associated with the use of indicator species come from lack of specific information about the relation of these organisms to the enrichment process. Much to the dismay of pollution experts who must use indicator organisms as legal evidence, biologists are seldom able to demonstrate that indicators are specific for the enrichment process. Only in rare instances are we able to demonstrate or surmise the basis of sensitivity (Macan, 1962). Judging from the small amount of data on hand, it would seem that sensitivity is based both on competitive interaction between species and on physiological and behavioristic responses. The list of sensitive indicator organisms is small, and ubiquitous organisms, unfortunately, are seldom the sensitive ones. We become aware of indicator species only from rare investigations in which biological data have been gathered concurrently with major shifts in enrichment level. This technique of discovering indicator species is exceedingly inefficient, to say the least. We urgently need quasi-experimental facilities that will enable us to improve our understanding of eutrophication by adding to the list of species sensitive to the enrichment process and by determining the physiological, behavioristic, or ecological basis for sensitivity.

PRODUCTIVITY

A number of aspects of production and elaboration of organic matter in natural waters may be considered as indices of eutrophication. The simplest index, of course, is direct measurement of rate of carbon fixation by phytoplankton, encrusting algae, and aquatic macrophytes. Detection of eutrophication through measurement of primary production is discussed elsewhere in this volume. An index based on changes in rate of primary production might be somewhat ubiquitous and adaptable to many, if not all, aquatic ecosystems; it might have considerable sensitivity if data were obtained over periods long enough to make them independent of seasonal fluctuation. Somewhat better indices might be those incorporating the activities of consumers. Imbalance between production and consumption has been proposed by Odum (1961) as characteristic of a high level of enrichment. Certain organic wastes apparently do not support production of micro-crustacea and other zooplankton consumers; this permits the accumulation of large algal blooms and large standing crops. Odum (1961) suggests that an open ecological niche of some sort exists at high levels of enrichment, permitting accumulation of crop. The example frequently cited, in which consumer organisms were seemingly unable to utilize algae produced by phyotosynthesis, is the one in Moriches Bay, Long Island, reported by Ryther (1954). In this instance, organic nutrients from duck excreta produced algae, which were not efficiently utilized by consumers. Whether blooms persist because of the lack of consumers capable of utilizing these species or because blooms produce toxic materials inhibiting consumers, the result is the same: crops of phytoplankton accumulate and await bacterial breakdown. In this way, grazers are eliminated, although at some stage of decomposition, bacterial and detrital feeders may become abundant (Edmondson, 1957). The net effect, however, is almost certainly a shortening of food chains.

Besides being convenient, indices based on crops of phytoplankton or chlorophyll reflect the activity or inactivity of consumers. Since they indicate the primary production not utilized by consumers, these indices indirectly reflect food-chain efficiency and the "bloom effect" produced at a given enrichment level. To those interested in predicting nuisance blooms and changes in aesthetic qualities, crop measurements may be a more useful index than carbon fixation rates.

The ratio of photosynthesis to respiration (both expressed in grams per day) has been used by Odum (1961) to characterize and classify communities. This ratio indicates the relative dominance of autotrophic and heterotrophic processes. It appears to differentiate between communities receiving allochthonous organic debris as their primary source of energy and those having an excess of photosynthesis over respiration, thus being able to report organic matter (Nelson and Scott, 1962). Although this ratio should change as eutrophication proceeds, it certainly does not differentiate between organics arising from cultural eutrophication and humic materials, which must be considered in certain situations. Thus, this ratio would be interpreted somewhat differently in each situation.

DIVERSITY AND STABILITY

Indicator species have proved useful. But more valuable, and perhaps more reliable, criteria of enrichment are indices that utilize population size (Gaufin and Tarzwell, 1956). By "diversity" an ecologist means a measure of the number of species. In high diversity, many kinds of animals or plants are ordinarily present. In low diversity there is a monotony of a few species. In severe environments such as hot springs (Brues, 1932) or saline lakes (Rawson and Moore, 1944), the intensity of one or more environmental factors tends to limit the number of species and thus to lower diversity.

Yount (1956) and Margalef (1961) have pointed out that increases in nutrient supply may bring about a simultaneous decrease in diversity. Thus, an increase in nutrient level acts in a way similar to increases in the intensity of other factors found in any severe environment. Margalef (1961) suggests that increases in diversity are associated with increases in nutrients because species increase in numbers to take full advantage of the increased supporting capacity of the environment. Since various species differ in their capacity for increase in numbers, some multiply more rapidly than others and manifest dominance, thereby decreasing diversity. Changes in diversity taking place with an increase in nutrients, therefore, may be the result of some form of interspecific competition.

The variety of mathematical expressions relating number of species (S) to number of individuals (N) have been proposed to calculate an index diversity (d). The simplest is the ratio S/N. The slope of a curve relating the number of species to the logarithm of the number of individuals has been proposed as a more suitable index (Fisher $et\ al.$, 1943). The expression $d = (S-1)/ln\ N$ is more commonly used (Margalef, 1958). More recently, Patten (1962) proposed

$$d = \sum_{i=1}^{s} n_i \log_2 \frac{n_i}{N} \text{ and } \overline{d} = \sum_{i=1}^{s} \frac{n_i}{N} \log_2 \frac{n_i}{N},$$

where n_i = the number of individuals per species and $\overline{d}$ is the diversity per individual. J. L. Wilhm (unpublished data)* calculated indices for a number of stations along an Oklahoma stream that was receiving organic wastes. He found that $\overline{d}$ described changes in community composition not reflected by graphs showing the number of individuals or the number of species.

It is beyond the scope of this paper to evaluate the utility of various diversity indices. All may depend on sample size (Hairston, 1959). The use of biomass units, rather than numbers of individuals, may eliminate difficulties with diversity indices arising from differences in the size of various species. It seems also that existing indices will not be useful in situations in which species composition changes but in which the ratio of the number of species to the number of individuals remains constant. Situations of this sort may, however, prove to be rare.

---

*Species Diversity of Benthic Macroinvertebrates in a Stream Receiving Domestic and Oil Refinery Effluents. Ph.D. thesis, Oklahoma State University, Stillwater (1965).

Indices of diversity proposed so far have been based chiefly on studies of natural communities. Further studies are needed to identify more precisely the relation of diversity indices to changes in nutrient level. Identifying the relation of diversity indices to changes in nutrient level can be attacked experimentally in replicated stream and pond environments by observing changes in numbers and kinds of organisms with changes in nutrient level. Experiments being carried out in a replicated series of ponds at Cornell University by Professors Donald Hall and William Cooper promise to provide useful information regarding the response of diversity to nutrient level. In these ponds the diversity of emerging insects was always lowest at the highest nutrient level, and bottom fauna showed a similar decrease in diversity at higher nitrogen and phosphorus levels.

In summary, diversity appears to be a useful and sensitive index because, like indicator species, it tends to integrate changes over a period of time and thereby reduce the sampling effort required to monitor environments. The degree of specificity that various diversity indices have for eutrophication is worth considerably more study. If we restrict the index to certain segments of the ecosystem, we might achieve a high degree of specificity. A proper diversity index should measure relatively subtle changes in the environment, such as those noted by Macan (1962) in Fort Wood Beck. In Macan's investigation, a seemingly minor change in amount of organic enrichment led to a major shift in composition of the community. This change would have gone undetected if there had not been data on the relative abundance of various species. Indicator organisms would have failed in this instance, since the list of species before and after enrichment change was almost the same.

A widely accepted ecological concept is that communities with a large number of species (that is, with high diversity and presumably with a large number of links in the food chain) will have a high stability (MacArthur, 1955). By "stability" an ecologist means resistance to adverse environmental factors—in this instance, high enrichment levels. Presumably, stability arises from homeostatic mechanisms operating within the ecosystem. If we accept this premise, it follows that an index based on ecological diversity may predict change in stability and thus may forecast critical phases of cultural eutrophication; that is, it may forecast major shifts in community structure and perhaps loss of aesthetic features of interest to man.

RELATION TO OTHER INDICES OF ECOSYSTEM CHANGE

All the indices discussed here measure processes occurring at all stages of enrichment; they are not highly specific for either normal or cultural eutrophication. Only when indices measure changes over a considerable period do they gain specificity. Acceleration thus appears as a change in intensity or rate of exisiting processes, and only by detecting accelerations can we distinguish normal and cultural processes. This distinction is a sensitive and practical point to government administrators who are forced to assign responsibility for cultural changes. Computer science and technology and the systems approach may refine the sensitivity and specificity of currently used indices. I have left this problem to my colleague Frederick Smith, who has referred to it in his paper (this volume, page 631).

REFERENCES

Beeton, A. M. 1965. Eutrophication of the St. Lawrence Great Lakes. Limnol. Oceanog. 10:240–254.

Brues, C. T. 1932. Further studies on the fauna of North American hot springs. Proc. Amer. Acad. Arts and Sci. 67:185–303.

Deevey, E. S. 1942. Studies on Connecticut lake sediments. III. The biostratonomy of Linsley Pond. Amer. J. Sci. 237:691–724.

Edmondson, W. T. 1956. The relation of photosynthesis by phytoplankton to light in lakes. Ecology 37:161–174.

Edmondson, W. T., G. C. Anderson, and D. R. Peterson, 1956. Artificial eutrophication 76:225–245.

Edmondson, W. T., G. C. Anderson and D. R. Peterson. 1956. Artificial eutrophication of Lake Washington. Limnol. Oceanog. 1:47–53.

Fisher, R. A., A. S. Corbet, and C. E. Williams. 1943. The relation between the number of individuals and the number of species in a random sample of an animal population. J. Anim. Ecol. 12:42–58.

Gaufin, A. R., and C. M. Tarzwell. 1956. Aquatic macro-invertebrate communities as indicators of organic pollution in Lytle Creek. Sewage Ind. Wastes 28:906–923.

Hairston, N. G. 1959. Species abundance and community organization. Ecology 40:404–416.

Hayes, F. R., J. A. McCarter, M. L. Cameron, and D. A. Livingston. 1952. On the kinetics of phosphorus exchange in lakes. J. Ecology 40:202–216.

Hutchinson, G. E., and V. T. Bowen. 1950. Limnological studies in Connecticut. IX. A quantitative radiochemical study of the phosphorus cycle in Linsley Pond. Ecology 31:194–203.

Hutchinson, G. E., and A. Wollack. 1940. Studies on Connecticut lake sediments. II. Chemical analysis of a core from Linsley Pond, North Branford. Amer. J. Sci. 238:493–517.

Macan, T. T. 1962. Biotic factors in running water. Schweizerische Zeit. für Hydrol. 24:386–407.

Margalef, R. 1957. La teoria de la informacion en ecologia. Mem. Real Acad. Ciencias y Artes de Barcelona 32:373–449.

Margalef, R. 1958. Temporal succession and spatial heterogeneity in phytoplankton, p. 323–349 In A. A. Buzzati-Traverso [ed] Perspectives in marine biology. Univ. Calif. Press, Berkeley.

Margalef, R. 1961. Communication of structure in planktonic populations. Limnol. Oceanog. 6:124–128.

MacArthur, R. 1955. Fluctuations of animal populations, and a measure of community stability. Ecology 36:533–536.

Nelson, D. J., and D. C. Scott. 1962. Role of detritus in the productivity of a rock-outcrop community in a piedmont stream. Limnol. Oceanog. 7:396–413.

Odum, E. P. 1961. Factors which regulate primary productivity and heterotrophic utilization in the ecosystem, p. 65–71 *In* Algae and metropolitan wastes (transactions of the 1960 seminar). U.S. Dept. Health, Education, and Welfare. Robert A. Taft Sanitary Engineering Center, Cincinnati, Ohio.

Odum, H. T. 1956. Primary production in flowing waters. Limnol. Oceanog. 1:102–117.

Patten, B. C. 1962. Species diversity in net phytoplankton of Raritan Bay. J. Mar. Res. 20:57–75.

Patrick, R. 1957. Diatoms as indicators of changes in environmental conditions, p. 71–80 *In* Biological problems in water pollution (transactions of seminar on biological problems in water pollution). U.S. Dept. Health, Education, and Welfare. Robert A. Taft Sanitary Engineering Center, Cincinnati, Ohio.

Rawson, D. S., and J. E. Moore. 1944. The saline lakes of Saskatchewan. Canad. J. Res. 22:141–201.

Rigler, F. H. 1966. Radiobiological analysis of inorganic phosphorus in lakewater. Verh. Int. Verein. Limnol. 16:465–470.

Ryther, J. H. 1954. The ecology of phytoplankton blooms in Mariches Bay and Great South Bay, Long Island, New York. Biol. Bull. 106:190–209.

Sylvester, R. O., W. T. Edmondson, and R. H. Bogan. 1956. A new critical phase of the Lake Washington pollution problem. The Trend in Engineering 8:8–14.

Yount, J. L. 1956. Factors that control species number in Silver Springs, Florida. Limnol. Oceanog. 1:286–295.

JOHN LANGDON BROOKS
Yale University, New Haven, Connecticut

Eutrophication

and Changes in the Composition

of the Zooplankton

If the adjective "oligotrophic" means having scant nutrition, and "eutrophic," having good nutrition, then "eutrophication" of a lake can mean only a bettering of the nutritional state of a lacustrine ecosystem. The inadvertent addition to lakes of soluble compounds of nitrogen and phosphorus by human activities has been called cultural, or artificial, eutrophication. This is opposed to natural eutrophication, the enrichment that is assumed to be the fate of lacustrine ecosystems.

An increase in the dissolved nutrients, however, is only one of a series of changes that a lake may undergo during its existence. Morphometric changes resulting from deposition of both autogenic and allogenic material, introduction of new biotic elements, and alterations in the climate of the region can all produce changes in the biotic composition of the ecosystem in addition to, or independently of, chemical enrichment. It seems preferable, therefore, to refer to the changes in a lake during its existence by a general, noncommittal term—"maturation," for example—instead of "eutrophication," a term that has sometimes been applied to these non-nutritional changes.

Obviously, then, we must look to ecosystems undergoing cultural eutrophication when seeking to answer the primary question that this paper asks: What effect does adding chemical nutrients to a lake have on the species composition of the zooplanktonic herbivores? Answers derived from these ecosystems, in which chemical enrichment is the change of overriding significance, might then be useful in our interpretation of data relating to natural maturation, where enrichment is usually only one of a complex of simultaneous, overlapping, or alternating changes.

TROPHIC DYNAMICS INVOLVING PLANKTONIC HERBIVORES

Before examining ecosystems to assess the effects of enrichment on zooplankton, we must review briefly the trophic dynamics involving animal plankton. To simplify this review, we will consider only those zooplankters that, by one semiautomatic mechanism or another, collect the fine particles—algae, bacteria, detritus—suspended in the surrounding medium. These particle-collectors will be referred to as "herbivores." Thus, we are eliminating the following from immediate consideration:

1. Various large, but relatively scarce, predaceous Cladocera—*Leptodora kindtii, Polyphemus pediculus,* and the European *Bythrotrephes longimanus.*
2. Cyclopoid copepods. The food habits of the numerous species are imprecisely known, but many seem more given to seizing small animals than to collecting fine particles.
3. Planktonic rotifers. Although most qualify as herbivores, they are so small that they do not play a prominent trophic role in most of the ecosystems that we will examine.

These disqualifications leave us the crustacean herbivores: Cladocera and calanoid Copepoda. Although some (probably most) large calanoids can seize small zooplankters, the collection of fine particles seems a prevalent mode of nutrition (Fryer, 1957; Mullin, 1966).

These crustacean herbivores, the rotifers, and many ciliated protozoa utilize—and therefore compete for—fine particles that are more than 1μ and less than about 15μ in long dimension. The larger Cladocera filter much more effectively than do their smaller relatives (Brooks and Dodson, 1965; Burns and Rigler, 1967), and they also take those larger algae that the smaller zooplankters cannot manage. This greater feeding effectiveness probably explains the oft-observed competitive suppression of populations of smaller zooplankters by populations of larger species (Brooks and Dodson, 1965).

The crustacean herbivores constitute one of the chief trophic links between the algae (and bacteria) and the fish that dominate the higher trophic levels of lacustrine ecosystems. Nearly all fish subsist on zooplankton shortly after they hatch and commence feeding. Some continue to find much food in the open-water zooplankton. There are fewer obligate planktivores in freshwaters than there are in the sea, undoubtedly because only scattered lakes can continuously provide an adequate supply of plankton. Most freshwater fish (excluding freshwater populations of anadromous marine species) are facultative planktivores; they feed on zooplankton when large forms (especially *Daphnia*) are plentiful, but switch to some other food source (small fish or benthic invertebrates) when the supply of large zooplankters fails. An important characteristic of all planktivorous fish is the high degree to which they are selective in their choice of food items from the variety of sizes and shapes presented by lacustrine zooplankters. In general, large zooplankters are preferred to small ones, and Cladocera are preferred to calanoid copepods of the same visual size. Cyclopoids are an intermediate choice (Berg and Grimaldi, 1965, 1966a; Brooks, 1968; Ivlev, 1961).

More will be said of this selectivity in the next section. It is sufficient now to emphasize that this trophic system of the open waters—a system in which a large number of closely competitive planktonic herbivores are subject to highly selective predation—has no obvious counterpart in substrate-dominated ecosystems.

In such an open-water system, where predation by the planktivorous fish is principally size-dependent, the following relations prevail (Brooks and Dodson, 1965):

1. Planktonic herbivores all compete for the fine particulate matter (1μ to 15μ in length) of the open waters.

2. Larger zooplankters compete more efficiently and can also take larger particles.

3. When predation is of low intensity, the small planktonic herbivores will be competitively eliminated by large forms (dominance of large Cladocera and calanoid copepods).

4. But when predation is intense, size-dependent predation will eliminate the large forms, allowing the small zooplankters (rotifers, small Cladocera) that escape predation to become the dominants.

5. When predation is of moderate intensity, it will, by falling more heavily upon the larger species, keep the populations of these more effective herbivores sufficiently low to prevent elimination of slightly smaller competitors.

One other aspect of this system is of potential relevance to the subject of this paper. Because of the differential ability of small zooplankters (*Bosmina longirostris*, rotifers), on the one hand, and large *Daphnia*, on the other, to utilize the size-spectrum of available particles, one would expect quantitative differences in the standing crop of particles at both ends of the usable spectrum in the presence of a population of small or large zooplankters. Let us decide to call the small particles nannoseston and the large particles net phytoplankton. Large *Daphnia*, then, should be able to collect some of the net phytoplankton so that the standing crop of net phytoplankton in the presence of an equilibrium population of *Daphnia* should be smaller than that in the presence of an equilibrium population of zooplankters that are too small to ingest these large algae. Furthermore, because the filtering rates of Cladocera (at low concentrations of algae) increase with at least the square, and sometimes the cube, of the body length (Burns and Rigler, 1967), the standing crop of nannoseston should be much scantier in the presence of a population of large *Daphnia* than in the presence of a population of small *Bosmina*. Hrbáček *et al.* (1961) demonstrated precisely this relation between standing crop of algae and size of dominant zooplankters. When a pond (Poltruba Pond, Czechoslovakia, 0.18 ha; mean depth, 2.77 m) contained a large fish population, *Bosmina longirostris* and rotifers were the dominant zooplankters (see 1955 in Table 1). After the fish had been poisoned, a large species of *Daphnia* became the dominant zooplankter (1957). Hrbáček separated the algae with a 50µ mesh, so that his "nannoseston" contained forms several times larger than µ-alga as defined by Lund (1961). Net phytoplankton here included any alga larger than 50µ. Table 1 gives approximate mean values for the various parameters that were measured by Hrbáček and co-workers from May to October. During 1957, the fish stock in Poltruba Pond was very low, and large *Daphnia* dominated the zooplankton. The standing crop of nannoseston and net phytoplankton during this period can be compared both with the crop in Poltruba Pond during the same months 2 years previously and with the crop in Procházka Pond (0.15 ha; mean depth, 1.95 m), a similar pond several hundred yards away from Poltruba Pond into which the water from Poltruba Pond drains. In both of these control situations, the heavy fish stock kept the size of the zooplankters small, with *Bosmina longirostris,* rotifers, and ciliates predominating. A dwarf form of *Daphnia cucullata,* about the length of the *Ceriodaphnia* sp. depicted in Figure 1, also occurred when the fish stock was heavy, but only then.

It is apparent from Table 1 that the standing crop of net phytoplankton is very low in the presence of *Daphnia* (Poltruba Pond, 1957) but abundant when *Bosmina longirostris* predominates (Poltruba, 1955; Procházka, 1957). Similarly, the organic nitrogen and chlorophyll content of the standing crop of nannoseston in the presence of *Daphnia* is about one third that in the presence of equilibrium populations of the small zooplankters. Comparison of rates of photosynthesis and respiration in Poltruba and Procházka Ponds in 1957 made it clear that a large part of what Hrbáček chose to call nannoseston is truly nannoplankton algae. As would be expected from these differences in population density of algae of all sizes, transparency of the water when planktonic herbivores are small differs strikingly from the transparency when the herbivores are large. When *Daphnia* predominated, the water was much clearer; the mean depth at which the Secchi disc disappeared was 3.5 m. In the presence of *Bosmina* and rotifers, it disappeared from view in about a meter. Thus, as Hrbáček (1962a) emphasized, when a large species of *Daphnia* is the dominant planktonic herbivore (fish stock low), a pond has the classic characteristics of "oligotrophy." But when small-bodied plank-

TABLE 1 Fish Predation and Composition of Plankton in Bohemian Ponds[a]

Pond	Total N (mg/l)	Total P (mg/l)	Fish Stock	Zooplankton				
				Organic N (mg/l)	Daphnia	Bosmina l.	Rotifers	Ciliates
Poltruba								
1955	1.5	—	High[c]	0.25	—	Abundant	Few	Common
1957	1.1	0.14	Low	0.13	Abundant (80%)	Rare	Few	Few
Procházka								
1957	1.2	0.11	High	0.1	Rare	Abundant	Abundant	Few

Pond	Net Phytoplankton		Nannoseston[d]				
	Diatoms	Dinobryon	Organic N[b]	Chlorophyll	Photosynthesis[e]	Respiration[e]	Transparency[f]
Poltruba							
1955	Abundant	—	0.35	0.13	—	—	1.0
1957	—	—	0.14	0.05	100	100	3.5
Procházka							
1957	Some	Abundant	0.47	0.22	870	275	1.0

[a] Data calculated from Hrbáček et al. (1961).
[b] Milligrams per liter.
[c] 907 kg, forage fish; 18.6 kg, pike.
[d] Smaller than 50 μ.
[e] Percent.
[f] Meters.

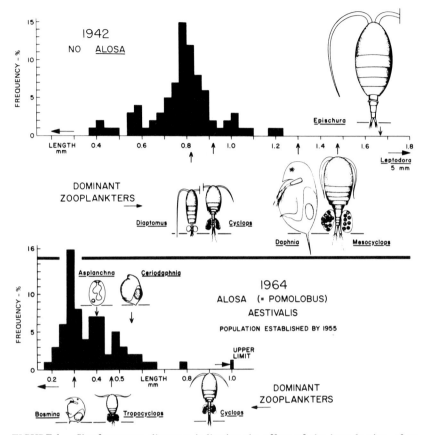

FIGURE 1   Size-frequency diagrams indicating the effect of the introduction of an obligate planktivore on the size—hence, species composition—of the crustacean zooplankton of a small lake. Composition of crustacean zooplankton of Crystal Lake (Stafford Springs, Connecticut) before (1942) and after (1964) a population of *Alosa aestivalis* had become well established. Each square of the histogram indicates that 1 percent of the total sample counted was within that size range. The larger zooplankters are not represented in the histograms because of the relative scarcity of mature specimens. The specimens depicted represent the mean size (length from posterior base lines to the anterior end) of the smallest mature instar. The arrows indicate the position of the smallest mature instar of each dominant species in relation to the histograms. The predaceous rotifer *Asplanchna priodonta* is the only noncrustacean species included; other rotifers were present but were not included in this study. (From Brooks and Dodson, 1965.)

tonic herbivores predominate (fish stock heavy), the same pond has the classic characteristics of "eutrophy." By eliminating fish stock in four other ponds, which were up to 3.8 ha, Hrbáček and Novatná-Dvořáková (1965) confirmed these shifts in both body size of zooplankton and standing crop of algae.

Although the control of the standing crop of algae by planktonic herbivores of different body size has not yet been experimentally demonstrated in a large lake, there is every reason to expect that the sustained dominance of *Bosmina longirostris* in a large lake would have an effect similar to that in a small pond. A shift in dominance from large-bodied to small-bodied planktonic herbivores (as a result of an increase in size-dependent predation) should cause a lake to become more "eutrophic."

SELECTIVE PREDATION ON THE ZOOPLANKTON

Any relationship between predation on the zooplankton and the density of the standing crop of algae is germane to a discussion of eutrophication. We must, therefore, briefly examine the nature of prey selection by various freshwater planktivores.

As pointed out earlier, few primary freshwater fish species feed entirely on zooplankton throughout their lives. The most complete dependence on zooplankton shown by a North American planktivore is, perhaps, that of *Alosa pseudoharengus,* the alewife, and its close relative *A. aestivalis,* both of which are recent invaders from the sea. Freshwater populations have become established in lakes and ponds on streams along the Atlantic Coast that marine *Alosa* ascend each spring to spawn. In recent years, however, the alewife has become conspicuous in numerous lakes, including the Laurentian Great Lakes, because it has been extensively introduced both accidentally and purposefully (as forage for game fish). The alewife is pelagic in habit, shunning the littoral zone and the bottom waters and finding its food only among offshore or limnetic zooplankton. Experiments have indicated that when offered typical limnetic zooplankton, *Alosa* first select *Daphnia,* then large copepods. As the supply of these becomes exhausted, the fish take progressively smaller zooplankton, finally eating copepod nauplii, even smaller (200μ) than the *Bosmina longirostris,* on which natural populations are known to subsist (Brooks, 1968). *Alosa* can be designated an obligate planktivore.

Planktivorous species that have evolved from ancient freshwater stocks are seldom as completely dependent upon zooplankton as *Alosa* is. Species from such divergent stocks as the yellow perch (*Perca flavescens*), various trout (species of *Salmo* and *Salvelinus*), and the characin *Alestes baremose* are similar in the restriction of their planktivory to large *Daphnia* over 1.2 mm long (O. M. Brynildson,* unpublished data; Galbraith, 1966; Green, 1967).

Lime Treatment of Brown Stained Lakes and their Adaptability for Trout and Largemouth Bass. Ph.D. dissertation, Univ. of Wisconsin (1958).

If *Daphnia* of this size are not readily available, these fish, as analyses of stomach contents indicate, switch to a nonplanktonic food source, for example, small fish or benthic invertebrates. They do so despite the fact that equally large copepods are available and despite the fact that the spacing between their gill rakers is fine enough to retain zooplankters smaller than those they take (Galbraith, 1966).

Populations of certain planktivores, especially coregonids, are intermediate between these two conditions in degree of trophic dependence on the zooplankton. The steps from the extreme facultative state involve, generally speaking, inclusion of (1) such large littoral plankters as *Sida crystallina* and cyclopoid copepods and (2) limnetic calanoid copepods.

Obviously, differential predator selection of zooplankters for species and size will substantially affect the relative abundance of the various zooplankters. For example, the standing crop of *Daphnia* in a lake dominated by a facultative planktivore that selects only large *Daphnia* will be smaller, relative to the abundance of its competitors, the unselected calanoid copepods, than if the planktivore were absent.

The extent to which planktivores influence the composition of zooplankton can best be appreciated by observing the effects of introducing a new planktivorous species into a lake. Galbraith (1966) observed the changes in the zooplankton in a small Michigan lake after populations of three planktivorous species were established. Rainbow trout (*Salmo gairdneri*) were the most numerous; the other species were smelt (*Osmerus mordax*) and fathead minnows (*Pimephales promelas*). Because of poisoning, the lake had been without fish for the preceding 4 years. During these years, a population of large *Daphnia pulex,* maturing when about 1.8 mm long and commonly reaching 3 mm, had, predictably, established itself as the dominant planktonic herbivore. In the summer after the rainbow trout were introduced, *Daphnia* longer than 2 mm were rare; and in the summer 4 years later, none were seen with head and carapace longer than 1.5 mm. The mean length of *Daphnia* in the plankton decreased from 1.4 mm before rainbow trout were introduced to 0.8 mm 4 years later. In the face of this severe predation on large specimens, too few *Daphnia pulex* reached maturity (at least 1.2 mm) to maintain the species. *Daphnia pulex* was replaced by *Daphnia galeata* and *Daphnia retrocurva.* (The first species is about the same size at maturity as *Daphnia catawba* in Figure 1.) *Daphnia retrocurva* is slightly smaller at maturity than is *D. galeata. Bosmina* (sp.?) also became more numerous, as did *Chydorus* (probably *C. sphaericus*?). Copepods, not being preyed upon by the rainbow trout, became somewhat more numerous.

The introduction of an obligate planktivore, *Alosa aestivalis,* had an even more profound effect on the composition of the zooplankton of a small Connecticut lake (Brooks and Dodson, 1965). Re-examination of the crustacean plankton of Crystal Lake about 10 years after the planktivore population was known to have become well established revealed that all the characteristic large limnetic zooplankton—*Leptodora kindtii, Epischura nordenskioldi, Mesocyclop edax, Daphnia catawba,* and *Diaptomus minutus*—had been eliminated from dominance in the plankton. Their dominance had been usurped by *Bosmina longirostris* and two small cyclopoids. (*Cyclops bicuspidatus thomasi* probably persisted because its adults often concentrate near the shore and bottom, places where *Alosa* does not feed.) The zooplankton association thus resulting from the establishment of this obligate planktivore is precisely that association characteristic of coastal lakes in which populations of *Alosa pseudoharengus* had become naturally established. One of these coastal lakes, Linsley Pond, appears to have long been a *"Bosmina"* Lake, to judge from the microfossils in its sediments (Deevey, 1942). Evidence for this early establishment of *Alosa* is considered in the final section of this paper. One might also recall that the plankton associations in these *Alosa* lakes is remarkably like that of the Bohemian ponds and backwaters that are very heavily stocked with fish of a variety of species, as Hrbáček and his co-workers have demonstrated.

The several planktivores of Lago Maggiore in northern Italy can provide us with some insight into the feeding habits of long-established planktivores that are about as complete in their trophic dependence on zooplankton as freshwater fish ever are. The indigenous freshwater populations of the Mediterranean shad, *Alosa fictus,* and a species of the cyprinid planktivores, *Alburnus,* have been supplemented by the introduction of two distinct populations of *Coregonus* (Berg and Grimaldi, 1966b). The two coregonids have similar food habits. The smaller of the two was introduced several decades ago, long after the other had become established. This smaller "bondella," sustaining less fishing pressure than its larger congener and enjoying greater reproductive success, is slowly becoming the more abundant *Coregonus.* The food habits of bondella in relation to the zooplankton deserve close scrutiny.

The zooplankton of Lago Maggiore, a large, deep, elongate lake, which becomes isothermal during the 3 winter months at about 6°C, is dominated by the calanoid copepod *Diaptomus vulgaris.* Tonolli (1962), in his careful and complete study of the zooplankton, determined the mean annual abundance of the various limnetic zooplankters and demonstrated that adult *Diaptomus* make up about 45 percent of the zooplankton. If the numerous copepodites and nauplii are included, it is quite clear that far more than half of all zooplankters encountered were life stages of this calanoid copepod. The next most abundant species was *Daphnia hyalina,* which made up 35 percent of the zooplankton when all life stages were included.

These two dominants are remarkably different in the seasonal distribution of their abundance. Whereas *Diaptomus,* although exhibiting fluctuations, is represented throughout the year, *Daphnia* is markedly seasonal in its occurrence. It is virtually absent from the plankton during the winter months (December through April). Shortly thereafter it becomes common, with an annual peak of abundance in June–July, followed by a sharp decline (due to excessive predation) to a low level in middle and late summer. Numbers then rise in the fall months. *Cyclops strenuus* (somewhat smaller than the *Mesocyclops* of Figure 1) resembles *Diaptomus* in maintaining a fluctuating presence the year round, contributing about 10 percent of the annual mean. These findings by Tonolli (1962) provide the background for the analyses of stomach content performed by Berg and Grimaldi (1965, 1966 a and b).

The percentage composition of the organisms found in the stomachs of bondella from April to December 1963 is presented in Figure 2. It is

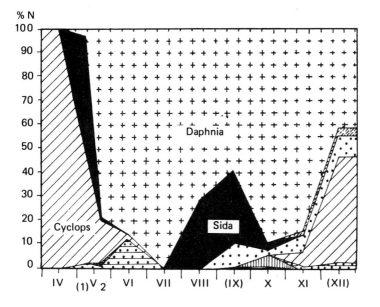

FIGURE 2   Seasonal variation in the relative abundance of the major zooplanktonic elements in the diet of the planktivore *Coregonus* ("bondella") in Lago Maggiore, Italy. The relative numbers of each planktonic species found in the fish stomachs from April (IV) to December (XII) are given on the vertical axis. Note that in April the stomachs contained only cyclopoid copepods (*Cyclops*), while in the middle of July, all contained almost nothing except *Daphnia*. (See text for further explanation.) *Daphnia = Daphnia hyalina, Sida = Sida crystallina, Cyclops* = various species of cyclopoid copepods, mostly *Cyclops strenuus.* Calanoid copepods, almost entirely *Diaptomus (Eudiaptomus) vulgaris,* are represented by the crosshatched area with large dots. The large predatory cladoceran *Bythrothrephes longimanus* is represented by the dotted (stippled) area (August to December). (After Berg and Grimaldi, 1965, 1966a.)

immediately apparent that *Daphnia* is the principal food item during the period of its occurrence in the lake. Between mid-May and mid-July, the period of maximum abundance of *Daphnia, Coregonus* makes its best growth, increasing its body weight by 2 percent each day. In late summer, despite the great decline in the abundance of *Daphnia* in the lake, it is still selected, making up over 50 percent of the items eaten. The other major component during this period is the large cladoceran *Sida crystallina*. Since Tonolli never found more than an occasional individual of *Sida* in the midlake plankton, it is apparent that *Coregonus* had moved inshore to eat the *Sida,* which are usually most abundant just off weed beds. These cladocerans (larger than the *Daphnia* of Figure 1) often hang suspended from the littoral vegetation, especially from *Potamogeton*. During the late fall and winter months, when all Cladocera become scarce, the *Coregonus* eats *Cyclops*. Berg and Grimaldi (1966b) present data from winter feeding in other years, indicating that the reliance upon *Cyclops* evident in April (Figure 2) is characteristic of the preceding 3 months as well. It is particularly noteworthy that even during the winter months, when *Diaptomus* (horizontal cross-hatching with dots in Figure 2) is the most abundant zooplankter, it is not eaten. June is the only month when *Diaptomus* contributes as many as 10 percent of the items in the stomach of this *Coregonus*. Its obvious preference for large cladocerans is further evidenced in the noticeable number of large (several millimeters) predatory polyphemid, *Bythrotrephes longimanus* (the dotted area of Figure 2), even though the latter was extremely rare in the plankton in 1963.

Although the shad, *Alosa fictus,* of Lago Maggiore depends more on copepods when *Daphnia* are of low abundance (and not on *Sida*), it is *Cyclops* that is eaten in quantity, not *Diaptomus*. This extremely low electivity for *Diaptomus* by both genera of planktivores, compared with that for *Daphnia,* undoubtedly has a profound influence on the relative proportions of the two dominant zooplankters. It seems reasonable to conclude, therefore, that relative dominance among species of zooplankters can be, and frequently is, largely determined by the food preferences of the planktivorous fish.

GENERAL EFFECTS OF ENRICHMENT ON ZOOPLANKTON

A general, and not unexpected, consequence of the enrichment of lake ecosystems has been the increase in the standing crop of the phytoplankton and (although to a lesser degree) of the planktonic herbivores. Possibly the recent increase in the standing crop of both plant and animal plankton of Lake Erie is one of the most dramatic demonstrations of this effect (Beeton, 1965; Bradshaw, 1964; Chandler, 1940; Davis, 1964). Hrbáček (1962b) considered the direct effects of sewage pollution on planktonic Crustacea. He concluded that if the deleterious effects of toxic substances such as heavy metals and insecticides were excluded, the primary effects of pollution on zooplankton would be mediated through its effect on abundance of the nannoplanktonic algae that are the principal food of zooplankton. Laboratory studies of amenable planktonic herbivores have demonstrated that the size of equilibrium populations in systems without predators is determined by the amount of particulate food available (Slobodkin, 1954). Predation on the zooplankton in the natural ecosystems undergoing enrichment may account for the fact that the zooplankton does not show an increase in standing crop proportionate to the apparent increase in its food supply.

As we have noted previously, however, the phytoplankton includes a complex array of organisms that differ in their utility as food for planktonic herbivores. For example, increases in the nannoplanktonic algae would benefit all planktonic herbivores; increases in the abundance of larger diatoms, on the other hand, would directly benefit only those herbivores large enough to collect and ingest them. Among the array of phytoplanktonic organisms, the blue-green algae have made the most noticeable response to enrichment. Minder (1938) first called attention to the increase of *Oscillatoria rubescens* in the Zürichsee. Since then, cultural eutrophication has commonly been observed to enhance especially the growth of this species and others. Apart from the possible chemical unacceptability of cells of certain blue-greens to herbivores (for this controversial subject, see Hrbáček, 1964, and Monakov and Sorokin, 1961), the large colonies cannot be utilized by true planktonic herbivores. These large clumps, however, produce food and substrate for *Chydorus sphaericus*, a widely dispersed bottom-dweller. This species, otherwise absent from the open waters, appears in often-enormous numbers during a bloom of blue-green algae. Its appearance provides one of the most characteristic changes in plankton samplings of strongly enriched waters. Strictly speaking, however, *Chydorus sphaericus* is not a true member of the plankton.

Extensive utilization of nannoplanktonic algae by planktonic herbivores may even encourage the development of these blue-greens, which they do little to crop. They could have this effect by so depressing the abundance of nannoplanktonic algae that these small forms remove nutrients less rapidly from the medium, offering the blue-greens more unrestricted use of any nutrients they require. Hrbáček (1964) has examined the associations of four species of various sizes that occur in association with blooms of *Aphanizomenon flos-aquae* and *Microcystis aeruginosa* in the ponds and backwaters of the Elbe in Czechoslovakia. He states that certain species of *Daphnia* tend to occur more frequently in association with blue-green blooms than do others. Hrbáček suggested that the different frequencies of association may be due to differing abilities of *Daphnia* species to graze down nannoplanktonic algae that compete with the blue-greens for nutrients. In view of the greatly increased filtering rates of large species, compared with those of related but smaller species, it is of considerable interest to note that the larger species of *Daphnia* are more frequently associated with blooms. This association makes Hrbáček's interpretation consistent with what is known of comparative filtering abilities (Brooks and Dodson, 1965). These *Daphnia* populations would, however, also graze down the populations of the bacteria that Thomas (this volume, page 29) has shown to produce substances acting as growth stimulants for the blue-greens.

The greatly increased production of both plant and animal plankton in the epilimnia of stratified lakes that are undergoing enrichment often results in severe depletion of hypolimnetic oxygen in late summer. This oxygen depletion can have a devastating effect on the salmonids and coregonids common in the well-oxygenated hypolimnia of deep north temperate lakes (Beeton, 1965; Hasler, 1947; Minder, 1938). While the decimation or elimination of these important facultative planktivores will cause a decrease in predation pressure on the zooplankton, this decrease might well be transitory. Various species of warmth-tolerant planktivores can quickly utilize this food source once monopolized by the coregonids and salmonids (Beeton, 1965; Hasler, 1947; Verduin, 1964). However, the fact that different planktivores have different selective preferences may cause some shifts in relative abundance of the zooplankters.

CHANGES IN LAKES ERIE AND MICHIGAN

It is manifestly impossible to analyze changes in the zooplankton of all the lakes that have been reported to have undergone eutrophication, chiefly because the necessary quantitative data are lacking. A general impression, from examination of available data on various lakes that have been enriched, is that the proportions of the various zooplanktonic species have been altered somewhat but that the composition of the species has not changed drastically. The oft-noted disappearance of *Bosmina coregoni* in the course of cultural eutrophication and the appearance of *Chydorus sphaericus* give an impression of instability that, in my opinion, is misleading. The qualitative changes in zooplanktonic composition of large lakes undergoing eutrophication (Lake Washington, for example) do not seem greater than the changes noted in the zooplankton of Lago Maggiore over a comparable period (Baldi, 1951; D'Ancona, 1955; Tonolli, 1962), although the changes seem to be in opposite directions.

Our attention will be restricted to two of the largest lakes that have undergone cultural enrichment: Lake Michigan and Lake Erie. These lakes differ in many ways (Chandler, 1964). In addition, the populations of planktivores of each have undergone different kinds of alterations.

The phytoplankton of Erie, the shallower and more southern of the two lakes, has increased greatly in abundance over the last few decades. Erie now appears thoroughly "eutrophic" (Beeton, 1965; Davis, 1964). Neither nutrient content nor phytoplankton of Lake Michigan has increased at Lake Erie's rate during the same period, and the lake can still be considered "oligotrophic" (Beeton, 1965; Damann, 1945, 1960). A quantitative increase in the standing crop of zooplankton in Lake Erie can be demonstrated (Bradshaw, 1964), but comparable quantitative data are lacking for Lake Michigan. However, probably valid shifts in the relative dominance of some species of limnetic Cladocera can be derived from the literature; these are indicated in Table 2. These shifts are discussed below.

In both lakes, the planktivore populations have undergone remarkable changes in composition during this period. Lake Erie once supported a sizable population of coregonids (Beeton, 1965). Their decrease apparently can be attributed to decreased ability to reproduce in a deteriorating environment and to great fishing mortality. These planktivores have been largely replaced by such warmth-tolerant planktivores as yellow perch and smelt. Similarly, the populations of some of the large predatory fish, such as the walleye (*Stizostedon vitreum*), have dropped greatly; thus, there are fewer restraints on the development of large populations of facultative planktivores. From our knowledge of the preference of facultative planktivores for large *Daphnia,* we might expect that the larger forms would be selectively removed and that this removal would make possible an increase in their slightly smaller competitors. Table 2 indicates that this may have occurred. The large *D.*

TABLE 2   Shifts in Relative Abundance of Cladocera in Lake Erie and Lake Michigan

Cladoceran	Size (mm)	Erie		Michigan	
		1938–1939[a]	1948–1949[b]	1927[c]	1954[d]
Daphnia pulex	1.7	+	−	−	−
galeata	1.1	+++	++	+	+++
retrocurva	0.9	++	+++	++	+++
Bosmina coregoni	0.6	−	−	+++	−
Diaphanosoma sp.	1.0	+	+++	+	+

[a] Data from Chandler (1940).
[b] Data from Bradshaw (1964).
[c] Data from Eddy (1927).
[d] Data from Wells (1960).

pulex is no longer found in Lake Erie, and the medium-sized *D. galeata* now appears to be exceeded in abundance by the small *D. retrocurva. Diaphanosoma,* which several studies, including those of Berg and Grimaldi (1965 and 1966a), have shown to be ignored by many planktivores, has increased in abundance. As we have noted, calanoids are relatively immune from predation by facultative planktivores. This relative immunity may have been a factor in both the rapid increase in the population of *Diaptomus siciloides* in Lake Erie and the growth of sizable populations of five other species of *Diaptomus* already present (Davis, 1962). But it should be noted that females of *D. siciloides* mature at about 1.0 mm, the size of *D. minutus,* the smallest of the previously existing species.

In Lake Michigan (but not in Lake Erie), the several species of once-numerous coregonids have suffered greatly from the predation (parasitism?) of the sea lamprey, which became established in 1940 (Smith, 1964). The salmonid and coregonid species that are large as adults have almost been eliminated. Only the smallest coregonid—the slow-growing *Coregonus hoyi,* which is too small to interest the lamprey—was numerous when the zooplankton study reported in Table 2 was made (1954). Since then, alewife and smelt have become abundant, and further changes can be expected in the zooplankton. Be that as it may, we probably can conclude that there has been an over-all drop in predation on zooplankton from 1927 (Eddy, 1927) to 1954 (Wells, 1960). Furthermore, Wells and Beeton (1963) have studied the stomach contents of *Coregonus hoyi,* which constituted about 90 percent of the coregonids caught in Lake Michigan at that time. Although the adult fish mostly fed on large benthic (and deep pelagic) *Pontoporeia affinis* and *Mysis relicta,* the small and intermediate-sized fish fed chiefly on zooplankton. In their midwater collections, the fish had been feeding about equally on cladocerans and copepods. Of the Cladocera in the stomachs, *D. galeata* was by far the commonest, although *D. retrocurva* and *Bosmina longirostris* had also been eaten. Of the copepods, *Cyclops bicuspidatus* was most commonly eaten in summer, and the larger species of *Diaptomus* were eaten in the fall. Table 2 reports Wells' finding that in the lake during 1954, *D. galeata* and *D. retrocurva* were roughly equal in abundance. The larger was obviously being selected preferentially by the one remaining species of *Coregonus.* Yet, in spite of this, *Daphnia galeata* has increased in abundance, whereas the once-abundant *Bosmina coregoni* has disappeared from this still "oligotrophic" lake. This shift toward a dominance of larger competitors is to be expected, if the predation pressure on the zooplankton has indeed decreased (Brooks and Dodson, 1965). In this situation, the elimination of *Bosmina coregoni* could be an example of competitive exclusion, having nothing to do with alterations consequent upon enrichment.

As Wells (1960) discovered, *Diaptomus oregonensis* increased from rarity to dominance in Lake Michigan between 1927 and 1954, roughly the same period in which *D. siciloides* rose to dominance in Lake Erie. *D. oregonensis* is, however, a larger species, the females maturing at 1.25 mm. It is not only larger than *D. siciloides* but is one of the largest of the species of *Diaptomus* found in the Great Lakes. The relation between the body sizes of these emerging dominants in the two lakes is consistent with our conclusions about the trends of predation pressure in the two lakes.

This shift in Lake Michigan toward zooplankton that are larger and more effective in collecting particles may have been a factor in the preservation of the oligotrophic condition of that lake, despite some chemical enrichment. The downward shift in body size among planktonic herbivores in Erie implies a decreased ability to cope with the rising biomass of phytoplankton under the present regime of steady enrichment.

In recapitulation, it can be said that chemical enrichment, by promoting an increase in the standing crop of phytoplankton, can result in an increase in the biomass of zooplankton. But—to answer the question asked at the beginning of this paper—the relative composition of zooplankters apparently seldom suffers major modification by such enrichment alone. The evidence suggests that in most lakes the composition of zooplankton is largely controlled by the degree and kind of planktivory and by the size-related filtering efficiency of the planktonic herbivores themselves. Therefore, should enrichment be sufficient to lead to a disturbance in the populations of dominant planktivorous fish (such as the elimination of coregonids in temperate lakes), the composition of the zooplankton may well be seriously affected. The resultant shift in body size—and hence a shift in filtering effectiveness—of the planktonic herbivores, may either slow the change toward the eutrophic state that the chemical enrichment induces or accelerate it.

BOSMINA AS AN INDICATOR OF ENRICHMENT

In the preceding section we concluded that the composition of the dominant planktonic herbivores need not reflect enrichment. But there is always the possibility that some minor component of the zooplankton might be especially sensitive to some aspect of enrichment. Such a species might then be an animal indicator of enrichment, much as *Oscillatoria rubescens* has been taken to be among the phytoplankton (Minder, 1938).

Several studies have indicated that *Bosmina coregoni longispina* disappears as a lake is enriched and is replaced by the rather smaller *Bosmina longirostris*. In 1938, Minder recorded that *B. coregoni* disappeared from the plankton of the lower Zürichsee about 1911, after about 50 years of that lake's enrichment with Zürich's sewage. In a study of the sediments of Linsley

Pond (Connecticut), Deevey (1942) recorded that the microfossils of *B. coregoni* were replaced by *B. longirostris* at about the level of sediment deposited when the lake had attained the high productivity (judged by organic content of sediment) that has characterized it since (Hutchinson and Wollack, 1940). *B. longirostris* is a dominant zooplankter in the modern lake and appears to have remained the sole *Bosmina* subsequent to its initial replacement of *B. coregoni*. Similar but less simple and clear-cut shifts in *Bosmina* have been noted in other lakes.

Caution in the interpretation of shifts in *Bosmina* species must be urged, however, because, as we have seen, such shifts can result from changes in the level of predation on the zooplankton. If predation on zooplankton greatly increases, a population of *B. coregoni* might not be able to persist. *B. coregoni longispina*, 2 to 3 times larger than *B. longirostris*, disappeared from the zooplankton of Crystal Lake (along with the other large zooplankters depicted in the upper panel of Figure 1) after a population of *Alosa aestivalis* became established (Brooks and Dodson, 1965). *B. longirostris* became the dominant planktonic herbivore. A rapid shift from *B. coregoni* to *B. longirostris* will be recorded in the sediments. To be sure, this *Alosa* is an obligate planktivore and will exert far more predation upon *B. coregoni longispina* than would the facultative coregonids and salmonids in the presence of which it has evolved. But this makes the general point that a shift in abundance of microfossils from *Bosmina coregoni longispina* to *B. longirostris* need have nothing to do with enrichment.

This effect of the introduction of *Alosa aestivalis* in Crystal Lake raises the more specific question of whether the shift from *B. coregoni* to *B. longirostris*, as recorded in the microfossils of Linsley Pond, might have been an immediate effect of the natural introduction of *Alosa pseudoharengus*. There is reason to believe that this obligate planktivore is responsible for the dominance of *B. longirostris* in several coastal lakes of New England at present. It seems not unreasonable to speculate that the complete and irreversible loss of *B. coregoni*—a unique circumstance among lakes whose histories are known—is similarly determined. But why should an *Alosa* population have become established just at the time the lake attained the high level of production that has been maintained, more or less, to the present? Why not before? When it is recalled that *Alosa pseudoharengus* is entirely dependent upon open-water plankton, it can be appreciated that the earliest stages of Linsley Pond did not support a dense enough crop of zooplankton to provide a continuing supply of food for a population of *Alosa pseudoharengus*. Anadromous spawners could have, and probably did, enter the lake during its earlier stages, but a population of obligate planktivores could become permanently established only when the lake had become much more productive than it had been originally. Thus, the elimination of *B.*

coregoni longispina and its replacement by *B. longirostris* might well have been an indirect result of enrichment but a direct result of the establishment of a planktivore population.

The elimination of *B. coregoni* from dominance in Lake Michigan (Table 2), on the contrary, occurred when other, larger planktonic herbivores were becoming more abundant. This could be interpreted as suppression by competitors rather than by predators. But whatever force may be causing a shift in the compositon of the zooplankton, it is only the microfossils of *Bosmina* that will reflect past changes with a reasonable degree of reliability. Changes in the specific representation of other genera of Cladocera and copepods undoubtedly also occur during the maturation of most temperate lakes. But *Bosmina* has an exoskeleton sufficiently resistant to the natural forces of decomposition to cause its subfossilized remains to reflect with considerable accuracy the abundance of the living zooplankters (Deevey, 1964). In this respect it is unique among the zooplankters. The usually scant remains of other limnetic zooplankters, when carefully sought and studied, can give further indications of past changes in the composition of the zooplankton (Goulden, 1964).

Some of the recorded shifts in the *Bosmina* species of the open waters of lakes now undergoing enrichment may also be due to changes in the fish populations that have been induced by the enrichment. In the "classic" case of the lower Zürichsee, Minder has reported (see Hasler, 1947) that the fish population has been seriously modified during the period of enrichment. The coregonid and salmonid planktivores were largely replaced by even more numerous cyprinid planktivores during the period of replacement of *B. coregoni longispina* by *B. longirostris*. Thus, in both artificial enrichment and natural maturation there is reason to believe that a primary effect of chemical enrichment has been an alteration in the populations of planktivorous fish and that this change has in turn affected the composition of the zooplankton.

This paper is based in part on research supported by National Science Foundation Grants GB 1207 and GB 6004.

REFERENCES

Baldi, E. 1951. Stabilité dans le temps de la biocénose zooplanktique du Lac Majeur. Verh. Internat. Verein. Limnol. 11:35–40.

Beeton, A. M. 1965. Eutrophication of the St. Lawrence Great Lakes. Limnol. Oceanogr. 10:240–254.

Berg, A., and E. Grimaldi. 1965. Biologia delle due formi di coregene (*Coregonus* sp.) del Lago Maggiore. Mem. Ist. Ital. Idrobiol. 18:25–196.

Berg, A., and E. Grimaldi. 1966a. Biologia dell'agone (*Alosa ficta lacustris*) del Lago Maggiore. Mem. Ist. Ital. Idrobiol. 20:41–83.

Berg, A., and E. Grimaldi. 1966b. Ecological relationships between planktophagic fish species in the Lago Maggiore. Verh. Internat. Verein. Limnol. 16:1065–1073.

Bradshaw, A. S. 1964. The crustacean zooplankton picture: Lake Erie 1939–49–59; Cayuga 1910–51–61. Verh. Internat. Verein. Limnol. 15:700–708.

Brooks, J. L. 1968. The effects of prey-size selection by lake planktivores. Syst. Zool. 17:272–291.

Brooks, J. L., and S. I. Dodson. 1965. Predation, body size, and composition of plankton. Science 150:28–35.

Burns, C. W., and F. H. Rigler. 1967. Comparison of filtering rates of Daphnia in lakewater and in suspensions of yeast. Limnol. Oceanogr. 12:492–502.

Chandler, D. C. 1940. Limnological studies of western Lake Erie. I. Plankton and certain physical-chemical data of the Bass Islands Region, from September 1938 to November 1939. Ohio J. Sci. 40:291–336.

Chandler, D. C. 1964. The St. Lawrence Great Lakes. Verh. Internat. Verein. Limnol. 15:59–75.

Damann, K. E. 1945. Plankton studies of Lake Michigan. I. Seventeen years of plankton data collected at Chicago, Illinois. Amer. Midland Natur. 34:769–796.

Damann, K. E. 1960. Plankton studies of Lake Michigan. II. Thirty-three years of continuous plankton and coliform bacteria data collected from Lake Michigan at Chicago, Illinois. Trans. Amer. Micro. Soc. 79:397–404.

D'Ancona, U. 1955. The stability of lake planktonic communities. Verh. Internat. Verein. Limnol. 12:31–47.

Davis, C. C. 1962. The plankton of the Cleveland Harbor area of Lake Erie, 1956–57. Ecol. Monogr. 32:209–247.

Davis, C. C. 1964. Evidence for the eutrophication of Lake Erie from phytoplankton records. Limnol. Oceanogr. 9:275–283.

Davis, C. C. 1966. Plankton studies in the largest great lakes of the world with special reference to the St. Lawrence Great Lakes of North America, p. 1–36 In Univ. Michigan Great Lakes Res. Div. Pub. No. 14.

Deevey, E. S., Jr. 1942. Studies on Connecticut lake sediments. III. Biostratonomy of Linsley Pond. Amer. J. Sci. 240:233–264, 313–338.

Eddy, S. 1927. The plankton of Lake Michigan. Bull. Ill. Nat. Hist. Survey 17:203–232.

Edmondson, W. T., G. C. Anderson, and D. R. Peterson. 1956. Artificial eutrophication of Lake Washington. Limnol. Oceanogr. 1:47–53.

Fryer, G. 1957. The feeding mechanism of some freshwater cyclopoid copepods. Proc. Zool. Soc. London 129:1–25.

Galbraith, M. G., Jr. 1966. Size-selective predation on Daphnia by rainbow trout and yellow perch. Trans. Amer. Fish. Soc. 96:1–10.

Goulden, C. E. 1964. The history of the cladoceran fauna of Esthwaite Water (England) and its limnological significance. Arch. Hydrobiol. 60:1–52.

Green, J. 1967. The distribution and variation of Daphnia lumholzi (Crustacea: Cladocera) in relation to fish predation in Lake Albert, East Africa. J. Zool. 151:181–197.

Hasler, A. D. 1947. Eutrophication of lakes by domestic drainage. Ecology 28:383–395.

Hrbáček, J. 1962a. Species composition and the amount of the zooplankton in relation to the fish stock. Rozpravy ČSAV, Řada Mat. a přír. Ved 72(10):1–116.

Hrbáček, J. 1962b. Relations of planktonic Crustacea to different aspects of pollution, p. 53–57 In Biological problems in water pollution. Robert A. Taft Sanitary Engineering Center, U.S. Public Health Service, Cincinnati, Ohio.

Hrbáček, J. 1964. Contribution to the ecology of water-bloom forming blue-green algae–*Aphanizomenon flos aquae* and *Microcystis aeruginosa.* Verh. Int. Verein. Limnol. 15:837–846.

Hrbáček, J., M. Dvořáková, V. Kořínek, and L. Procházková. 1961. Demonstration of the effect of the fish stock on the species composition of zooplankton and the intensity of metabolism of the whole plankton association. Verh. Int. Verein. Limnol. 14:192–195.

Hrbáček, J., and M. Novatná-Dvořáková. 1965. Plankton of four backwaters related to their size and fish stock. Rozpravy ČSAV, Řada Mat. a Přír. Ved 75(13):1–64.

Hutchinson, G. E., and A. Wollack. 1940. Studies on Connecticut lake sediments. II. Chemical analyses of a core from Linsley Pond, North Brandford. Amer. J. Sci. 238:493–517.

Ivlev, V. S. 1961. Experimental ecology of the feeding of fishes [trans. from Russian]. Yale Univ. Press, New Haven, Conn. 302 p.

Livingstone, D. A. 1957. On the sigmoid growth phase in the history of Linsley Pond. Amer. J. Sci. 255:364–373.

Lund, J. W. G. 1961. The periodicity of μ-algae in three English lakes. Ver. Internat. Verein. Limnol. 14:147–154.

Minder, L. 1938. Der Zürichsee als Eutrophierungs-phänomen. Geol. Meere Binnengewässer 2:284–299.

Monakov, A. V., and Yu. I. Sorokin. 1961. Quantitative data on the feeding of *Daphnia* [in Russian]. Trav. Inst. Biol. Vodokhr. 4:251–261.

Mullin, M. M. 1966. Selective feeding by calanoid copepods from the Indian Ocean, p. 545–554 *In* Some contemporary studies in marine science. Allen and Unwin, Ltd., London.

Slobodkin, L. B. 1954. Population dynamics in *Daphnia obtusa* Kurz. Ecol. Monogr. 24:69–88.

Smith, S. H. 1964. Status of the deepwater cisco population of Lake Michigan. Trans. Amer. Fish. Soc. 93:155–163.

Tonolli, V. 1962. L'attuale situazione del popolamento planctonico del Lago Maggiore. Mem. Ist. Ital. Idrobiol. 15:81–134.

Verduin, J. 1964. Changes in western Lake Erie during the period 1948–1962. Verh. Int. Verein. Limnol. 15:639–644.

Wells, L. 1960. Seasonal abundance and vertical movements of planktonic Crustacea in Lake Michigan, p. 343–369 *In* U.S. Fish and Wildlife Service Bull. 60.

Wells, L., and A. M. Beeton. 1963. Food of the bloater, *Coregonus hoyi,* in Lake Michigan. Trans. Amer. Fish. Soc. 92:245–255.

P. A. LARKIN AND T. G. NORTHCOTE
The University of British Columbia, Vancouver, Canada

Fish as Indices of Eutrophication

A substantial literature, dating from antiquity, confirms that fishes respond to changes in the trophic nature of aquatic environments, whether these changes are natural or the result of human activity. Man's interest in bodies of water is often concerned with production of fish, and in consequence, intense effort has been directed at increasing yields of edible fish.

Long before practitioners of the pond culture arts appreciated why, they had used many of the techniques of the science of "deliberate eutrophication" (Neess, 1949). In addition, the possibilities of increasing production of fishes in lakes by fertilization were extensively discussed prior to widespread concern about the unintentional fertilization with which we now largely associate the word "eutrophication." For example, Hasler's review (1947) of case histories of domestic eutrophication is an important gathering point in the literature, and it is relevant that this review was presented at a symposium on fertilization of aquatic areas. The object of this paper was "to examine the consequences of inadvertent lake fertilization in an effort to forecast the outcome of deliberate fertilization projects." Many authors, both before and since, have drawn attention to the mutual relevance of studies of "intentional and inadvertent fertilization," particularly with regard to fish production (see for example, Hayes, 1951). Finally, much of the concern over the accelerating pace at which freshwater is becoming culturally influenced stems from the effects on fish populations. There is little doubt that fish are a prime source of motivation for studies of nutrient enrichment in aquatic communities, and that the literature of pond fish culture and lake fertilization is as relevant to our subject as that concerned with cultural influence.

While fish are a focal point of human interest, they may not be particularly good indicators of eutrophication. Being at the top of a pyramid of production, they are the last, and perhaps the least, affected (Maciolek, 1954). In the words of Neess (1949), "the functional relationship between the amount of fertilizer applied and the weight of fish produced is blurred by intervening processes." Because fish are mobile, they may respond to changes by moving from the scene, to which they may return when conditions are more auspicious. Having growth and survival rates that fluctuate widely, fish pose a difficult substrate of variability against which to measure minor change. Being flexible in their food habits and being associated with other species in a complex interaction on food organisms, they may persist by exploiting alternatives (Larkin, 1954; Keast, 1965). For all these reasons, fish have a reputation as poor indicators of eutrophication (Hynes, 1960). Nevertheless, there are some things to be said in their favor. Because of interest in them, they are under closer public scrutiny; they attract more attention, especially when dead; and many statistics document their population dynamics. Moreover, much more is known about the physiology of fishes than about many other aquatic organisms, so that in the laboratory they may have particular immediate usefulness as calibrated indicators of changed conditions. On balance, although fish are far from ideal as indicators, from an economic and political point of view they will certainly be exploited for some time to come as one of the most relevant indicators.

From the viewpoint of fisheries biology, eutrophication has been considered both as the *process* of adding nutrients and as the *effect* on environments of adding nutrients to natural waters. In consequence, there are two kinds of effects on fish to be considered: first, those that arise directly and immediately from the process of adding nutrients of any kind; and, second, those that are mediated through the various pathways of the metabolism of aquatic communities. The literature on direct effects is evidently enormous, because it concerns the potential immediate influence on fish of a wide range of substances (Hynes, 1960; Jones, 1962).

Paralleling the distinction between direct and indirect effects, the separation is fundamental between pond and lake environments on the one hand and stream environments on the other. For streams, much of the literature on "eutrophication" is concerned largely with exclusion of fish by the direct effects of the added nutrients. By contrast, for lakes, where the term "eutrophication" has its strictest limnological derivation, the literature shows that the main effects on fish are indirect. They are indirect as a consequence of increased production at all levels, which renders lakes more "eutrophic" in their several physical, chemical, and biological characteristics. The same is true of pond fish culture, which is a group of techniques concerned with concentrating aquatic production of fish by manipulations of the environment that could be termed "recurrent eutrophication" (Neess, 1949; Hasler and Einsele, 1948).

It has obviously been necessary for us to confine this review to particular facets of the whole nexus of questions concerned with the direct and indirect effects on fish of adding nutrient substances to water. We have chosen to treat streams separately, focusing attention on (1) the direct effects on fish of pollutants that are commonly responsible for eventual nutrient enrichment of aquatic environments, and (2) the ecological consequences of addition of nutrients to streams. Ponds and lakes are treated together, with emphasis only on the many effects on fish (such as growth, feeding habits, distribution, mortality, reproduction, and population biology) that result indirectly from changes in trophic status.

STREAMS

DIRECT EFFECTS ON FISH OF COMMON NUTRIENT-RICH POLLUTANTS

Characteristic of many organic, nutrient-rich pollutants is their high immediate oxygen demand. A substantial literature documents the lethal effects of such pollutants on fish. The effects are especially lethal when the pollutants are accidentally discharged in large quantities into streams (Hynes, 1960; Klein, 1962; Jones, 1964). Effluent from sewage plants has direct toxic effects (Pruthi, 1927; Herbert, 1965), some of which are enhanced at reduced oxygen concentrations (Allan et al., 1958). Fish may thus be absent from the vicinity of outfalls even when oxygen concentrations are high (Katz and Gaufin, 1953; Rasmussen, 1955; Pentelow et al., 1938). Carbon dioxide concentrations, which strongly influence respiration (Black, 1940), may also be a factor in rendering regions of active organic decay uninhabitable for fishes, particularly in warm weather (Jones, 1952). Sedimentation and growths of sewage fungi may smother spawning areas (Rasmussen, 1955) as well as fish-food organisms (Katz and Gaufin, 1953). Effects of low oxygen concentrations on embryonic development and hatching of fish are reviewed by Doudoroff and Warren (1965) and Doudoroff and Shumway (1967). Diel fluctuations in oxygen concentration may depress feeding and growth of fish. Where low oxygen concentrations are not lethal, they may influence the vigor of fish and so affect competition between species (Davison et al., 1959).

In most cases of chronic sewage pollution, one might expect seasonal movements of fishes away from the immediate outfall, with avoidance rather than mortality the common occurrence (Doudoroff and Warren, 1965; Höglund, 1961). The avoidance reactions of fish may well provide one of the most sensitive indices of changed conditions, particularly where highly toxic substances are involved.

In some instances sewage effects on streams may be mediated through lakes. For example, Mackenthun and Herman (1948) reported a heavy mortality of fish resulting from decomposition of algae that entered a river from Lake Kegonsa (Wisconsin). Although sewage was not mentioned in their paper, Domogalla (1935) records the same lake as influenced by sewage. Presumably the eutrophication of the lake was responsible ultimately for the fish mortality in the stream.

Detergents are common sources of phosphates as nutrients, and they may have toxic effects on fish (Klein, 1962). For example, they may have detrimental effects on trout sperm and on fertilization and development of eggs (Mann and Schmid, 1961). Also, the chemical senses of *Ictalurus natalis* are affected by detergents at low, sublethal concentrations (Bardach *et al.*, 1965).

Such insidious effects, which impair each fish to some degree, may well be more serious to populations than factors that kill a certain number of fish outright. A growing literature suggests that such effects may be more common than is realized. For example, Fujiya (1961) reports decreased liver glycogen and pancreatic RNA, kidney malfunction, and degeneration of intestinal mucosa of *Sparus macrocephalus* exposed to kraft-mill wastes. Chavin (1964) indicates that goldfish show endocrine changes in response to very slight environmental changes. The influence of sublethal levels of some pollutants may be reflected in changes of plasma or tissue proteins (Bouck, 1965). The amount of low-mobility protein in the blood is associated with general pollution tolerance; salmonids have little mobility protein, and carp have much (Bouck and Ball, 1967). It is probable that many of these sublethal effects will find increasing use as indicators of stress such as may be induced by eutrophication.

As a direct measure of contamination (and presumably enrichment) of natural waters by sewage and animal manure, intestinal tracts of fish may be sampled for the presence of mammalian enteric bacteria and associated bacteriophages. A review of literature is provided by Guelin (1962). Viscera of fish of many species taken in coastal marine waters may contain enteric bacteria of man and other warm-blooded animals, their presence indicating a pollution of the waters the fish have been frequenting. The contamination of the fish by *Escherichia coli* and by coliphages is apparently not of long duration, the former being undetectable 7 days after experimental contamination, and the latter after 8 to 12 days. The presence of these organisms is thus evidence of relatively recent contamination.

The degrees of occurrence of various helminth parasites in freshwater fish may reflect pollution by human sewage (Van Duijn, 1962), but it would seem difficult to use information of this type as a reliable indicator when many alternative warm-blooded hosts are commonly available for the parasites.

Even when fish are not contaminated with human bacteria and parasites, they may be rendered unpalatable by exposure to various pollutants. Most of the chemical substances affecting taste are not generally considered as nutrients, but sewage fungi and algal growth may indirectly contribute to unpleasant taste and odor of fish (Klein, 1962; Baldwin *et al.*, 1961; Huet, 1965). Detergents may intensify unpleasant taste effects (Mann, 1962).

ECOLOGICAL EFFECTS OF ADDITION OF NUTRIENTS

Directly deleterious effects of nutrients (chiefly sewage) are usually ameliorated within a short distance downstream, where one might expect to see increases in production associated with the addition of nutrients. Fish generally reappear in the *"Cladophora-Asellus* zone" downstream (Hynes, 1960). Salmonids seem generally more sensitive than "coarse fishes" (usually cyprinids and centrarchids). The survey conducted by Ellis (1937) of 982 stream sites in the United States led him to the general rule that a "good mixed fish fauna" would be found only where oxygen concentrations remained above 5 mg per liter during warm weather. The dissipation of deleterious effects may extend various distances downstream, the distance depending on the quantity and characteristics of the effluent and the stream. Katz and Gaufin (1953) reported a complete fish fauna 5 miles downstream from a sewage-pollution outfall, with the number of both species and individuals decreasing close to the sewage source.

The literature on ecological effects on fish of enrichment of streams by nutrients contains rather few references to the mechanisms of interaction involved, although effects on summer feeding and growth rate are evidently important. Brown trout in a Michigan stream fed heavily throughout summer on an abundant supply of *Asellus* in a section enriched with domestic sewage. But they were obliged to rely more on terrestrial organisms in an unenriched section when *Asellus* and other aquatic invertebrates became scarce in midsummer (Ellis and Gowing, 1957). Fish from the enriched section consistently showed a higher and less variable coefficient of condition than those from the less productive area. In a study of feeding habits of cutthroat trout in a small stream experimentally enriched with sucrose, Warren *et al.* (1964) found that food consumption was about doubled compared with that in unenriched sections. Increased consumption was from aquatic food sources, largely chironomid larvae associated with a heavy *Sphaerotilus* growth resulting from the sugar enrichment. Trout gained weight throughout most of the year in the enriched sections, but in unenriched sections the nutrient supply was barely above maintenance requirements, and gains were sporadic.

Katz and Howard (1955) showed that in the lower recovery part of a small Ohio stream receiving domestic sewage, growth rate of creek chub (*Semotilus a. atromaculatus*) remained high throughout summer, not showing the sharp midsummer retardation evident in unenriched sections upstream or downstream. Growth rate of the chub was reduced in the upper recovery section of the stream, possibly because the fish were restricted to small pools immediately below riffles where the oxygen content of the water was adequate. Enrichment may have caused both a retardation and an enhancement in growth of fish in different sections of the stream.

Shifts in distribution or elimination of fishes in streams and rivers as a result of nutrient enrichment and pollution are well known. Pentelow (1953), Katz and Gaufin (1953), and Krumholz and Minckley (1964) give typical examples from different regions. It is difficult to generalize, because stream systems are branched environments characterized by continuous change from one habitat type to the next. Changes induced by enrichment influence the proportion of the total stream length inhabited by particular associations of fish, the general pattern being to move the downstream communities farther upstream. In streams, eutrophication is a condition requiring virtually continuous enrichment.

LAKES AND PONDS

The common pattern of response of lakes and ponds to nutrient enrichment is an immediate increase in plant material, particularly of blue-green algae. The consequences to fish populations may be almost immediate, because blue-greens not only cause sharp oxygen deficiencies but may also have toxic effects on fish. This increase in algae is a common frustration of pond fertilization (Bennett, 1952; Krumholz, 1952; Smith and Swingle, 1939; Saila, 1952). For an extensive review, see Maciolek, 1954.

With proper techniques of management, ponds can demonstrate the heightened fish production that results from addition of nutrients. The draining, dry-fallowing, and liming techniques of pond fish culture are a form of cyclical enrichment. Together with fertilization techniques, they provide a basis for an extensive aquiculture in various parts of the world (Swingle, 1952; Maciolek, 1954; Mann, 1961; Tamura, 1961). Sewage may be used very effectively in producing pond fish, and, of course, it has been for centuries. Kisskalt and Ilzhöffer (1937) report that ponds to which sewage was added produced 300 to 500 lb per acre of carp and tench. Hey (1955) reports similar results in South Africa. Although there is an abundant literature on pond fish culture and fertilization, it is apparent that most present knowledge is empirical, there being far from sufficient understanding of the processes involved.

The eutrophication of lakes affects fish in many ways that devolve primarily from the increase in production, the consequent deoxygenation of the hypolimnion, and the change of many other features of the biological environment that determine abundance of various species of fish.

GROWTH AND FEEDING HABITS

In early stages of eutrophication, one of the first apparent changes may be enhancement in growth of fish. Since many freshwater fish are quite catholic in their diet, and since their growth is elastic, changes in growth, both seasonal and annual, may provide a good summative and sensitive index to progressive eutrophication of their environment. Thus, Nümann (1964) suggests that fish may provide a better means than plankton for expressing the total effect of all factors that have been responsible for changes in the Bodensee (Lake Constance) during the course of its eutrophication. He also notes that fish may be less subject to local variations or to the spatial and seasonal variations that are often associated with chemical indices of eutrophication.

A long series of lake-fertilization experiments has shown that fish can respond with higher growth rate fairly rapidly (Juday and Schloemer, 1938; Hasler and Einsele, 1948; Frost and Smyley, 1952; Smith, 1955; Weatherley and Nicholls, 1955; Nelson, 1959; Munro, 1961). In the Munro study, for example, the growth rate of brown trout increased during the same year that fertilization was employed in four Scottish lochs. The increased growth rate was evident in most of the age groups (1+ to 4+).

The changes in growth rate and associated phenomena in a fish population subject to progressive eutrophication have been carefully documented for "Blaufelchen" (*Coregonus wartmanni* Bloch) in the Bodensee by Nümann (1962, 1963, 1964). In the 1930's only minor changes in growth rate were apparent, probably resulting from fluctuations in year class strength, and even up to 1950, no significant change occurred. In the 1950's, however, growth rate began to increase and was first evident in young of the year 1954–1955. By the late 1950's, fish were attaining in 3 years the length they had previously attained in 5. There was also a marked effect on age class composition in the catch. Whereas previously some individuals attained an age of 8 to 10 years, and most of the adults were 4 to 6 years old, by 1964 most of the adults had only completed their second or third year. Previously, the catch was composed entirely of individuals more than 2 years old; in 1964, 1-year-olds made up a third of the catch, the remainder being age 2.

In the Grosser Plöner See, which also has undergone strong eutrophication, plankton is concentrated in upper layers of the lake during the summer as a result of severe oxygen depletion in the hypolimnion (Morawa, 1958).

Consequently, the marane (*Coregonus maraena*), utilizing the denser food supply, now have a higher growth rate.

A spectacular increase in growth rate of kokanee in Kootenay Lake, British Columbia (R. A. H. Sparrow, unpublished data) may in part be associated with early stages of eutrophication of the lake caused by phosphate enrichment. Kokanee from the west arm of the lake, which in 1951 rarely exceeded 200 g at maturity (Vernon, 1957), weighed more than 1 kg at spawning time in 1963–1967. Summer standing crops of most zooplankton species have increased manyfold since 1949 (E. R. Zyblut,* unpublished data) and provide a rich food source for plankton-feeding fish. Utilization of the introduced *Mysis relicta,* which are abundant in some areas of the lake (Sparrow *et al.,* 1964; E. R. Zyblut,* unpublished data), is also probably involved in the increase growth rate of the kokanee.

The changes in abundance of food organisms caused by eutrophication may or may not be reflected in the feeding habits of indigenous fish. The intentional enrichment of Crecy Lake brought about an appreciable increase in its benthic fauna, but many of the abundant forms were not directly fed upon by the brook trout population (Smith, 1961). Morgan (1966) also discusses the production of unsuitable components of benthos for fish following fertilization of Scottish lochs. Perch in the Obersee (Nümann, 1964) switched to an entirely planktonic diet after this food became abundant. Price (1961) suggests that several species of Lake Erie's fishes that formerly fed heavily on mayflies have changed to midge larvae and pupae. The mayfly populations had declined greatly by the 1960's (Britt, 1961).

As might be expected, accelerated growth rates of fish in enriched waters often have concomitant effects on their condition. Consistent differences may be found in the fat and water contents of fish from environments showing various degrees of enrichment. The marane from the Grosser Plöner See, the most highly eutrophic of the three lakes studied by Morawa (1958), have the highest as well as the strongest seasonal variation in fat content.

In lakes of northeastern Wisconsin, eutrophic environments force ciscoes (*Coregonus artedii*) to live under undesirable conditions of temperature and dissolved oxygen. They fail to thrive even in the presence of abundant food (Hile, 1936).

SPATIAL DISTRIBUTION AND MORTALITY

Although shifts in horizontal and vertical distributions of fish are probably a common response to changes associated with eutrophication, there are few convincing demonstrations in the literature. One example is provided by the

*Temporal and Spatial Changes in Distribution and Abundance of Macro-zooplankton in a Large British Columbia Lake. M.S. thesis, Univ. British Columbia (1967).

three indigenous species of salmonids in Lake Erie that now are restricted to the eastern basin during most of the year (Beeton, 1965). Again, in the Obersee, the perch and whitefish (*Coregonus fera*) have moved from the littoral zone into the pelagic zone, a region where they were never found before eutrophication (Nümann, 1964). Although not due to induced eutrophication, the depth restriction of cisco during summer in lakes of northeastern North America, such as reported for Lake Nipissing, Ontario (Fry, 1937), seems to be the sort of thing that would be expected.

This restriction of coldwater fishes to a thin layer between the oxygen-deficient hypolimnion and the warm epilimnion may lead eventually to mortalities. Hasler (1947) reports the observation of Brutschy and Güntert (1924) of probable exclusion of coregonines from the hypolimnion of the Hallwilersee in Switzerland. The gradual disappearance of ciscoes from Lake Mendota, Wisconsin, which is now nearly complete (A. D. Hasler, personal communication), may well be attributable to similar causes. Tanner (1960) states that the elimination of oxygen from the hypolimnion is a milestone in eutrophication.

REPRODUCTION

The many changes involved in eutrophication may have their largest effects on fish populations by influencing the survival of the young. The evidence is surprisingly limited. Reproductive failure has been cited as the major factor in decline of commercial fishes in Lake Erie, which has undergone marked eutrophication (Beeton, 1965). (See also page 150, this volume). Nümann (1964) refers to a reduction in the number of Blaufelchen spawners in the Bodensee caused indirectly by its eutrophication, and he notes that both the gonads and the eggs are now smaller. In addition, early mortality of the embryos is appreciably higher (50 percent as opposed to 20 percent). Algal growth and increased turbidity, associated in part with eutrophication, may seriously degrade spawning grounds in littoral regions of lakes (Trautman, 1957; Davis, 1961). In the Bodensee, spawning places of winter spawners are covered by decaying matter that suffocates the eggs (Kriegsmann, 1955). By contrast, summer spawners profit from the increased weediness of shore regions. Presumably there are many examples that have not been extensively reported.

CHANGES IN YIELD AND IN COMPOSITION OF FISH POPULATIONS

While evidence is scanty as to the reasons for changes in fish populations as a result of eutrophication, there is abundant documentation of its occurrence. Hasler (1947) provides one of the better early summaries, indicating the general pattern of change from coregonines to coarse fish as eutrophication proceeds. In North America the outstanding example is usually taken to be the dramatic changes in abundance of fishes in the Great Lakes, although it must be observed that a variety of factors other than eutrophication may be involved. For instance, Langlois (1954) presented an extensive argument about changes in the fish fauna of the western end of Lake Erie, including nutrient effects but placing greatest emphasis on turbidity (resulting from land-use practices) as a factor influencing survival of young fishes. He did not consider overfishing as responsible, and in this respect, he differs vigorously from Van Oosten (1948). The current view seems to be that eutrophication is responsible for at least some of the changes in Lake Erie. Beeton (1965) reports that during most of the year, salmonids are restricted to the eastern end of the lake, and even there warmwater fishes now predominate. (See also page 150, this volume.) Changes in the relative abundance of species in Lake Erie have been dramatic. Catches of ciscoes dropped from 20 million lb per year to 7,000 lb in 1962; whitefish from 2 million to 13,000 lb in the same year; sauger from 1 million lb in 1946 to a few thousand; walleye from 15 million lb in the 1940's and 1950's to less than 1 million; blue pike from 15 million lb to only 1,000 in 1962. The total catch has remained near 50 million lb, but the catch has become progressively more concentrated on the warmwater species, freshwater drum, carp, yellow perch, and smelt. A major feature in the decline of several once-valuable species has been the failure of reproduction. Verduin (1964) has recorded summer mortalities of several species.

Equally dramatic changes in relative abundance have occurred in other of the Great Lakes (Moffett, 1957; Smith, 1964), but the introduction of the sea lamprey (and more recently the alewife) has apparently been a major factor in these changes. Most recently, Pacific salmon have been introduced. So many changes have taken place in Great Lakes fish populations, and they have taken place in such rapid sequence as a result of so many different kinds of factors, that it seems unlikely that their causes will ever be understood. Perhaps the only safe conclusion with respect to eutrophication is that it has been an important contributing factor.

Rapid changes have also been taking place in European lakes, especially since 1950. Müller (1966) refers to several instances of eutrophication in lakes of the Democratic Republic of Germany in which the classification of the lakes for fisheries purposes has changed. He cites as an impressive example the largest lake of North Germany, Lake Muritz. Large catches of *Coregonus albula* were being made as late as 1950 to 1953, but from 1957 to 1966 the species was not taken. In the early 1960's, the main fishery of the pelagic zone was for the "Zander" (*Lucioperca lucioperca*), which was stocked. Müller (1966) also refers to the possibility that the effects of civilization are likely to result in further changes in many of the lakes in which oxygen concentrations may be critical seasonally.

The change in fish populations with eutrophication of the Bodensee is fully discussed by Kriegsmann (1955), who has carefully considered the various factors that have influenced the catch statistics for the lake. The Obersee was classified as unproductive as a fish producer (under 10 kg per ha) by Schäperclaus (1936), but in 1953 the yield was 13.5 kg per ha. The Untersee was classified as a minor producer (between 10 and 20 kg per ha), but in 1953 it produced 23 kg per ha. The increase in yield was attributed to the effects of sewage-induced eutrophication.

The proportions of different fish species have also changed over the period 1910 to 1953. The fine fishes (edelfische), which include the whitefish, trout, pike-perch, and eel, were more than 80 percent of the 1910–1914 catch in the Obersee and 53 percent in the Untersee. The percentages in 1949–1953 were 65 percent in the Obersee and 38 percent in the Untersee. During the same period, the coarse fish catches (perches and cyprinids) increased in both lakes, particularly in the Obersee. The yield in the Obersee in the later period was close to the early yields in the Untersee, which suggests that the sequences for the two lakes may be "spliced" to give a continuous picture of the change in species composition in the transition from most oligotrophic (early years in the Obersee) to most eutrophic (late years in the Untersee). It would seem from this comparison (Figure 1) that the switch from predominantly fine fishes to predominantly coarse fishes takes place over a relatively narrow range of mesotrophy. Thus, the changes in the Obersee have taken place with increasing speed as eutrophication has progressed. Because the Obersee has attained eutrophic status, it would be anticipated that yields could continue to increase but that the species composition would be relatively more stable, as has been the case in the Untersee.

This pattern of change is quite in conformity with what one would expect from a comparative study of lakes. More than 40 years ago, A. S. Pearse studied several lakes in Wisconsin, and his review on the ecology of lake fishes summarizes major differences in the quantity and species composition among the various lake types (Pearse, 1934). Increasing eutrophy is associated with greater production. The largest oligotrophic lakes are dominated by salmonids and coregonines, whereas smaller oligotrophs support centrarchids in abundance as well as coregonines. Such eutrophic lakes as Mendota, in Pearse's day, produced large quantities of perch, largemouth bass, white bass, rock bass, carp, and buffalofish. The shallow Lake Wingra (maximum depth, 4.3 m) produced large quantities of carp, crappie, sunfish, dogfish, and perch. In the words of Pearse, "Each lake presents a type in which one or more species of fishes may be at their best and become dominant." It is scarcely surprising that with the changes attendant upon eutrophication, changes in the fish populations should ensue.

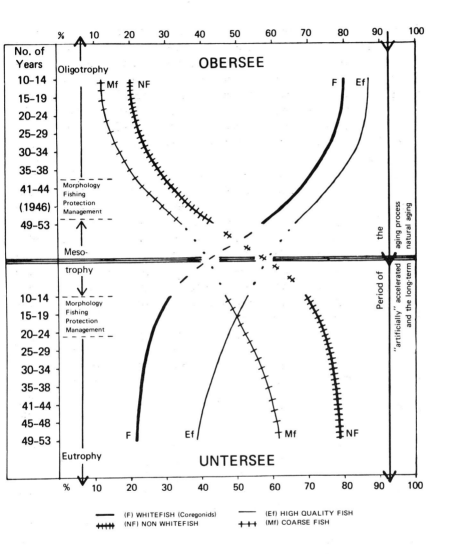

FIGURE 1   Representation of changes in associations of fishes in the Bodensee in the transition from most oligotrophic conditions (Obersee, 1910–1914) to most eutrophic conditions (Untersee, 1949–1953). (Kriegsmann, 1955.)

TAKING ADVANTAGE OF EUTROPHICATION

There can be little doubt that the increasing world population will involve greater cultural influence on aquatic environments. In these circumstances, it would seem wise to devote increasing attention to the techniques of guiding nutrient enrichment into greater production of aquatic foods. The emphasis should be not only on correcting eutrophication but also on exploiting it. Hynes (1960) devotes the last chapter of his book to assessing prospects for the future, drawing attention to the "appalling waste" of nutrients in sewage disposal and the opportunities for their recirculation for production of human food.

Among the ways of recovering fertility is the culture of fish in ponds to which sewage is added (Hey, 1955; Kisskalt and Ilzhöffer, 1937). It is equally apparent that the nutrient enrichment of lakes and streams offers avenues for greater food production. With knowledge of the processes involved in eutrophication, we may be able to regulate the rate of nutrient flow, modify its qualitative characteristics, and manipulate the species composition in lakes so as to produce the types of changes in aquatic communities that are most desired. Also, it may prove simpler to produce plankton in some ponds and lakes for feeding of fish in others (Smith, 1932). Many other possibilities can be envisaged and are worth exploration.

It is well to bear in mind that much of the experience of lake fertilization has been discouraging, partly because the rate of production in colder climates is not sufficient to justify the cost, but also because the fish produced must be close enough to the consumers and in sufficient demand to make the venture economically sound (Maciolek, 1954). Despite the many problems, research aimed at harnessing the quantities of waste nutrients would seem to have great value to mankind.

Similarly, there have been suggestions for increasing estuarial and littoral production in the sea by the use of nutrient enrichment. Gaarder and Bjerkan (1934) are reported by Smith (1959) as advocating the use of waters receiving drainage from cultivated land for oyster culture in Norway. The possibilities for increasing lobster production by sewage effluent have also been mooted (D. A. Wilder, personal communication). And, as is well known, lagoon production in many parts of the world is already considerably enhanced by sewage effluent and land drainage rich in fertilizers.

An argument of Thomas (1962) is highly relevant to the prospects for taking advantage of eutrophication; that is, a good way to "rectify" eutrophication is to remove nutrients in the form of fish. One might suppose that if the fish are consumed locally, one might hope for the elusive "perpetual motion," so far as phosphate ions are concerned.

It is appropriate to conclude this review with reference to the remarks of

Thienemann in his paper of 1952 entitled "Limnologie und Wasser-wirtschaft." Thienemann warned of the then-approaching catastrophe of stream regulation and waste pollution in German rivers, and he stressed the need for intelligent development of all aquatic systems. He emphasized the importance of the mutual dependence and interrelations of land and water resources and the value of pursuing limnological studies in the broad framework of the ecology of a whole area. Eutrophication is evidently one facet of the study of the impact of man on the total economy of nature. Because freshwater fish are of particular interest as a source of food and recreation, it is to be hoped that the effects of eutrophication on fish populations will be increasingly well documented and studied. At present, our understanding does not seem commensurate with the significance of the rapid changes in aquatic environments that are taking place.

REFERENCES

Allan, I. R. H., D. W. M. Herbert, and J. S. Alabaster. 1958. A field and laboratory investigation of fish in a sewage effluent. Fishery Investigations, series I, 6(2):1–76. H.M.S.O., London.

Baldwin, R. E., D. H. Strong, and J. H. Torrie. 1961. Flavor and aroma of fish taken from four freshwater sources. Amer. Fish. Soc., Trans. 90:175–180.

Bardach, E., M. Fujiya, and A. Hall. 1965. Detergents: Effects on the chemical senses of the fish *Ictalurus natilis* (Le Sueur). Science 148:1605–1607.

Beeton, A. M. 1965. Eutrophication of the St. Lawrence Great Lakes. Limnol. Oceanogr. 10:240–254.

Bennett, G. W. 1952. Pond management in Illinois. J. Wildlife Manag. 16:249–253.

Black, E. C. 1940. The transport of oxygen by the blood of freshwater fish. Biol. Bull. 79:215–229.

Bouck, G. R. 1965. Influence of sub-lethal pollution on the protein composition of fish and its ecological implications. Abstracts 16th Int. Cong. Soc. Int. Limnol. (Warsaw), p. 15.

Bouck, G. R., and R. C. Ball. 1967. Distribution of low-mobility proteins in the blood of fishes. J. Fish. Res. Bd. Canada 24:695–697.

Britt, N. W. 1961. Extended limnological studies in Western Lake Erie sponsored by the Natural Resources Institute of the Ohio State University. Publ. No. 7, Great Lakes Res. Div., Univ. Michigan, p. 158.

Brutschy, A., and A. Güntert. 1924. Gutachen über der Rückgang des Fischbestandes im Hallwilersee. Arch. für Hydrobiol. 14:523–571.

Chavin, W. 1964. Sensitivity of fish to environmental alterations. Publ. No. 11, Great Lakes Res. Div., Univ. Michigan, p. 54–67.

Davis, C. C. 1961. The biotic community in the Great Lakes with respect to pollution. Conf. on Water Pollution and the Great Lakes, Proc. (DePaul Univ., Chicago), p. 80–87.

Davison, R. C., W. P. Breese, C. E. Warren, and P. Doudoroff. 1959. Experiments on the dissolved oxygen requirements of coldwater fishes. Sewage Ind. Wastes 31:950–966.

Domogalla, B. P. 1935. Eleven years of chemical treatment of the Madison Lakes: Its effect on fish and fish foods. Amer. Fish. Soc., Trans. 65:115–121.

Doudoroff, P., and C. E. Warren. 1965. Dissolved oxygen requirements of fishes, p. 145–155 *In* Biological problems in water pollution. Third seminar, 1962. U.S. Public Health Service Environmental Health Series.

Doudoroff, P., and D. L. Shumway. 1967. Dissolved oxygen criteria for the protection of fish, p. 13–19 *In* a symposium on water quality criteria to protect aquatic life. Spec. Publ. No. 4. Am. Fish. Soc.

Ellis, M. M. 1937. Detection and measurement of stream pollution. Bull. U.S. Bur. Fish. 48:365–437.

Ellis, R. J., and H. Gowing. 1957. Relationship between food supply and condition of wild brown trout, *Salmo trutta* Linnaeus, in a Michigan stream. Limnol. Oceanogr. 2:299–308.

Frost, W. E., and W. J. P. Smyly. 1952. The brown trout of a moorland fishpond. J. Anim. Ecol. 21:62–86.

Fry, F. E. J. 1937. The summer migration of the cisco, *Leucichthys artedi* (Le Sueur), in Lake Nipissing, Ontario. Univ. Toronto Stud. Biol. Ser. 44, Publ. Ont. Fish. Res. Lab., 55, 91 p.

Fujiya, M. 1961. Effects of Kraft pulp wastes on fish. J. Water Pollut. Control Fed. 33:968–977.

Gaarder, T., and P. Bjerkan. 1934. Osters og Osterskultur i Norge. John Griegs Boktrykkeri, Bergen. 96 p.

Guelin, A. 1962. Polluted waters and the contamination of fish, p. 481–502 *In* Fish as food, vol. 2, Academic Press, New York.

Hasler, A. D. 1947. Eutrophication of lakes by domestic drainage. Ecology 28:383–395.

Hasler, A. D., and W. G. Einsele. 1948. Fertilization for increasing productivity of natural inland lakes. Trans. Thirteenth North American Wildlife Conf., p. 527–555.

Hayes, F. R. 1951. On the theory of adding nutrients to lakes with the object of increasing trout production. Can. Fish. Cult. 10:32–37.

Herbert, D. W. M. 1965. Pollution and fisheries, p. 173–195 *In* G. T. Goodman, R. W. Edwards, and J. M. Lambert [ed] Ecology and the industrial society. Blackwells, Oxford.

Hey, D. 1955. A preliminary report on the culture of fish in the final effluent from the new disposal works, Athlone, S. Africa. Verh. Int. Ver. Limnol. 12:737–742.

Hile, R. 1936. Age and growth of the cisco, *Leucichthys artedii* (Le Sueur), in lakes of the northeastern highlands, Wisconsin. Bull. U.S. Bur. Fish. 48:211–217.

Höglund, L. B. 1961. The reactions of fish in concentration gradients. Inst. Freshwater Res. (Drottningholm), Rep. 43. 147 p.

Huet, M. 1965. Water quality criteria for fish life, p. 160–167 *In* Biological problems in water pollution. Third seminar, 1962. Public Health Service Publ. 999-WP-25. Robert A. Taft Sanitary Engineering Center, U.S. Public Health Service, Cincinnati, Ohio.

Hynes, H. B. N. 1960. The biology of polluted waters. Liverpool Univ. Press, Liverpool. 202 p.

Jones, J. R. E. 1952. The reactions of fish to water of low oxygen concentration. J. Exp. Biol. 29:403–415.

Jones, J. R. E. 1962. Fish and river pollution, p. 254–310 *In* L. Klein [ed] River pollution. II. Causes and effects. Butterworths, London.

Jones, J. R. E. 1964. Fish and river pollution. Butterworth, Inc., Washington. 203 p.

Juday, C., and S. L. Schloemer. 1938. Effects of fertilizers on plankton production and on fish growth in a Wisconsin lake. Progr. Fish Cult. 40:24–27.

Katz, M., and A. R. Gaufin. 1953. The effects of sewage pollution on the fish population of a midwestern stream. Amer. Fish. Soc., Trans. 82:156–165.

Katz, M., and W. C. Howard. 1955. The length and growth of 0-year class creek chubs in relation to domestic pollution. Amer. Fish. Soc., Trans. 84:228–238.

Keast, A. 1965. Resource subdivision amongst cohabiting fish species in a bay, Lake Opinicon, Ontario. Publ. No. 13, Great Lakes Res. Div., Univ. Michigan, p. 106–132.

Kisskalt, K., and H. Ilzhöffer. 1937. Die Reinigung von Abwasser in Fischteichen. Arch. Hyg. Berl. 118:1–66.

Klein, L. 1962. River pollution. II. Causes and effects. Butterworths, London. 456 p.

Kriegsmann, F. 1955. Der Wechsel in der Vergesellschaftung der Fischarten des Ober-und Untersees und die Veränderungen des See-Reagierens. Arch. für Hydrobiol., suppl. 22:397–408.

Krumholz, L. A. 1952. Management of Indiana ponds for fishing. J. Wildlife Manag. 16:254–257.

Krumholz, L. A., and W. L. Minckley. 1964. Changes in the fish population in the upper Ohio River following temporary pollution abatement. Amer. Fish. Soc. Trans. 93:1–5.

Langlois, T. H. 1954. The western end of Lake Erie and its ecology. J. W. Edwards, Ann Arbor, Mich. 479 p.

Larkin, P. A. 1954. Interspecific competition and population control in freshwater fish. J. Fish. Res. Bd. Canada 13:327–342.

Maciolek, J. A. 1954. Artificial fertilization of lakes and ponds. U.S. Fish and Wildlife Service, Spec. Sci. Report: Fisheries No. 113. 41 p.

Mackenthun, K. M., and E. F. Herman. 1948. A heavy mortality of fishes resulting from the decomposition of algae in the Yahara River, Wisconsin. Amer. Fish. Soc. Trans. 75:175–180.

Mann, H. 1961. Fish cultivation in Europe, p. 77–102 *In* Fish as food (vol. 1). Academic Press, New York.

Mann, H. 1962. Increase by detergents of effects on taste in fish. Fischwirt 12:237–240.

Mann, H., and O. L. Schmid. 1961. Der Einfluss von Detergenien auf Sperma, Befruchtung und Entwicklung bei der Forelle. Int. Rev. Ges. Hydrobiol. Hydrog. 46:419–426.

Moffett, J. W. 1957. Recent changes in the deep-water fish populations of Lake Michigan. Amer. Fish. Soc. Trans. 86:393–408.

Morawa, F. W. F. 1958. Einege Beobachtungen über die Schwankungen des Fett- und Wassergehaltes von Fischen aus verschiedenen Umweltverhaltnissen. Verh. Int. Ver. Limnol. 13:770–775.

Morgan, N. C. 1966. Fertilization experiments in Scottish freshwater lochs. 2. Effects on the bottom fauna. Freshwater Salm. Fish. Res., 36. H.M.S.O., Edinburgh. 19 p.

Müller, H. 1966. Eine fischerei wirtschaftliche Seenklassifizierung Norddeutschlands und ihre limnologischen Grundlagen. Verh. Int. Ver. Limnol. 16:1145–1160.

Munro, W. R. 1961. The effect of mineral fertilizers on the growth of trout in some Scottish lochs. Verh. Int. Ver. Limnol. 14:718–721.

Neess, J. C. 1949. Development and status of pond fertilization in central Europe. Amer. Fish. Soc., Trans. 76:335–358.

Nelson, P. R. 1959. Effects of fertilizing Bare Lake, Alaska, on growth and production of red salmon (*O. nerka*). Fish. Bull. (U.S. Fish and Wildlife Serv.) 159:59–86.

Nümann, W. 1962. Schnelleres Wachstum, grössere Fangerträge, jungere Jahrgänge und Frühreife bei den Bodenseeblaufelchen als Folge der Düngung des Sees durch Abwässer. Allgm. Fischerei Zeitung 87:114–116 (as cited by Nümann, 1964).

Nümann, W. 1963. Die Auswirkung der Eutrophierung auf den Eintritt der Reife, aud die Eizahl und Eizrösse bein Bodenseeblaufelchen (*Coregonus wartmanni*). Allgm. Fischerei Zeitung 88:8.

Nümann, W. 1964. Die Veränderangen im Blaufelchenbestand (*Coregonus wartmanni*) und in der Blaufelchen-fischerei als Folge der künstlichen Eutrophierung des Bodensees. Verh. Int. Ver. Limnol. 15:514–523.

Pearse, A. S. 1934. Ecology of lake fishes. Ecol. Monog. 4:475–480.

Pentelow, F. T. K., R. W. Butcher, and J. Grindley. 1938. An investigation of the effects of milk wastes on the Bristol Avon. Fishery Investigations, series I, 4(1):1–80. H.M.S.O., London.

Pentelow, F. T. K. 1953. Pollution and fisheries. Verh. Int. Ver. Limnol. 12:768–771.

Price, J. W. 1961. Food habits of some Lake Erie fish. Publ. No. 7, Great Lakes Res. Div., Univ. Michigan. p. 160.

Pruthi, H. S. 1927. Preliminary observations on the relative importance of the various factors responsible for the death of fishes in polluted waters. J. Marine Biol. (U.K.) 14:729–739.

Rasmussen, C. J. 1955. On the effect of silage juice in Danish streams. Verh. Int. Ver. Limnol. 12:819–822.

Saila, S. B. 1952. Some results of farm pond management studies in New York. J. Wildlife Manag. 16:279–282.

Schäperclaus, W. 1936. Eine neue Klasseneinteilung für die Naturalroheträge mittel-europaischer Binnenseen. Zeit. für Fischerei 34 (as cited by Kriegsmann, 1955).

Smith, E. V., and H. S. Swingle. 1939. The relationship between plankton production and fish production in ponds. Amer. Fish. Soc., Trans. 68:309–315.

Smith, M. W. 1932. The fertilization of water in experimental ponds. Amer. Fish. Soc., Trans. 62:317–322.

Smith, M. W. 1955. Fertilization and predator control to improve trout angling in natural lakes. J. Fish. Res. Bd. Canada 12:210–237.

Smith, M. W. 1959. Phosphorus enrichment of drainage waters from farm lands. J. Fish. Res. Bd. Canada 16:887–895.

Smith, M. W. 1961. Bottom fauna in a fertilized natural lake and its utilization by trout (*Salvelinus fontinalis*) as food. Verh. Int. Ver. Limnol. 14:722–726.

Smith, S. H. 1964. Status of the deepwater cisco population of Lake Michigan. Amer. Fish. Soc., Trans. 93:155–164.

Sparrow, R. A. H., P. A. Larkin, and R. A. Rutherglen. 1964. Successful introduction of *Mysis relicta* Loven into Kootenay Lake, British Columbia. J. Fish. Res. Bd. Canada 21:1325–1327.

Swingle, H. S. 1952. Farm pond investigation in Alabama. J. Wildlife Manag. 16:243–249.

Tamura, Tadashi. 1961. Carp cultivation in Japan, p. 103–120 *In* Fish as food (vol. 1). Academic Press, New York.

Tanner, H. A. 1960. Some consequences of adding fertilizer to five Michigan trout lakes. Amer. Fish. Soc., Trans. 89:198–205.

Thienemann, A. 1952. Limnologie und Wasserwirtschaft. Sartryck ur Vatlenhygien 2:25–43.

Thomas, E. A. 1962. The eutrophication of lakes and rivers, cause and prevention, p. 299–305 *In* Biological problems in water pollution. Third seminar, 1962. Public Health Service Publ. 999-WP-25. Robert A. Taft Sanitary Engineering Center, U.S. Public Health Service, Cincinnati, Ohio.

Trautman, M. B. 1957. The fishes of Ohio. Ohio State University Press, Columbus. 683 p.

Van Duijn, C., Jr. 1962. Diseases of freshwater fish, p. 573–593 *In* Fish as food (vol. 2). Academic Press, New York.

Van Oosten, J. 1948. Turbidity as a factor in the decline of Great Lakes fishes with special reference to Lake Erie. Amer. Fish. Soc., Trans. 75:281–322.

Verduin, J. 1964. Changes in western Lake Erie during the period 1948–1962. Verh. Int. Ver. Limnol. 15:639–644.

Vernon, E. H. 1957. Morphometric comparison of three races of kokanee (*Oncorhynchus nerka*) within a large British Columbia lake. J. Fish. Res. Bd. Canada 14:573–598.

Warren, C. E., J. H. Wales, G. E. Davis, and P. Doudoroff. 1964. Trout production in an experimental stream enriched with sucrose. J. Wildlife Manag. 28:617–660.

Weatherley, A., and A. G. Nicholls. 1955. The effects of artificial enrichment of a lake. Australian J. Marine Freshwater Res. 6:443–468.

PÉTUR M. JÓNASSON
University of Copenhagen, Hillerød, Denmark

Bottom Fauna and Eutrophication

Bottom dwellers are among the most common of freshwater animals, and they form an important link in the food chain of lakes. Many are microphagous, feeding on either phytoplankton or organic mud constituents. In their turn, they are eaten by many aquatic predators.

They show a great variety of adaptations to their semisessile mode of life. They adapt to the substratum, to the amplitude and rhythm of physical and chemical factors, and to the exploitation of food resources.

The number of bottom fauna is usually underestimated; thus, techniques for sampling, sieving, and sorting have been defective and inefficient (Jónasson, 1955). The time-consuming nature of this type of work has discouraged the study of the freshwater bottom fauna, and there are few papers of ecological value.

Eutrophication may affect rivers as well as lakes, but since the most reliable information on the bottom fauna has been obtained for lakes, the present discussion is restricted to this habitat.

ENVIRONMENT

To understand the influence of eutrophication on the bottom fauna, one must understand the basic environmental factors and their rhythm. These factors are described for a transect of a well-known eutrophic lake: Lake Esrom, in Denmark.

The transect may be divided into different zones (Figure 1C):

- Upper littoral zone. With the surf zone, it consists of sand, stones, and algae.

- Lower littoral zone. It is characterized by subsurface weeds that extend from the lower limit of the stony surf zone.
- Sublittoral zone. This zone is without macrophytic vegetation: mollusks extend to it.
- Profundal zone. In a eutrophic lake, this zone is characterized by mud of increasing thickness; it is without mollusks except *Pisidium*.

The lake stratifies during summer and during winter when there is ice cover. Spring and autumn overturns also occur. Temperatures in the epilimnion during summer may increase to 20° or 25° C, whereas the hypolimnion fauna survives in a "refrigerator" at a temperature of 8° to 10° C. The stratification of oxygen follows the same trend as the temperature. This means that the hypolimnion is devoid of oxygen for several months (Figure 1, A and B).

Phytoplankton is an important food for bottom invertebrates; an estimate of the production of organic matter by the phytoplankton is desirable, therefore, together with its vertical and seasonal distribution. The thickness of the primary producing layer and its seasonal changes are described for the moderately eutrophic Lake Esrom throughout the year 1963 and expressed as of g of C/m^3 per day (Figure 2, A and B). During spring a coffee-brown diatom plankton dominates. Transparency is low so that production goes on in a thin layer of about 2 m during the spring maximum. In May and June the green algae produce less food in a layer up to 10 m thick. In summer (July through September) the plankton consists of blue-greens, which build up the summer maximum and occupy an upper layer of 3 to 4 m. In autumn the diatoms dominate again, and primary production declines (cf. Jónasson and Kristiansen, 1967, Table 4).

Figure 1C roughly repeats the seasonal trend in the thickness of the primary producing layer and the sequence of algal groups. The combined effects of the primary production at the surface, the vertical temperature, and oxygen content result in the following conditions for bottom fauna (Figure 1, A, B, and C): The diatoms are rapidly distributed to the bottom during spring and autumn (and during winter in an ice-free lake); the green algae are mainly circulated to the bottom; the blue-green algae circulate within the epilimnion and sink slowly through the pycnocline, decay in the hypolimnion, and reach the bottom invertebrates in that condition (Jónasson and Kristiansen, 1967; Lund et al., 1963; Likens and Hasler, 1962; Likens and Ragotskie, 1965). The value of phytoplankton as food for bottom fauna is thus very different in different seasons and depends on lake rhythm. At the height of the summer stagnation, pH increases in the surface layers because of photosynthesis; alkalinity declines because of the precipitation of $CaCO_3$. In the hypolimnion, conditions are the reverse: pH is low and alkalinity is high. This is again very important for the bottom fauna, since essential biological processes are linked to photosynthesis and the carbon dioxide cycle of a lake (Figure 1C).

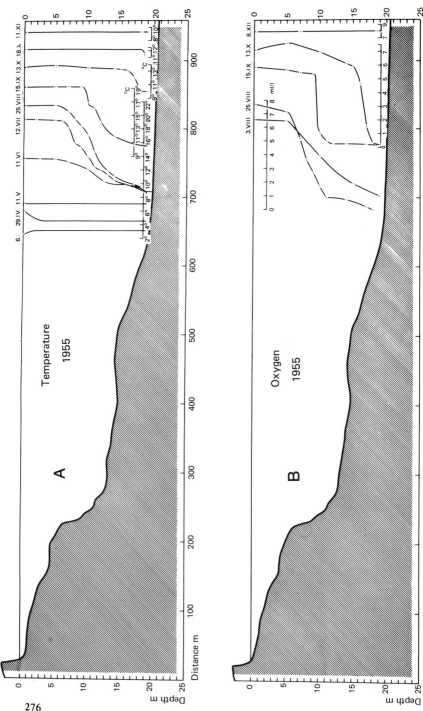

276

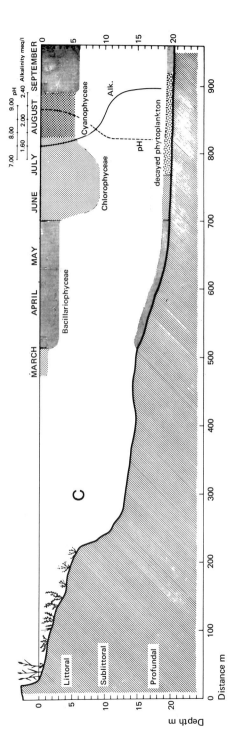

FIGURE 1 *A* and *B*, Vertical distribution of temperature (*A*) and oxygen (*B*) during 1955 along transect through western part of Lake Esrom. Littoral, sublittoral, and profundal zones are included. Macrophytes descend to a depth of about 6 m. *C*, Main seasonal occurrence of phytoplankton, approximate thickness of the primary producing layer in various seasons, and the condition of algae arriving at the bottom. The vertical variation in pH and alkalinity at the summer stagnation peak (August 25, 1955) is shown. Line and stipple patterns represent dominating forms: Bacillariophyceae (diatoms), Chlorophyceae (green algae), and Cyanophyceae (blue-green algae). Algae that were alive when they reached the bottom are represented by a repetition of the surface patterns. Decayed phytoplankton at the bottom are represented by dots.

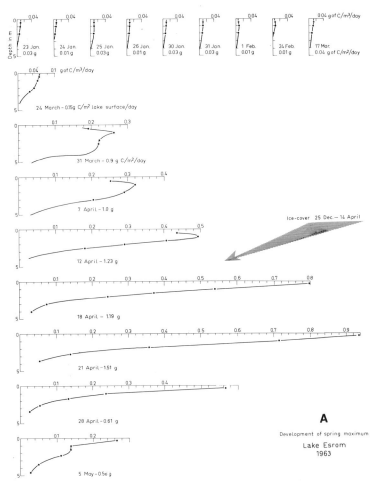

FIGURE 2   Vertical distribution of primary gross production during 1963. Each minor figure shows an *in situ* experiment, and rate of gross production at various depths is measured as mg of C/m^3 per day. *A*, Until April 14, experiments were carried out below ice. The very high volumetric values below ice in March and April and the very thin producing layer during the spring maximum in April are remarkable. *B*, The summer maximum in August has a thicker productive layer and thus higher values per square meter of lake surface. A transitional situation of low volumetric values and a thick producing layer are seen in June.

A comparison between the morphometry of the basin, temperature, and oxygen content shows that throughout extensive areas of a stratified lake the bottom temperature is less than half the surface temperature during the period July to November. During this period, the oxygen content in the bottom layers is so low that only a few species are able to survive.

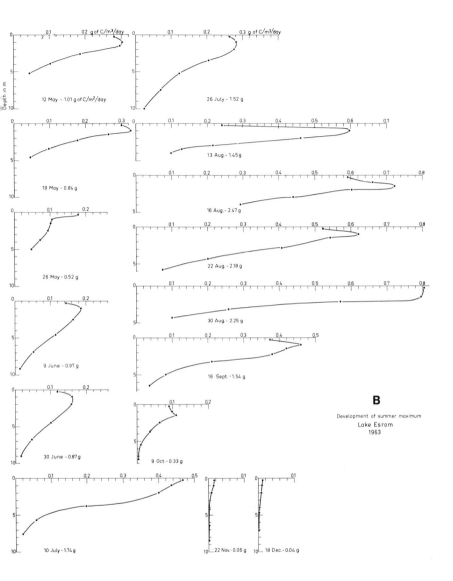

B

Development of summer maximum
Lake Esrom
1963

EFFECT OF INCREASE IN NUTRIENTS
ON THE BOTTOM ENVIRONMENT

An increase in the supply of nutrients changes the qualitative composition of phytoplankton and the amount of organic matter produced. Figure 3A

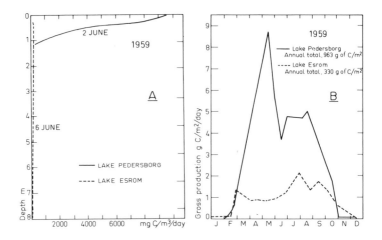

FIGURE 3   Comparison of phytoplankton and primary production in moderately eutrophic Lake Esrom and highly eutrophic (and polluted) Lake Pedersborg. *A*, daily values (June 2 and 6); *B*, annual values.

shows the daily vertical primary production in June in the clear, eutrophic Lake Esrom and in Lake Pedersborg, a lake rich in nutrients. Each shows a maximal production but at different levels. In the clear lake, maximal volumetric production, 200 mg of C/m^3, occurs at a depth of 1 m; in the nutrient-rich lake, the maximum volumetric production occurs at the surface and is 50 times greater. In the clear lake, primary production extends to a depth of 8 m; in the nutrient-rich lake, to only 1 m. In extremely rich lakes the thickness may be reduced to 0.20 m. Transparency in the clear lake was 4.70 m; in the nutrient-rich lake, 0.20 m. Thus, in the clear, eutrophic lake, primary production occurs in a relatively thick layer; in the very rich lake, it is reduced to a thick soup of algae at the surface (Jónasson and Mathiesen, 1959; Mathiesen, 1962; Jónasson, 1964).

A comparison on an annual basis is shown in Figure 3*B*. In the clear lake, a spring maximum is followed by a spring minimum, when nutrients are reduced. Then a summer maximum occurs in August; during autumn, primary production by phytoplankton declines rapidly concurrently with the decreasing light. In the very rich lake, the primary production increases in early spring in the same way as in the clear, eutrophic lake, but, in contrast to the latter, it continues at a high level all through summer until the autumn decline. In the lake rich in nutrients, the annual total production was 3 times that of the clear, eutrophic lake.

When the amount of nutrients entering a lake increases, the following changes in production of organic matter by phytoplankton occur:

• Primary production (expressed in volumetric units) increases tremendously.

• Seasonal variation in primary production (expressed in units per square meter of lake surface) becomes atypical.

• Annual primary production (expressed in square meters of lake surface) may increase several times.

• A qualitative change of planktonic algae occurs; such a change is shown by the "compound index" for Lake Fure, in Denmark, in the period 1911 to 1951 (Nygaard, 1949, 1958).

The influence of eutrophication on macrophytic vegetation was investigated in the eutrophic Lake Fure in 1911 and 1912 and again in 1950 and 1951. Figure 4 clearly shows the reduction in depth distribution of four plant species during 40 years as a result of increasing eutrophication and reduction

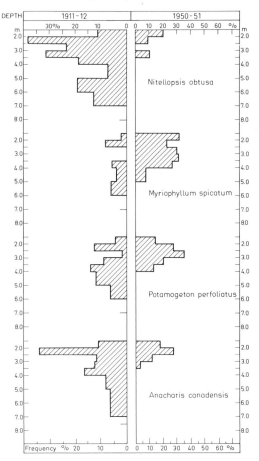

FIGURE 4   Depth distribution of four plant species in Lake Fure in 1911–1912 and 1950–1951. Line patterns indicate frequency in percentage. The absolute depth limit in 1911–1912 is indicated by horizontal lines for each species.

TABLE 1   Relation between Floristic Lake Types and Gross Production by Phytoplankton[a]

Floristic Lake Type	Bottom Vegetation — Maximum Depth Distribution of Vegetation (m)	Bottom Vegetation — Dominant Forms	Phytoplankton — Annual Total (g of C/m²)	Phytoplankton — Daily Maximum (g of C/m³)
Drepanocladus	2–4	*Drepanocladus, Fontinalis,* and *Sphagnum*		
Lobelia	>10	*Lobelia, Isoëtes,* and *Littorella*	60– 70	—
Lobelia-Potamogeton	5–10	*Lobelia-Elodeids*[b]	60– 70	0.1[c]
Potamogeton	2–6	*Elodeids*[d]	60– 70	1.0
Potamogeton-Nuphar	1–5	Few *Elodeids* and *Nymphaeidae*	60– 140	
Nuphar		*Nymphaeidae*[e]	160– 500	5.0
			200–1,200	40.0

[a]Modified from Mathiesen (1966).
[b]For example, *Myriophyllum.*
[c]Approximate.
[d]*Potamogeton* and *Myriophyllum.*
[e]No *Elodeids.*

282

in subsurface illumination. Other species, among them *Ranunculus circinatus* Sibth., *Ceratophyllum demersum* L., *Potamogeton pectinatus* L., and *P. lucens* L., show a similar trend (Christensen and Andersen, 1958).

Recent research in Danish lakes has shown a close relation between primary production by phytoplankton and the physiognomy and depth distribution of macrophytic vegetation. Table 1 shows clearly that many floristically different lakes have low values of primary production, but all possess a submerged vegetation extending to a depth of 5 to 10 m. In contrast, no submerged vegetation is found if the annual total primary production per square meter of lake surface is high and if the volumetric values are high during the growing season. The lake becomes devoid of submerged vegetation and the lake bottom becomes covered with mud.

It is true that this classification is based on Danish lakes and is especially characteristic of the northern European area (southern Baltic). Nevertheless, these lakes include a wide variety of types well known in the Temperate Zone of the Northern Hemisphere (Boreal Zone).

This section has been concerned with environment, submerged vegetation, and production of phytoplankton. The following conclusions may be stated:

• Primary production occurs in the surface layers, since light is the main limiting factor.
• In the absence of stratification, phytoplankton is evenly distributed throughout the lake volume.
• These facts are related to the seasonal cycle of physical and chemical conditions.
• Amount of nutrients, size of primary production by phytoplankton, and submerged vegetation are closely related. These factors also influence the type of substratum. The precipitation of algae to the bottom increases with nutrients, as does the thickness of mud, whereas the bottom becomes a uniform muddy area.
• The distribution of macrophytic vegetation becomes restricted to shallower depths because of a decrease in light ultimately caused by increased nutrients.

BOTTOM FAUNA: ITS COMPOSITION AND QUANTITY

In the previous section, the influence of increasing eutrophication on the environment was outlined. This section attempts to show the effects on the bottom fauna of increasing eutrophication. These effects may include both qualitative and quantitative changes. Unfortunately, there have been no quantitative investigations of bottom fauna similar to the studies of the effects of increasing eutrophication on macrophytic vegetation and plankton

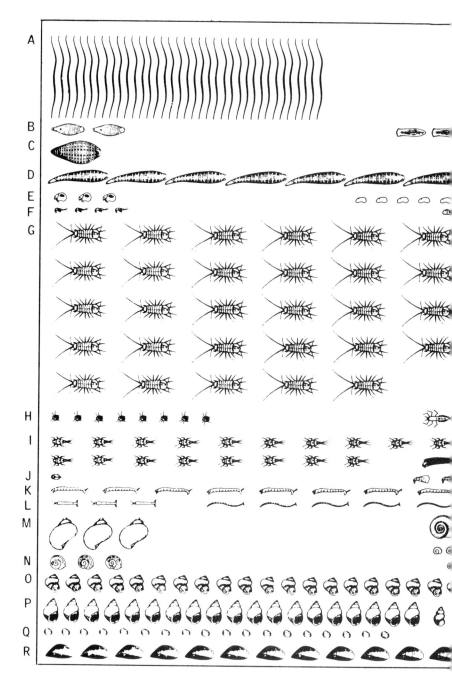

in Lake Fure. Because of lack of data, the effect of increasing eutrophication on the bottom fauna is described as a change in which a typical littoral fauna becomes more like a profundal fauna. The influence of different environmental conditions results in different composition and quantity of the bottom fauna, as shown in Figures 5 and 6.

In a calcareous eutrophic lake, the depth distribution of macrophytes seems to determine the occurrence and quantity of some bottom fauna groups; the vegetation provides a valuable substratum with optimal physical and chemical characteristics. The bottom fauna of the subsurface weeds is shown in Figure 5; the total is 10,810 individuals per m^2, with some 33 groups and species represented. It is clear that the primary consumers are mainly represented by snails living on the vegetation and by various groups living in the thin but rich mud (*Oligochaeta, Asellus, Caenis, Pisidium, Chironomidae*). Other groups (e.g., *Dreissena*) need a firm substratum. The richness of the fauna in the littoral zone is emphasized by the many predators—for example, the leeches, the *Hydracarina,* the *Ostracoda,* and the *Tanypodinae.* The zone is thus characterized by many microhabitats occupied by a varied fauna.

The profundal fauna of the same lake shows a great contrast (Figure 6). Figure 6 appears very uniform in comparison with Figure 5, only five species being represented. There are three primary consumers: *Ilyodrilus hammoniensis, Chironomus anthracinus,* and *Pisidium casertanum.* The predators are represented by *Corethra flavicans* and *Procladius pectinatus.* These are the specialists, and they thrive well in this habitat, with 20,441 individuals per m^2. At the same time, a marked change in ecology occurs. The animals burrow into the bottom instead of living on or near the surface. From a

FIGURE 5   Average number of individuals per 1/50 m^2 in unpolluted, eutrophic Lake Esrom from the littoral zone with macrophytes at a depth of 2 m. The drawings are life-size except those of *Eurycercus, Canthocamptus, Candona,* and *Cypris,* which are more than life-size. For simplicity, specimens of the same species are depicted as having the same size. The figure represents average populations over a period of 15 months. The populations are coincident. In the list that follows, the number of individuals per 1/50 m^2 is given in parentheses. A, *Oligochaeta* (37). B, *Helobdella stagnalis* (2). C, *Glossiphonia complanata* (1). D, *Herpobdella octoculata, H. testacea* (7). E, *Eurycercus lamellatus* (3). F, *Canthocamptus staphylinus* (4). G, *Asellus aquaticus* (29). H. *Hydracarina* (8). I, *Caenis moesta* (18). J, *Micronecta minutissima* (1). K, *Chironomidae* (8). L, *Tanypodinae* (3). M, *Lymnaea pereger* (3). N, *Gyraulus albus* (3). O, *Valvata piscinalis* (20). P, *Bithynia tentaculata* (19). Q, *Pisidium* (18). R, *Dreissena polymorpha* (12). S, *Polycelis tenuis* (2). T, *Candona candida, C. neglecta* (5). U, *Cypris vidua* (1). V, *Centroptilum luteolum* (1). W, *Leptoceridae* (1). X, *Oxyethira costalis* (1). Y, *Ceratopogon vermiformis* group. Z, *Planorbis planorbis* (1). A[′], *Gyraulus crista* (2). B[′], *Valvata cristata* (1). C[′], *Bithynia leachii* (2). Total number of individuals per 1/50 m^2: 219. Total per m^2: 10,810. (Data from Berg, 1938.)

temperature viewpoint, this fauna lives in a "refrigerator." During certain months the substratum is devoid of oxygen, low in pH, and high in alkalinity; the animals also live in complete darkness.

The communities shown in Figures 5 and 6 were carefully selected to correspond with the littoral and profundal peaks indicated in Figure 7, which shows the vertical distribution of the bottom fauna in relation to certain environmental factors: vegetation, oxygen, temperature, and mud deposits. Such information was originally provided by Lundbeck (1926) for the Plöner See. Later, Berg (1938) depicted his quantitative data of Lake Esrom similarly. Data for the profundal fauna during the period 1954–1956 were obtained by using more-adequate techniques for such sampling. The community structure shown in Figure 6 was based on these data.

The ecological background for understanding the change in the fauna from diversity to uniformity caused by eutrophication is determined by three main factors: dependence on substratum, respiratory adaptation of bottom fauna, and utilization of food as a function of lake rhythm.

DEPENDENCE ON SUBSTRATUM

We have discussed the change of substratum in the littoral zone from several microhabitats to a uniform mud surface, noting that the change is caused by increased production of phytoplankton and, consequently, by a decrease in subsurface light. Since no firm substratum exists, the fauna no longer has an opportunity to climb on submerged weeds, and it is reduced to burrowing into the mud. This causes a material change from an oxygen-rich to an oxygen-poor environment; it also causes changes in the chemical factors, in food, and in feeding habits. Such changes in the environment cause many groups to disappear and new ones to appear.

RESPIRATORY ADAPTATION OF BOTTOM FAUNA

The physiological properties of the various types of bottom fauna are undoubtedly one of the most effective factors determining both concentra-

FIGURE 6 Average number of individuals per $1/50$ m^2 in unpolluted, eutrophic Lake Esrom from the muddy profundal zone at a depth of 20 m. The drawings are life-size. For simplicity, specimens of the same species are depicted as having the same size. The figure represents average populations over a period of 2 years. The populations are coincident. In the list that follows, the number of individuals per $1/50$ m^2 is given in parentheses. A, *Ilyodrilus hammoniensis* (110). B, *Ilyodrilus hammoniensis* cocoons (15). C, *Corethra flavicans* larvae (35). D, *Chironomus anthracinus* larvae (179). E, *Procladius pectinatus* larvae (6). F, *Pisidium casertanum* (65). Total number of individuals per $1/50$ m^2: 410. Total per m^2: 20,441.

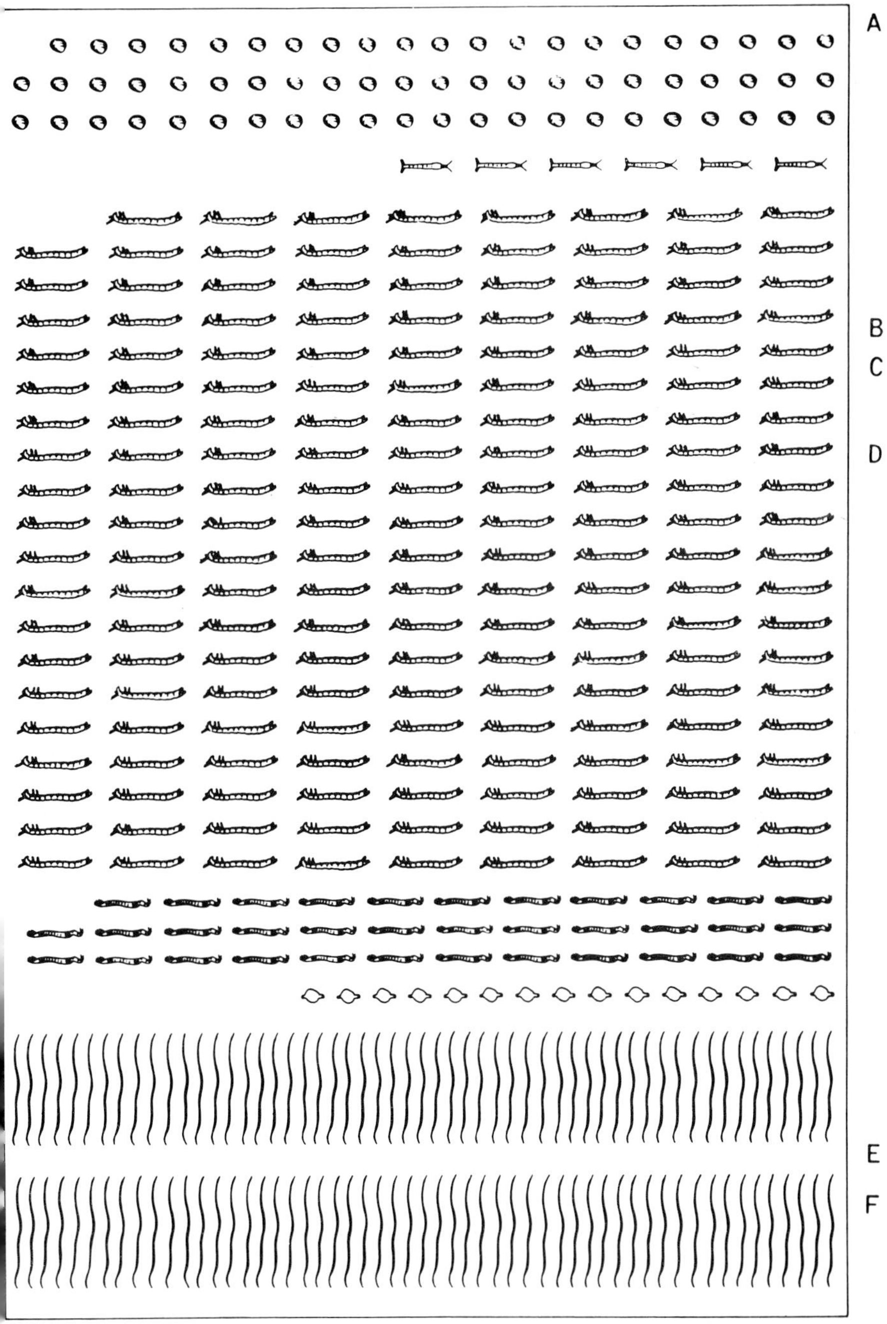
A
B
C
D
E
F

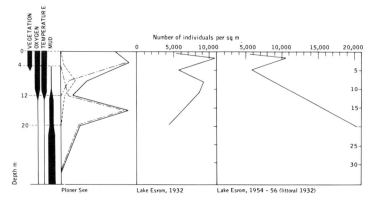

FIGURE 7   Benthic distribution of bottom fauna in typical eutrophic lakes in relation to depth distribution of vegetation, thickness of mud layers, temperature, and oxygen content of lake water. The solid line shows the total bottom fauna. The broken lines show littoral, sublittoral, and profundal animals, in the order named. (Redrawn and modified from Lundbeck, 1926; Berg, 1938; and Jónasson, 1961.)

tions of animals in a transect of a lake and differences between lakes of various degrees of eutrophication. Any competition between these groups or species of bottom fauna seems confined mainly to oxygen requirements. Figure 8 illustrates the respiration of the littoral fauna (A) and profundal fauna (B) in relation to decreasing oxygen content of the lake water (Berg and Ockelmann, 1959; Berg et al., 1962; Berg and Jónasson, 1965). The littoral fauna, here presented by snails, shows two kinds of adaptation to the environment.

1. Oxygen consumption decreases when the oxygen declines (*Lymnaea palustris* and *Bithynia tentaculata*).

2. A critical point and a corresponding drop in oxygen consumption are found when oxygen supply decreases in the following species.

Species	Oxygen Content of Lake Water (%)
Bithynia leachi	14–15
Lymnaea auricularia	11
Valvata piscinalis	9–10

The limits at which oxygen consumption becomes critical for these species range between 10 and 15 percent oxygen and apparently coincide with the oxygen content of lake water at night.

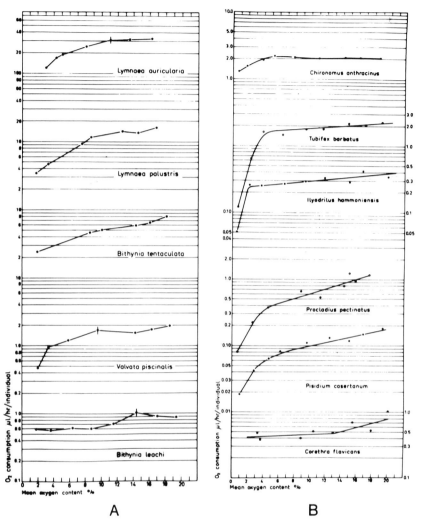

A   **B**

FIGURE 8   *A*, Oxygen consumption in relation to oxygen content of the water for five littoral species. The ordinate shows the consumption in microliters per hour per standard individual. The abscissa shows the oxygen percentage of a gas mixture in equilibrium with the water. Vertical lines crossing the curves indicate the approximate critical points of oxygen supply. *B*, The same data for five profundal and one sublittoral species. Note that critical points are much lower.

Figure 8*B* shows the profundal animals adapted to an environment that is characterized by very low oxygen tensions. The three best-adapted species are shown on top: *Chironomus anthracinus, Tubifex barbatus,* and *Ilyodrilus* *hammoniensis.* Their critical point in oxygen consumption is reached when the oxygen content of lake water is between 2 and 5 percent.

Species	Oxygen Content of Lake Water (%)
Chironomus anthracinus	5
Tubifex barbatus	4
Hyodrilus hammoniensis	2

The best-adapted species, *C. anthracinus,* respires at 75 percent of normal respiration even when the oxygen content is as low as 1 percent. Even very small differences in critical limits have a big ecological influence. *T. barbatus* has a limit of 4 percent and is found only in the sublittoral zone of Lake Esrom. *I. hammoniensis* has a limit of 2 percent and is found only in the profundal zone of Lake Esrom.

These two species thus divide the deeper part of the lake between them according to their tolerance of oxygen. They and *C. anthracinus* are able to utilize oxygen at very low tensions because their blood contains hemoglobin.

The adaptation of *Corethra (Chaoborus) flavicans* is quite different from that of the five other species. The decrease in oxygen consumption is steepest at 20 to 13 percent oxygen saturation, that is, at the highest oxygen concentration of lake water; further decrease does not affect it, since the species is able to maintain its respiration at this lower level. This might be the ecological explanation of its existence, because it migrates between the bottom, where it lives during the day, and the pelagic water layers, where it spends the night hours.

Pisidium casertanum and *Procladius pectinatus* show a trend in oxygen consumption that is similar to that shown by *Lymnaea auricularia,* which lives in the littoral zone. This trend agrees with the ecology of *P. pectinatus* larvae, which occur only in the profundal zone during autumn, winter, and spring, when the bottom layers are rich in oxygen. A physiological explanation of the presence of *Pisidium,* which lives permanently in this biotope, has not been found.

Figure 8 shows that the animals have various respiratory adaptations to their environment. One of the most effective adaptations is possession of hemoglobin in the blood; much discussion has been devoted to the function of hemoglobin. In *Chironomus* the main significance seems to have three aspects:

1. Hemoglobin acts in oxygen transport at very low oxygen concentrations, thereby making possible continued respiratory irrigation.

2. It enables the larva to maintain the active process of filter-feeding when relatively little oxygen is present.

3. It greatly increases the rate of recovery from periods of oxygen lack, making such recovery possible even under adverse respiratory conditions (Walshe, 1950).

However, one cannot assume that the respiratory responses of the bottom fauna indicate a direct sensitivity to declining oxygen pressure. The low oxygen content in lakes is accompanied by a drop in pH owing to the production of carbon dioxide and other metabolic products. Thus, when the animal pauses between irrigation periods in its tube, it may be affected by both factors. Experiments made by Walshe (1950) on *C. plumosus* indicate that irrigation is not initiated by oxygen deficit alone. The behavior of larvae at a high carbon dioxide pressure causes lengthening of the irrigation period. Walshe states:

The animal has normally a permanent slight oxygen debt, above which additional accumulation of metabolites initiates irrigation. Elimination of these must become increasingly difficult at progressively lower oxygen concentrations until, finally, with poor incoming oxygen supplies the animal must be fighting a losing battle against anaerobic products.

The decreased rate of irrigation and final immobility below 3 percent air saturation may be the result of their excessive accumulation. Brundin (1949, 1951) has shown by his extensive studies of bottom fauna in Swedish lakes

... that the order in which the larvae are grouped in the table represents a series with decreasing body size and that this order directly corresponds to the proved ability of the larvae to manage under unfavourable conditions. (See Table 2.)

TABLE 2   Relation between Body Size and Surface Area in Various Chironomids[a]

	Length (mm)	Width (mm)	Body Surface (mm^2)
Chironomus plumosus	28.0	2.0	176
Chironomus anthracinus	18.0	1.4	77
Sergentia coracina	14.0	0.8	35
Stictochironomus Rosenschöldi	11.5	0.7	27
Tanytarsus sp. (medium size)	7.0	0.5	11

[a]Modified from Brundin (1951).

Undoubtedly the ability of the larvae to break through a microstratification of the bottom layers with their irrigation currents increases with increasing size. The smallest larvae are therefore found in the most oligotrophic lakes, and the larval size increases with increasing eutrophication.

UTILIZATION OF FOOD AS A FUNCTION OF LAKE RHYTHM

The previous section has shown the occurrence of various adaptations to the environment and to habitats with increasing eutrophication. The adapted species must be able to grow and breed in the habitat.

Two species of the genus *Chironomus* dominate the profundal mud-living fauna of eutrophic lakes of our latitudes: *C. anthracinus* and *C. plumosus* (Thienemann, 1922). *C. anthracinus* might be taken as a model to illustrate the growth and food utilization of a species under increasing eutrophication, because it has shown an ecological relationship between food (i.e., primary production by the standing crop of phytoplankton) and the production of organic matter by *C. anthracinus* (Jónasson, 1964, 1965). Both species are filter feeders or deposit feeders, and their food consists of all particles that can be brought down into the tube by the irrigation currents of the larvae or by a collection around the mouth of the tube (Figure 9). Systematic experiments on the food uptake of C. *anthracinus* under well-defined experimental conditions are desirable. Experiments have been made at our laboratory on the food uptake of the closely related species *C. plumosus*. The *C. plumosus* larvae are filter feeders. Walshe (1947, Figure 2) studied the feeding mechanism of the larvae in specially built tubes. The larvae live in U-shaped tubes through which an intermittent anteroposterior irrigation current is maintained by dorsoventral undulations of the body. Three types of feeding were observed:

1. Inside the tube the larvae make a shallow conical net that collects detritus and plankton.
2. The larvae scrape the walls of their tubes, which, being lined with salivary secretion, also trap particles.
3. Occasionally they feed from the mud surface. In doing so, they extend the body out of the tube entrance, retaining contact with it by means of the posterior prolegs, and spread a net of salivary secretion over the mud, which is then dragged down with its attached particles for consumption in the seclusion of the tube. This last type of feeding seems to be the main feeding habit of *C. anthracinus*, shown in Figure 9B.

With this background, we now turn to the growth of C. *anthracinus*. Figure 10A shows the variation in wet and dry weight of C. *anthracinus* through a

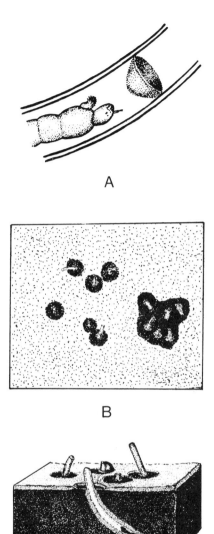

A

B

C

FIGURE 9   Feeding behavior of *C. plumosus* (*A*) and *C. antracinus* (*B* and *C*).

2-year period in relation to temperature, oxygen (summer stagnation period and autumn overturn), and primary production of the phytoplankton (spring and summer maximum and autumn decline). (See Figure 10*B*.) No measurements of primary production were made during 1960, and the curve for June to December of that year is an average curve for 8 years. The curves for 1961 and 1962 are actual.

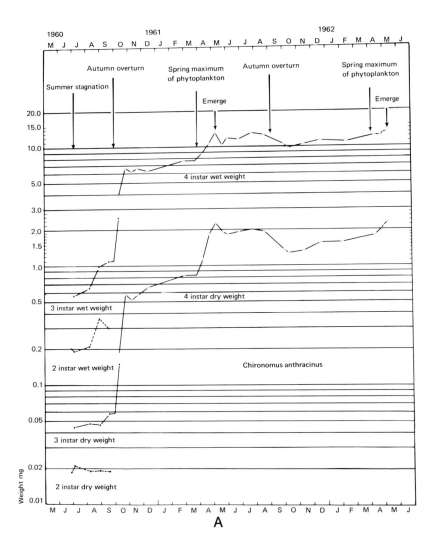

FIGURE 10   A, Variation in wet and dry weight in the larvae of C. antracinus through a 2-year period in relation to temperature and oxygen (summer stagnation period and autumn overturn) and primary production of the phytoplankton (i.e., spring and summer maximum and autumn decline). C. anthracinus has four larval instars. The weights of the second, third, and fourth instars are shown in the figure. Many larvae have a 2-year life cycle, but part of the population has a 1-year life cycle (cf. Jónasson, 1961, Figure 6). Growth takes place in limited periods, mainly just after autumn overturn of the first year and during the spring maximum. The diagram shows average weight per larva, obtained by weighing 8,000 larvae. B, Seasonal changes in factors controlling growth of C. anthracinus:

   a.   Temperature of the bottom layers shows low values during winter, rapid increase

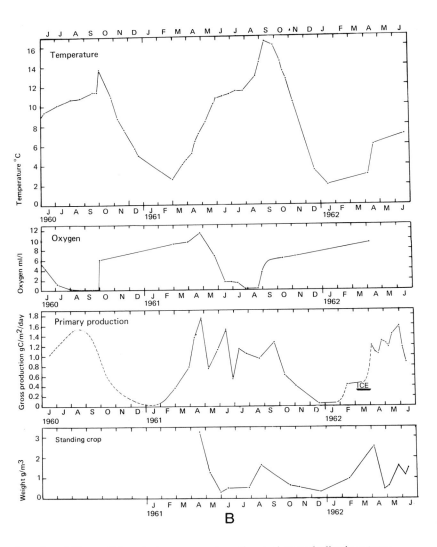

B

in spring, stable conditions during summer stagnation, and steep decline in autumn.

b. Oxygen at the bottom shows decreasing values to zero during summer stagnation. Other periods show sufficient amount even below ice.

c. Primary production of phytoplankton is expressed in grams of carbon per square meter per day. The numbers show actual measurements except for June to December, 1960; average values for 8 years are used for this period (broken line). A spring (April) peak and a summer (August) peak are easily recognized. Decline is rapid during autumn, and values are low in winter.

d. Standing crop of phytoplankton, expressed in grams per cubic meter, shows the same general trend as primary production: peak values in spring (April) and summer (August).

C. anthracinus has four larval instars. The individual weights of the second, third, and fourth instar larvae are shown in Figure 10*B* on a logarithmic scale. As already mentioned, many larvae have a 2-year life cycle (23 months), but part of the population has a 1-year life cycle (11 months). The diagram is based on weights of 8,000 larvae.

The figure shows clearly that larval growth is restricted to two short periods of the year: (1) in autumn, immediately after an overturn and (2) in spring, during the spring maximum of phytoplankton.

The eggs hatch in May and June. Figure 10*A* shows an increase in wet weight of the second and third instars during the summer stagnation period. However, the dry weights remain constant in both instars at this time. Increase in water content, which takes place during the last part of the period, simply reflects starvation; that is, in spite of the relatively high rate of respiration at 1 percent oxygen, shown in Figure 8, the larvae are not able to utilize the summer maximum of phytoplankton as food. This inability is not due to the inability of the larvae to ingest or digest the blue-green algae, which cause the summer peak, because they thrive on this diet in shallower water.

After the autumn overturn, the oxygen content increases, and the bottom temperature shows peak values. These conditions are the most favorable for the larvae. Between October 1 and 10, the second and third instar larvae moult; 80 percent of the population now consists of fourth instar larvae.

Wet weight increases are 255 and 750 percent in second and third instars, respectively.

Dry weight increases are 210 and 605 percent in the second and third instars, respectively.

Between October 10 and 25, the fourth instar larvae increase their wet weight by 75 to 160 percent and their dry weight by 220 to 285 percent. This means an average total increase within 4 weeks of 900 percent in both dry and wet weight. However, the increase is most remarkable in the third instar immediately after the overturn. On September 26 the average dry weight of the second instar is 0.0189 mg per individual, whereas on October 25 the average dry weight of the fourth instar is 0.584 mg. The maximal increase is thus about 3,000 percent.

After this outburst, growth stops suddenly in a period of optimal temperature and oxygen conditions. The reason seems to be lack of food, caused by the rapid decline in October of phytoplankton production and standing crop, as shown in Figure 10*B*. This situation is maintained during winter, but by March 1 a weight increase due to the early phytoplankton production in an ice-free lake can be demonstrated. Final growth takes place in April and early May, simultaneously with the spring maximum of phytoplankton. It seems reasonable to emphasize the parallel course of the primary production and weight curves.

During the first half of May, one group of larvae increases its average weight to 13.14 mg and emerges, for at this time the individuals have accumulated enough organic matter. The average weight of larvae not emerging is 10.36 mg. These larvae are left behind because the spring maximum has passed, and they are unable to accumulate organic matter rapidly enough. During June and July, 1961, they increased their weight to 13.10 mg but lost weight after the autumn overturn. At the same time, their water content increased.

The larvae thus have both a 1- and a 2-year life cycle, the length of the cycle depending on the effect of external factors on their life history and population dynamics (Jónasson, 1961). The most important external factors are the influence of increasing eutrophication on the food and feeding of larvae and on lake rhythm.

The *Oligochaeta* feed in a quite different way. Sorokin (1966) showed that *Limnodrilus hoffmeisteri* feeds on bacterial food in a layer 2 to 3 cm below the mud surface. This information coincides with observations on growth of *Ilyodrilus hammoniensis* in Lake Esrom, which is completely different from the growth of surface feeders (Jónasson, unpublished data).

DISCUSSION

We assume that phytoplankton is the primary source of food for bottom fauna, very little being known about the bacteria of muds. The condition of planktonic algae when they reach the bottom as food for bottom invertebrates is very important and is closely linked to the physical and chemical cycle of the lake. It appears that the bottom invertebrates receive fresh food during spring and autumn and that the food partly decays during summer. The food supply is very sparse during autumn and winter.

Larvae eat most of the spring peak of phytoplankton. That they do so is shown by their growth, population size, oxygen uptake, and the amount of primary production. This statement is based on the following data. During the spring maximum of phytoplankton in April 1961, production was about 1.5 g of C/m^2 per day, which is equivalent to 3.75 g of organic matter/m^2 per day. At this time Bacillariophyceae are the dominant algae. Table 3 shows the chemical composition of this type of plankton. We notice that the fat content is very high, one of the highest in the literature, but it is similar to that found by Lund (1965) for *Asterionella,* which is common in Lake Esrom in the spring. (See Jónasson and Kristiansen, 1967, Table 3; see also Wesenberg-Lund, 1918, and Birge and Juday, 1922.)

Larvae were present on April 17, 1961, at a density of 5,000 per m^2, and their mean weight was 12.2 mg. Thus, the biomass was about 61 g/m^2, and from measurements of oxygen uptake (Berg *et al.,* 1962), the consumption by this population was estimated at 2.4 liters of O_2/m^2 per day.

TABLE 3   Composition of a Diatom Plankton from Lake Esrom[a]

Constituent	Fresh Phytoplankton (% of total)	Dry Matter, Ash Included (% of total)	Dry Matter, Ash Free (% of total)
Water	88.4	–	–
Crude fat	2.8	24.1	50.9
Crude protein	2.3	19.8	41.8
Other organic components (carbohydrates)	0.4	3.5	7.3
Ash	6.1	52.6	–

[a]From Jónasson and Kristiansen (1967); analysis made by A. Krogh.

Net production is estimated at 75 percent of the gross production. Available organic matter is thus 2.8 g/m^2 per day. Let us assume that the carbohydrate, fat, and protein are utilized in the ratio in which they are present and that the amounts of O_2 required for physiological combustion are as follows:

Component	O_2 Required (liters)
1 g of carbohydrate	0.83
1 g of fat	2.0
1 g of protein	1.0

With these assumptions, we find that the 2.8 g of plant material will require 4.17 liters of O_2 (cf. Nielsen, 1949). Thus, the oxygen consumed by the larvae in respiration is about 60 percent of the oxygen required for the combustion of the entire plankton production at the spring maximum. Moreover, the larvae will consume an additional amount of food that is not respired but is incorporated in the tissues in growth. Clearly, food requirements of the larvae for most of the year at least equal the phytoplankton production.

The intermediate period between the spring and summer peaks of phytoplankton is characterized by lower production. During this period, larvae that have not yet emerged (the 2-year generation) increase in weight; also the first and second instar larvae of a new (1-year) generation breed during this period.

However, the summer peak of phytoplankton (Figures 2 and 3) is not utilized by the larvae in the deep profundal zone. The summer stagnation period implies oxygen lack in the hypolimnion, and the phytoplankton therefore reaches the mud surface partly decayed and relatively unsuitable as food for the profundal organisms.

Experiments on the respiratory behavior of *C. plumosus* show alternating periods of filter feeding and complete immobility. In well-aerated water, about 50 percent of the time is occupied by respiratory irrigation, 35 percent by filter feeding, and the remainder by periods of rest. As the oxygen concentration in the water drops, progressively less time is occupied by filter feeding and immobility and more by respiratory irrigation. When air saturation of the water is below 30 percent, filter feeding is reduced; and when it is below 10 percent, the larvae no longer feed (Walshe, 1950, Figure 2). Further, a decrease in oxygen and pH causes a lengthening of the irrigation period in *C. plumosus* (Walshe, 1950, Figure 4). If this lengthening is related to the increasing CO_2 content during summer stagnation in the hypolimnion of Lake Esrom, which is indicated by increasing alkalinity and decreasing pH in Figure 1C (see also Jónasson and Mathiesen, 1959, Figure 2), it also suggests less time for feeding activity.

A comparison of these results obtained by laboratory experiments on *C. plumosus* with measurements of oxygen uptake and growth in *C. anthracinus* show the same thing from two approaches. The very low oxygen saturation of lake water during the summer stagnation period reduces or prohibits oxygen uptake and thus implies low metabolism and slow or negligible growth of *C. anthracinus* larvae. At the height of summer stagnation, the larvae have difficulty in surviving, and the question of food utilization hardly arises.

Owing to their ability to utilize oxygen at low tensions, the larvae increase in weight when the stratification begins to break down. The cessation of growth at 1 to 2 percent oxygen saturation agrees well with the experiments on *C. plumosus* by Walshe (1950), who showed that the larvae are immobile at 3 percent air saturation and do not feed below 10 percent. Fox (1945) and Leitch (1916) observed no oxyhemoglobin absorption bands when the external oxygen pressure fell to 13 and 2.9 mm, respectively. The studies by Weber (1965) on the hemoglobin of *C. plumosus* have confirmed this.

The termination of the summer stagnation period at the autumn overturn allows warmer lake water, rich in oxygen and fresh algal food, to reach the mud surface. The growth response in first-year larvae is immediate and very considerable.

During both autumn and winter, the low primary production is a direct function of the low intensity of the light. We have learned from the dry and wet weights of the larvae that growth depends on the limited amount of available food during this season and not, as usually explained, or temperature (Jónasson and Kristiansen, 1967, Tables 8 and 9). Keeping this in mind, it seems reasonable to suggest a lower limit of primary organic production of 0.2 to 0.3 g of C/m^2 per day at which larval growth stops because of food shortage.

The growth of *C. anthracinus* has also been related to the seasonal cycles and renewal rate (turnover) of phytoplankton (Jónasson and Kristiansen, 1967, Table 13; cf. Rodhe, 1958).

The two most important periods of growth of *C. anthracinus* thus coincide with the high renewal coefficient of planktonic algae during spring and autumn. The high renewal coefficient of algae and their even vertical distribution in a homothermous lake will provide the bottom invertebrates with fresh food that is immediately utilized. The vertical distribution of the standing crop of *Asterionella formosa* found in relation to certain chemical and physical conditions in Windermere (Lund *et al.*, 1963) strongly supports the results obtained in Lake Esrom, that is, a continuous vertical distribution in the unstratified lake and a discontinuous one in the stratified lake. The vertical counts by Nauwerck (1963) in Lake Erken may be considered also to support these results.

Living conditions at the mud surface in the profundal zone of Lake Esrom, however, also interfere with the utilization of the relatively high primary production of organic matter during the summer stagnation period, as described; similar conditions may occur in the ice-covered lake.

In summary, growth of *C. anthracinus* is limited to very short periods: during the spring maximum of phytoplankton, during the period of green algae in June, and, for first-year larvae, immediately following the autumn overturn. It is at these times that plankton algae reach the bottom as fresh food and, coinciding with a high oxygen content of bottom water, thus allow the growth of *C. anthracinus*. It therefore seems clear that events in the bottom fauna depend not only on the size of standing crop and annual gross production but also on having sufficient amounts of plankton algae reaching the bottom as fresh food when the oxygen content of the bottom water is high enough to enable the animals to utilize the algae.

It may well be that algae are in part consumed by bacteria and that *Chironomus* ingests the bacteria. Kuznetsov (1958, 1959) has shown that there is a large increase in the bacterial population in the thermocline after the phytoplankton maximum, and that in the mud of certain eutrophic Russian lakes the wet weight of bacteria may reach 1.5 kg/m^2, of which 1 kg may be alive at the surface. If the plankton production is first consumed by bacteria, the total amount of food available to *Chironomus* is correspondingly reduced, and it is almost certain that the available biomass of phytoplankton limits the growth of the *Chironomus*.

Dark fixation in the epilimnion of Lake Esrom is usually 1 percent of primary production, the same size that is mentioned for chemosynthesis in the epilimnion of lakes (Kuznetsov, 1959). This means for Lake Esrom approximately 0.01 g of C/m^2 per day during the maxima, which is much less than that at which growth of *Chironomus* stops during autumn (0.2 to 0.3 g of C/m^2 per day). The chemosynthesis of the hypolimnion must thus be 15 to 20 times greater than that of the epilimnion before *Chironomus* is able to increase its weight on bacterial food.

Jónasson (1965, Table 1) has further shown that *C. anthracinus* larvae accumulate fat in their bodies during the spring maximum of phytoplankton and before pupation. There seems to be little doubt that the larvae store fat as fuel for the flight of the imago. Fat has an advantage over glycogen as fuel for flight: higher energy content per volume. At other seasons, the biomass of *C. anthracinus* larvae also fluctuates synchronously with changes in environmental conditions (Figure 10*B* and Jónasson, 1965, Figure 6). It now seems clear that environment has a dominating influence on the growth and life cycle of the bottom fauna as described for *C. anthracinus*. This influence explains much of the seasonal rhythm of chironomids because they emerge only during or immediately after the spring and autumn overturn. In a shallow, rich lake with continuous circulation, development is much quicker—two generations per year as described for *C. plumosus* (Hilsenhoff, 1966).

The use of phytoplankton as food for the filter-feeding zooplankton is known (e.g., Berg and Nygaard, 1929; Krogh and Berg, 1931; Storch, 1924a and b, 1925). Accumulation of zooplankton in the same layers as the phytoplankton would be expected (Woltereck, 1908, 1913). However, an investigation of the vertical distribution of both phytoplankton and zooplankton in 11 deep lakes of the eastern Alps by Ruttner (1920, 1937) did not give clear support to this hypothesis, if average values are used. For selected species of phytoplankton and zooplankton, accumulations in the same layer have been observed, but only during the maxima (Ruttner, 1937, page 290).

Estimating the feeding of zooplankton in order to estimate how much food is left over for the bottom fauna is difficult. Nauwerck (1963) has shown that in Lake Erken about 50 percent of the phytoplankton is of limited importance or of no importance as direct nourishment for the zooplankton. Anyway, only the nannoplankton can be taken by the zooplankton. The zooplankton, certainly, to a great extent, feeds on bacteria (with standing crop of the same order as that of the phytoplankton, but with a much higher multiplication rate). G. W. Saunders, Jr., has reviewed these aspects, giving special attention to the relative importance of detritus and phytoplankton as food for zooplankton (see page 556, this volume).

Despite the "grazing" of the zooplankton, we can assume that a great part of the phytoplankton (at any rate, most of the net plankton) sinks and becomes available for the sedentary bottom fauna. The problems are more complicated for the profundal bottom invertebrates than for the zooplankton since the food is produced in the upper few meters near the surface and is consumed at the surface of the mud. The life of the bottom invertebrates is thus intricately related to both food and lake rhythm.

A simultaneous investigation on the zooplankton is missing. However, Berg and Nygaard (1929) investigated the periodicity of standing crops of both phytoplankton and zooplankton in the lake of Frederiksborg Castle, a shallow eutrophic lake 5 km from Lake Esrom. Owing to its small body of water and its sheltered position, the lake warms earlier than Lake Esrom. The peaks of phytoplankton and zooplankton thus may develop earlier than in Lake Esrom. Only one species, *Bosmina longirostris,* has a peak occurrence in May. *Cyclops strenuus* has peaks in both spring and summer, but even at a peak, the organisms are few. The most important zooplankton species *Daphnia cucullata, Diaphanosoma brachyurum, Chydorus sphaericus,* and *Diaptomus graciloides* all show only one peak in August and September. These facts, combined with the observations of Wesenberg-Lund (1904) on Lake Esrom, support the previous arguments that in the unstratified lake in spring the bottom invertebrates receive the bulk of freshly produced food since the standing crop of zooplankton is present at a very low level. The same is true in the unstratified lake after autumn overturn, because the zooplankton peaks are markedly reduced or obliterated.

On the other hand, despite the various theories on the food and feeding of zooplankton, it seems reasonable to suggest that a considerable part of the organic matter produced by the phytoplankton during the summer maximum is used by the zooplankton and bacteria.

In Lake Erken, nevertheless, Nauwerck (1963, Figures 51 and 52) has shown that zooplankton has both a spring (April) and a summer (July) peak of abundance. While the spring peak coincides with the maximum of standing crop and primary production of phytoplankton in April, the summer peak does not, since the peaks of zooplankton abundance and primary production of phytoplankton occur in July and August, respectively.

CONCLUSIONS

The bottom fauna fits into an ecological pattern set by primary production of algae, submerged macrophytes, and physical and chemical factors of a lake. The physiochemical factors affect the size and seasonal trend of primary production. In turn, the primary production determines the range of various physical and chemical factors at the bottom.

Phytoplankton production, vertical distribution of macrophytes, and abundance of macrophytes are important determinants of oxygen regime, alkalinity, pH, and other factors and are especially important in relation to vertical and temporal variations of these factors.

The bottom fauna is correspondingly affected in quality, in quantity, and in depth distribution. Above the thermocline is a rich fauna with high oxygen demands; below the thermocline are a few specialists tolerating the low oxygen tension.

These qualitative contrasts are paralleled by quantitative ones. Two peaks of abundance occur with depth, one in the littoral zone (epilimnion) and another in the profundal zone (hypolimnion). Increasing the supply of nutrients to the epilimnion causes increases in the standing crop and in production of phytoplankton. Thereupon, transparency decreases; the subsurface light dwindles. The macrophytes climb the shelf to compensate for decreased illumination. In the end, the lake changes into one devoid of submerged vegetation but dominated by species of Nymphoids and showing only the lower peak of animal biomass.

Increasing the supply of nutrients to the hypolimnion causes a rapidly developing oxygen deficit; periods with lack of oxygen become more prolonged; pH decreases and alkalinity increases. This trend in environmental factors lowers the respiratory activity, reduces the periods of growth, and has other adverse effects on the bottom fauna and its survival. One such effect is to increase the life cycle from 1 year to 2 years.

During the final stage of eutrophication (and pollution), the number of sessile mud-dwellers may drop to almost zero or become restricted to *Oligochaeta.*

REFERENCES

Berg, K. 1938. Studies on the bottom animals of Esrom Lake. K. Danske Vidensk. Selsk. Skr. Nat. Math. Afd. 9 Rk. 8:1–225.

Berg, K., and G. Nygaard. 1929. Studies of the plankton in the lake of Fredriksborg Castle. K. Danske Vidensk. Selsk. Skr. Nat. Math. Afd. 9 Rk. 1:227–314.

Berg, K., and K. W. Ockelmann. 1959. The respiration of freshwater snails. J. Exp. Biol. 36:690–708.

Berg, K., P. M. Jónasson, and K. W. Ockelmann. 1962. The respiration of some animals from the profundal zone of a lake. Hydrobiologia 19:1–40.

Berg, K., and P. M. Jónasson. 1965. Oxygen consumption of profundal lake animals at low oxygen content of the water. Hydrobiologia 26:131–144.

Birge, E. A., and C. Juday. 1922. The inland lakes of Wisconsin. The plankton. I. Its quantity and chemical composition. Wisconsin Geol. and Nat. Hist. Survey. Bull. 64. Sci. Ser. 13:1–223.

Brundin, L. 1949. Chironomiden und andere Bodentiere der südschwedischen Urgebirgsseen. Rep. Freshw. Res. Drottningholm 30:1–914.

Brundin, L. 1951. The relation of O_2-microstratification at the mud surface to the ecology of the profundal bottom fauna. Rep. Freshw. Res. Drottningholm 32:32–43.

Christensen, T., and K. Andersen. 1958. De større vandplanter i Furesø, p. 114–128 *In* K. Berg *et al.* Furesøundersøgelser 1950–54 (Investigations on Fure Lake 1950–54. Limnological studies on cultural influences). Folia Limnol. Scand. 10.

Fox. H. M. 1945. The oxygen affinities of certain invertebrate haemoglobins. J. Exp. Biol. 21:161–165.

Hilsenhoff, W. L. 1956. The biology of *Chironomus plumosus* in Lake Winnebago, Wisconsin. Ann. Entomol. Soc. Amer. 59:465–473.

Jónasson, P. M. 1955. The efficiency of sieving techniques for sampling freshwater bottom fauna. Oikos 6:183–207.

Jónasson, P. M. 1961. Population dynamics in *Chironomus anthracinus* Zett. in the profundal zone of Lake Esrom. Verh. Int. Ver. Limnol. 14:196–203.

Jónasson, P. M. 1964. The relationship between primary production and production of profundal bottom invertebrates in a Danish eutrophic lake. Verh. Int. Ver. Limnol. 15:471–479.

Jónasson, P. M. 1965. Factors determining population size of *Chironomus anthracinus* in Lake Esrom. Mitt. Int. Verein. Limnol. 13:139–162.

Jónasson, P. M., and H. Mathiesen. 1959. Measurements of primary production in two Danish eutrophic lakes, Esrom sø and Furesø. Oikos 10:137–167.

Jónasson, P. M., and J. Kristiansen. 1967. Primary and secondary production in Lake Esrom. Growth of *Chironomus anthracinus* in relation to seasonal cycles of phytoplankton and dissolved oxygen. Int. Revue Ges. Hydrobiol. 52:163–217.

Krogh, A., and K. Berg. 1931. Über die chemische Zusammensetzung des Phytoplanktons aus dem Frederiksborg-Schlosssee und ihre Bedeutung für die Maxima der Cladoceren. Int. Revue Ges. Hydrobiol. 25:204–219.

Kuznetsov, S. I. 1958. A study of the size of bacterial populations and of organic matter formation due to photo- and chemosynthesis in water bodies of different types. Verh. Int. Verein. Limnol. 13:156–169.

Kuznetsov, S. I. 1959. De Rolle der Mikroorganismen im Stoffkreislauf der Seen. Berlin. VEB. Deutsche Verlag der Wissenschaften. 301 pp.

Leitch, I. 1916. The function of haemoglobin in invertebrates with special references to Planorbis and Chironomus larvae. J. Physiol. 50:370–379.

Likens, G. E., and A. D. Hasler. 1962. Movements of radiosodium (Na^{24}) within an ice-covered lake. Limnol. Oceanogr. 7:48–56.

Likens, G. E., and R. A. Ragotskie. 1965. Vertical water motions in a small ice-covered lake. J. Geophys. Res. 70:2333–2343.

Lund, J. W. G. 1965. The ecology of freshwater phytoplankton. Biol. Rev. 40:231–293.

Lund, J. W. G., F. J. H. Mackereth, and C. H. Mortimer. 1963. Changes in depth and time of certain chemical and physical conditions and of the standing crop of *Asterionella formosa* Hass. in the North Basin of Windermere in 1947. Roy. Soc. (London), Phil. Trans. Ser. B. No. 731. 246:255–290.

Lundbeck, J. 1926. Die Bodentierwelt norddeutscher Seen. Arch. Hydrobiol. Suppl. 7:1–473.

Mathiesen, H. 1962. Measurements of the production of the organic matter by the phytoplankton. p. 128–131 *In* P. Johnsen, H. Mathiesen, and U. Røen. Sorø-søerne, Lyngby sø og Bagsværd sø (The Sorø-lakes, Lake Lyngby sø and Lake Bagsværd sø. Limnological studies on five culturally influenced lakes in Sjælland (Zealand)). Dansk Ingeniørforening. Spildevandskomiteen. Skrift nr. 14.

Mathiesen, H. 1966. Om planteplanktonets bruttoproduktion og bundvegetationens forekomst i nogle danske søer (Midtjylland). Vattenhygien 22:102–104.

Nauwerck, A. 1963. Die Beziehungen zwischen Zooplankton und Phytoplankton im See Erken. Symb. Bot. Upsal. 17:1–163.

Nielsen, C. O. 1949. Studies on the soil microfauna. II. The soil inhabiting nematodes. Nat. Jutlandica 2:1–131.

Nygaard, G. 1949. Hydrobiological studies on some Danish lakes and ponds. II. K. Danske Vidensk. Selsk. Biol. Skr. 7:1–293.

Nygaard, G. 1958. Furesøens planteplankton, p. 109–114 *In* K. Berg *et al.* Furesøundersøgelser 1950–54. (Investigations on Fure Lake 1950–54. Limnological studies on cultural influences). Folia Limnol. Scand. 10.

Rodhe, W. 1958. Primärproduktion und Seetypen. Verh. Int. Ver. Limnol. 13:121–141.

Ruttner, F. 1930. Das Plankton des Lunzer Untersees. Int. Revue Ges. Hydrobiol. 23:1–138, 161–287.

Ruttner, F. 1937. Limnologische Studien an einigen Seen der Ostalpen. Arch. Hydrobiol. 32:167–319.

Sorokin, Ju. I. 1966. Carbon-14 method in the study of the nutrition of aquatic animals. Int. Revue Ges. Hydrobiol. 51:209–224.

Storch, O. 1924a. Morphologie und Physiologie des Fangapparates der Daphniden. Ergebn. Fortschr. Zool. 6:125–234.

Storch, O. 1924b. Der Phyllopoden-Fangapparat I. Int. Revue Ges. Hydrobiol. 12:369–391.

Storch, O. 1925. Der Phyllopoden-Fangapparat II. Int. Revue Ges. Hydrobiol. 13:78–93.

Thienemann, A. 1922. Die beiden *Chironomusarten* der Tiefenfauna norddeutscher Seen. Arch. Hydrobiol. 13:609–642.

Walshe, B. M. 1947. Feeding mechanism of *Chironomus* larvae. Nature 160:474–476.

Walshe, B. M. 1950. The function of haemoglobin in *Chironomus plumosus* under natural conditions. J. Exp. Biol. 27:73–95.

Weber, R. E. 1965. On the haemoglobin and respiration of *Chironomus* larvae with special references to *Chironomus plumosus plumosus* L. Drukkerij Pasmans. S – Gravenhage. 91 p.

Wesenberg-Lund, C. 1904. Studier over de danske søers plankton. København. 223 pp.

Wesenberg-Lund, C. 1918. Om planktonolien og dens eventuelle betydning i fedtstoffattige tider. Nat. Verden 2:9–19.

Woltereck, R. 1908. Die natürliche Nahrung pelagischer Cladoceren und die Rolle des "Zentrifugenplanktons" im Süsswasser. Int. Revue Ges. Hydrobiol. 1:871–874.

Woltereck, R. 1913. Über Funktion, Herkunft und Entstehungsursachen der sogenannten "Schwebe-Fortsätze" pelagischer Cladoceren. Zoologica 67:473–547.

J. W. G. LUND

Freshwater Biological Association, Ambleside, England

Phytoplankton

Since this paper was written before the International Symposium on Eutrophication was held, some of the views expressed in it may need modification in relation to what was said at Madison. Other articles in this volume describe what is known about eutrophication as a whole and what may be done to control it. Nevertheless, in considering phytoplankton, the practical aims of the symposium must be paramount. If it is agreed that eutrophication means enrichment, in this case by planktonic algae, then consideration can be given to what limits the growth of phytoplankton. However, eutrophication here refers mainly to excessive or undesirable enrichment. Therefore, a discussion of what limits growth in waters whose algal production is small, or of what causes moderate increases in production, may seem to be of little importance. Nevertheless, it is from such conditions that eutrophication can start, and there is no general agreement on what is meant by undesirable or excessive enrichment. What is desirable in a commercial fishpond can be undesirable in waterworks practice, and what is undesirable in waterworks practice may be acceptable in a recreational area.

It is necessary to make a distinction between what is produced and the rate at which it is produced. In general it is what is present in a given period that is objectionable. It is the sizes of the populations in reservoirs and recreational areas that are of prime importance.

The rate at which populations are produced affects the length of time during which excessive growths are present and the rate at which the decline of one population may be followed by the rise of another. The rate of production also gives insight into the factors causing these growths. If the causes of massive eutrophication are known, then the probable success of methods of water treatment can be assessed and their cost can be evaluated.

The emphasis here is placed on what is not understood. There is an immediate need for careful evaluation of what is known, for research on what can be done on the basis of our present imperfect understanding of the ecology of the phytoplankton, and for more fundamental research on the many unsolved problems. No attempt is made to review present information because several accounts have been published recently (Jackson, 1964; Fogg, 1965; Goldman, 1965; Lund, 1965; Hutchinson, 1967).

Table 1 compares the largest observed populations of diatoms and the maximal and minimal concentration of phosphate phosphorus, nitrate nitrogen, and silica in seven English waters. Diatoms are used as an example because it is easier to illustrate their relationship with increasing eutrophication than other groups. The forms of silicon that diatoms utilize are estimated

TABLE 1 Diatom Populations in Waters Rich or Poor in Nutrients

Body of Water and Location	Year	Nutrient[a]						Diatoms (Max)[b]
		SiO_2		$NO_3.N$		$PO_4.P$		
		Max	Min	Max	Min	Max	Min	
Southeast England								
River Lee	1963	18	2	12	1	3	0.2	40
Queen Elizabeth II Reservoir	1966	11.5 ($>$15)*	5.0	6.8	4.7	1.33 ($>$2.25)*	0.89	10
King George VI Reservoir	1966	2.4 ($>$7)*	0.5	7.5	2.7	1.40 ($>$2.17)*	0.04	7
Midlands								
Crosemere	1966	5.5	1.4	2.3	0.08	0.08	0.006	11
Northwest England								
Blelham Tarn	1963	2.7	0.50	0.62	0.05	0.0081	$<$0.00010	4
	1964	2.6	0.02	0.60	0.01	0.0138	$<$0.00011	17
	1965	2.3	0.01	0.53	0.07	0.0091	$<$0.00010	18
	1966	2.5	0.01	0.60	0.11	0.0135	0.00026	14
Windermere (north basin)	1963	1.8	0.10	0.41	0.15	0.0042	$<$0.00010	5
	1964	1.9	0.02	0.42	0.11	0.0037	$<$0.00015	7
	1965	1.7	0.02	0.47	0.19	0.0041	$<$0.00015	4
	1966	1.8	0.13	0.43	0.15	0.0036	$<$0.00011	2
Buttermere	1964	1.9	1.6	0.16	0.05	0.00090	0.00011	$<$0.1
	1965	1.8	1.6	0.14	0.04	0.00160	$<$0.00015	$<$0.1
	1966	1.8	1.5	0.13	0.04	0.00026	$<$0.00011	$<$0.1

[a]Data are expressed in mg per liter.
[b]Millions of cells per liter.
*For Queen Elizabeth II and King George VI reservoirs, the period covered is from February to September. The nutrient concentrations (in parentheses) are midwinter values (J. E. Ridley, unpublished data).

by the standard methods used for water analysis, and in the vast majority of waters, abiogenic loss of silica is insignificant (Tessenow, 1966). Therefore, production can be assessed by estimating the decrease in the content of dissolved silica in the water. Care must be taken that the estimates are not invalidated by other causes of decrease, such as utilization of silica by diatoms in the inflows or, in the present case, by nonplanktonic forms. An experiment in which utilization by epiphytes occurred is described later.

The data for the River Lee, Queen Elizabeth II reservoir (hereinafter Q.E. II), King George VI reservoir (hereinafter K.G. VI), Crosemere, and Buttermere cover 1 to 3 years between 1963 and 1966. Data for Blelham Tarn and Windermere are available for all 4 years and are given for that period.

The maximal concentrations of nitrate nitrogen and phosphate phosphorus rise from low or very low values in the highly oligotrophic Buttermere to very high values in the eutrophic waters in southeast England (London region). In silica content, Buttermere and Windermere are the same, and Blelham Tarn is somewhat higher. Crosemere has about half the average silica content of the world's rivers (Livingstone, 1963) and the River Lee about 5 mg per liter more. It appears that the amounts of other nutrients (not necessarily nitrates and phosphates) in Buttermere are insufficient for significant growth of diatoms and consequent utilization of silica. In the other bodies of water, increasing amounts of phosphates, nitrates, and silica are coupled with large or very large populations of diatoms. However, the efficiency with which the silica is utilized does not match the increasing amounts of these three nutrients. Apart from 1963, Blelham Tarn's silica is almost completely utilized, and that of Windermere is depleted in nearly the same degree. The silica in Crosemere, Q.E. II, and the River Lee is relatively inefficiently utilized, for the minimum concentrations recorded are close to, or exceed, the maximum concentrations in Blelham Tarn and Windermere. Even though the maximum population in the River Lee is higher than in Blelham Tarn, the maximum populations of K.G. VI, Q.E. II, and Crosemere are not, and the question is: Why is this so?

EXAMPLES OF EUTROPHICATION

The recent history of the phytoplankton of Blelham Tarn is well documented but largely unpublished. It is a small lake (Macan, 1949) that has been mildly enriched since 1952–1953, when piped water became generally available for the inhabitants of a small village (Macan, 1962) and the farmers began to apply more fertilizer to the land. The village has 50 to 100 inhabitants, more in summer than in winter. The maximum concentration of phosphate phosphorus in the lake has increased, but the concentrations of nitrate nitrogen and silica have not (Figure 1, Table 2). In the same period (1954 and after), the minimum concentrations of silica have decreased in relation to the increased production by diatoms (Figure 2, Table 2). The mean maximum concentration of phosphate has increased threefold, and the mean minimum concentration of silica has decreased threefold. The dominant diatom at the time of the vernal maximum usually is *Asterionella formosa* Hass. Six times in the last 14 years, more than 10^7 cells per liter of this diatom have been

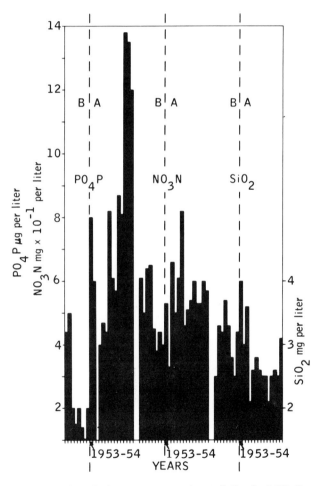

FIGURE 1   Maximum concentrations of dissolved PO_4P, NO_3N, and SiO_2 in Blelham Tarn between 1945 and 1967. B = before onset of eutrophication. A = after onset of eutrophication.

TABLE 2   Nutrients and Diatom Populations in Relation to Mild Eutrophication
(Summary of Data in Figures 1 and 2)

Years[a]	PO_4P Max[b]		NO_3 N Max[b]		SiO_2 Max[c]	
	Mean	Range	Mean	Range	Mean	Range
1945–1953	2.4	1.0–5.0	509	380–650	3.1	2.5–3.7
1954–1967	7.9	4.0–13.8	556	330–820	2.8	2.1–4.0

Years[a]	SiO_2 Min[c]		Time Max[d]		Diatom Population[e]	
	Mean	Range	Mean	Range	Mean	Range
1945–1953	0.58	0.24–1.10	182	147–225	3.5	1.1–7.6
1954–1967	0.20	0.01–0.64	140	110–190	9.5	1.3–19.0

[a]Data collected during the period October through March.
[b]Micrograms per liter.
[c]Milligrams per liter.
[d]Number of days after preceding December 31 to date of diatom maximum.
[e]Millions of cells per liter.

produced (maximum 1.8×10^7). In the 9 years before eutrophication started, the maximum number of cells did not exceed 5×10^6 per liter. The time of the vernal maximum has become earlier (Figure 2, Table 2). It seems that the diatoms now grow faster, though the number of cells at the winter minimum has to be taken into consideration. The more numerous the cells at the start of the vernal increase, the earlier the maximum will be reached if the population always grows at the same rate.

It is noteworthy that the increase in phosphate in the last 3 years (Figure 1) cannot be ascribed to sewage, because no appreciable increases in the population of the village or of summer visitors have taken place. It can be explained only by the increasing amount of fertilizers used by the farmers. It seems that this increase in phosphorus, so well known as a major factor in eutrophication, is responsible for the enhanced utilization of silica by the diatoms. Nevertheless, there must remain a doubt as to whether this is so or whether the effect of phosphorus is direct or indirect, because Mackereth (1953) has shown that 1 μg of P per liter can theoretically produce some 1.6×10^7 cells of Asterionella, and he obtained such a population in a culture. The maximum phosphate phosphorus in the years before eutrophication started varied between 1 and 5 μg per liter. On the other hand, the uptake of phosphorus by the diatoms depends on the concentration of phosphorus at the time the diatoms start their vernal increase and the proportion of phosphorus utilized by other algae. There are insufficient data to give exact answers to these questions.

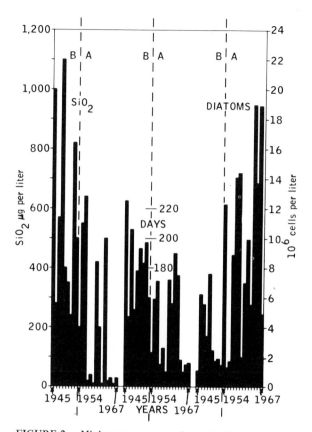

FIGURE 2 Minimum concentration of dissolved SiO_2, time of maximum diatom population expressed as days after the preceding December 31, and size of this maximum in Blelham Tarn between 1945 and 1967. B = before onset of eutrophication. A = after onset of eutrophication.

If phosphorus is the major factor determining the increase in the number of diatoms since 1954, then it should be possible to produce a similar degree of eutrophication in Buttermere by fertilizing it with phosphate. (Buttermere, too, is in the English Lake District and has basically the same chemistry as Blelham Tarn. Despite its more mountainous environment, the nature of the rocks is not fundamentally different; they are harder and less easily leached.) Several experiments to test this theory have been carried out, the investigator using untreated or treated Buttermere water and, for the purpose of biological assay, the diatom *Asterionella formosa,* which does not occur in Buttermere but usually is the dominant species in Blelham Tarn and the north basin of Windermere. The clones used came from one or the other of these two lakes. They were first grown in Windermere water, which had been boiled to kill all other algae and to which approximately 1 mg of extra silica per liter was added in order to obtain as large a population as possible.

In one such experiment (Figures 3 and 4), four polythene tubes, each 1 m wide and 15 m long, were filled with Buttermere water and suspended vertically in the lake. To one of these tubes 10 μg per liter of phosphate phosphorus was added, approximately 10 times the maximum concentration in Buttermere. The concentration of 1.6 μg per liter recorded once in 1965 is

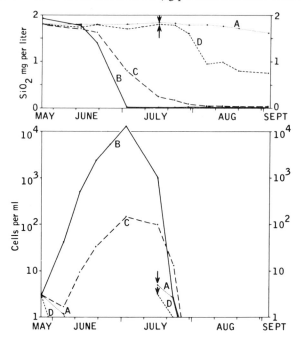

FIGURE 3   The effect of fertilizing water from Butter-mere, English Lake District. The water was enclosed in polythene tubes (about 11m³). A, tube to which no fertilizer was added until July 16 (see arrows), when 6.0 mg CaCO₃ and 20 mg EDTA per liter were added. B, tube to which 20 liters of soil extract was added. C, tube to which 10 μg per liter of PO₄P was added. D, tube to which 100 mμ per liter of vitamin B₁₂ was added and, on July 16 (see arrow), 10 μg per liter of PO₄P. Asterionella formosa was added to all four tubes at the start of the experiment and again to tubes A and D on July 16. Upper part of figure, changes in the concentrations of silica in tubes A–D; lower part of figure, changes in the number of Asterionella cells per milliliter in tubes A–D.

unique. Usually Buttermere's phosphate does not exceed 1 μg of P per liter; in 1966 the maximum concentration was 0.3 μg of P per liter (Table 1). The added phosphate led to a massive development of diatoms (Figure 4, line C) and a reduction in the silica content of the water to a level equivalent to that found in Blelham Tarn in recent years (Figure 2). In a tube not receiving any addition of nutrients (Figures 3 and 4, line A), no significant multiplication of diatoms or reduction in the dissolved silica occurred. *Asterionella* increased in the tube to which phosphate was added, but only to a degree equivalent to the uptake of about 0.2 mg of silica [10^6 cells contain approximately 140 μg

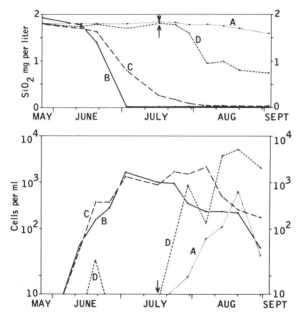

FIGURE 4   The effect of fertilizing water from Buttermere, English Lake District. The water was enclosed in polythene tubes (about 11 m³). *A*, tube to which no fertilizer was added until July 16 (see arrows), when 6.0 mg of $CaCO_3$ and 20 mg of EDTA per liter were added. *B*, tube to which 20 liters of soil extract was added. *C*, tube to which 10 μg per liter of PO_4P was added. *D*, tube to which 100 mμ per liter of vitamin B_{12} was added and, on July 16 (see arrow), 10 μg per liter of PO_4P. *Asterionella formosa* was added to all four tubes at the start of the experiment and again to tubes *A* and *D* on July 16. Upper part of figure, changes in concentrations of silica in tubes *A−D*; lower part of figure, changes in the total number of diatoms (i.e., *Asterionella* and indigenous epiphytic diatoms) expressed as cells per milliliter in tubes *A−D*.

of SiO$_2$ (Lund, 1965; Table 3)] out of a total uptake of about 1.8 mg of SiO$_2$. The major portion of the silica was utilized by endemic epiphytic diatoms, mainly species of *Achnanthes, Nitzschia,* and *Synedra.* To a third tube (Figure 3, line B) was added soil extract that contained phosphate and numerous other known or unknown substances. In this tube, *Asterionella* outgrew the attached diatoms. The utilization of silica was similar to that which took place when only phosphate was added, but the rate of utilization was faster (Figures 3 and 4). The fourth tube (Figures 3 and 4, line D) was fertilized with vitamin B$_{12}$ (100 mμ per liter) but no significant uptake of silica or growth of diatoms took place.

After the experiment had lasted 51 days, the tube that had not been fertilized (Figures 3 and 4, dotted line) received 6 mg of calcium carbonate and 24 mg of ethylenediaminetetraacetic acid (EDTA) per liter. A new inoculum of *Asterionella* was added also. *Asterionella* did not grow, but by the end of the experiment there had been a slight decrease in the silica concentration and some increase in attached diatoms. The possible reason for the change is discussed later. The tube to which vitamin B$_{12}$ had been added (Figures 3 and 4, line D) received 10 μg of PO$_4$P per liter on the 51st day (July 16) and a new inoculum of *Asterionella.* The added phosphorus did not promote the growth of *Asterionella,* the result being different from the one obtained at the beginning of the experiment when the same amount of phosphorus was added to the tube (Figure 3, line C). In similar experiments not described here, additions of 8–10 μg of PO$_4$P per liter either supported a slight increase of *Asterionella* or resulted in a decrease in the number of *Asterionella.* In the present experiment, the attached diatoms grew in this tube (line D), after the addition of 10 μg of PO$_4$P on the fifty-first day, as vigorously as they had done in the tube (line A) to which this amount of phosphorus was added on the first day of the experiment.

To summarize, if phosphate or soil extract is added, it is possible to produce increases in diatoms (and so decreases in the dissolved silica) in Buttermere that are as great as in the mildly polluted Blelham Tarn. This supports the view that the increase in phosphate that has taken place in Blelham Tarn since 1952–1953 is the cause of the increased growth of diatoms. However, in Buttermere, phosphate alone stimulates the growth of attached diatoms occurring naturally in the lake to a greater extent than the growth of *Asterionella;* sometimes the *Asterionella* does not grow at all. Soil extract causes the added *Asterionella* to grow faster than the attached diatoms (this result, too, has been confirmed in other experiments). The "absence" of *Asterionella* in Buttermere is caused by insufficient amounts of one or more substances present in soil extract. One of these substances could be phosphate, since its addition to Buttermere water sometimes leads to some growth of *Asterionella,* and the tubes containing soil extract generally show an increase in dissolved phosphate. The word "absence" in the sentence above was placed in quotation marks because there can scarcely be doubt that *Asterionella* is sometimes present in Buttermere, though in such small numbers that it has not been detected. This is because Buttermere is only separated from the next lake in a chain of three lakes, Crummock Water, by a narrow delta, which was formed in postglacial times. Moreover, *Asterionella* occurs in abundance in Loweswater, the third lake in the chain.

The many descriptions now available of the development of plankton in new reservoirs show that *Asterionella* and other algae will soon appear after the reservoirs are filled if the water is of the right kind to support their growth. Although we do not have much knowledge of how plankton algae are distributed, there is no doubt that they can be carried over land barriers. The history of the changes in Blelham Tarn and the results of the experiments in Buttermere support the general view that phosphate plays a key role in the early stages of eutrophication. However, there remains the question of why the addition of 8 to 10 μg of phosphate per liter to Buttermere water does not result in the massive growth of *Asterionella,* when it has been shown that 1 μg per liter is sufficient for some 1.6×10^7 cells (Mackereth, 1953).

Finally, it is necessary to consider the relatively small increase of attached diatoms produced by the addition of EDTA and calcium carbonate to the tube originally fertilized with vitamin B_{12} (Figure 4, line A, after July 16). It was later discovered that one of the bottles in the batch from which the EDTA was taken contained 30.1 μg of PO_4P per g of EDTA. As a result, the EDTA added could have increased the phosphorus content of the water in this tube by 0.72 μg per liter. As the acid was added in a liter of distilled water to the top of the tube, the main part of the attached diatom population, which was on the uppermost 2 to 3 m of the tube, was exposed for a time to high concentrations of phosphate. The total amount present as a contaminant in the EDTA added could have been 8 mg. Clearly, if such experiments are to be carried out in relation to substances such as phosphorus and trace elements, which can be utilized at very low concentrations, only chemicals of very high quality can be used.

Leaving aside several other matters of limnological interest arising from these observations and experiments (Tables 1 and 2, Figures 1–4), there are a number of points relating specifically to the theme of the symposium. It may seem unsatisfactory to place so much emphasis on diatoms when, in so many eutrophic waters, the most abundant algae are Cyanophyta (blue-green algae), which are also among the most troublesome. Nevertheless, the lessons to be learned from studies on diatoms are relevant to eutrophication. In reservoir practice they are usually the first algae that become troublesome when enrichment of the drainage area takes place. The diatom *Asterionella* is one of the worst pests of British reservoirs (Lund, 1955). The differences in the quantity and quality of the diatom populations are paralleled in other algal groups, notably Cyanophyta. Therefore, when considering how to study or combat eutrophication, the facts discussed here can be borne in mind. Other contributors discuss the blue-green algae from various points of view.

QUANTITATIVE AND QUALITATIVE
ASPECTS OF EUTROPHICATION

It is clear, as shown in Tables 1 and 2 and Figures 1–4, that a given degree of nutrient enrichment does not lead, without exception, to an equivalent increase in algae populations. Therefore, though analyses of water can give warning that troublesome growths of algae are probable, the exact nature and severity of these troubles cannot be predicted. There can be situations where increased nutrient inflow will have no significant effect because some other factor retards the growth of algae. High turbidity and color of the water are obvious examples, producing such absorption of incident radiation that photosynthesis is restricted to a shallow depth. Those concerned with commercial fishponds are well aware that there are interrelations between the quality and quantity of the algae present and the various kinds of animals, including fish. One of the reasons blue-green algae are so often the dominant algae in the plankton of eutrophic waters is their relative freedom from grazing, compared with smaller green algae that could flourish under similar conditions (Vinberg and Lyakhnovich, 1965; Lund, 1967). Although there is some uncertainty about the degree to which blue-green algae are grazed (e.g., Sorokin *et al.*, 1965; Smirnov and Feoktistova, 1965), the prolonged blooms often seen in carp ponds are evidence for this belief. Very large populations of small algae can be decimated by the zooplankton (see, for example, Pennington, 1941). There are, of course, other factors of vital importance controlling the development of water blooms. Apart from certain *Oscillatoria* spp., planktonic Cyanophyta nearly always form blooms in warm weather. Their ability to remain alive and actively photosynthesize on the surface in summer or in tropical climates (Talling, 1965) suggests that they are often little harmed by intense illumination and high temperatures.

The chemical factors permitting the growth and, sometimes, the prolonged maintenance of massive populations of Cyanophyta are not properly understood (Hutchinson, 1967). It is well known that lakes that commonly have water blooms do not always have them, nor do the blooms always arise in the expected period of the year. There is a need for observations over several years, together with as many physical, chemical, and other biological supporting data as can be produced.

The possibility should not be overlooked that such data exist but have not been examined from this point of view. In a different sphere, that of water movements, Mortimer (1963, 1965) discovered that a vast amount of valuable data existed in the records of certain water undertakings. The temperature measurements concerned had not been taken with any idea of their use for studying internal seiches. Records exist from the Great Lakes that have permitted Damann (1960) and Davis (1964, 1966) to record the lakes' increasing eutrophication.

Can such data give a clue as to why Cyanophyta are more common in one year than in another? Although water blooms commonly arise in waters rich in inorganic phosphorus and nitrogen, the concentrations of phosphates and, especially, of nitrates may be very low by the time the blooms arise. The probable reason is that the phosphorus and nitrogen have been incorporated into the cells of the algae. The evidence from work with pure cultures (Gerloff and Skoog, 1954, 1957 a and b; Staub, 1961; Fogg, 1965; Gorham, 1964) suggests that the species concerned grow as vigorously in purely inorganic media, apart from the need for chelating agents, as in media enriched with organic compounds. However, it does not follow from this that organic substances may not be important in nature at times when the inorganic salts have been reduced to very low concentrations.

If the inorganic salts of nitrogen and phosphorus are the main enriching agents in relation to the growth of planktonic Cyanophyta, then enriching a mixed plankton with such compounds should lead to large increases in the species present. Since Cyanophyta, apart from certain species of *Oscillatoria,* are most abundant in summer or in the warmer regions of the world, these experiments should not be carried out in cold, poorly illuminated water. Such experiments have been made on water from Windermere that was stored in 100-ml bottles that had either been kept under suitable light and temperature regimes in the laboratory or suspended at various depths in the lake. The Cyanophyta have not increased significantly and, in time, their place has been taken by other algae, notably small Chlorophyta. It may be that the results of such experiments are not reliable guides to the ecology of these algae. If lake water is enclosed in a bottle, with or without added nutrients, the planktonic flora often changes in a similar manner. Possibly a better insight into the changes in the natural populations could be obtained by using large plastic containers, such as those used in the experiments conducted on Buttermere water described earlier. It would be valuable to carry out such studies, perhaps with even larger containers, in relation to the problems of eutrophication.

It has been shown, in relation to data in Table 1, that increasing amounts of phosphate, nitrate, and silica in certain waters were not accompanied by proportionately larger populations of diatoms. As a result, the utilization of silica was not as efficient as in less-eutrophic waters, or even in oligotrophic waters, for example, Blelham Tarn and Windermere. The fact that the minimum silica concentrations in the River Lee, Q.E. II reservoir, and Crosemere are similar to the maximum contents in Blelham Tarn and Windermere seems to exclude this nutrient as a limiting factor in the three water bodies named first. The minimum values in the Lee, Q.E. II reservoir, K.G. VI reservoir, and Crosemere are of the order of 10 to 100 times the maximum values in Blelham Tarn and Windermere. The relatively inefficient use of silica in the richer waters is paralleled by even less efficient use of phosphorus. Nitrogen is also less efficiently utilized in the richer waters, with the exception of Crosemere. Moreover, the minimal values in the more eutrophic waters are the extremes for a year. During a given diatom bloom, all three nutrients may remain at even higher levels than the annual minima. For example, in the spring of 1967, 1.2×10^7 *Asterionella* cells per liter were present in the phytoplankton of Crosemere. The water was isothermal, and loss by outflow was small. Yet, at the end of the diatom bloom, the concentrations of silica, nitrate nitrogen, and phosphate phosphorus, respectively, were 3.5, 0.75, and 0.05 mg per liter. Since no other algae were present in large numbers, the subsequent decline of *Asterionella* cannot be related to the inhibitory effects of their extracellular products; nor could the decrease in this diatom be ascribed to fungal parasitism or grazing.

Other examples could be given illustrating that the factors controlling populations of phytoplankton are often difficult to identify. I do not mean to suggest that decreases in the amounts of nitrogen, phosphorus, or silicon are not known to control algal populations in eutrophic waters. Ohle (1965) and Tessnow (1966) have found, in several eutrophic north German lakes, that silica is reduced to concentrations as low as those found in Windermere and Blelham Tarn (e.g., 0.02 mg per liter). In these eutrophic lakes, unlike the English ones in Table 1, high concentrations of nitrogen, phosphorus, and silicon are followed by very large increases in the diatom populations and correspondingly low concentrations of silica. In summer, nitrate nitrogen can also be reduced to concentrations too small to estimate accurately (Ohle, 1965).

As far as diatoms are concerned, the observations on Blelham Tarn show that populations in excess of 1×10^7 cells per liter can be produced in a water containing less than 10 μg of phosphorus per liter. The maximum phosphorus content is usually found between November and February and in the period of minimum algal growth. On the basis of Mackereth's estimations (1953), the phosphorus, in most years, should be sufficient for populations of the order of 10^8 cells per liter. The reason Blelham Tarn and Windermere do not produce more diatoms is the lack of sufficient silica. Since diatoms such as *Asterionella* are likely to become troublesome in a waterworks when they exceed about 10^6 cells per liter, it may be difficult to reduce concentrations of phosphorus in the inflow of a eutrophic body to such an extent that diatom troubles are overcome.

The figures for Crosemere, K.G. VI, and Q.E. II reservoirs raise the possibility that waters less rich in nutrients than they are may be more favorable for the growth of diatoms, a matter of importance in relation to reservoir practice. One reason that the diatom populations are so large in Blelham Tarn is the earliness of the vernal maximum, notably in recent years (Figure 2), so that thermal stratification is absent. Then, virtually the total utilization of silica is present in suspension in the form of diatoms. In Windermere, the utilization of silica in the epilimnion may be as great as in Blelham Tarn, but the diatom populations are smaller (Table 1) because of the continuous loss of cells to the hypolimnion and bottom deposits. Then, the total utilization of silica is never present in suspension in the form of diatoms. This effect of density stratification of the water is considered further in relation to controlling eutrophication (p. 326).

Though some waters richer than Blelham Tarn in phosphorus, nitrogen, and silica do not produce more diatoms every year, they commonly have larger maxima of other algae. For example, in Crosemere, populations of flagellates (such as *Cryptomonas* and *Ceratium*) and of Cyanophyta and certain Chlorophyta exceed those in Blelham Tarn, and total production per year is considerbly higher in Crosemere. Many waters in southeast England, notably certain London reservoirs, produce, as might be expected, large blooms of blue-green algae. Therefore, the growth of diatoms may be inhibited by these or other algae. There is plentiful evidence that some algae can produce substances that inhibit the growth of other algae (Lund, 1965; Hutchinson, 1967). But the extent to which inhibition determines the succession of algae in nature is still uncertain.

It is also not known why increasing eutrophication leads to changes in the quality of the phytoplankton, although many examples are recorded. Once again, the changes in Blelham Tarn may be taken as an example. As has been described (Figure 1, Table 2), the only increase in nutrients that has been detected is the rise in the concentration of phosphate from an average of 2.4 μg before the start of eutrophication to 7.9 μg per liter of phosphorus after the start.

This history of the lake's eutrophication also illustrates how "resistant" a water can be to the introduction of new species before eutrophication starts. From 1945 to 1949, samples were taken by net on the same morning each week from Blelham Tarn and the nearby Esthwaite Water. The net was first used in Esthwaite Water, and after being washed with lake water, it was used in Blelham Tarn. As a result, contamination of Blelham Tarn with organisms from Esthwaite Water must have taken place. Tests showed that it was virtually impossible to remove all the algae entangled in the meshes of the wet net. However, no changes in Blelham Tarn's phytoplankton were observed that could have been caused by this contamination. Among the algae present in Esthwaite but never observed in Blelham Tarn were *Aphanizomenon flos-aquae* Ralfs ex Born. et Flah., *Cyclotella pseudostelligera* Hust., *Staurastrum chaetoceros* (Schroed.) Smith, *Fragilaria crotonensis* Kitton, and *Anabaena solitaria* Kleb., all of which are common in eutrophic lakes.

In 1949 it was decided that this contamination of Blelham Tarn should cease. Since then a different net has been used for each lake, and these nets have not been used for collecting phytoplankton from any other lakes. In 1954, 2 years after the beginning of eutrophication from sewage and 1 year after the beginning of increased fertilization of the land, *Aphanizomenon flos-aquae* was detected in Blelham Tarn. In 1960 *Cyclotella pseudostelligera*, in 1962 *Fragilaria crotonensis*, in 1964 *Staurastrum chaetoceros*, and in 1967 *Anabaena solitaria* appeared and had become sufficiently numerous to be recorded in counts of the algae in 1 ml of Blelham Water. In 1958 the maximum count of *Aphanizomenon* exceeded 500 filaments per ml; in 1961 the maximum count of *Cyclotella* exceeded 3,000 cells per ml; in 1966 the maximum count of *Anabaena* exceeded 10 filaments per ml; and in 1967 the maximum count of *Fragilaria* exceeded 800 cells per ml.

It is impossible to be certain that these algae were not in Blelham Tarn between 1945 and 1953, but it is probable that they were not present for at least the greater part of the period. Samples of phytoplankton collected by net were examined each week. Until September 1953, three people were studying the phytoplankton, and thereafter, two people. It is certain that all these algae were extremely rare, because large amounts of water are filtered in collecting by net for qualitative studies of phytoplankton. It is equally clear that mild eutrophication soon produced both qualitative and quantitative changes in the flora.

It is not known whether these qualitative changes were caused by the increase in phosphate phosphorus. If they were, it would be possible to carry out an experiment with the water of Buttermere, similar to those described earlier (pp. 311–315), that would cause species to grow in it that do not do so at present. It has already been seen that additions of phosphate and, especially, of soil extract can so alter the water in Buttermere that *Asterionella formosa* can grow in it. One such experiment has been carried out. Phytoplankton from Windermere containing at least 15 species not present in Buttermere, and including green, blue-green, and diatom algae, was added to about 11 m^3 of Buttermere water enclosed in a polythene cylinder. Phosphate phosphorus was also added to the Buttermere water at the rate of 8 μg per liter. This is the average concentration in Blelham Tarn since its eutrophication. None of the algae not present naturally in Buttermere increased significantly; the majority rapidly decreased. Too much emphasis should not be placed on a single experiment, but, once again, the result of this experiment raises the question of whether we have much understanding of the detailed effects of eutrophication.

It can be accepted that the main causes of the increased production of phytoplankton in eutrophic lakes is the enrichment with nitrogen and phosphorus. But the effect of self-shading by algae must also be allowed for when probable effects of reversing the eutrophication are considered (Talling, 1960 and 1965; the latter paper is discussed also in Lund, 1967). The River Lee (Table 1) may be taken as an example. The water commonly contains about 1 mg of phosphate phosphorus and 3 mg of nitrate nitrogen per liter (maxima 3 and 12 mg per liter, respectively). The river is about 2 m deep and, in view of the series of weirs and the river's flow, there seems to be little likelihood that the distribution of phytoplankton in depth is markedly uneven.

The major species during large algal maxima is *Stephanodiscus hantzschii* Grun., whose cells are kept in suspension by relatively little turbulence (E. M. F. Swale, unpublished data,* and personal observations). The other common algae are either small or as in the case of *Chlamydomonas,* they are motile. The maximum number of cells per liter during the period of Swale's (unpublished data* and 1964) survey was about 5×10^7, of which 4×10^7 were *Stephanodiscus hantzschii.* Swale's experiments show that 1 mg of phosphorus is sufficient for more than 2×10^8 cells, and 3 mg of nitrogen is sufficient for more than 10^9 cells of *Stephanodiscus.*

Therefore, even a reduction of phosphorus and nitrogen to a tenth of their concentrations could still permit very large populations to develop. E. M. F. Swale (unpublished data* and 1963) emphasizes that for most of the year, fluctuations in the concentrations of these nutrients cannot be the factors determining the numbers of algae. She lays emphasis on rates of flow and detrital turbidity as major factors limiting algal production. With 5×10^7 cells per liter, there are 10^{11} cells under 1 m² of the surface. If we assume that the dry weight of each cell is the same as that of a cell of *S. hantzschii* (E. M. F. Swale, unpublished data* and 1964) and that its chlorophyll content is 1 percent of the dry weight, there will be about 130 mg of chlorophyll per m² at the time of the algal maximum recorded by Swale (unpublished data* and 1964). This figure is an underestimate. The chlorophyll content is likely to be more than 1 percent of the dry weight of *Stephanodiscus* and other diatoms present and will be higher in the green algae than in the diatoms. The total dry weight is also likely to be an underestimate. At this population level, the rate of growth will be limited by light penetration because of self-shading (Talling, 1960, 1965; Steeman Nielsen, 1962). Steeman Nielsen (1962) has reckoned that the maximum amount of chlorophyll per square meter in the euphotic zone—the depth to which 1 percent of the incident solar radiation between the wavelengths 400 to 700 mμ penetrates—would be 300 mg for a

*Studies in the Phytoplankton of a Calcareous River. Ph.D. thesis, Univ. London (1962).

diatom population. In fact, this maximum value is very unlikely to be reached, for other light-absorbing particles are likely to be present, and the water is likely to contain some coloring matter. In this connection it may be mentioned again that in the Lee the timing of the end of the spring maximum is controlled by turbidity, and that the periods of greatest algal growth are those when the rate of flow is slow, because only then is detrital turbidity low (Swale, unpublished data,* and 1964). The maximum amount of euphotic chlorophyll found in natural waters is about 200 mg per m^2 (e.g., Ichimura, 1956, for Japanese waters; Talling, 1965, for African waters). Talling (1965) has estimated the maximum chlorophyll value for a population of the diatom *Asterionella formosa* to be 185 mg per m^2. Although there are no light measurements for the River Lee, the estimated amount of chlorophyll present, actually an underestimate, is so large that limitation of the algal rate of increase can scarcely be doubted.

My conclusion, from this consideration of the effects on light penetration of very large algal populations, is that a reduction in the nutrients entering the Lee would not necessarily be accompanied by a reduction in algal populations. The reduction in nutrients would be effective only when their rate of supply controlled the growth of algae instead of light penetration, a situation often encountered in physiological investigations on cultured algae. In the Lee the reduction in the phosphorus concentration would have to be about 80 percent and that of nitrogen 95 percent to reduce the potential maximum population of *S. hantzschii* below the observed maximum, judging by Swale's (unpublished data* and 1963) experiments. Anyone familiar with water blooms in lakes, fishponds, or sewage oxidation ponds knows that algal populations in very eutrophic waters often are so dense that the euphotic zone is extremely shallow. In such waters, self-shading may be expected to limit the rate of production. Therefore, as in the Lee, a reduction in the inflow of nutrients may not lead to the expected decrease in algal blooms.

How are we to determine the nutrient levels at which algal growth will be reduced? There is no simple answer to this question which is of such vital importance. If we find the minimal amounts of nutrients per cell or per unit volume of water that will support growth, they are likely to be so small that our new knowledge will have limited value in development of methods for reducing the nutrient income of highly enriched waters. An example is the relationship between the external and internal concentrations of phosphorus and the growth of *Asterionella* (p. 310, and Mackereth, 1963). In the author's view, the best approach to the problem is that of Gerloff and Skoog (1954, 1957a, b), whose work was carried out with special reference to lakes in the Madison, Wisconsin, area. They paid special attention to the cell contents of algae (e.g., *Microcystis*), comparing the results of laboratory experiments with

*See footnote on page 321.

analyses of algae in lakes and the concentrations of nutrients in these lakes. They pointed out that the natural ecosystem is so complex that their results might not be directly applicable to natural conditions. Nevertheless, their results offer a factual basis for devising experiments with lake water.

It is doubtful whether limits can be determined for the concentrations of various nutrients causing algal troubles. However, more data of the types collected by Sawyer (1947) would be valuable. The types of algae, and the quantities thereof, that are considered undesirable depend to a great extent on what the water is to be used for and how much money can be spent on remedial measures. As a result, great care must be taken not to erect arbitrary regulations concerning water quality. Regulations that could be valuable in one situation could impose unjustifiable economic burdens on a community in another. It might be better to decide first what is an undesirable amount of algal matter in a given set of circumstances and then demand that certain remedial measures be carried out in order to reduce, if possible, the eutrophication of the water body concerned. A convenient measure of the size of an algal population is the amount of chlorophyll present. Despite the fluctuations in the amount of chlorophyll per cell in different kinds of algae and under diverse environmental conditions, it does not seem that the more extreme variations recorded from laboratory experiments often arise under natural conditions.

REDUCING EUTROPHICATION

Before we consider some theoretical aspects of reducing or combating the increases in phytoplankton caused by eutrophication, a summarizing discussion of the viewpoints expressed so far will be helpful. The previous matter has been concerned largely with unsolved problems; any measures proposed for controlling algal blooms must take these problems into account. Certain English waters known to me have been used as examples of the problems involved. In agreement with most workers, I believe that nitrogen and phosphorus are of prime importance in the process of eutrophication. I suggest that eutrophication has become so great in some waters that the concentrations of these elements are no longer the major factors limiting the growth of phytoplankton. If this is so, it is also true that we do not know which elements or substances do limit production. I also suggest that, during periods when weather conditions are suitable for rapid algal growth (e.g., summer in temperate lands), the major factor limiting production can be light. If such suitable weather conditions arise, then reductions in the nutrient supply may not be followed by comparable decreases in the phytoplankton. An alga that is troublesome when water is used for one purpose is not necessarily troublesome when the water is used for another. Except for such generalities as illustrated by the statement "Eutrophic waters are likely to have more planktonic Cyanophyta than oligotrophic waters," we do not know what controls the qualitative composition of the phytoplankton. If a person had no knowledge of a particular body of water other than estimates about physical and chemical variables, it is doubtful that he would be very successful in estimating the constitution of the phytoplankton of the water body. It is doubtful that he would be successful even if very detailed records, by present standards, had been kept of the physical and chemical variables.

Techniques for reducing fertility of water and other methods of combating eutrophication are not strictly within the subject of this paper. They are discussed elsewhere in this volume. Nevertheless, from a theoretical viewpoint, it is permissible to consider ways in which algal growth might be controlled, even though such a discussion may overlap to some extent the descriptions of methods known to be of practical application.

There are two matters to be discussed: First, methods for which there is factual evidence of efficacy; and second, suggestions for water treatments that, though they may prove to be impracticable, relate to matters discussed in this paper. The need for further research on the ecology of phytoplankton in relation to eutrophication is obvious and will not be discussed further. In addition, the literature already in existence, which is very extensive, should be reviewed critically. Finally, some of the possible methods suggested will not be considered in detail because they have been discussed elsewhere (Lund, 1966, 1967) or because they are not my suggestions but those of my colleague Mr. F. J. H. Mackereth.

Considerable research has been carried out to devise methods for reducing the nitrogen and phosphorus contents of sewage effluents (e.g., Wuhrmann, 1964; Ambühl, 1964; Hanisch, 1964; Thomas, 1965). The most promising seem to be dentrification and the use of alum or ferric chloride, with or without the addition of lime. The diversion of sewage from Lake Washington, U.S. (W. T. Edmondson, unpublished data,* including discussion of other examples), and certain Swiss and German lakes (Thomas, 1965; Ohle, 1965) should give valuable information on how soon the removal of nutrients from the inflows of these lakes leads to a decrease in the phytoplankton, and to what extent. The problem of reducing enrichment by agricultural fertilizers, a steadily increasing cause of eutrophication, is more difficult because such a large part of the drainage area of a lake may be involved that the enrichment is not necessarily confined to one or a few of the sources of supply.

The nature of the treatments that may be effective can also vary in relation to the use to which a water is put. For example, reduction in the

*Water Quality Management and Lake Eutrophication: The Lake Washington Case. Presented at seminar, Univ. Washington (1964).

number of diatoms in a recreational lake may be of little or no importance compared with reduction in a reservoir. One of the main troubles in a recreational water body is the production of obnoxious superficial scums of Cyanophyta. These blooms often are distributed unevenly. Any method of reducing the blooms may solve the problem, even if over-all production by other algae remains high. Therefore, spraying the affected areas with suitable algal poisons or mixing flocculants into the surface layers might be effective, if there were no undesirable side effects. Copper sulfate is often used to reduce algal growths in reservoirs, but the price of copper has risen considerably recently. Other algal poisons probably could be used, but the lack of evidence that they are harmless to human beings has prevented their being used. Even if good evidence were obtained that other substances are harmless, the history of the objections to the use of fluorides suggests that it might be difficult to convince the users of the water that they are harmless.

In theory, the addition of flocculants directly to the water in the lake or reservoir has much to recommend it, and experiments should be carried out to discover whether it is possible in practice and not too expensive. Coagulants such as alum have advantages over algicides, for they can remove both algae and nutrients. In waterworks, flocculants are commonly used in filtration and clarification plants after the water has left the reservoir. If flocculants were added directly to the water in the reservoir or the inflow, part of the algal population would be lost to the sediments, and the treatment plant could be less extensive than if they were added after the water had left the reservoir. Moreover, the present system does not permit their nutrient-reducing properties to be utilized. The experience with flocculants in sewage treatment, mentioned earlier, shows that both phosphorus and organic matter may be removed. In addition, their ability to adsorb nutrients may well extend to trace elements, which, in waters containing large amounts of nitrogen and phosphorus, may be the main chemical factors limiting growth. Finally, the difficulty of reducing the fertility of waters enriched by agricultural fertilizers would be overcome, because the flocculants would be added to the water body into which the agriculturally enriched waters had flowed.

The importance of light has already been alluded to in relation to the River Lee. It is obvious that if light could be excluded, algal growth would be impossible. Therefore, the question of whether light exclusion is possible arises. As far as reservoirs are concerned, the use of aluminized underwater plastic screens has been suggested (Lund, 1966). This method, even if it is practical for reservoirs of moderate size, is impractical for recreational areas.

In England, the Metropolitan Water Board is experimenting with controlling the thermocline, and so limiting the size of hypolimnion (Taylor, 1966). Experiments have also been undertaken to examine the effect of destroying stratification (discussed in Lund, 1966; Taylor, 1966).

Destratification of a temperate lake in summer or a tropical lake at any time of year might be expected to increase algal production, for the nutrients throughout the water body would be available to the algae. Instead of some of the algae from the epilimnion passing into a hypolimnion, which is partly or wholly below the photic zone, all the algae would receive sufficient light for growth from time to time. The productive capacity of the whole water mass could be utilized, subject to the effects of self-shading referred to earlier. That this can happen naturally is shown by events in Blelham Tarn in recent years (Figures 1 and 2). The largest populations of diatoms arise in years when the maximum is reached before thermal stratification starts (e.g., during April). Then the whole water mass is depleted of silica.

The opposite situation usually occurs in Windermere, the lake into which the outflow of Blelham Tarn passes. Stratification arises before the silica is fully utilized; there follows a loss of cells by sedimentation through the thermocline into depths where photosynthesis is negligible or impossible. Thus the utilization of silica in the epilimnion does not match the observed size of the population of diatoms (compare Windermere and Blelham Tarn, 1965 and 1966, Table 1). In 1947, the concentration of silica in Windermere in the layers of water below the mean depth of the lake (26 m) altered by only 0.4 mg per liter, at the most, from that existing before the vernal increase of diatoms started (Lund $et\ al.$, 1963, Figure 7). The number of cells per unit volume over this depth, expressed as the amount of silica they contained, did not equal the observed decrease in dissolved silica. During 20 days in June, the loss of silica from the epilimnion was equivalent to the observed increase of the population plus the passage of some 8×10^{10} cells through each square meter of the thermocline; that is, 4×10^9 cells per m^2 per day.

Data such as these support the view that stratification can be advantageous in reducing algal troubles, although it is often pointed out that it is disadvantageous because of remineralization and release of nutrients in the hypolimnion. It is important to remember that whatever may happen in a lake over a certain period, nutrients are not produced there but are lost to the deposits and outflow or recycled. The vital factor in relation to eutrophication is what happens in its drainage area, for it is from there that the nutrients come. In the case of a lake such as Windermere, the maximum size of a population of diatoms is reduced by stratification, and much of the silica in the lake is never incorporated into algal cells. When the lake overturns, the short days and, to a lesser extent, low temperatures prevent any rapid increase of algae. As far as silica is concerned, the total amount incorporated into the diatom is lost to the productive cycle forever because there is virtually no resolution in the water of this lake or in the aerobic superficial deposits (Lund $et\ al.$, 1963; see especially Figure 9).

On the basis of facts such as these, the production of a permanent hypolimnion has been advocated as a method of reducing fertility (Lund, 1966). Permanent hypolimnia have arisen naturally in meromictic lakes. Large amounts of nutrients can be locked up in them. Findenegg (1965) points out that the rate of eutrophication of the meromictic Austrian Klopeiner See has been reduced by such losses of nutrients from the productive zone. Artificial meromixis could be achieved by so increasing the normal density difference between hypolimnion and epilimnion that thermal stratification produces. The use of flocculants and adsorbents, such as alum or ferric chloride, would be especially beneficial if a lake were meromictic. The material so removed would be retained in the hypolimnion. Superficial blooms of Cyanophyta could probably be flocculated. If, like the Klopeiner See in August (Findenegg, 1965, Table 3), the nitrogen and phosphorus in the upper layers of an artificially produced meromictic reservoir were almost entirely in organic form, the value of sedimenting the organisms out of the epilimnion would be obvious. At first sight, such a suggestion seems to be paradoxical, for the algal fertility of so many eutrophic lakes is bound up with microbiological processes in the hypolimnion. However, the stronger the meromixis, the lower the quantity of nutrients that will be recycled when thermal stratification breaks down.

Against the view outlined above must be set the experience of the Metropolitan Water Board of London. The reservoirs Q.E. II and K.G. VI have had their potential stratification in summer reduced by use of jet-type inlets or axial flow pumps (Taylor, 1966). As stated earlier (Table 1), the utilization of nutrients, although it can be considerable, is not efficient because only a small percentage of the nutrients are utilized by algae. Why this is so is not known. The possible mechanical effects on the algae of pumping or jet-type inflows cannot be disregarded. Ridley *et al.* (1966) have discussed the possible advantages and disadvantages of deepening the epilimnion in relation to reservoir practice. As far as algae are concerned, a certain amount of increased production might not be harmful because it would be distributed over a greater depth. The same could apply to recreational water bodies with blooms of blue-green algae. In practice, possible means of reducing algal production or the distribution of algae within a water body would have to be related to their effects on other organisms. Whatever effect artifical mixing of a water has on the phyto-plankton, it is likely to be beneficial where fishing is concerned. It has been found that overturning a lake in summer leads to less algal production, judging by the absence of blooms and the increased depth to which a Secchi disc could be seen (Wirth and Dunst, 1967). Meromixis could have deleterious effects.

I am very grateful to several people for information and advice. Mr. E. G. Bellinger and Dr. J. Ridley of the Metropolitan Water Board, and Mr. C. S. Reynolds, who is investigating the plankton of Crosemere, have permitted me to use the unpublished data in Table 1. They and colleagues at the Windermere Laboratory of the Freshwater Biological Association, especially Mr. F. J. H. Mackereth, Dr. E. M. F. Swale, and Dr. J. F. Talling, have discussed many matters with me.

REFERENCES

Ambühl, H. 1964. Die Nährstoffelimination aus der Sicht des Limnologen. Schweiz. Z. Hydrol. 26:569–594.

Damann, K. E. 1960. Plankton studies of Lake Michigan. 2. Thirty-three years of continuous plankton and coliform bacteria data collected from Lake Michigan at Chicago, Illinois. Amer. Microsc. Soc., Trans. 79:397–404.

Davis, C. C. 1964. Evidence for the eutrophication of Lake Erie from phytoplankton records. Limnol. Oceanogr. 9:275–283.

Davis, C. C. 1966. Plankton studies in the largest great lakes in the world with special reference to the St. Lawrence Great Lakes of North America. Publ. 14 (p. 1–36), Great Lakes Res. Div., University of Michigan.

Findenegg, I. 1965. Die Eutrophierung des Klopeiner Sees. Öst. Wasserw. 17:175–181.

Fogg, G. E. 1965. Algal cultures and phytoplankton ecology. University of Wisconsin Press, Madison. 126 p.

Gerloff, G. C., and F. Skoog. 1954. Cell contents of nitrogen and phosphorus as a measure of their availability for growth of *Microcystis aeruginosa.* Ecology 35:348–353.

Gerloff, G. C., and F. Skoog. 1957a. Availability of iron and manganese in southern Wisconsin lakes for the growth of *Microcystis aeruginosa.* Ecology 38:551–556.

Gerloff, G. C., and F. Skoog. 1957b. Nitrogen as a limiting factor for the growth of *Microcystis aeruginosa* in southern Wisconsin lakes. Ecology 38:556–561.

Goldman, C. R. [ed]. 1965. Primary production in aquatic environment. Mem. Ist. Ital. Idrobiol., suppl. 18. 471 p.

Gorham, P. R. 1964. Toxic algae, p. 307–336 *In* D. F. Jackson [ed] Algae and man. Plenum Press, New York.

Hanisch, B. 1964. Technische und wirtschaftliche Überlegungen zur Frage der Stickstoff- und Phosphorelimination. Schweiz. Z. Hydrol. 26:559–568.

Hutchinson, G. E. 1967. A treatise on limnology. John Wiley & Sons, Inc., New York. 2 vol.

Ichimura, S. 1956. On the standing crop and production structure of phytoplankton community in some lakes in Central Japan. Bot. Mag. (Tokyo) 69:7–16.

Jackson, D. F. [ed]. 1964. Algae and man. Plenum Press, New York.

Livingstone, D. A. 1963. Chemical composition of rivers and lakes. *In* M. Fleischer [ed] Data of geochemistry. U.S. Geol. Surv. Prof. Paper 440-G. U.S. Govt. Printing Office, Washington. 63 p.

Lund, J. W. G. *In* C. H. Oppenheimer [ed.], Marine Biology 2. New York Academy of Sciences.

Lund, J. W. G. 1955. The ecology of algae and waterworks practice. Soc. Water Treat. Exam., Proc. 4:83–99.

Lund, J. W. G. 1965. The ecology of the freshwater phytoplankton. Biol. Rev. 40:231–293.

Lund, J. W. G. 1966. Limnology and its application to potable water supplies. J. Brit. Waterworks Ass. 49:14–26.

Lund, J. W. G. 1967. Planktonic algae and the ecology of lakes. Sci. Progr. (Oxford). 55:401–419.

Lund, J. W. G., F. J. H. Mackereth, and C. H. Mortimer. 1963. Changes in depth and time of certain chemical and physical conditions and of the standing crop of *Asterionella formosa* Hass. in the north basin of Windermere. Phil. Trans. Roy. Soc. B. 246:255–290.

Macan, T. T. 1949. Corixidae (Hemiptera) of an evolved lake in the English Lake District. Hydrobiologia 2:1–23.

Macan, T. T. 1962. Biotic factors in running water. Schweiz. Z. Hydrol. 24:386–407.

Mackereth, F. J. H. 1953. Phosphorus utilization by *Asterionella formosa* Hass. J. Exp. Bot. 4:296–313.

Mortimer, C. H. 1963. Frontiers in physical limnology with special reference to long waves in rotating basins. Publs. Great Lakes Res. Inst. 10:9–42.

Mortimer, C. H. 1965. Spectra of long surface waves and tides in Lake Michigan and at Green Bay, Wisconsin. Publs. Great Lakes Res. Inst. 13:304–325.

Ohle, W. 1965. Nährstoffanreicherung der Gewässer durch Düngenmittel und Meliorationen. Münchn. Beitr. 12:54–83.

Pennington, W. 1941. The control of the number of freshwater phytoplankton by small invertebrate animals. J. Ecol. 29:204–211.

Ridley, J. E., P. Cooley, and J. A. P. Steel. 1966. Control of thermal stratification in Thames Valley reservoirs. Soc. Water Treat. Exam., Proc. 15:225–244.

Sawyer, C. N. 1947. Fertilization of lakes by agricultural and urban drainage. J. New Engl. Waterworks Ass. 61:109–127.

Smirnow, N. N., and O. I. Feoktistova. 1965. Vliyanie sinezelenȳkh vodorosleĭ na vodnȳkh zhivotnȳkh i racteniĭ, p. 212–223 In L. P. Brazinckiĭ [ed] Ekologiya i Fiziologiya Sinezelenȳkh Vodorosleĭ. Moscow, Leningrad.

Sorokin, Yu. I., A. V. Monakov, E. E. Mordukhaĭ-Boltovskaya, E. A. Tsikhon-Lukanina, and R. A. Rodova. 1965. Onȳt primeneniya radiouglerodnogo metoda dlya izucheniya troficheskoĭ roli sinezelenȳkh vodorosleĭ, p. 235–240 In L. P. Brazinskiĭ [ed] Ekologiya i Fiziologiya Sinezelenȳkh Vodorosleĭ. Moscow, Leningrad.

Staub, B. 1961. Ernährungsphysiologisch-autökologische Untersuchungen an der planktischen Blaualge *Oscillatoria rubescens* DC. Schweiz. Z. Hydrol. 23:83–198a.

Steeman, Nielsen, E. 1962. On the maximum quantity of plankton chlorophyll per surface unit of a lake or the sea. Int. Rev. Ges. Hydrobiol. Hydrograph 47:333–338.

Swale, E. M. F. 1963. Notes on *Stephanodiscus hantzschii* Grun. in culture. Arch. Mikrobiol. 45:210–216.

Swale, E. M. F. 1964. A study of the phytoplankton of a calcareous river. J. Ecol. 52:433–446.

Talling, J. F. 1960. Self-shading effects in natural populations of a planktonic diatom. Wett. Leben 12:235–242.

Talling, J. F. 1965. The photosynthetic activity of phytoplankton in East African lakes. Int. Rev. Ges. Hydrobiol. Hydrograph 50:1–32.

Taylor, E. W. [ed]. 1966. Forty-first report on the results of the bacteriological, chemical and biological examination of the London waters for the years 1963–1964. Metropolitan Water Board, London.

Tessenow, U. 1966. Untersuchungen über den Kiesselsäurehaushalt der Binnengewässer. Arch. Hydrobiol. Suppl. Band 32:1–136.

Thomas, E. A. 1965. The eutrophication of lakes and rivers, cause and prevention, p. 299–305 *In* Biological problems in water pollution (3rd seminar, 1962), Robert A. Taft Sanitary Engineering Center, U.S. Public Health Service, Cincinnati, Ohio.

Vinberg, G. G., and V. L. Lyakhnovich. 1965. Udobrenie prudov. Moscow. 271 p.

Wirth, T. L., and R. C. Dunst. 1967. Limnological changes resulting from artificial destratification and aeration of an impoundment. Departmental Research Report 22 (Fisheries), Wisconsin Conserv. 15 p.

Wuhrmann, K. 1964. Stickstoff- und Phosphorelimination; Ergebnisse von Versuchen im technischen Massstab. Schweiz. Z. Hydrol. 26:520–558.

ELIZABETH McCOY AND WILLIAM B. SARLES
University of Wisconsin, Madison

Bacteria in Lakes:
Populations and Functional Relations

Bacteria in Wisconsin lakes have long been objects of curiosity and of serious study by Wisconsin bacteriologists. It is fitting, therefore, to begin this presentation of our current knowledge of bacterial populations by citing contributions of early workers.

The names of four pioneers come to mind: H. L. Russell, E. B. Fred, A. T. Henrici, and E. A. Birge, men who for more than five decades provided the stimulus to our later studies. Both authors of this paper have been privileged to know these pioneers. Much of their work was addressed to the same problems of lake bacteriology that challenge us today. Furthermore, it may be said that the bacteriological problems of Wisconsin lakes were not unique then, nor are they today. Although this paper cites local data and local interpretations of the relations of lake bacteria, it actually speaks for relations of bacteria to freshwater lakes in temperate regions anywhere in the world.

Since we consider eutrophication as being accelerated by man's use of lakes, early papers on lake bacteriology can provide us with some useful reference points. Russell in 1889 was concerned by the public use of ice from Lake Mendota and Lake Monona, which received, untreated, "all the filth and waste" from Madison, then a city of 13,000 persons, plus that from a state institution on the shore of Lake Mendota. His plate counts on 27 samples of lake water ranged from 3 to 3,167, with an average of 683 bacteria per ml. He recognized a natural flora of "water bacteria" (the quotation marks are his) but he was unable to demonstrate any actual pathogens. Nevertheless, he pointed out the danger of false security and stressed that where sewage exists, "safety is not absolute."

Two papers by Fred in the early 1920's also provide insight into the bacterial flora of Lake Mendota at that time. The first, "Distribution and Significance of Bacteria in Lake Mendota" (Fred *et al.*, 1924), gives data on numbers of bacteria as affected by depth of the water, season, and influx of storm-sewer water. The second paper, by Snow and Fred (1926), presents a laboratory study of a collection of chromogens (mostly yellow, orange, pink, and red forms), but it is cited also for its comparison of plate counts on different culture media, including those currently popular for soil bacteria. Out of that comparison came the recommendation for the use of the nutrient sodium caseinate, which is still widely used in media for plating of water bacteria.

In the 1930's Henrici made his contributions to the bacteriology of lakes: the Henrici slide technic (Henrici, 1933, 1936), the discovery of stalked bacteria (Henrici and Johnson, 1935), and, at the invitation of Birge, the comparative study of oligotrophic, eutrophic, and dystrophic lakes in southern and northern Wisconsin. His paper, "The Distribution of Bacteria in Lakes" (Henrici, 1940), sums up his findings and is enlivened by his own ideas about the lake bacteria.

THE PLACE OF BACTERIA IN THE ECOSYSTEMS OF LAKES

Let us now consider the functions of bacteria in relation to the organic systems of the lake. To do so, it will be useful if we state, and then defend, a working hypothesis with which Birge used to challenge the bacteriologists in his group. In Birge's concept the bacteria *stand at the base of the fertility of lakes,* as they do also of soil. They have two functions:

1. Bacteria are the prime agents of the return of dead organic matter (plant and animal bodies) to the soluble state. If this were not so, the carbon of the world would soon be tied up in a complex organic state in dead tissues, and photosynthesis would stop for lack of CO_2.
2. Bacteria are the prime agents for the return of soluble organic nutrients to the biological system. By using such nutrients in their own growth, heterotrophic bacteria become particulate matter and are food for protozoa, which in turn are food for higher metazoa, leading ultimately to fish. Recovery of soluble organic matter is a remarkable feat when one considers the extreme dilution concerned. Even in mesotrophic and eutrophic lakes, the range of concentration is in milligrams per liter.

The ideal situation would be to have just enough bacterial activity to release organic nutrients from dead organic matter, to consume much of the soluble products so as to benefit higher animals through the food chain, and to benefit plants by release of some C as CO_2 and by mineralization of organic N as NH_4^+ or NO_3^-, and organic P as PO_4^-. Such an ideal balance would be found in a truly "eutrophic" lake, if one defines eutrophism as "good growth." Perhaps today the term of choice would be "mesotrophic," and "eutrophic" would be reserved for the advanced stages of enrichment resulting in choking of the lake with unbalanced growth and, finally, death of the lake.

Dr. Birge's concept can be developed further to cover the features of eutrophication, as considered in this conference. As shown in Figure 1, the nutrient and organismal balance of a lake can be diagrammed as a unilateral triangle. If the nutrients are restricted, as in an oligotrophic lake, the triangle is small; all forms of life are numerically restricted, but in balance. As the lake develops to progressively higher levels of nutrients, the triangle becomes larger but is still in balance. Ultimately, in late eutrophication, that balance is lost, and nutrients accumulate in excess of the annual turnover by bacteria. The imbalance consists of *too much,* not necessarily of unnatural kinds of bacterial nutrients in the lake. It is our contention that the imbalance occurs because, for physical reasons of the habitat, there is a limit to the bacterial population, and thus the bacterial population fails to cope with the amounts of plant and animal tissues synthesized annually. In a north temperate climate the rate of bacterial activity is reduced by lowered temperature of the water for about two thirds of the year. The flush of bacterial activity in summer (one third of the year), although efficient for organic decomposition, might well be detrimental to animals because of excessive consumption of oxygen. The summer flush might also be detrimental to plants because of failure to return mineralized nutrients early enough in the season for the high demands of the food chain—from bacteria, to plants, to animals. One might speculate further, but it is sufficient to say that progressive eutrophication is represented by extending the triangle beyond the point of balance of plant and animal populations that the limited bacterial population can support.

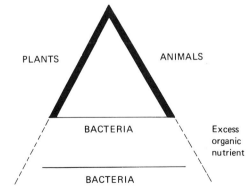

FIGURE 1   Organismal and nutrient balance in lakes undergoing eutrophication.

THE BACTERIAL POPULATIONS IN LAKES

It is difficult to describe the bacterial population in a lake. Admittedly, our understanding of the equilibrium determining the numbers and kinds that make up the population at any given time is imperfect for many reasons. Among the reasons are limitation of the counting methods, lack of identification of the types of bacteria in many studies, and lack of appreciation of the ecological relations.

It is a truism that the lake microflora is comparable to, but different in some respects from, the microflora of soil. Presumably, the lake flora is derived from soil flora by elimination of soil types that cannot compete in the lake habitat. In general, the high population of Gram-positive bacteria in soil gives way to a high population of Gram-negative bacteria in lakes. Among the latter, the proportion of species classed in *Pseudomonadales* is conspicuous. This order, of course, includes a great variety of forms, including spirilla, sulfur bacteria, strongly proteolytic pseudomonads, and others strongly oxidative and capable of using a tremendous range of carbon compounds, including the aromatic.

Another feature of the lake flora is the high percentage of chromogens belonging to several genera in both *Pseudomonadales* and *Eubacteriales*. The reason for the greater incidence of chromogens is not certain, but it has been suggested that certain pigments offer protection against the sun's radiation. Among the chromogens in lakes, *Micromonospora* spp. are often found in very high proportion to the total chromogenic count (Colmer and McCoy, 1944; Umbreit and McCoy, 1941). This is in contrast to the *Streptomyces* population, which is much lower in the lakes than in the surrounding soil. Stalked bacteria and other sessile forms are conspicuously high in the lake flora (Henrici and Johnson, 1935). So are representatives of *Chlamydobacterales* and *Beggiatoales*.

Lastly, the filamentous fungi in lakes are lower both in numbers and in variety than in soil. This statement may not hold true for all lakes. While no extensive study has been made, it is our impression that the dystrophic lakes of northern Wisconsin may have rich fungal flora.

Our concluding statement on the kinds of bacteria in the lake flora is that "Most Probable Number" counts for important physiological groups show that a well-balanced mixture is present to carry out degradation of chitin, cellulose, pectins, proteins, and other complex organic compounds.

The numerical aspect of lake populations of bacteria is difficult to document. No single medium can be recommended for poured plates, although our choice usually has been the sodium caseinate formula of Henrici and McCoy (1938). For organic nutrient it contains 0.5 g each of sodium caseinate, peptone, starch, and glycerol. It was noted recently that the

Canadian workers (Hayes and Anthony, 1959) used a modification of this medium; that is, they added liver, vitamin B_{12}, and traces of Fe. British investigators use a similar formula (Taylor, 1940; Collins and Willoughby, 1962). Counts for autotrophic bacteria are seldom attempted, probably for lack of adequate techniques.

Direct microscopic methods have been favored by some investigators. One great limitation has been the need first to concentrate the bacteria to make possible counts on very-low-count waters. Among the approaches to the problem have been the following:

Precipitation of the bacteria by $Al(OH)_3$ flocculation (Snow and Fred, 1926)

Evaporation of the water (Kuznetzov and Karzinkin, 1931, and as applied to Wisconsin lakes, Bere, 1933)

Attachment to surfaces such as Henrici slides (Henrici, 1936)

Membrane filters (Jannasch, 1958)

TABLE 1   Comparison of the Plate Method with the Direct Microscopic Method of Counting Bacteria in Lake Water[a]

Date	Depth (m)	Plate Count	Direct Count	Increase	Times Increased
Feb. 26	10	160	2,200	2,040	12
Feb. 26	15	150	1,480	1,330	9
Mar. 3	20	90	740	650	7
Mar. 12	10	160	2,200	2,040	12
Apr. 24	15	570	6,600	6,030	11
May 7	5	1,130	9,600	8,470	7
June 22	15	3,300	32,600	29,300	9
July 6	10	3,400	29,000	25,600	8
Aug. 3	5	2,200	23,700	21,500	10
Aug. 30	20	620	5,900	5,288	9
Sept. 14	10	530	5,100	4,570	9
Nov. 4	15	210	2,200	1,990	9
Dec. 18	5	400	3,700	3,300	8
Jan. 7	15	160	1,480	1,321	8
Feb. 2	20	180	1,480	1,301	7
Mar. 11	10	50	740	690	14
Apr. 8	5	110	1,480	1,370	12

Number of Bacteria in 1 ml of Water

[a]From Snow and Fred (1926).

In general, the direct methods give much higher counts, often in the range of 10 times higher than plate counts. In addition to being concerned about the difficulty in counting, investigators have been concerned about finding the sites of bacterial populations within lakes. The numerical data, and in some instances the data on dominance of groups, are quite different for samples taken from deep-water stations, shallow bays (with and without rooted plants), bottom muds, and beaches. From experience we can say that there is strong evidence of the importance of surfaces in the ecology of lake bacteria. Some of the apparently high concentrations of lake bacteria may, in fact, be surface-induced.

The magnitude of populations of lake bacteria in relation to location in the lake can be illustrated with selected data from some Wisconsin studies (Tables 1, 2, 3, and 4).

TABLE 2   Comparison of Numbers of Bacteria in Different Lakes[a]

Lakes	Periphytic Bacteria[b]	Plate Counts, Water[c]	Plate Counts, Bottom Deposit[d]	Microscopic Counts, Water[e]
Brazelle	3,853	2,963	44,600	2,000,000
Eutrophic				
Boulder	711	–	47,000	98,000
Alexander	526	675	144,240	–
Little John	402	505	39,050	64,500
Mendota	375	–	609,300	975,000
Muskellunge	197	133	10,930	400,000
Mean	442	438	170,100	384,400
Oligotrophic				
Weber	183	132	2,350	45,000
Trout	177	66	29,790	85,500
Crystal	63	80	2,160	36,000
Mean	141	93	11,400	55,500
Dystrophic				
Helmet	377	380	120,300	394,000
Mary	24	58	39,450	745,500
Mean	200	219	79,880	569,750

[a]From Henrici (1940).
[b]Bacteria per mm² per day deposited on slides.
[c]Colonies per ml of water on agar plates.
[d]Bacteria per ml on bottom mud on agar plates. (From Henrici and McCoy, 1938.)
[e]Bacteria per ml of water computed from counts of evaporated samples. (From Bere, 1933.)

TABLE 3   Bacterial Population in Mud, Lake Mendota, Station 1[a]

Mud Depth (cm)	July 7, 1936	July 13, 1936
0	182,000	148,000
2	336,000	108,000
4	1,300,000	126,000
6	820,000	17,800
8	414,000	15,000
10	256,000	19,000
12	490,000	19,800
14	172,000	10,200
16	23,000	–
18	–	14,600
20	25,600	–
22	–	21,500
24	4,820	–
26	–	9,400
28	2,220	–
30	–	7,500
32	1,670	–

[a]From Henrici and McCoy (1938).

TABLE 4   Bacterial Population in Shallow Bays[a]

Station		Depth (m)	Bacteria per ml	
Location	Description		Minimum	Maximum
LITTLE JOHN LAKE				
Resort bay	Algae, very weedy	0	1,320	10,800
		1	1,880	28,200
		2	34,600	129,000
Entrance bay	Some weeds, some algae	0	284	9,950
		1	746	8,800
		2	136	13,500
Deep water	Weedless, algae	0	88	5,000
		1	78	5,500
		2	236	5,000
LAKE ALLEQUASH				
Near outlet	Weedy, algae	0	2,380	54,000
		0.5	3,700	16,750
Deep water	Weedless, algae	0	494	2,222
		1	181	4,200
		2	105	4,000

[a]From Stark and McCoy (1938).

CONCLUSION

This paper has considered only the so-called natural microflora of freshwater lakes, and has attempted to generalize on the probable function of bacteria in the biological systems within the lakes. The flora resulting from sewage pollution has been omitted intentionally, although it may be important in many lakes; certainly it is important from the public health standpoint. We omit it only because of lack of data on its place in eutrophication.

In considering the problems of bacteria and eutrophication, the following questions are naturally raised:

Does eutrophication involve a change in numbers or kinds of bacteria within the so-called natural microflora?

What is the fate of pollution bacteria in a lake in an advanced state of eutrophication?

What needs to be studied on the nitrogen cycle in a eutrophic lake? How important are nitrification and nitrogen fixation in the light of recent studies on these bacterial processes?

It is important to the resolution of the eutrophication problem to find answers to these questions, for without such knowledge there is little that can be done.

REFERENCES

Bere, R. 1933. Numbers of bacteria in inland lakes of Wisconsin as shown by the direct count microscopic method. Int. Rev. Ges. Hydrobiol. Hydrol. 29:248–263.

Collins, V. G., and L. G. Willoughby. 1962. The distribution of bacteria and fungal spores in Blelham Tarn with particular reference to an experimental overturn. Arch. Mikrobiol. 43:294–307.

Colmer, A. R., and E. McCoy. 1944. Micromonospora in relation to some Wisconsin lakes and lake populations. Wis. Acad. Sci. Arts Letters, Trans. 35:187–220.

Fred, E. B., F. C. Wilson, and A. Davenport. 1924. The distribution and significance of bacteria in Lake Mendota. Ecology 5:322–339.

Hayes, F. R., and E. H. Anthony. 1959. Lake water and sediment. VI. The standing crop of bacteria in lake sediments and its place in the classification of lakes. Limnol. Oceanogr. 4:299–315.

Henrici, A. T. 1933. Studies of freshwater bacteria. I. A direct microscopic technique. J. Bacteriol. 25:277–287.

Henrici, A. T. 1936. Studies of freshwater bacteria. III. Quantitative aspects of the direct microscopic method. J. Bacteriol. 32:265–280.

Henrici, A. T. 1940. The distribution of bacteria in lakes. Amer. Ass. Adv. Sci., Pub. 10, 39–64.

Henrici, A. T., and D. Johnson. 1935. Studies of freshwater bacteria. II. Stalked bacteria, a new order of *Schizomycetes.* J. Bacteriol. 30:61–93.

Henrici, A. T., and E. McCoy. 1938. The distribution of heterotrophic bacteria in the bottom deposits of some lakes. Wis. Acad. Sci. Arts Letters, Trans. 31:323–361.

Jannasch, H. W. 1958. Studies on planktonic bacteria by means of a direct membrane filter method. J. Gen. Microbiol. 18:609–620.

Kuznetzov, S. I., and G. S. Karzinkin. 1931. Direct method for the quantitative study of bacteria in water and some considerations on the causes which produce a zone of oxygen-minimum in Lake Glubokoje. Zentbl. Bakt. Parasit. Infektionskrank., 2 Abt. 83:169–174.

Russell, H. L. 1889. Preliminary observations on the bacteria in ice from Lake Mendota, Madison, Wis. Medical News, Aug. 17, 1889, 15 p.

Snow, L. M., and E. B. Fred. 1926. Some characteristics of the bacteria of Lake Mendota. Wis. Acad. Sci. Arts Letters, Trans. 22:143–154.

Stark, W. H., and E. McCoy. 1938. Distribution of bacteria in certain lakes of northern Wisconsin. Zentbl. Bakt. Parasit. Infektionskrank., 2 Abt. 98:201–209.

Taylor, C. B. 1940. Bacteriology of freshwater. I. Distribution of bacteria in English lakes. J. Hyg. (London) 40:616–640.

Umbreit, W. W., and E. McCoy. 1941. The occurrence of actinomycetes of the genus *Micromonospora* in inland lakes. Symposium Hydrobiol., U. Wis. 1940, 106–114.

C. H. MORTIMER

University of Wisconsin, Milwaukee

Physical Factors with Bearing on Eutrophication in Lakes in General and in Large Lakes in Particular

Four main interdependent categories of physical factors control biological production—and therefore eutrophication and its effects—in natural waters. They are associated with (1) radiant energy input, (2) nutrient input and loss, (3) oxygen supply, and (4) the interactions of morphometry and motion. However, because present knowledge of eutrophication is fragmentary, any comments under the title of this paper must be fragmentary also.

For the purposes of this review, the borderline between physics and physical chemistry, normally indistinct, has been so drawn that adsorbtion, pH, colloid properties, clay mineral–solution equilibria, base exchange, and electrochemical aspects of dissociation and redox equilibria lie in the realm of chemistry and are not discussed here. On the other hand, the dynamics of diffusion (including turbulent diffusion), of oxygen balance (including redox control), and of nutrient gains to and losses from the lake system are included as part of the following subject matter. The main emphasis is on aspects of fluid dynamics that regulate exchange and transport mechanisms, illustrated where necessary by occasional excursions into chemistry.

Some well-known results will be mentioned briefly, and some recent findings will be discussed concerning (1) factors influencing the mobilization and loss of phosphorus and (2) the pattern of motion in large lakes, insofar as patterns of motion can influence the rates at which the effects of eutrophication come to be felt in different regions of the lake.

While these effects of eutrophication, accelerated by pollution and linked with it, have serious consequences in small lakes, their economic and social impact threatens to be much greater in large, urbanized lakes—for instance, the Laurentian Great Lakes. These large lakes are the only adequate source of

water for nearby cities and must also serve as a sink for treated and sometimes untreated wastes. For much of the year, mechanisms are in operation in
these lakes that impede the free exchange between inshore and offshore water
masses, and this leads to higher rates of eutrophication near shore.

Figure 1 presents, in the form of a simplified flux diagram, the elements of
the eutrophication process and its effects. Naumann's original (1919) definition of "eutrophy" and "oligotrophy" emphasized the nutrient status of
"rich" and "poor" lakes as the key to differences between their plankton
communities—differences that are both quantitative and qualitative. To avoid

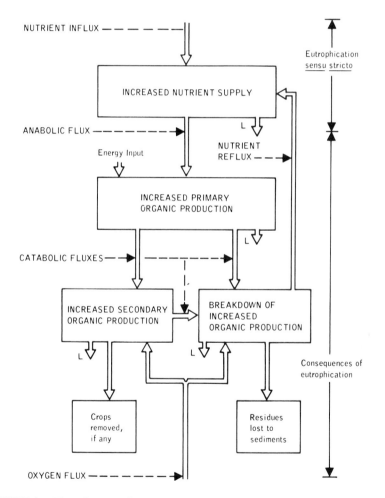

FIGURE 1   Flux diagram of eutrophication and its consequences. L indicates a leak
through the lake outflow.

the wide confusion that has come to be associated with the concept of eutrophication (for instance, it is frequently equated with pollution), it is helpful to distinguish between eutrophication *sensu stricto* and the effects of eutrophication, which may be variously expressed in different lakes.

Essentially, eutrophication is an increase in nutrient supply from soil and lake processes and from human activities in the drainage basin. If the results are desirable (a greater fish yield from a pond, for example), the increase is looked upon as a fertilizing enrichment. More frequently, however, the results are undesirable, and eutrophication receives the connotation of pollution. Nutrients, of course, include not only the major inorganic elements but also organic micronutrients and probably some nutrients yet to be discovered. This points to the need for much more critical research on the nutrient requirements of algae, and of algae in association with bacteria, before eutrophication can be adequately understood and perhaps eventually controlled. Such research may also explain the tendency toward planktonic monoculture (for instance, in algal blooms) and the changes in species composition at all trophic levels so often associated with eutrophication.

NUTRIENT AND ENERGY SUPPLY

It is evident from Figure 1 that physical factors exert their main influences on the fluxes, but in the present state of knowledge it is not possible to give detailed examples of physical control in each case.

Although they receive only brief mention in this review, the influences of *water temperature* and the *availability of light energy* are, of course, overriding. Temperature will control rates of the anabolic and catabolic fluxes, of the nutrient reflux, and of many physiological processes. Temperature also occupies a central position in physical limnology as the main density-determining factor, exerting a fundamental influence on motion and stratification.

Photosynthetic rates are regulated by temperature and by availability of light, and the latter frequently controls the initiation of growth and the rate of growth of algae during the early phases of seasonal production cycles. Except for self-shading at very high production levels, or inhibition of photosynthesis at very high light levels, and except for extreme temperatures, light and temperature usually are not, in themselves, the chief limiting factors that bring a production cycle to an end or slow it down. These factors are usually chemical or biochemical, and are therefore not the main subject of this review.

Availability of light depends not only on incoming radiant energy and water transparency but frequently also on the ratio of "stirred depth" to "illuminated depth." Stirred depth is the depth of the layer throughout which algal cells are maintained in random vertical motion by turbulence (typically the whole water column in homothermal conditions, and the column above the thermocline in stratified conditions). Illuminated depth is the depth of the layer in which photosynthetic gain exceeds respiration loss. This ratio can affect not only production rates but also the timing of the onset of growth in spring. (For an example, see Figure 2.) It is also obvious that a high ratio (stirred depth to illuminated depth) must impose a condition of morphometric oligotrophy, in which primary production is limited not by radiation received in the illuminated layer but by limitations on the time that an algal cell may spend in that layer. This is one of many examples of the way in which the influence of one physical factor (light energy) is modified or overridden by others (morphometry and turbulence).

The *nutrient influx* will be influenced, qualitatively and quantitatively, by *rock-weathering, soil formation,* and *soil processes* (some physical) in the drainage basin. Some of these processes are purely mechanical. One example is erosion. Another is rock-splitting by ice, which exposes fresh rock surface to weathering and mobilization of nutrients (silicate, for instance).

From the historical record provided by chemical evidence from cores from lakes in the English Lake District, Mackereth (1966, page 180) recognizes two processes that are, to a certain degree, opposed.

These are (1) maturation by leaching, in which some components of the soil mineral material are removed in solution, thus leaving the residue *in situ* impoverished in the leached components, and (2) erosive removal of the soil from the site of leaching and subsequent deposition of the eroded material in the lake sediment where it is effectively protected from further leaching.

In all cores examined by Mackereth there was a close correlation between concentrations of sodium and potassium—elements particularly sensitive to leaching attack—and the over-all percentage of mineral matter; this suggests that the mineral content at each level is directly related to the intensity of erosion in the drainage basin at the time the sediment was laid down. The correlation also suggests that "under given climatic conditions, the potential internal productivity of a lake basin is determined at least in part by the topography of the drainage system which sets the erosion rate characteristic of that basin and determines the position of the equilibrium between soil formation and leaching, on the one hand, and erosion and soil removal, on the other" (Mackereth, 1966, page 183).

A regime of more intense erosion does not favor high biological productivity in lakes; a larger proportion of nutrient material becomes locked away in the sediments in unavailable unleached mineral form. During periods of

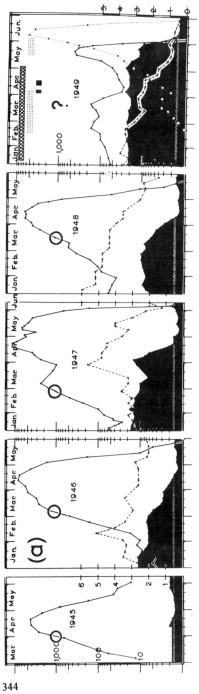

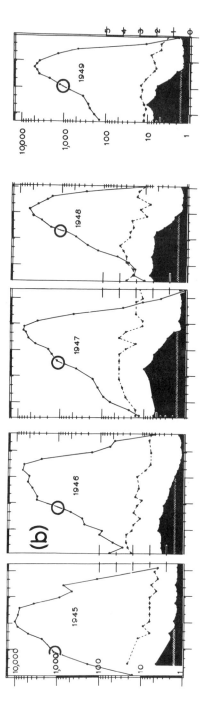

344

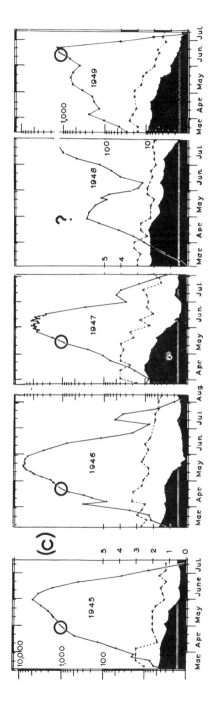

FIGURE 2 Spring growth of *Asterionella formosa* in the upper 5 meters of three lakes in the English Lake District. (a) Esthwaite Water; (b) Windermere, southern basin; and (c) Windermere, northern basin. Maximum depths are, respectively, 15 m, 33 m, and 63 m.

Cell numbers are given (unbroken lines) on a logarithmic scale, 1 to 10,000 per ml. Concentrations of dissolved silicate (as SiO_2, black area) and nitrate (as N, broken line) are shown on a linear scale, 0 to 5 mg per liter. The horizontal shaded lines within the black areas indicate the critical silicate concentration, 0.5 mg per liter.

In the top right-hand diagram (Esthwaite Water, 1949), the hatched area indicates abundance of the blue-green alga *Oscillatoria;* the dotted areas indicate severe parasitization of *Asterionella* by fungi; the black rectangles indicate floods; and the dotted lines show cell numbers of other diatoms, *Fragilaria* and *Tabellaria.* For further details, see Lund (1950).

The circles show the dates on which the *Asterionella* population had reached 1,000 cells per ml, and the question marks indicate exceptional years in which the "normal" spring increase did not take place. After Lund (1950) and reproduced from Mortimer (1959).

345

less-intense erosion, the mineral particles are held longer in the soils, and more of the potential nutrient material enters the lake in solution, favoring higher productivity. Sediment composition will, therefore, reflect not only the direct effect of erosion but also secondary effects associated with the higher contribution of organic matter from the lake itself. Mackereth, however, cites evidence that the lake-produced organic matter is consumed with relative rapidity and that the residues in the sediments are in a very stable form, largely derived from the soils.

Having considered briefly the controlling factors associated with radiant energy and nutrient input, we may turn to the categories associated with interactions between morphometry and motion and with oxygen supply. One example, the control of the rate of primary production by the ratio of illuminated depth to mixed depth, has already been mentioned. It represents a particular expression of a more general problem: exchange between subsurface and surface layers. The controlling mechanisms are those associated with turbulent diffusion, shear flow in stratified fluids, and, in large basins, the earth's rotation. These controlling mechanisms, applicable to the economy of lakes in general and to the regulation of eutrophication rates in particular, may be classed as diffusion control, stratified shear flow control, and Coriolis control. Each type of control will be illustrated by a detailed discussion of one selected example with which the writer happens to be familiar.

CONTROL THROUGH DIFFUSION

Consider the following well-known facts:

1. "The amount of oxygen in the lower water depends not merely on the length of time that the bottom water is cut off from the external air, but it depends also upon the amount of decomposable material discharged into it by the upper water and on the volume of the lower water, which, in turn, depends on the depth of the lake." (Birge, 1904, 1906.)

2. Respiration "goes on at all depths, consuming oxygen and liberating carbon dioxid. But as both animals and plants sink after death, most of the decomposition goes on in the deeper water, and especially, close to and at the bottom. . . . After the oxygen is used up, anaerobic decomposition continues, with evolution of carbon dioxid, methane and carbon monoxid. The lower water of a lake forms a *zone of decomposition*, whose processes are most vigorous at the bottom and decrease in intensity upward." (Birge and Juday, 1911.)

3. Under aerobic conditions, and at pH values representative of most

lakes, iron and manganese are present only in sparingly soluble form and in this form enter into precipitated complexes with other substances, notably the phosphate ion and organic solutes (for instance, see Shapiro, 1966). In their reduced (ferrous and manganous) forms, on the other hand, these elements are soluble; under anaerobic conditions, a reduction of the precipitated complexes liberates iron, manganese, phosphate, and other materials into solution. Shapiro (1967) has demonstrated that microorganisms also can liberate phosphate under reducing conditions and take it up again when aerobic conditions return.

4. In lakes, the "interface" between predominantly oxidizing conditions and reducing conditions usually lies a few millimeters below the sediment surface, as long as the water in contact with this surface contains oxygen. In one example (Mortimer, 1942) this interface coincided with a redox potential level of about 0.2 volt at pH 7, later referred to as the "0.2 isovolt." When the hypolimnetic oxygen supply becomes exhausted, the 0.2 isovolt usually ascends relatively rapidly to thermocline level, a direct result (as will be shown) of control through diffusion. Incidentally, the correlation between the liberation of ferrous iron from the sediment surface and the ascent of the 0.2 isovolt through that surface (see Figure 3) suggests that an iron system may have exerted some poising action in Mortimer's example.

5. The "average intensity" of turbulent stirring in the hypolimnia of lakes is generally positively correlated with lake dimensions, including depth.* This important empirical result suggests that the time that must elapse before the effect of any change in bottom water will be detected at upper hypolimnion levels does not differ greatly from lake to lake, irrespective of depth. Exceptions are lakes with abnormally large or small wind exposures.

The interplay of these five general results is illustrated by seasonal events (Figure 3) in Esthwaite Water, a shallow lake of medium to high productivity in the English Lake District. The principles involved have wide application to the "metabolism" of lakes in general and therefore to the eutrophication process.

To assist in interpreting the observations presented in Figure 3, certain technical details must be mentioned. The observations were made on short, relatively undisturbed cores of surface sediment and overlying water. The cores were taken in the deep region of the lake (but not at the deepest point) with the Jenkin Surface Mud Sampler described by Mortimer (1942). Two cores were taken on each occasion; one was used to provide a sample of water in contact with the sediment (typically a mixed sample from the bottom 10- to 20-cm water layer) for the determinations illustrated in the lower part of

*For an example of a method of measuring the average intensity, see Mortimer (1942), Figure 36.

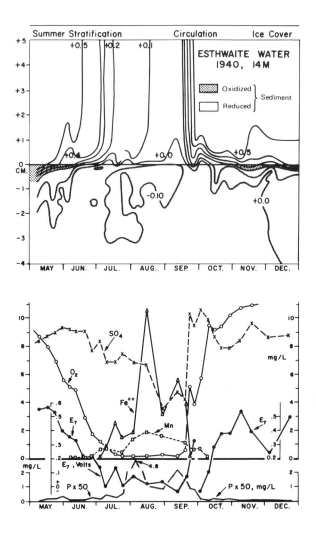

FIGURE 3   Esthwaite Water (English Lake District), May to December, 1940 (redrawn from Mortimer, 1942).

Upper portion: distribution of redox potential (E_7, adjusted to pH 7.0 from observed potential and pH, and equating 1 pH unit to 58 mV) above and below the sediment surface (electrodes normally placed at 50, 10, and 1 mm above and 1, 2, 3, 5, 7.5, 10, 13, 16, 19, 30 and 40 mm below the sediment surface).

Lower portion: redox potential (E_7) at 1 mm above, and concentrations (mg per liter) of certain dissolved substances at roughly 10 cm above the sediment surface.

Figure 3; the other was returned to the laboratory for the potential measurements illustrated in the upper part of the figure. Usually within 30 minutes after sampling, the top lid of the coring tube was opened, and an array of platinum electrodes was gently lowered into the core so that (normally) one electrode was introduced at each of the depths, relative to the sediment—water interface, indicated in the legend to Figure 3. The electrodes were 10 mm long and 2 mm high and could, therefore, indicate only whether an environment of those dimensions was, on the average, predominantly oxidizing or reducing. There was also some disturbance at insertion, including the introduction of oxygen. As a result, the electrode potentials usually fell after insertion. The rate of fall was greatest during the first hour, becoming progressively less thereafter. Routinely, readings were taken 2 hours after insertion. The higher potentials (i.e., in the presence of oxygen) were unstable and poorly reproducible. In such a mixed and nonhomogeneous environment, the potentials, even if clearly defined, were difficult to interpret. The particular value of the potential measurements in this application lay in the means they provided of locating the interface between regions that showed other (chemical) evidence of oxidation or reduction. This interface was very close to the 0.2 isovolt. With modern oxygen electrodes, suitably miniaturized, it should now be possible to make *direct* oxygen estimations at small depth intervals.

Returning to Figure 3, which begins with the onset of thermal stratification in the lake, the 0.2 isovolt was initially some 8 mm below the sediment surface. At that time, the potential at 1 mm above the surface (illustrated in the lower half of the figure) was of the order of 0.5 volt (at pH 7) and unstable. With the progressive fall in dissolved oxygen in the bottom water as a result of continued stratification, the 0.2 isovolt had risen close to the sediment surface by mid-June, and traces of manganous salts had appeared in the water. By the beginning of July, the oxygen concentration in the bottom water (modified Alsterberg method) had fallen to below 2 mg per liter; the 0.2 isovolt had risen above the sediment—water interface; and appreciable quantities of manganous salts and traces of ferrous salts had been liberated into the water. As the summer progressed, and as the oxygen content of the water fell to near zero, the concentrations in the water of the reduced forms of iron and manganese rose sharply, as did phosphate. The 0.2 isovolt climbed rapidly to thermocline level, and the whole hypolimnion turned anaerobic. In late summer the potential at the sediment surface fell to its lowest value, and the simultaneous decrease of ferrous iron and sulfate concentrations in the water suggests that ferrous sulfide was formed and precipitated at this low potential.

At the fall overturn, with the reintroduction of oxygen to the sediment surface, ferrous iron virtually disappeared from the water, the phosphate concentration fell to a low value, and, later, manganous salts disappeared. The potential at the sediment surface showed considerable fluctuations at this time, but after mid-October it remained above the 0.2-volt level for the rest of the year. The depth of the 0.2 isovolt in the sediment was also subject to considerable fluctuations after the fall overturn, and it was some weeks before it stabilized at levels comparable with those found during May. These fluctuations may have been a result of intermittent turbulent stirring of the upper few millimeters of sediment, as postulated from indirect evidence by Gorham (1958) and not sufficiently considered by Mortimer (1942). The lag in the establishment of a surface layer of oxidized sediment comparable in depth with that found early in the spring may have been the result of the slowness of oxidation of the accumulated products of summer reduction, notably ferrous sulfide.

The events just described are interpreted in terms of changes in the oxidized surface layer of the sediment and the contrast between slow molecular diffusion rates below this layer and the much more rapid turbulent diffusion rates in the water. As long as there is a ready supply of oxygen to the sediment surface, iron and manganese are precipitated and accumulate at the surface as insoluble oxidized complexes. These complexes effectively scavenge the environment (water and lower sediment layers) for phosphate and other materials with an intensity that depends on the relation between the ambient pH and the amphoteric points of the complexes concerned. In this case, the "environment" includes, as Mackereth (1966) has shown, the whole drainage system of the lake, the soils of which are the original source of iron and manganese. These materials are mobilized in reducing soils and coprecipitated with phosphate in the lake. At the same time, there is biological precipitation of phosphate onto the sediment in the form of dead algae and other organic materials.

The thickness of the oxidized surface layer depends on the degree of oxygen penetration into the sediments, and this penetration in turn is regulated (1) by the texture of the sediment surface, (2) by the balance between the rates of oxygen demand by the sediment and the oxygen supply to the surface (the factor emphasized by Mortimer, 1942), and (3) by any turbulent stirring that may take place in surface layers (Gorham, 1958). These factors combine to produce the maximum thickness of the surface-oxidized layer during periods of isothermal mixing.

During stratification, typically accompanied by a reduction in turbulence and oxygen supply at the sediment surface, the thickness of the layer readjusts itself, but perhaps not immediately if there is a lag in reducing the accumulated iron and manganese complexes. Figure 3 (upper portion) presents a picture, for the months of May and June, of the surface-oxidized layer being reduced in thickness from below. Gorham's later direct observations

(1958) on cores from the deepest point in Esthwaite Water show that deposition of dead plankton onto the sediment surface can produce a situation in which the layer reduces from the top downward. It should also be noted that the dimensions of Mortimer's electrodes (10 x 2 mm), and perhaps the disturbance and oxygen introduction at their insertion, would smooth out much of the detail noted by Gorham, namely, discrete sites of reduction, which showed up as spots or bands of black ferrous sulfide in the otherwise brown surface layer, associated with settlement or turbulent plowing-in of freshly decomposing plankton.

But by whichever route the surface-sediment layer becomes reduced, the effects on the lake economy, and on the properties of the subthermocline water mass in particular, are profound; and these effects are associated not only with the reduction and mobilization of iron and manganese and the unlocking of adsorbed materials, including phosphate, but also with the resultant removal of a barrier to unimpeded diffusion of materials across the sediment—water interface. On passing through this interface, the mobilized materials pass from a region of slow molecular diffusion into a region of much more rapid turbulent diffusion. They become disseminated throughout the hypolimnion, with the thermocline as the upper boundary, acting as a partial barrier to further upward diffusion. Some properties of the thermocline barrier are described in a later section.

DIFFUSIONAL CONTROL IN THE LIGHT OF HISTORICAL EVIDENCE FROM SEDIMENTS

An important question is: How much of the nutrient material (notably phosphate) mobilized into the hypolimnion by the processes described can enter the surface-illuminated layer and take part in the cycle of biological production? Some nutrient material may well do so—by mechanisms described in the next section—particularly at the fall overturn, before it is reprecipitated onto the sediments or lost through the outflow. However, the productivity of the lake as a whole, over the year as a whole, will greatly depend on the ratio of outflow losses to inflow gains of nutrients, including phosphate, from soil processes in the drainage basin. Anaerobic conditions in the hypolimnion produce a temporary enrichment of phosphate in that layer. But there is growing evidence that in many cases the onset of these conditions leads to an over-all loss of phosphate (coupled with loss of iron and manganese) from the lake system and from the sediments, as documented from lake-sediment cores by Mackereth (1966).

With so many variables at play, it is impossible to give a general answer to the question of how anaerobiosis influences over-all productivity, and no doubt this question will attract considerable future research. But the qualitative effects on production at various trophic levels is profound, and it is evident that diffusional factors are important. Whether, or at what stage in its history, a lake gradually or suddenly turns anaerobic below the thermocline depends not only on the intensity of biological production but also (as Birge noted in 1904) on the total reserve of subthermocline oxygen, which is a morphometrically determined factor.

Historical evidence has been sought from the organic remains and the chemical profiles in cores of lake sediments and will be increasingly sought from these sources. Mackereth's pioneer investigation (1966) of the chemistry of lake sediments, already mentioned, throws light on the influences of soil erosion and leaching. His general conclusion is that the composition of the residues eventually incorporated in the sediment can "be accounted for in terms of the rate of erosion of the drainage basin rather than in terms of changing rates of organic productivity either on the drainage basin or in the lake waters." But he demonstrates that the distribution in the cores of those elements (for instance, iron and manganese) that migrate more readily under reducing conditions do throw some light on events in the lakes themselves. For instance, the distribution of iron and manganese may be used "to deduce the redox conditions in the soils of the drainage basin in past times, or in the muds or hypolimnetic waters of the lakes themselves."

But correlations between sediment chemistry and biological events are likely to be largely indirect. Even in the case of phosphorus (a notably biophilic element), coprecipitation with iron and manganese appears to have "more influence on deposition efficiency than does incorporation into biological tissues." The hope of finding a simple, direct, historical, chemical record of the eutrophication process will, therefore, be small in many instances. For example, the carbon profiles of the English Lake District cores, which Mackereth examined, and those of Linsley Pond, Connecticut, strongly point to the soils of the drainage basin as the main source of organic matter and are all remarkably similar when reduced to the same time scale. (See Mackereth, 1966, Figures 7 and 8.) By the time the organic matter has become incorporated in the sediments, it has already reached "a state of considerable stability towards further oxidation." It shows little influence of any changes in biological productivity (or "eutrophication") that can be attributed to the possession by two of the lakes (Esthwaite Water and Linsley Pond) of anaerobic hypolimnia for long periods during their histories. The carbon profiles suggest, rather, that climatic and erosional factors existed on a wider-than-regional scale, a view also supported by the distribution of halogens and boron.

Of particular interest in connection with the historical origins of the situation illustrated in Figure 3 is Mackereth's interpretation of the distributions of iron, manganese, and phosphorus in the Esthwaite Water cores. These distributions are illustrated in Figure 4 and are consistent with the following postulations:

1. During the earliest phase of postglacial sedimentation (starting at a core depth of 460 cm), iron and manganese were mobilized in reduced soils in the drainage area. This mobilization, combined with reprecipitation in a lake with no appreciable reduced phase in the hypolimnion, led to an accumulation in the sediments of iron and manganese, and also of phosphorus, which was coprecipitated with them, reaching peak concentrations of all three elements at about 420-cm depth and corresponding to a low Fe/Mn ratio.

2. A key to the understanding of subsequent changes with time of the iron and manganese concentrations and to the Fe/Mn ratios in the core is provided by the present-day seasonal cycles of those elements in Esthwaite Water (lower half of Figure 3). In the lake environment, manganese is reduced and mobilized at a higher redox potential than iron is. Therefore, in a seasonal cycle with a phase of increasing reduction followed by a phase of reoxidation (for instance, summer stratification followed by the fall overturn), manganese goes into solution earlier and is reprecipitated later than iron is. Manganese therefore remains in solution and is transported by water movements for a considerably longer period than iron is. Manganese will consequently be more vulnerable to loss at the outflow, particularly during and after the fall overturn. While such seasonal changes are not usually reflected in the sediments, long-term changes are. During a progression from oxidizing to reducing conditions in the lake, manganese loss from the sediments begins before iron loss; and during the reverse process (change from reducing to oxidizing conditions), iron begins to accumulate in the sediment before manganese does.

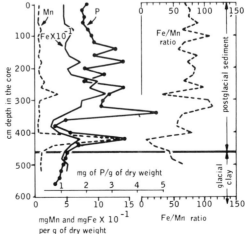

FIGURE 4   Esthwaite Water. Depth distribution of concentrations of manganese, iron, and phosphorus (milligrams per gram of dry weight), and the iron—manganese ratio, in a sediment core.

Redrawn from Figures 17 and 21 in Mackereth (1966).

For instance, the time interval corresponding to a core-depth range of 420 to 370 cm in Figure 4 is interpreted as one of increasing reduction in the hypolimnion of the lake, characterized first by severe loss of manganese, leading to a high Fe/Mn ratio, followed by a less severe loss of iron and phosphorus. This development is seen as the result of higher organic production (in soils and lake) leading to the onset of reducing conditions. The iron concentration fell to a minimum at the 370-cm level, which is marked by a medium Fe/Mn ratio. This was followed by an interval of increasing oxidation, in which iron and phosphorus concentrations increased, leading to a peak in the Fe/Mn ratio at 320 cm, followed by accumulation of manganese and a falling Fe/Mn ratio, culminating in a peak concentration of manganese and a minimum Fe/Mn ratio at 270 cm. After this the trend reverted to reducing conditions, and the sequence of rapid manganese loss (high Fe/Mn ratio) followed by a slower and partial loss of iron was repeated. The minimum concentration of iron was reached at 140 cm, after which there occurred a reversion to a weak oxidizing phase with a small increase in iron concentration (and rising Fe/Mn ratio) peaking at 80 cm. Iron loss thereafter represented a swing to the reducing conditions observed in the hypolimnion today (Figure 3).

The iron, manganese, and phosphorus profiles in Figure 4 support, in Mackereth's view, a hypothesis (1) of oscillations between oxidizing and reducing conditions in which the variations in sedimentary concentrations were brought about mainly by variation in redox conditions in the lake, particularly at the sediment surface, which controlled the rates of loss of these elements from the lake. Mackereth argues that the alternative hypothesis (2), which supposes that variation in sedimentary concentration was brought about by variation in supply of the elements from the drainage system while the redox conditions in the lake remained relatively constant, is ruled out by a consideration of the Fe/Mn ratios mentioned so frequently in the previous paragraph. For instance, it will be noted in Figure 4 that the peaks in iron concentration at 410 and 270 cm corresponded to distinct minima in the Fe/Mn ratio. If, according to hypothesis (2), these peaks were wholly or largely produced by increased supply of ferrous iron from the drainage system, then manganese loss from the soil must already have been maximal before the reduction had reached the point of iron mobilization. Increased reduction would, therefore, lead to an increased supply of iron but to little or no increase in the rate of supply of manganese. This would, in turn, lead to an increase in the Fe/Mn ratio in the sedimenting material, and the peaks of iron concentration in the profile should, under hypothesis (2), correspond to high Fe/Mn ratios; but this is the opposite of what is observed.

The conclusion is, therefore, that although variations in the rates of supply from the drainage basin are not entirely without influence, the iron and manganese profiles in the core reflect the control of redox conditions in the lake on the efficiency of precipitation and retention of iron and manganese and that this retention efficiency increases or decreases as conditions become more or less oxidizing. In the first phase of increasing oxidation, iron precipitation occurs before that of manganese, producing a rise in the Fe/Mn ratio. As oxidation increases further, manganese is precipitated as well as iron, and the ratio falls. Reversal of this process toward more reducing conditions leads, first, to a reduction in the rate of manganese accumulation and an increase in the Fe/Mn ratio and, later, to a fall in this ratio, when the efficiency of iron retention is diminished by increasing reduction. This is why the peaks in the Fe/Mn ratio occur in the sedimentary profile (Figure 4) both before and after the period of maximum oxidation.

CONTROL THROUGH STRATIFIED SHEAR FLOW

The redox cycles described in the previous section operate within the hypolimnion, but the hypolimnion is not an entirely closed system. The extent to which it is open to exchange with overlying oxygenated layers must influence the rate and degree of reduction (and, presumably, biological productivity) through transfer of nutrients to the illuminated layers. The hypolimnion participates, with the rest of the lake, in forced and free motions induced by wind stress. The development of a "steady state" by prolonged wind stress on a stratified lake that is small enough for the effects of the earth's rotation to be neglected is illustrated in Figure 5. (The different situation in large lakes, in which the earth's rotation introduces important terms, is discussed in the next section.) There is an initial shift of the epilimnion downwind and of the hypolimnion upwind, with the production of (a) a circulating wedge of epilimnion water at the downwind end of the basin, (b) a gradient on the lake surface, (c) a surface wind-driven current, and (d) a gradient-driven return current running upwind along the top of the thermocline in the lower part of the epilimnion. This is a special case of a type of flow—shear flow in a stratified fluid—very widespread in nature and with properties that exert a fundamental regulatory influence. If the shear is expressed as a velocity gradient, (du/dz), and the stratification is expressed as a density gradient, $(d\rho/dz)$—both, in this instance, in terms of the vertical axis z—then the controlling nondimensional parameter is the Richardson number: $R_i = g(d\rho/dz)/\rho(du/dz)^2$, in which g is acceleration of gravity, ρ is the density, u is the velocity, and z is the depth. The Richardson number expresses the balance between the stabilizing force of the density gradient and the disturbing force of the shear flow. In the light of investigations discussed by Phillips (1966, pp. 186–187), a condition sufficient for stability

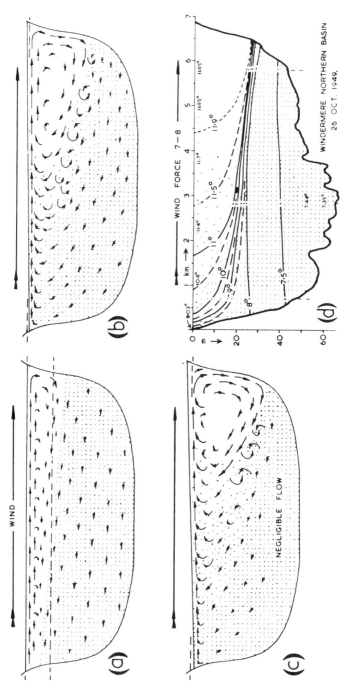

FIGURE 5   Motion induced by a strong sustained wind blowing along the axis of a stratified lake. Drawing (d) is an example after 12 hours of wind at Beaufort Force 7-8; (a)–(c) illustrate stages in the development of (d). Broken lines in (a)–(c) represent initial equilibrium levels of thermocline and water surface; the hypolimnion is strippled. Speed and direction of flow are roughly indicated by arrows. From Mortimer (1961).

of the flow is for the local R_i everywhere to be greater than 1/4. This should not be regarded as a unique "critical" number, because a range of stability may exist when $R_i < 1/4$, but it is a useful, if not universal, criterion for the onset of instability.

As long as the flow remains stable, disturbances are suppressed by the buoyancy forces associated with the density gradient; mixing across the density gradient, and friction, remain small. If $R_i < 1/4$, the supply of energy from the shear flow, proportional to $(du/dz)^2$, is greater than the loss of energy experienced by the turbulent motions in doing work against gravity on the density gradient; flow becomes unstable; wavelike disturbances grow into large vortices, which result in stirring and mixing across the density gradient in the manner described by Mortimer (1961). The horizontal wavelength of the most rapidly growing disturbances is usually of the order of the thickness of the shear layer. Unstable flow and resultant mixing lead, in turn, to the formation of a new, thicker shear layer with density and velocity gradients less steep than those initially present. Because the "disturbing term" in the R_i ratio is proportional to $(du/dz)^2$, the result of unstable flow in a thermocline situation will often be a reversion to a condition of neutral stability, with a density gradient that is just sufficient to stabilize the disturbance at the time of maximum shear. In other words, the interactions of shear flows, density gradients, and turbulence—and sometimes also internal waves—play a major role in determining the shape and depth of the thermocline density profile.

For instance, wind-induced shear flow in the stratified lake represented by Figure 5, being greatest at the surface, breaks down all density gradients that may have been previously present and, as is well known, produces a mixed homothermal epilimnion. The sharpness of the thermocline at the downwind end of the basin is, in this hypothetical picture, associated with the return current. As long as the shear developed by current does not lead to subcritical Richardson numbers, the thermocline remains a slippery surface with very little mixing across it. However, in the situation represented by (c) and (d) in Figure 5, a plausible interpretation of the fan-shaped distribution of the isotherms is that large eddies or vortices associated with the Richardson instability at the thermocline give rise to large-scale mixing across the thermocline and to an upwind drift of mixed water. This drift acts rather like a carpenter's plane, removing "shavings," in the form of vortices indicated by the curling arrows in the figure, and carrying them upwind. A sharp thermocline is left at the downwind end, which just balances the local shear to produce a neutrally stable Richardson number of the order of 1/4.

When the wind stress is removed, the water layers displaced or created in the manner just described are redistributed by gradient currents and progressive and standing internal waves until a new equilibrium is achieved. The eventual shape of the density profile at any point in the lake must, therefore, be regarded as a combined result of a number of processes operating during and after a wind disturbance and in widely separated regions of the basin. It is often found that when the disturbance and subsequent oscillations have subsided, the mean shape of the density profile is similar to what it was before the disturbance, but it has been pushed a little deeper.

If Richardson instability is localized (for instance, at the nodes of internal standing waves after a strong disturbance), isolated and circumscribed masses of mixed water are formed; these are "pancakes" of small-scale turbulence (Phillips, 1966, page 188), As Schooley (1967) has shown, the masses are initially subject to turbulent growth and later to collapse and decay by lateral spreading. Because the characteristic collapse time for such mixed regions is approximately 1/2 cycle of an important internal wave period (the Brunt-Väsälä period associated with stability waves on the thermocline and with periods of a few minutes under typical lake conditions), this may be an efficient mechanism of short-period internal wave generation.

The concepts just discussed are fundamental to physical limnology, and their relevance to the eutrophication process is general rather than particular. However, it is evident from Figure 5 that if the lake basin is long enough, or the wind disturbance is great enough, a portion of the hypolimnion will be raised into contact with the atmosphere at the upwind end and any oxygen taken up in this way will be distributed later to subthermocline levels. Also, some nutrient enrichment of the illuminated layer may result from the incorporation of hypolimnetic water into the surface wind drift. In a lake in which the hypolimnion becomes reduced and enriched and in which the events illustrated in Figure 5 are of frequent occurrence, these events could exercise some control over the rate of eutrophication. Such injections of nutrients into the epilimnion and of oxygen into the hypolimnion are probably a common feature of many lakes and, although complicated by gains and losses of heat through the water surface and by variability of the strengths and direction of wind impulses, they are repeated storm by storm throughout the stratified season until the final instability of the overturn.

CONTROL THROUGH THE CORIOLIS EFFECT

An observer attempting to formulate the equations of motion within his local geographical frame of reference (i.e., ignoring the fact that his frame is subject to rotation) has to invoke a deflecting force (the Coriolis force arising from the earth's rotation) to describe the motion. As explained in oceanographic textbooks, the horizontal component of the deflective force is directed at $90°$ to the right of the direction of motion in the northern hemisphere, and is given by the product of the velocity (u) and the Coriolis parameter (f, namely, $2\Omega \sin \phi$, where Ω is the angular velocity of the earth's rotation, and ϕ is the geographic latitude).

In the special case in which the motion of the water mass is unconstrained by boundaries and not subject to forces other than the Coriolis force, the (N hemisphere) track followed by the mass is a clockwise-turning circle (the inertial circle) of diameter u/f and with a rotational period (the inertial period) of $2\ \pi/f$, which is equivalent to $12/\sin \phi$ hours. If the external forces are not zero, but are of the same order of magnitude as the Coriolis force, or smaller, the circular track is modified. For instance, if u is slowly increasing or decreasing, the corresponding tracks will be expanding or contracting spirals.

If there is a lateral boundary (a shoreline) within a few inertial-circle diameters of the motion, the track will be modified or masked by gradient currents imposed by the boundary and frequently running parallel to it. The resultant track can then be one with loops, cusps, or meanders, of which Figure 8 (page 365) provides examples. In inshore regions, or in lakes and channels not wide enough for inertial motion to be developed, a common pattern is one of shore-parallel flow, relatively uniform in direction and velocity, in which the Coriolis force is balanced or nearly balanced by a gradient (surface, internal, or both) at right angles to the flow. If this balance is exact, the current is said to be geostrophic and in geostrophic equilibrium.

The preceding paragraphs outline a chapter in physical oceanography and limnology. Their relevance to eutrophication becomes apparent when motion and exchange of water masses are considered in large lakes.

Recent findings in the example discussed here (Lake Michigan during summer stratification) indicate that both the forced motions (wind-driven currents) and the free motions (internal waves) are characterized by the dominance of shore-parallel, quasigeostrophic flow near the shore and by the frequent occurrence of rotating (inertial or near-inertial) flow offshore. Under some circumstances, this combination leads to a partial or temporary separation of characteristic coastal and offshore water masses, which may result in "coastal entrapment" (Csanady, 1967a) of water with its nutrient or pollution load. The practical consequences of this behavior, still incompletely understood, are very great in the Laurentian Great Lakes, where large populations rely mainly on coastal intakes for water supply, dispose of processed and unprocessed wastes into the same water, and strive to maintain the lake beaches as a recreational "lung" for the cities.

In the end, decisions on how lakes are to be used and exploited must be based on enlightened social and political choices and must involve compromises and the expenditure of money. Aware of the urgency, as far as the Great Lakes are concerned, and of the lack of basic knowledge on water movements in those basins, the United States Government in 1960 appointed a group to undertake an extensive study of the physical mechanisms by which currents are driven and by which pollution-laden waters mix with and are transported by the current. A report on this study has been prepared by the Federal Water Pollution Control Administration, U.S. Department of the Interior. It describes a 2-year investigation of the currents of Lake Michigan, using a network of self-recording current meters and thermometers suspended from moored buoys.

The findings, a sample of which is shown in Figure 8, have confirmed a theoretical model assembled by Mortimer (1963) concerning the properties of internal waves in large basins, particularly in regions remote from shore constraint.

These properties may be outlined briefly by reference to the simple, rotating, constant-depth models illustrated in Figure 6 and assembled from figures in Mortimer (1963). The wave surfaces shown in Figure 6 can be regarded either as an air—water interface or (for the simplest example of stratification) as a "thermocline" interface between two homogeneous layers of differing density. In that case the current indications in the figures (highly generalized and associated with the waves) are shown for the lower layer (hypolimnion). The corresponding epilimnetic currents will be exactly opposed in direction, and the speed ratio of the upper and lower current systems will be the inverse depth ratio of the two layers.

In the narrow channel (Figure 6a) all currents, including geostrophic or quasigeostrophic currents, must run parallel to the sides; no motion transverse to the boundary is possible. The free-wave solution then takes the form of a Kelvin wave in which the response to rotation imposes an exponential decrease in wave amplitude and current speed along a line normal to the shore. For the internal variety of Kelvin waves and under conditions representative of Lake Michigan, these amplitudes fall to negligible values at a distance of 10 km or so from the shore.

Figure 6b represents the shoreline and a portion of a semi-infinite "ocean" of constant depth. At distances offshore great enough for boundary constraint to be neglected, the response to rotation takes the form of a Sverdrup wave, in which the wave crests remain level but in which the currents rotate (clockwise in the Northern Hemisphere).

Lake Michigan, however, has boundaries; and, by analogy with the well-known result for small lakes that standing internal waves (seiches) are mathematically equivalent to the combination of two progressive waves of equal amplitude traveling in opposite directions, so the boundary conditions in a large lake also impose a standing-wave pattern. In a rectangular basin of Lake Michigan dimensions, closed at the ends as well as at the sides, the boundary conditions for long wavelengths are met by two pairs of Sverdrup waves so combined that flow normal to all boundaries is zero. The result is a pattern of standing (Poincaré) wave "cells" with a discrete number of nodes

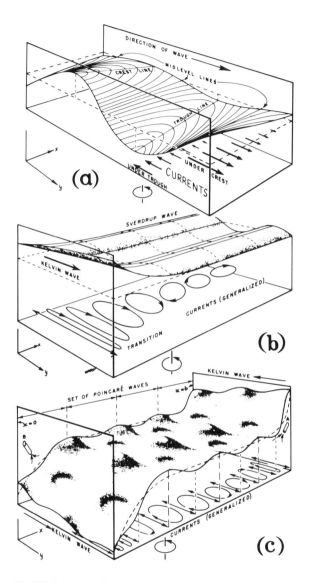

FIGURE 6   Qualitative representations of waves and associated currents in clockwise rotating rectangular models of
constant depth, assembled from Figures 1, 2, and 3 of
Mortimer (1963). (a) Kelvin wave in portion of an infinitely
long narrow channel; (b) portion of a semi-infinite sea with
Sverdrup waves offshore and part of a Kelvin wave inshore;
(c) portion of a wide channel or basin with Poincaré waves
offshore and with a part of a Kelvin wave on each shore.

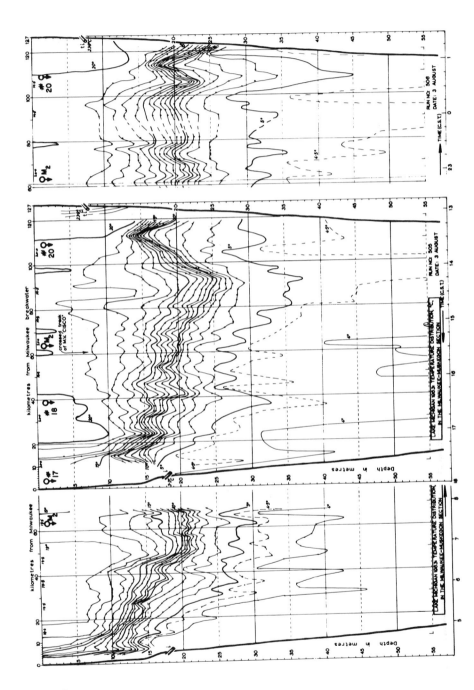

362

across the basin and also along the basin. A portion of such a pattern, with part of an "inshore" Kelvin wave superimposed, is illustrated in Figure 6c, which is regarded as descriptive of the thermocline topography in the central section of Lake Michigan.

In 1963, the writer attempted to check this hypothetical picture by making anchor-station measurements of the depth distribution of temperature and current in midlake and by repeated observations (from railroad ferries) of temperature distribution along the Milwaukee-Muskegon section. The Figure 6c model depicts a rather high number of transverse nodes. But the 1963 observations in the Lake suggest that a smaller number of transverse nodes (one or three or a combination of these) appears commonly. The preference for an odd low nodal number is the result of the interaction of wind stress and Coriolis force whereby surface water is transported to the right of the wind direction. This usually leads to depression of the thermocline on one shore and upwelling on the opposite shore. Figure 7 provides an example of this in which strong winds, predominantly from the southeast during the week preceding August 3, had produced a downward deflection of the thermocline on the right-hand shore and upwelling on the opposite (Wisconsin) shore. A striking example of the reverse situation, a result of strong northerly winds, is given by Mortimer (1963, Figures 16 and 17).

Figure 7 also illustrates the isotherm distribution on one crossing and two part-crossings by the railroad ferry on the Milwaukee-Muskegon route during a 21-hour interval selected as representative of a common summer situation. (In all, 80 ferry crossings were measured; and anchor-station observations of current and temperature were made on MV *Cisco,* with the aid of an Office of Naval Research contract.) The pictures in Figure 7 are, of course, not synoptic, as the ferry takes about 5.5 hours for the crossing. They show the result of previous (prevailing) winds with southerly component, with consequent downwelling and upwelling effects in the eastern and western

FIGURE 7   Distribution of temperature ($^{\circ}$C) in the Milwaukee-Muskegon cross section of Lake Michigan (course 80° true at long 87°, lat. 43°) during the 21-hour interval centered on 1500 hours, CST, August 3, 1963, obtained under Office of Naval Research contract and by means of bathythermographs operated from the Grand Trunk and Western Railroad Company's car ferry (central and right-hand portions of the figure) and from MV *Cisco* of the U.S. Bureau of Commercial Fisheries (left-hand portion). The "pipes" labeled *"i"* on the western shore and *"ti"* on the eastern shore represent the water intake at Milwaukee and a temporary intake at Muskegon, respectively.

Arrows at the top of the figure indicate relative positions of fixed measuring stations as follows (with distance in kilometers north and south of the Milwaukee-Muskegon section indicated in parentheses): FWPCA stations 17 (12 N) and 18 (wind only, 7 N); MV *Cisco* anchor station, M_2 (12 N); FWPCA station 20 (6 S). For current and temperature measurements at stations 17 and 20, see Figure 8.

10-km-wide coastal strips, respectively, and wavelike fluctuations in isotherm depth elsewhere. There is a suggestion of standing-wave behavior (for instance, indications of nodes at approximately 45 and 100 km from Milwaukee), and this impression is confirmed after comparison with preceding and succeeding ferry crossings and with temperature and current measurements at nearby fixed stations whose positions relative to the ferry track are indicated in the legend and at the top of Figure 7. Stations 17, 18 (wind only), and 20 were those established by the Federal Water Pollution Control Administration and referred to in Figure 8; M_2 was the anchor-station occupied by MV *Cisco* for observations of depth and of distribution of temperature and current.

The anchor-station observations, over a 45-hour interval between August 3 and 5 disclose (1) a wavelike change in thermocline depth with a main period and range of approximately 17 hours and 7 meters, respectively; (2) an epilimnetic current of close to 40 cm/sec and with similar speed and direction throughout the epilimnion; (3) a much smaller hypolimnetic current (about 10 cm/sec and also fairly uniform in depth) running in the opposite direction to the epilimnetic current; and (4) a clockwise rotation of the current vectors at all depths (epilimnion and hypolimnion) with the same periodicity as that of the internal wave. This is consistent with the predicted internal Poincaré wave pattern in Figure 6c, if we assume that the measuring station was not situated at an antinode (with zero or rectilinear currents) or at a nodal point (maximum rotary current but no thermocline displacement).

In the lake, however, as opposed to the simplified model in Figure 6c, a mixture of nodalities is commonly present (for instance, one, three, and five), and this ensures that any offshore station will show some rotary current contribution from one or more of these nodalities, associated with the corresponding internal waves. This surprising and previously unsuspected result— that the most conspicuous current patterns in offshore waters in summer bear no direct relation to wind direction—has been confirmed in great detail by the Federal Water Pollution Control Administration's investigation of the currents in Lake Michigan, as documented in their report. Most of the wind energy goes into the generation of internal waves, typically, standing Poincaré waves in the central region of the lake, and it is only when current direction is averaged over very long time intervals that general relationships with the wind emerge. By contrast, motion near shore, as already mentioned, is more directly related to the wind and commonly takes the form of shore-parallel currents, with the result that the dispersal of inshore waters into offshore regions may sometimes be impeded.

By way of example, Figure 8 (assembled from data supplied by the Federal Water Pollution Control Administration) presents a contrast between a predominantly "offshore" current pattern at station No. 20 (18 km from

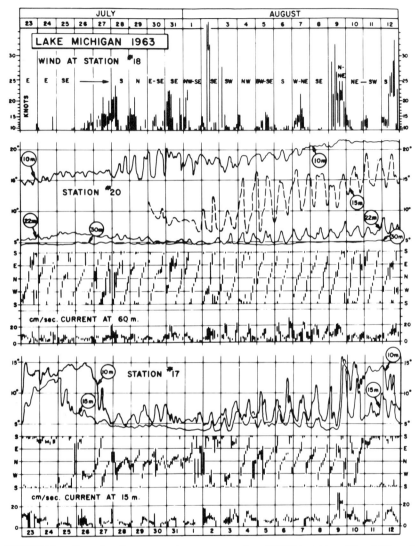

FIGURE 8   Records of current and temperature from FWPCA stations 17 and 20, and of wind from station 18, for the interval July 23 to August 12, 1963. Station positions are given in Figure 7 and its legend.

Comparison is made between (a) hourly readings of temperature at various depths (10, 15, 22, and 30 m at station 20; 10 and 15 m at station 17); (b) 2-hour envelopes (vertical line joining maximum and minimum values) of wind speeds above 9 knots at station 18 (plotted on a square-law scale); and (c) 2-hour envelopes of current speed and direction at 60-m depth at station 20 and 15-m depth at station 17. Current directions are those *toward* which the current flows; wind directions are those *from* which the wind blows.

shore; for position see Figure 7) and some features of "inshore" behavior at another station (No. 17, 4 km from shore). But before this contrast is considered, two points of general interest may be noted: (1) when temperature waves are present at various depths, they are in phase (this indicates an internal wave of the first vertical mode appropriate to a two-layered system); and (2) a relatively precise determination can be made of the internal wave period, for instance, 17.1 hours during August at station 20 is 2.5 percent less than the local inertial period of 17.55 hours. This is consistent with Poincaré wave theory for long wavelength and low nodality.

The current indications in Figure 8 are 2-hour "envelopes" (i.e., vertical lines joining the minimum and maximum speeds and the extremes of direction during each 2-hour interval). During most of the interval covered by the figure, the current direction at 60 m at station 20 (in the hypolimnion) showed a steady clockwise rotation, in phase with the internal wave displayed on the thermograph records. A suggestion of synchronous fluctuations in velocity is also evident and leads to the conclusion that the vector envelopes of the current were ellipses (i.e., not inertial circles), again consistent with Poincaré wave theory.

At station 17, on the other hand, there were long intervals when the current was either northgoing or southgoing (i.e., parallel to the shore), although even at this short distance from the shore rotational behavior associated with internal waves occurred occasionally, for example, during the period of low winds from August 4 to 8. [Indeed, although they may be obscured by near-shore motions, the internal waves of near-inertial period affect the whole of the thermocline surface; and the first hint of their existence in Lake Michigan was provided by water-intake temperature records at Milwaukee (Mortimer, 1963).] But it is evident that the current envelope is not a simple circle or ellipse; the current has a northgoing trend with loops in it, coupled to the internal wave, which is evident on the thermograph traces.

Stations 20 and 17 also showed contrasted responses to wind, for example, the southerly wind on July 27 and the northerly wind on August 9. At the "offshore" station (20) the influence on the thermocline was small. At the western "inshore" station (17), however, the southerly wind produced an upwelling and the northerly wind a downwelling of the thermocline.

It will be noted that the 15-m current measuring depth at station 17 was above the thermocline and showed a southgoing current on July 24 and 25. The upwelling, resulting from the southeastern winds of the following 2 days, raised the thermocline above the current meter, which then showed a northgoing meandering current, followed by a reversal to a southgoing trend for August 2 and 3, followed in turn by the internal wave influence already noted. This gave way, at the downwelling on August 9, to a steady southgoing current.

These interpretative remarks concerning the currents at station 17 have been couched in terms of wind-forcing. But free responses are also involved; and here it is relevant that the shore-parallel currents associated with a Kelvin wave response (see Figure 6a), or with the "current-like" and "wave-like" responses discussed by Csanady (1967b), are negligible outside a coastal strip of characteristic width (10 km). In particular, Csanady has demonstrated, on theoretical grounds, that a nonuniform distribution of wind stress over a large lake, on a scale that may often occur over the Great Lakes, can lead to a concentration of energy in a narrow "coastal jet" current, as one of the possible free responses of a large stratified basin, and perhaps as the interpretation of a persistent shore-bound current in Lake Superior (Ragotzkie, 1966).

Each of the possible mechanisms of coastal entrapment described here is associated with the Coriolis force and with the presence of a thermocline. There is another mechanism, the "thermal bar" (Rodgers, 1966), that can achieve the same result in certain circumstances during periods of spring warming or winter cooling. Because seasonal warming or cooling, with consequent stratification, takes place more rapidly in the shallow inshore water masses than in deeper isothermal waters offshore, there are times when the surface temperatures of the two water masses lie on either side of the temperature of maximum density. Mixing of the two masses at their boundary then produces a mixture more dense than either; the sinking of this mixture develops a region of convergence that impedes, for a time, the outward movement of the inshore water, in which, for a time, the inflow and pollutants from rivers and streams are trapped. This phenomenon may provide part of the explanation of the differences, now coming to light, in phytoplankton distribution between inshore and offshore waters of the Great Lakes. Nalewajko (1966) refers to the thermal bar as a likely explanation of the distribution of *Stephanodiscus tenuis* in Lake Ontario.

The recognition that, at certain times, one of several mechanisms can operate to delay or impede exchange between inshore and offshore water masses will have an important bearing on the interpretation to be placed on water-quality changes and on decisions concerning water-intake and waste-disposal points in the Great Lakes. These matters must now be related, not to whole-lake volumes or retention times, but to their inshore equivalents. This also means that the Great Lakes are more vulnerable to cultural changes than was, perhaps, previously supposed. The effects of such changes may be expected to appear first in the coastal strips of the characteristic width of 10 km or less.

This paper is Contribution No. 6, Center for Great Lakes Studies, University of Wisconsin—Milwaukee.

REFERENCES

Birge, E. A. 1904. The thermocline and its biological significance. Amer. Microbiol. Soc., Trans. 25:5–33.

Birge, E. A. 1906. The oxygen dissolved in the waters of Wisconsin lakes. Amer. Fish. Soc., Trans. 35:142–163.

Birge, E. A., and C. Juday. 1911. The inland lakes of Wisconsin; the dissolved gases of the water and their biological significance. Bull. Wis. Geol. Nat. Hist. Surv., No. 22 (Sci. Ser. No. 7). 259 p.

Csanady, G. T. 1967a. Large-scale diffusion experiments at Douglas Point. Internat. Ass. Great Lakes Res., Proc. 10th Conf., 1967, p. 274–276.

Csanady, G. T. 1967b. Large-scale motion in the Great Lakes. J. Geophys. Res. 72:4151–4161.

Gorham, E. 1958. Observations on the formation and breakdown of the oxidized microzone at the mud surface in lakes. Limnol. Oceanogr. 3:291–298.

Lund, J. W. G. 1950. Studies on *Asterionella formosa* Hass. II. Nutrient depletion and the spring maximum. J. Ecol. 38:1–35.

Mackereth, F. J. H. 1966. Some chemical observations on postglacial lake sediments. Phil. Trans. Roy. Soc. B, 250:165–213.

Mortimer, C. H. 1942. The exchange of dissolved substances between mud and water in lakes. III and IV. J. Ecol. 30:147–201.

Mortimer, C. H. 1959. The physical and chemical work of The Freshwater Biological Association, 1953–57. Adv. Sci. (London) 61:524–530.

Mortimer, C. H. 1961. Motion in thermoclines. Verh. internat. Ver. Limnol. 14:79–83.

Mortimer, C. H. 1963. Frontiers in physical limnology with particular reference to long waves in rotating basins. 6th Conf. Great Lakes Res. Pub. No. 10, Great Lakes Res. Div., Univ. Michigan, p. 9–42.

Nalewajko, C. 1966. Composition of phytoplankton in surface waters of Lake Ontario. J. Fish. Res. Bd. Canada 23:1715–1725.

Naumann, E. 1919. Några synpunkter angående planktons ökologi. Med. särskild hänsyn till fytoplankton. Svensk. bot. Tidskr. 13:129–158. (Swedish; German summary.)

Phillips, O. M. 1966. The dynamics of the upper ocean. Cambridge University Press. 261 p.

Ragotzkie, R. A. 1966. The Keweenaw Current, a regular feature of the summer circulation of Lake Superior. Univ. Wisconsin, Dep. Meteorol. Tech. Rep. No. 29.

Rodgers, G. K. 1966. The thermal bar in Lake Ontario, spring 1965 and winter 1965–66. 9th Conf. Great Lakes Res., Pub. No. 15, Great Lakes Res. Div., Univ. Michigan, p. 369–374.

Schooley, A. H. 1967. Wake collapse in a stratified fluid. Science 157:421–423.

Shapiro, J. 1966. The relation of humic color to iron in natural waters. Verh. Internat. Ver. Limnol. 16:477–484.

Shapiro, J. 1967. Induced rapid release and uptake of phosphate by microorganisms. Science 155:1269–1271.

V.

PREVENTIVE
AND CORRECTIVE
MEASURES

GERARD A. ROHLICH
University of Wisconsin, Madison

Engineering Aspects of Nutrient Removal

More than 35 years ago Naumann (1931) defined eutrophication as "an increase of the nutritional standards especially with respect to nitrogen and phosphorus." It has been recognized for several decades that natural drainage from forest areas, agricultural lands, and urban areas contributes these and other nutrients in varying amounts. It has been known also that in many locations the inorganic and organic constituents of municipal and industrial waste waters discharged to receiving streams and lakes are major sources of nutrients for rooted aquatic plants and algae.

In a rational approach to the design and construction of engineering works for the control of nutrients in bodies of water, the initial step is to conduct a survey to determine the origins and quantities of nutrients and to identify the point sources most readily subject to control. Such a survey should include chemical and biological investigations to determine which nutrients are limiting, so that maximum benefit may be realized from expenditures for nutrient control and removal. In defining the problem, it is desirable to know what nutrients have to be removed, how much of each must be removed, what the cost will be, and what results we can expect.

Other papers in this volume are concerned with problems associated with urban, agricultural, and forest drainage; management of aquatic plants and algae; dredging; mechanical removal of organic production; and shoreland corridor regulations. In this paper, discussion is directed principally to nutrient removal from municipal and industrial waste waters and is restricted to removal of phosphorus and nitrogen.

In conventional methods of waste water treatment, the principles of mechanical, hydraulic, chemical, and biological separation have been used to remove suspended solids and bacteria and either to remove the biochemical oxygen demand or to satisfy it. Until recently, laboratories of relatively few treatment plants have determined the nutrient elements that influence the productivity of the receiving bodies of water.

Although the early workers were not concerned with nutrient removal, Gleason and Loonam (1933, 1934), in developing the Guggenheim Process more than 30 years ago, used the removal of nitrogen as one measure of the effectiveness of the process. Unfortunately, they did not report on the removal of phosphorus. In their 1934 paper, they stated that the process consisted essentially of two operations:

1. Removal of the suspended matter and a major portion of the non-basic dissolved material by coagulation and precipitation with an iron salt and an alkali, preferably ferric sulphate and lime, and
2. Removal of the soluble basic compounds by passing the clarified sewage through a bed of base exchange zeolite.

They added:

Two auxiliary operations, first, filtration and incineration of the sludge with recovery of the iron for further use, and second, regeneration of the zeolite and recovery of the basic nitrogen, principally as ammonia, complete the process.

In a pilot plant having a capacity of about 30,000 gpd that was operated over a period of 7 months at the North Side Sewage Treatment Works in Chicago, the investigators reported the following removals.

	Average for 6-Month Period (%)	Range in Monthly Average (%)
Biological oxygen demand	91.2	88.3–95.0
Suspended solids	97.4	96.0–99.6
Organic nitrogen	79.3	74.0–83.5
Ammonia nitrogen	67.4	58.4–73.0

In 1934, construction costs for a 12-million-gpd plant using the process were estimated to be $40,900 per million gallons, and the operating cost per million gallons was estimated to be $36.16.

In discussing this paper, A. Potter observed:*

As to the wisdom of a more general adoption of sewage treatment plants capable of producing such highly satisfactory results as shown in the author's

*See Gleason and Loonam (1934), page 466.

paper, I am not prepared to state, but considering the activities of Izaak Walton Clubs . . . the time may come when the highest degree of purification possible will be demanded everywhere.

Other studies on the effectiveness of ion exchange for nitrogen removal have been reported by Eliassen *et al.* (1965), Culp and Slechta (1966), and E. J. Nesselson (unpublished data).* Nesselson reported these findings in laboratory studies:

1. Strong base anion exchangers, regenerated with common salt, perform satisfactorily for the removal of nitrate. A number of evaluations established that Amberlite IRA-410, in the treatment of trickling filter effluent, has an exchange capacity of 8.7 to 11.7 kilograms/cu. ft. as $CaCO_3$. It operates under an efficiency of 4.0 to 6.8 lbs. of $NaCl$/kilograin of anions removed. Nalcite SAR had respective values of 6.5 to 7.7 kgrs/cu. ft. as $CaCO_3$ and 6.0 to 9.9 lbs. $NaCl$/kgr of anions removed. Both media were regenerated to an end point of 1 ppm of nitrogen in the waste regenerant. A minimum volume of about 7 per cent of the influent feed was required for this operation.

2. The removal of ammonia nitrogen by nuclear sulfonic cation exchangers was investigated. Nalcite HCR has an exchange capacity of 16.0 to 22.0 kgrs/cu. ft. as $CaCO_3$ in the treatment of activated sludge effluent. It operates with an efficiency of 1.4 to 2.5 lbs. $NaCl$/kgr cations removed. Amberlite IR-120 has respective values of 13.0 to 17.0 kgrs/cu. ft. as $CaCO_3$ and 1.3 to 2.6 lbs. $NaCl$/kgr cations removed. A minimum volume of about 6 per cent of the influent feed was required to perform regeneration.

Eliasson *et al.* (1965) found that when an effluent from the activated sludge process, followed by sand filtration, was treated with a strong base anion exchanger, the average removal of nitrogen with the first 200 bed volumes was 84 percent, the initial nitrogen concentration of the feed to the exchange column being 18 ppm and the effluent concentration 2.9 ppm. Culp and Slechta (1966) showed reductions in ammonia nitrogen of 85 to 99 percent, with influent concentrations of ammonia nitrogen at approximately 20 to 30 ppm and effluent concentrations less than 1 ppm for throughputs of up to 400 bed volumes.

P. A. Kuhn (unpublished data†), after considering preliminary studies by Nesselson on air stripping in packed towers to remove ammonia nitrogen, showed that it is possible to remove ammonia nitrogen by desorption in a tower packed with Raschig rings.

Removal of Inorganic Nitrogen from Sewage Effluent. Ph.D. thesis, Univ. Wisconsin (1954).

†*Removal of Ammonia Nitrogen from Sewage Effluent.* M.S. thesis, Univ. Wisconsin (1956).

In the studies conducted, the optimum pH for stripping was 11.0. The optimum pH value was determined from a study of ammonia nitrogen removals through the range pH 8.0 to pH 12.0. Increased removals were obtained by increasing pH values. No significant difference existed between removals at pH 11.0 and pH 12.0.

The effect of air-to-liquid loading, expressed as cfm of air per gpm of liquid, was studied at ratios of 40, 59, 85, 230, and 447 cfm/gpm. Removals obtained at these ratios were 15.1, 28.5, 37.8, 67.0 and 78.7 percent, respectively. These results are consistent with theory, which predicts best removals at an air:liquid ratio of 453 cfm/gpm.

Ammonium nitrogen removals were studied on half-inch Raschig rings at depths of 2.5, 4.0, 5.0, 5.5, and 7.0 ft in an 8-in.-diameter column. At the optimum pH value, pH 11.0, removals of ammonia nitrogen from sewage effluent of 53.9, 71.1, 78.5, 82.2, and 92.3 percent, respectively, were obtained at the depths stated. Loadings for these removals were 52 to 55 cfm of air and 0.10 gpm of effluent. These results are consistent with the theory of design of desorption columns.

Culp and Slechta (1966) conducted similar studies, and with influents containing approximately 25 mg/liter of ammonia nitrogen they reported removals of 98 percent at a pH value of 10.8, the effluents containing less than 1 mg/liter of ammonia nitrogen.

Chemical methods for removing phosphorus from sewage and sewage effluents have received considerably more attention than studies on removing nitrogen.

More than 20 years ago, Sawyer (1944) called attention to the need for "further treatment of sewage plant effluents or auxiliary treatment of sewages to remove fertilizing elements" and discussed methods of removing nitrogen or phosphorus or both. He reported results of studies on chemical treatment of sewages with ferric chloride and showed that total phosphorus concentrations could be reduced to approximately 0.50 ppm and soluble phosphorus concentrations to values as low as 0.01 ppm with dosages of 50 ppm of ferric chloride.

Rudolfs (1947) conducted extensive studies on phosphates in sewage and sludge treatment. As part of these studies he demonstrated that phosphates in sewage are readily precipitated by lime. Owen (1953) reported that analyses of domestic sewage from nine communities in Minnesota, with populations ranging from 1,200 to 940,000, showed that the raw sewage of these communities at that time contained 1.5 to 3.7 g of elemental phosphorus per capita per day, with a median value of 2.3 g, and that the amount of phosphorus removed in conventional sewage-treatment plants varied from 2 to 46 percent. The lowest removal was in a primary plant in which the removal of biochemical oxygen demand was only 24 percent. Removals of 42 and 46 percent of phosphorus were obtained in plants showing BOD removals of 94 and 85 percent, respectively. In general, those plants with the highest 5-day BOD removals also had higher removals of phosphorus. On the basis of laboratory tests indicating phosphorus removals of 95 percent on samples of effluent from a trickling filter that were treated with 545 ppm of CaO, Owen conducted studies on an operating plant consisting of fine screens, high-rate trickling filter, secondary clarifier, low-rate trickling filter, final clarifier, and facilities for vacuum filtration of sludge.

Slaked lime when added to the influent of the final settling tank in a concentration of 545 ppm (CaO) reduced the phosphorus concentration from 7.4 ppm to 1.7 ppm, a reduction of 77 percent. Although the laboratory studies had indicated that at the same dosage the phosphorus concentration was reduced from 6.0 ppm to 0.3 ppm with 1 hour of settling, Owen concluded that the plant-scale tests would have more closely approached the laboratory results if mixing and flocculation had been adequate and settling had been improved. He estimated (1953) that the chemical costs for year-round treatment with unslaked lime would be approximately $7,600 for treating sewage flowing at a rate of 0.5 mgd.

Lea et al. (1954) conducted laboratory studies and pilot plant investigations at the Madison, Wisconsin, Nine Springs Sewage Treatment Plant. Various coagulants were used, but in the most intensive part of the study, aluminum sulfate in the form of "filter alum" was used as the coagulant. The laboratory studies indicated that at an alum dosage of 200 mg/liter, 96 to 99 percent of the soluble phosphates were removed from effluent samples from the plant. A process for the recovery of the coagulant was developed in which the aluminum hydroxide floc with the sorbed phosphate was pumped from the sedimentation tank to a recovery tank. Sodium hydroxide was added to the floc suspension to raise the pH to approximately 11.9. At this pH, the aluminum hydroxide is converted to soluble sodium aluminate, and the phosphate is converted to soluble sodium phosphate. Addition of calcium chloride at this point results in the formation of tricalcium phosphate, which is readily separated from the sodium aluminate. The comparatively phosphate-free sodium aluminate is then adjusted in strength and pumped back to the flocculation unit to be reused as a coagulant.

A pilot plant was operated at a flow rate of 10 gpm. Results of material balances in the pilot plant were presented by the authors. Tables 1, 2, and 3 summarize the results obtained.

As part of the study, chemical costs were estimated for operating the system for a sewage flow of 14.4 mgd, which was the average daily flow for Madison, Wisconsin, at the time of the study. The total chemical cost per year was estimated to be $106,450, or $20.25 per million gallons. If the tricalcium phosphate were salable, the chemical costs would be reduced to $14.90 per million gallons, based on 1954 chemical costs.

TABLE 1 Results of Removing Phosphate with 200 ppm of Aluminum Sulfate (Alum)[a]

Sampling Period (hr)	Phosphorus, P					Total Organic N[b]			Biological Oxygen Demand[b]		
	Influent[b] (ppm)	Effluent (ppm)		Removal (%)		Influent (ppm)	Effluent (ppm)	Removal (%)	Influent (ppm)	Effluent (ppm)	Removal (%)
		Unfiltered[b]	Filtered[c]	Unfiltered[b]	Filtered[c]						
24	4.08	0.578	0.088	86	98	1.078	0.281	74	–	–	–
26	4.45	0.80	0.024	82	99	1.112	0.290	67.6	–	–	–
38	5.90	0.80	0.022	86	99	1.190	0.50	58	36	11.7	67.5

[a]Flow rate, 10 gpm; test time, 108 hr.
[b]Unfiltered before analysis.
[c]Filtered before analysis.

TABLE 2 Results of Removing Phosphate with 200 ppm of Aluminate Solution[a]

Influent[b] (ppm)	PO$_4$, as P				Total Organic N[b]			Biological Oxygen Demand[b]		
	Effluent (ppm)		Removal (%)		Influent (ppm)	Effluent (ppm)	Removal (%)	Influent (ppm)	Effluent (ppm)	Removal (%)
	Unfiltered[b]	Filtered[c]	Unfiltered[b]	Filtered[c]						
4.62	0.60	0.024	87.0	99	1.46	0.384	74.0	70.85	5.0	76.0
4.10	0.60	0.036	85.5	99	1.85	0.622	66.4	17.9	3.34	80.7
4.58	0.64	0.018	86.0	99	1.344	0.608	54.8	13.1	4.05	69.0
4.51	0.99	0.001	78.0	99	1.114	0.496	56.5	16.15	7.25	55.0

[a]Flow rate, 10 gpm; test time, 96 hr.
[b]Unfiltered before analysis.
[c]Filtered before analysis.

TABLE 3 Results of Removing Phosphate with 200 ppm of Sodium Aluminate from Alum Recovery[a]

Sampling Period (hr)	PO$_4$, as P Influent[b] (ppm)	Effluent (ppm) Unfiltered[b]	Effluent (ppm) Filtered[c]	Removal (%) Unfiltered[b]	Removal (%) Filtered[c]	pH Influent	pH Effluent
23	4.31	0.600	0.013	86.1	97	7.8	7.9
27	4.73	0.506	0.142	89.3	97	7.75	7.95
24	4.37	0.830	0.016	81	96.4	7.8	8.00
22	4.51	0.992	0.032	78	93	7.8	8.1
29	4.40	1.01	0.031	77	93	7.8	8.1

[a]Flow rate, 10 gpm.
[b]Unfiltered before analysis.
[c]Filtered before analysis.

More recently, Malhotra *et al.* (1964) studied removal of phosphorus and nitrogen compounds in laboratory tests using lime, alum, and sodium hydroxide. Using prices for chemicals as of July 1962, they reported that, at 95 percent removal, the costs of lime, alum, and sodium hydroxide were $32, $73, and $330 per million gallons, respectively. A removal of 99 percent could be obtained with lime treatment only at a chemical cost of $41 per million gallons.

Biological methods by which living organisms incorporate the nutrients into protoplasmic structure, thereby removing nutrients from waste water or waste-water effluents, are also in various stages of investigation and development. A number of process flow systems have been investigated, both in the laboratory and in pilot plants.

Stabilization ponds have been used for many years, and data on BOD and solids-reduction performance of such ponds are more readily available than data on their nutrient removal. Fitzgerald and Rohlich (1958), however, in an evaluation of stabilization pond literature, pointed out that considerable amounts of plant nutrients are removed from sewage by the action in ponds. Only a small quantity of the nutrients is lost from ponds but is removed from solution and concentrated in the algal cells. Wide variations in removal of phosphorus and nitrogen are noted for ponds for which such data are available. Parker (1962) reported data on an Australian installation of eight ponds operated in series. In this series of ponds, organic nitrogen levels of 26.3 mg/liter in the raw sewage influent was reduced to 7.5 mg/liter in the effluent from the eighth pond. The raw sewage contained 32.4 mg/liter of ammonia nitrogen, and the eighth pond contained 18.9 mg/liter. The results were obtained in summer at temperatures of about 70°F. In winter, at temperatures of about 48°F, organic nitrogen decreased about 65 percent, and ammonia nitrogen increased.

Bush *et al.* (1961) reported on a pond following treatment of sewage by the activated sludge process. Phosphate removals ranged from 19 to 68 percent, ammonia nitrogen removals from 63 to 90 percent, nitrate from 27 to 60 percent, and organic nitrogen from 32 to 74 percent. Phosphate concentrations in the effluent were in the range 5 to 15 ppm and in the effluent, 2 to 9 ppm. Effluent concentrations of ammonia were usually less than 2.5 ppm, and nitrate concentrations were 1 to 4 ppm.

The wide variations in removals of nitrogen and phosphorus in ponds have also been noted by Mackenthun and McNabb (1961), Fitzgerald (1961), and others. Fitzgerald showed removals of about 70 percent of inorganic nitrogen under favorable summer conditions. Average removal of nitrogen throughout the year in an experimental pond, however, was about 30 percent. Phosphorus removals during periods of high algal growth coincided with high pH values and were probably the result of coagulation and adsorption. Bogan (1961) reports similar results.

At present, ponds in many locations cannot be considered dependable as biological process operations for removal of nutrients until improved operating and control procedures and satisfactory, low-cost methods of harvesting algae are developed. Engineering design criteria for BOD removal will continue to be used; at present, design criteria for nutrient removal are not well established.

Several workers in the laboratory and in field experiments have investigated removal of nitrogen and phosphate in the activated sludge process or in modifications of the process. Wuhrmann (1957) stressed the importance of the C:N:P ratio if maximum amounts of nitrogen and phosphorus are to be incorporated in cell tissue. Earlier, Sawyer (1944) reported on experiments in which glucose was added to sewage in varying amounts. He showed that the addition of carbonaceous matter to domestic sewage in the activated sludge process would result in increased amounts of activated sludge and in the conversion of more inorganic nitrogen and phosphorus to organic forms. If an industrial waste having a high carbohydrate content were available, therefore, a process could be devised in which the carbon source could be fed in controlled amounts in relation to available nitrogen and phosphorus. The biological sludge produced would, of course, have to be handled in such a manner that the nutrients would not be returned to the aquatic environment.

Elimination of nitrogen by controlled microbial nitrification–denitrification has been reported by Wuhrmann (1957, 1964), Ludzak and Ettinger (1961), Johnson and Schroepfer (1964), and others. Basically, the unit operations consist of extensive nitrification by aeration in the presence of activated sludge. The highly nitrified mixed liquor then is denitrified in the absence of dissolved oxygen in a suitable mixing tank; next, it is allowed to settle in a basin to remove the activated sludge. Wuhrmann (1964) reported that the denitrification process could be run without difficulty and with high reliability. He found that when influent nitrogen concentrations were in the range of 25 to 30 ppm, final effluents contained 2 to 4 ppm of total nitrogen. Ludzak and Ettinger (1961) reported nitrogen removals of up to 77 percent under favorable denitrification conditions, and Johnson and Schroepfer (1964) reported removals in the range of 63.3 to 69.2 percent.

Process flow diagrams for the denitrification process are given in the articles cited above.

Tenney and Stumm (1965) proposed another process of interest, basing their proposal on research that they had conducted on chemical flocculation of microorganisms in biological waste treatment. The process consists of primary treatment followed by aeration of short detention time sufficient to remove organic soluble material. The effluent from the aeration tank is then discharged to a rapid-mix and flocculation basin and finally to a settling tank. The chemical flocculation complements the biological waste-treatment process and substitutes for part of it by removing phosphates and effecting a phase separation of the microorganisms.

Operational control in the activated sludge process for maximum phosphate removal has been discussed by Vacker et al. (1966). Basing their conclusions on studies of phosphate removal through municipal waste-water treatment at San Antonio, Texas, they proposed the following design and operation considerations:

1. Facilities should be provided that will permit operation at a constant loading of BOD to aeration solids. Optimally, this should be about 50 lb of BOD to 100 lb of aeration solids. To avoid unduly long periods of retention of solids in secondary clarifiers, which occurs during periods of low flow, an aerated surge tank for return sludge should be provided. The rate of flow of raw waste should be controlled by providing an aerated surge tank for use during periods of peak flows.

2. The facilities for air supply should be sufficient to provide dissolved oxygen of at least 2 mg/liter in the first half of the aeration tank and about 5 mg/liter at the effluent end of the tank. Over-aeration, however, should be avoided because it results in excessive nitrification and aerobic digestion of solids.

3. Phosphate-rich, waste-activated sludge should be disposed of separately from primary sludge. Among the methods that may be used are lagooning, irrigation, filtration and drying, filtration or centrifuging followed by incineration, and aerobic or anaerobic digestion followed by chemical precipitation and recovery of phosphate from the digestor liquor.

4. To treat raw waste water in which the BOD-to-PO_4 ratio is low, it will be necessary to augment biological treatment by cationic precipitation of phosphate.

These authors reported that maximum phosphate removal was obtained at an average daily BOD loading rate of about 50 lb of BOD per 100 lb of aeration solids, with a sludge volume index of about 150, a dissolved-oxygen level of about 2.0 mg/liter at about the midpoint in the aeration tanks, and a dissolved-oxygen level of about 5 mg/liter at the effluent end of the aeration tanks.

Levin and Shapiro (1965) conducted studies (in the laboratory and on a full plant scale) on metabolic uptake of phosphorus by waste-water organisms and demonstrated "luxury uptake" of dissolved orthophosphate by activated sludge. They placed emphasis on the rate of aeration as a major factor in the control of the amount of phosphate uptake.

Other alternatives to be considered in reducing the nutrient input to receiving waters from point sources such as waste-water effluents are diversion and irrigation. In particular locations these may prove to be the most feasible and practicable after engineering analyses of alternates have been made.

A variety of physical, chemical, and biological processes for nutrient removal have been proposed from laboratory and pilot plant studies, and techniques for operational control have been developed. The improvement and implementation of these techniques in large-scale pilot plants and more efficient operation of existing sewage-treatment plants are needed to further the establishment of design criteria and costs.

REFERENCES

Bogan, R. H. 1961. The use of algae in removing nutrients from domestic sewage. Algae and Metropolitan Wastes, Trans. of the 1960 Seminar, Div. of Water Supply and Pollution Control, SEC TR W61-3.

Bush, A. F., J. P. Isherwood, and S. Rodgi. 1961. Dissolved solids removal from waste water by algae. J. Sanit. Engr. Div. Amer. Civil Engrs. SA3 87:39.

Culp, G., and A. Slechta. 1966. Nitrogen removal from sewage. Final Progress Report. USPHS Demonstration Grant 86-01, February, 1966. 38 p.

Eliassen, R., B. M. Wyckoff, and C. D. Tonkin. 1965. Ion exchange for reclamation of reusable supplies. J. A.W.W.A. 57:1113–1122.

Fitzgerald, G. P., and G. A. Rohlich. 1958. An evaluation of stabilization pond literature. Sewage and Industrial Wastes 30:1213–1224.

Fitzgerald, G. P. 1961. Stripping effluents of nutrients by biological means. Algae and Metropolitan Wastes, Trans. of the 1960 Seminar, Div. of Water Supply and Pollution Control, SEC TR W61-3.

Gleason, G. H., and A. C. Loonam. 1933. The development of a chemical process for the treatment of sewage. Sewage Works J. 5:61–73.

Gleason, G. H., and A. C. Loonam. 1934. Results of six months operation of a chemical sewage purification plant. Sewage Works J. 6:450–468.

Johnson, W. K., and G. J. Schroepfer. 1964. Nitrogen removal by nitrification and denitrification. J. Water Poll. Control Fed. 36:1015–1036.

Lea, W. L., G. A. Rohlich, and W. J. Katz. 1954. Removal of phosphate from treated sewage. Sewage and Industrial Wastes 26:261–275.

Levin, G. V., and J. Shapiro. 1965. Metabolic uptake of phosphorus by wastewater organisms. J. Water Poll. Control Fed. 37:800–821.

Ludzak, F. J., and M. B. Ettinger. 1961. Controlling operation to minimize activated sludge effluent nitrogen. J. Water Poll. Control Fed. 34:920–931.

Mackenthun, K. M., and C. D. McNabb. 1961. Stabilization pond studies in Wisconsin. J. Water Poll. Control Fed. 33:1234–1251.

Malhotra, S. K., G. F. Lee, and G. A. Rohlich. 1964. Nutrient removal from secondary effluent by alum flocculation and lime precipitation. Int. J. Air Water Poll. 8:487–500.

Naumann, E. 1931. Limnologische Terminologie. Urban and Schwarzenberg, Berlin-Wien, p. 153 and 413.

Owen, R. 1953. Removal of phosphorus from sewage plant effluent with lime. Sewage and Industrial Wastes 25:548–556.

Parker, C. D. 1962. Microbiological aspects of lagoon treatment. J. Water Poll. Cont. Fed. 34:149–161.

Rudolfs, W. 1947. Phosphates in sewage and sludge treatment. II. Effect on coagulation, clarification and sludge volume. Sewage Works J. 19:178–190.

Sawyer, C. N. 1944. Biological engineering in sewage treatment. Sewage Works J. 16:925–935.

Tenney, M. W., and W. Stumm. 1965. Chemical flocculation of microorganisms in biological waste treatment. J. Water Poll. Control Fed. 37:1370–1388.

Vacker, D., C. H. Connell, and W. N. Wells. 1966. Phosphate removal through municipal wastewater treatment at San Antonio, Texas. Paper presented at Annual Short Course Texas Water and Sewage Works Ass. March 9, 1966. Texas A. & M. University.

Wuhrmann, K. 1957. Die dritte Reinegungsstufe: Wege und bisherige Erfolge in der Eliminierung eutrophierender Stoffe. Schweiz Z. Hydrol. 19:409–427.

Wuhrmann, K. 1964. Nitrogen removal in sewage treatment processes. Verh. Int. Ver. Limnol. 15:580–596.

S. R. WEIBEL

U.S. Public Health Service, Washington, D. C.

Urban Drainage

as a Factor

in Eutrophication

In the broad sense, the topic "urban drainage" includes sewage as well as storm-water runoff and combined-sewer overflows. Since current research activities on the control of nutrients in connection with sewage treatment seem well documented, this paper is devoted to storm-water runoff and combined-sewer overflows as sources of water pollution, including nutrient contributions, an area of study where much work is needed.

The urbanization explosion means that more people, more demands for water for all purposes, more wastes, more storm-water runoff—all are impressed upon existing time, space, facilities, and habits already representing huge investments. It seems appropriate first to see where this dense urbanization is expected to develop and then to consider some of the drainage problems and efforts toward control.

URBANIZATION

Projections in the study by Pickard (1967) indicate that by the beginning of the twenty-first century some 77 percent of the 312 million population of our 48 mainland states and the District of Columbia will occupy about 11 percent of the land. These 240 million urban dwellers are expected to live in three major urban regions, six metropolitan areas, and 13 outlying urban regions, as indicated in Figure 1. These areas are identified as follows: the metropolitan belt (stretching from the Atlantic seaboard to the Wisconsin-Illinois complex) and the California and Florida regions; six metropolitan areas (St. Louis, Louisville, Memphis, Oklahoma City, Twin Cities, and

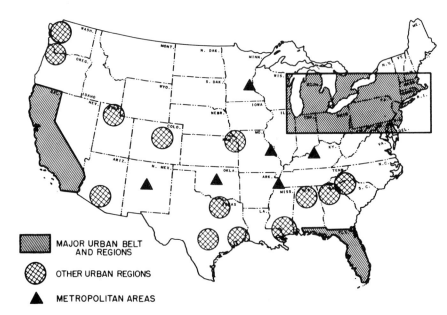

MAJOR URBAN BELT
AND REGIONS

OTHER URBAN REGIONS

METROPOLITAN AREAS

FIGURE 1   Urbanization in the year 2000. Sources: Pickard (1967) and Chicago Tribune (March 26, 1967. Section 3A, p. 1).

Albuquerque, each expected to have at least 1 million persons); and the 13 urban regions (Carolina Piedmont, north central Georgia, north central Alabama, central Gulf Coast, Texas-Louisiana Gulf Coast, north central Texas, south central Texas, Missouri-Kaw Valley, Salt Lake Valley, Colorado Piedmont, Puget Sound, Willamette Valley, and metropolitan Arizona).*

Other sources (National Academy of Sciences, Committee on Pollution, 1966) estimate that as much as 95 percent of the total population may live in urban environments at the turn of the century.

We had rather a good start on this by 1960. The population in urban areas then was about 125 million, or 70 percent of the total for the 48 mainland states plus the District of Columbia. About 98 percent of the urban population was served by community sewer systems at that time.

Although there has been no tabulation of the total number of communities ("communities" include housing developments with more than 100 residents), the number of communities having sewers in 1962 was 11,350 for the 50 states and the District of Columbia (U.S. Public Health Service, 1964). About 70 percent of our communities have populations of fewer than 5,000 persons. In 1960, sewered communities ranged in size from less than

*Chicago Tribune, March 26, 1967. Section 3A, page 1.

$1/2$ mi^2 to 454 mi^2 for Los Angeles (Weibel *et al.*, 1964). The most densely populated large city was New York, with 39 persons per acre. The metropolitan belt is expected to cover 159,000 mi^2 and have a population of 129 million persons by the year 2000.

WATER USE

Aside from future needs for increased supplies of suitable quality water for domestic, industrial, and agricultural uses, demands for recreational waters will probably outrun population growth. The demographic projections into areas generously endowed with surface water are consistent with this. In addition, recreation is already a large industry and will continue to grow. An important item in the control task ahead will be the coordination of political and technical elements in the broad planning so necessary for economy and effectiveness.

RAINFALL

Rain contributes constituents directly and indirectly to surface waters. It also serves as an eroding, dissolving, and transporting agent in storm-water runoff and in combined-sewer overflows. Rainfall averages 30 in. annually over the United States, ranging from less than 2 in. in the Southwest to some 150 in. in the Northwest.

The sources of rainwater constituents are the earth's soil and the by-products of man's activities. These become atmospheric pollution and wash out with rainfall. Lightning is claimed to be a minor factor in the fixation of rainfall nitrogen (Gambell and Fisher, 1964; Junge, 1963). Rainwater may further eutrophication through the direct contribution of constituents such as phosphorus to surface waters.

During the storm-water runoff studies at Cincinnati, many summer storms were highly local, sometimes with a downpour at one location and no rain $1/2$ mile away. Thus many times only parts of a city sewer system receive runoff.

Constituent concentrations in rainfall at the urban storm-water runoff study site at Cincinnati are shown in Table 1. Concentrations of solids, chemical oxygen demand, and nutrients are low compared with those of storm-water runoff, or combined-sewer overflows, or sewage, as shown in Tables 3, 5, and 7. Nonetheless, they warrant consideration. For instance, inorganic nitrogen (as N) averages 0.69 mg per liter, and total acid-hydrolyzable phosphate (which can eventually become soluble phosphate), called total phosphate P, and expressed as P in this paper, averages 0.08 mg per liter. These are in excess of Sawyer's (1952) nutrient thresholds of 0.3 and 0.01 mg per liter for inorganic nitrogen N and inorganic phosphorus P, respectively, in lake waters containing algae blooms.

The inputs of total phosphate P into Lake Erie have been estimated as shown in Table 2 (U.S. Department of the Interior, 1967).

Annual rainfall averages 32 in. at several large cities around Lake Erie and Detroit. If we assume that this is the annual rainfall on Lake Erie and, further, that its total phosphate P is the 0.08 mg per liter observed at Cincinnati, the rainfall phosphate contribution to the 9,910-mi^2 lake might be 10,000 lb per day, or about 6.5 percent of the total load. On the other hand, if the rainfall total phosphate P were the 0.027 mg per liter found in rain at Coshocton,

TABLE 1   Rainfall Constituents Measured at Cincinnati, Ohio, August 1963 to December 1964

Constituents	Average Concentration (mg/liter)	Amount[a] (1,000 lb/mi^2 per year)[b]
Suspended solids	13	57
Chemical oxygen demand	16	70
Total N[c]	1.27	5.5
Inorganic N[d]	0.69	3.0
Total PO$_4$ (as P)[e]	0.08	0.35

[a]Computations are based on assumption that annual rainfall is 30 in. (U.S. average).
[b]Multiply by 0.175 for 1,000 kg per km^2 per year.
[c]Σ four forms as N.
[d]Σ three forms as N.
[e]Total acid-hydrolyzable phosphate.

TABLE 2   Total Inputs of Phosphate P to Lake Erie, by Sources

Sources	PO$_4$ (as P) (lb/day)[a]
Lake Huron	< 20,000
Rural runoff	20,000
Municipal	
Detergents	70,000
Human excreta	30,000
Urban land	6,000
Industrial (direct)	6,000
Total	152,000

[a]Multiply by 0.454 for kilograms per day.

Ohio (Weibel *et al.*, 1966a), which approximates the 0.025 mg per liter suggested total phosphate P goal for western Lake Erie (U.S. Department of the Interior, 1967), the direct rainfall contribution would be 2 percent of the total load.

MUNICIPAL SEWAGE

Municipal sewage is a major source of waterborne pollution, including plant nutrients. Unfortunately, routine analyses and characterizations of sewage have rarely included information on nutrients, particularly on phosphates; therefore, data are scarce and fragmentary. In general, sewage derives from excreta, ground garbage, laundry and other cleaning wastes from households, wastes from commercial and industrial processes, and cleanup and wash operations from innumerable industrial activities. Sewage may carry infectious and toxic agents, plant nutrients, persistent organics, oxygen-demanding substances, petrochemicals, oil, radioactive substances, heat, acids, and sludges, to name several.

In the separate sewer system, sewage is independently collected and transported in a pipe system to treatment before discharge. The storm-sewer system, or subsystem, independently carries storm-water runoff to convenient watercourses. The two functions are combined, rather poorly, in the combined-type system, wherein the sewage, which is the only flow in the large sewers during dry weather, is diverted from that sewer to an intercepting sewer before reaching the outlet, and is led to the sewage-treatment plant. During wet weather, the diversion-to-interceptor arrangement accepts an amount of the sewage-and-storm-water mixture equal to 1½ to 5 times the sewage flow during dry weather, the amount depending on design and other circumstances. The rest of the mixture, which may have flows of 50 to 200 times the sewage flow during dry weather, overflows directly to a watercourse.

Past studies, mostly studies of flow relationships and dilutions rather than amounts of constituents, have yielded estimates of the annual loss of sewage through overflows at about 3 percent, although during storms the losses might be as high as 60 to 90 percent, the amount depending on the storm intensity and diversion capacity (Camp, 1963). Thus, we are interested in the character and amount of (1) raw sewage escaping untreated through combined-sewer overflows, (2) storm water involved, and (3) sewage filth (stranded during dry weather because of the low velocities) awaiting the scouring effect of storm-water flow. Storm-water constituents and stranded filth were not included in the 3-percent annual-loss estimates. In fact, as will be shown later, urban storm-water runoff itself, without any sewage, may contain annually quantities of some of the same contaminants, amounting to more than 3 percent of those of raw sewage.

A modest review of literature, for a general characterization of community raw sewage, including plant nutrients, was rather disappointing because of the incomplete, out-of-date, and ambigious nature of much of the information. Rather than repeat this, I present in Table 3 what I believe, from a review and discussions with other workers, to be a fair and up-to-date characterization of community sewage (mostly domestic).

The amounts of constituents produced annually per unit area for the various sources are compared later in this paper. In passing, note that at 10 persons per acre, which is the Cincinnati city population density, 3-percent annual loss of raw sewage nutrients (Table 3) would amount to 1,700 lb per mi^2 of inorganic nitrogen N and 600 lb per mi^2 of total phosphate P. Thirty inches of rainfall annually at concentrations observed at Cincinnati (Table 1) contain 3,000 lb of inorganic nitrogen N per mi^2 and 350 lb of total phosphate P per mi^2.

Research is continuing on methods for removal of nutrients, as well as other constituents, from sewage and from effluents from some types and stages of conventional sewage-treatment processes. These are documented in the literature under authorship of such authorities as Sawyer and Rohlich (see Mackenthun, 1965). A recent publication entitled "Eutrophication and Water Pollution" includes a review of treatment (Martin and Weinberger, 1966). It is concluded that methods for substantial removal of nutrients from sewage are available, for a price.

TABLE 3   Generalized Composition of Community Raw Sewage (Largely Domestic)

Constituents	Average Concentration[a] (mg/liter)	Amount Produced (1,000 lb/mi² per year)[b]	
		10 persons per acre	50 persons per acre
Suspended solids	200	390	1,900
Chemical oxygen demand	350	680	3,400
Biochemical oxygen demand	200	390	1,900
Total N[c]	40	78	390
Inorganic N[d]	30	58	290
Total PO₄ (as P)[e]	10	20	97

[a]Assuming 100 gpcd.
[b]Multiply by 0.175 for 1,000 kg per km^2 per year.
[c]Σ four forms as N.
[d]Σ three forms as N.
[e]Total acid-hydrolyzable phosphate (which can eventually become soluble phosphate).

It should be emphasized that the sewage-treatment plant, however sophisticated, can treat only the sewage it receives. Uncontrolled escape of significant portions of the raw sewage from combined-sewer overflows, or from overloaded or infiltrated separate sanitary sewers, may seriously discount plant performance.

SEWER SYSTEMS

The two general patterns of drainage systems for urban sewage and storm-water runoff in the United States are called "separate" systems and "combined" systems. The first type includes a separate sanitary-sewer system of closed conduits generally designed to take advantage of the natural slopes of the local topography to enable gravity flow to a treatment plant near a receiving stream, lake, or estuary. This is usually one system, but there may be more where a community straddles drainage divides. The storm-water runoff is collected and transported in separate pipes, gutters, or ditch-type storm drains to handy local waterways. There is usually more than one system of storm sewers in a community.

The combined system attempts to do the whole job by connecting the wastes as well as the roof leaders, area and footing drains, catch basins, and other components, to the single large street sewers. These are designed to accommodate storm-water runoff from storms up to a selected size, since the storm-water runoff is so large compared with the dry-weather flow, which is sewage. The combined-sewer subsystems (usually several) are connected to the sewage-treatment plant by a system of interceptor sewers designed to carry from 1½ to 5 times the sewage flow. In dry weather the large combined sewer carries sewage to a diversion point, which prevents it from flowing out of the local storm-water outlet. At the diversion point the sewage drops through a hole in the pipe to the interceptor, or does it by some other means, such as weirs or gates, and goes to the treatment plant. During wet weather, a portion of the sewage and storm-water mixture, equivalent to 1½ to 5 times the dry-weather sewage rate of flow, is diverted to the treatment plant. The bulk of the mixture, not so intercepted, goes out the storm-sewer outlet to the local watercourses. The bulk of the mixture may discharge to the local watercourse at a rate 50 to 200 times that of the dry-weather sewage rate of flow, the rate of discharge depending on population density, storm size, and other factors.

Some 11,350 communities in the 50 states and the District of Columbia have sewer systems of one of the two types, or combinations of both. Some of these systems are badly overloaded or otherwise abused because of failure to keep up with the pace of today's urbanization with its rapid population and industrial growth and because of the conversion of vegetated surfaces to paved surfaces. Other common causes are leaky and broken pipes and joints in street sewers, blanked stubs and house sewers, and misuse of sanitary sewers because of deliberate or unintentional connection of roof and other storm drains. Some of the items warrant strict control over the quality of installation and maintenance—more control than many communities have been willing to pay for.

STORM-WATER RUNOFF, SEPARATE SYSTEMS

Of the 11,350 sewered communities in the 50 states and the District of Columbia, 9,013 have separate-type sewer systems, 1,305 have combined systems, 618 have both types, and the remaining 414 are unidentified. So, some 9,631 communities, 88 percent of the identifiable total, are wholly or partly sewered by separate systems (U.S. Public Health Service, 1964a).

With growing awareness of the pollutional significance of raw sewage included in combined-sewer overflows, the complete separation of storm water from sanitary sewage and adequate treatment of the latter have gained support. Storm-water runoff itself has been viewed with suspicion. This is not surprising. One wonders what happens to residues from vehicle wear and drippage, garbage from careless handling by workers and householders, animal droppings, construction-job silt, runoff from generously treated gardens and lawns, dustfall, and so on. For instance, atmospheric dustfall can average about 300,000 lb per mi^2 per year in a city, and in some parts it can reach 80,000 lb per mi^2 per month. Sampling of street runoff by Palmer (1950, 1963) in Detroit, Sylvester (1960) in Seattle, and Wilkinson (1956) and others in England show that urban storm-water runoff is far from clean. (See also U.S. Department of the Interior, 1966.)

In 1961, Dr. R. L. Woodward, Chief of Engineering Research, Division of Water Supply and Pollution Control, U.S. Public Health Service, inspired support in that agency for study of urban storm-water runoff. A 27-acre, separately sewered, residential–light-commercial section of Cincinnati was chosen for study. The storm sewer there was instrumented to gauge runoff and automatically collect samples. This was a relatively clean type of urban land use. The area included single- and multiple-family dwellings, stores, local streets, part of a public park, parking lots, and residential lawns and gardens. Impervious areas, such as roofs, streets, and parking lots, totaled 37 percent, and the coefficient of runoff, based on the ratio of summed amounts of runoff and rainfall, was also 37 percent. The nature of this runoff is shown in Table 4 by average concentrations and unit annual yields for several parameters, including plant nutrients. This runoff carried suspended-solids concentrations exceeding those of raw community sewage and biochemical oxygen demand equal to the concentration in effluent from secondary sewage-treatment plants. The nutrient contents, with inorganic nitrogen N averaging 1.0 mg per liter and total phosphate P averaging 0.36 mg per liter, are considerably in excess of Sawyer's (1952) thresholds for trouble. There was evidence (Weibel et al., 1966b) that concentrations of suspended solids in storm-water runoff varied with rate of runoff, peaking at runoff peaks. Since runoff was very responsive to rainfall rates on the 27-acre area studied, peaks did not always occur early in a storm. Consequently, the heavy solids were not always discharged early in the storm. There was sometimes more than one peak, and each was accompanied by an increase in suspended solids.

Because all of this is so closely related to recreational waters, the microbiological quality of the same runoff is presented in Table 5. In many places in this country the limit on total coliform content of waters used for

TABLE 4 Constituents of Urban Storm-Water Runoff from 27-Acre Residential–Light-Commercial Area Having Separate Storm and Sanitary Sewers[a]

Constituent	Average Concentration[b]	Amount[c]
Suspended solids	227	366
Chemical oxygen demand	111	178
Biological oxygen demand	17	27
Total N (four forms as N)	3.1	5
Inorganic N (three forms as N)	1	1.6
Total Phosphate (as P)	0.36	0.6

[a]Computations are based on assumption that annual rainfall is 30 in. and that runoff is 37 percent.
[b]Data are expressed in milligrams per liter.
[c]Data are expressed in thousands of pounds per square mile per year. Multiply by 0.175 for thousands of kilograms per square kilometer per year.

TABLE 5 Bacterial Constituents of Urban Storm-Water Runoff from a 27-Acre Residential–Light-Commercial Area Having Separate Storm and Sanitary Sewers

Constituent	Bacterial Population Level Exceeded in Designated Percentage of Samples (Numbers per 100 ml)		
	90%	50%	10%
Total coliforms	2,900	58,000	460,000
Fecal coliforms	500	10,900	76,000
Fecal streptococci	4,900	20,500	110,000

full recreation—that is, full body contact as in swimming—is 1,000 per 100 ml. Since 90 percent of the runoff samples contained more than 2,900 total coliforms per 100 ml, this water would be condemned for full recreational use. Salmonella have been found in this runoff.

It is unlikely that the urban runoff from a residential—light-commercial area, characterized in Tables 5 and 6, is representative of runoff from high-density residential sections, industrial sections, market areas, railroad or trucking operation areas, parks, zoos, or downtown city sections. It is likely that these differ in both quality and quantity.

COMBINED-SEWER OVERFLOWS

Combined-sewer overflows are individually unique and variable. In addition to variables attributable to the source of the storm water are those of the escaping sewage and those of the sewer system. Unsatisfactory hydraulic design permits low flow velocities in dry weather to strand sewage solids in the relatively large combined sewers. These accumulations are dislodged, mixed, and partly lost in the overflow during storms. The pipe system, if it has a short flow-concentration time, may show a first flush of highly polluted overflow, followed by a tapering off. If the system is extensive, with long flow-concentration time, the dislodged materials may arrive sequentially from various distances. The result, then, is a prolonged overflow of high-strength mixture. Classic studies at Buffalo bring this out (Riis-Carstensen, 1955).

The uniqueness of combined-sewer overflows is evident from field data (Table 6), which characterize the raw sewage (dry-weather flow) and the sewage and storm-water mixture (wet-weather flow) from three urban combined-sewer systems. The suspended-solids concentrations in dry-weather sewage at East Bay, California (U.S. Public Health Service, 1964b), and at Twelve Town Drainage District, Michigan (G. E. Hubbell, unpublished data, 1966), were higher than those of their wet-weather combined-sewage flows, whereas those at Buffalo, New York (Riis-Carstensen, 1955), were the reverse. A reversal in the biochemical-oxygen-demand dry- and wet-weather flow concentrations is also evident for East Bay and Buffalo versus Twelve Town Drainage District. It is obvious from these that there are wide differences, in at least these parameters, between cities.

Studies of Connor Creek combined-sewer overflows at Detroit, Michigan, and of the Allen Creek storm-water drain at Ann Arbor, Michigan (Vaughn and Harlow, 1965; Benzie and Courchaine, 1966), provide nutrient information of interest. The concentrations and estimated amounts for the combined-sewer overflows at Connor Creek are shown in Table 7. The report of Benzie and Courchaine (1966) indicated that nitrogen and phosphate discharges

TABLE 6   Concentration of Constituents in Dry-Weather (Sewage) and Wet-Weather (Sewage and Storm Water) Flows

	Concentration of Suspended Solids (mg/liter)		Concentration of Biochemical Oxygen Demand (mg/liter)	
Location	Dry	Wet	Dry	Wet
East Bay, California[a]				
South Interceptor	336	162	449	178
North Interceptor	–	162	285	195
Buffalo, New York[b]				
Bird Avenue	158	554	162	100
Bailey	126	436	127	121
South Oakland County, Michigan				
Twelve Town Drainage District[c]	198	176	52	140

[a]U.S. Public Health Service (1964b).
[b]Riis-Carstensen (1955).
[c]G. E. Hubbell (unpublished data, 1966).

TABLE 7   Constituents of Combined-Sewer Overflow

Constituents	Average Concentration (mg/liter)	Amount[a] (1,000 lb/mi^2 per year)[b]
Suspended solids	150	162
Organic N	0.37	0.4
Ammonia N	3.25	3.5
Total PO$_4$ (as P)	3.0	3.2

[a]Computations are based on assumption that concentrations apply to reported 4¼ billion gal discharged during 1 year from an area of 32.8 mi^2 (Vaughn and Harlow, 1965; Benzie and Courchaine, 1966).
[b]Multiply by 0.175 for 1,000 kg per km^2 per year.

from the combined system (Connor Creek, Detroit) were 4 and 3 times greater, respectively, on the average than those of the separate storm-sewer system (Allen Creek, Ann Arbor).

Unit amounts from the foregoing tables for the various sources are listed in Table 8. As previously indicated, there is good reason to believe that concentrations and amounts may vary widely from one environment and its particular sewer system to another. Table 8 is merely a rough indication of what some of these values may be. It is offered in the hope that it may

TABLE 8   Amounts of Pollution from Urban Drainage[a]

Source of Pollution	Constituent				
	Suspended Solids	Biochemical Oxygen Demand	ΣN	Inorganic N	Total PO$_4$ (as P)
Rainfall[b]	57	–	5.5	3.0	0.4
Separate-system runoff[c]	366	27	5.0	1.6	.6
Combined-system overflow	162	–	3.9	3.5	3.2
Community sewage (mostly domestic); 10 persons per acre					
Raw sewage	390	390	78	58	20
3-percent loss	12	12	2.3	1.7	0.6
5-percent loss	20	20	3.9	2.9	1.0
10-percent loss	39	39	7.8	5.8	2
Community sewage (mostly domestic); 50 persons per acre					
Raw sewage	1,900	1,900	390	290	97
3-percent loss	57	57	12	8.7	2.9
5-percent loss	95	95	20	14	4.8
10-percent loss	190	190	39	29	9.7

[a]Data are expressed in thousands of pounds per square mile per year. Multiply by 0.175 for thousands of kilograms per square kilometer per year.
[b]30 in. per year.
[c]30 in. rainfall per year and 0.37 storm-water runoff coefficient.

stimulate others to generate factual data for study of the significance of sources of pollution of this nature. In addition to the amounts of constituents to be found in raw community sewage at the two population densities of 10 and 50 persons per acre, amounts represented by 3 percent of the raw sewage are included to give an idea of what a 3-percent annual loss of raw sewage to combined-sewer overflows would be. Similar data are calculated for the 10- and 5-percent losses in effluents from 90- and 95-percent removal by treatment from community sewage. As shown in Table 8, urban sources of phosphate such as combined-sewer overflows and possibly separate storm-water runoff become increasingly significant.

DISTRIBUTION OF U.S. COMMUNITIES WITH COMBINED SYSTEMS

About 1,920 U.S. communities (18 percent of the communities with identifiable systems) are partly or wholly sewered by combined-type sewer systems (U.S. Public Health Service, 1964a). Many of our older and larger cities are served by combined sewers. The geographic distribution is also of interest; 98 percent of the communities having combined systems are in a band of states across the northern part of the country and the West Coast. The distribution is tabulated in Table 9.

The geographical distribution of numbers of communities with "combined" or "both" types of sewer systems tabulated in Table 9 is shown in Figure 2. A comparison of Figure 2 with Figure 1 shows that the expected metropolitan belt (see page 283) will blanket possibly one third to one half of all the communities with combined systems in the country. Several cities in other parts of the country that are expected to be involved in areas of high-density urban occupancy and that now have combined systems are Seattle, Tacoma, Portland (Oregon), Pueblo, Kansas City, St. Louis, Minneapolis-St. Paul, San Francisco, Eureka (California), Atlanta, and Louisville (Pickard, 1967).

In addition to the waste-control challenges posed by existing combined systems in the path of urban intensification, many of these areas are generously endowed with surface waters for which future water recreation demands will be enormous. Add to this the occurrence of eutrophication in some of these areas now, and one gets some idea of the immensity of the preventative problems ahead.

The problems facing a particular city or area are unique to that area, and the preventive program may have to be tailored to its needs. Past shortcomings include lack of understanding or application of adequate urban hydrology, underestimated rate of growth in design, failure to utilize modern business tools to win adequate appropriations, reluctance to write off old investments in long-obsolete works and to modernize, and failure to buy the goods and services necessary to proper operation and maintenance of existing facilities.

PREVENTIVE MEASURES FOR CONTROL

Urban storm-water runoff is considered by some to be a resource out of place. An example of the usefulness of storm water as a resource is the old rain barrel, or the current versions, the cistern and the pressure water systems in fringe-area homes.

Intercepted on roofs, parking lots, streets, and similar impervious surfaces, storm water is conveniently distributed for a variety of uses, including local storage, cooling, watering lawns and gardens, recharging to ground, filling ornamental ponds, and flushing toilets. It can also be stored for delayed release to reduce flow peaks in storm sewers or combined sewers.

Runoff from some types of surface or land use may not be suitable for all uses. A suitable use may then be selected for the water, or the water may be modified to make it suitable for the desired use, for example, treatment including disinfection, if body contacts are to be involved in its use.

SEDIMENTATION OF STORM-WATER RUNOFF

Studies of storm-water runoff in the Cincinnati area indicated that less than 1 hour of settling did not substantially reduce chemical oxygen demand, biochemical oxygen demand, nitrogen, phosphates, or solids. Total phosphate removals were 15 percent in 1 hour and 30 percent after 4 hours of quiescent batch settling (Evans *et al.*, 1968).

TABLE 9   Distribution of Communities with Combined Systems

State	Number[a]	Percent[b]
STATES WITH MORE THAN 10 SYSTEMS EACH		
Pennsylvania	271	32
Indiana	205	71
Illinois	183	34
Ohio	179	41
Michigan	162	57
New York	102	23
Wisconsin	85	20
Maine	70	69
West Virginia	66	46
Washington	55	31
New Hampshire	55	79
Massachusetts	51	30
North Dakota	48	27
Oregon	43	29
Vermont	41	89
Kentucky	32	20
Minnesota	30	7
Iowa	28	7
South Dakota	27	15
Missouri	26	7
Connecticut	24	34
Nebraska	18	6
New Jersey	18	7
Alaska	13	62
California	12	2
Idaho	12	13
Maryland	11	10
Montana	11	10

TABLE 9–*Continued*

State	Number[a]	Percent[b]
STATES WITH FEWER THAN 10 SYSTEMS EACH[c]		
Delaware	7	35
Tennessee	6	5
Virginia	5	3
Georgia	5	2
Nevada	4	11
Kansas	4	1
Rhode Island	3	12
Arkansas	2	2
Florida	2	1
Texas	2	< 1
Arizona	1	2
Colorado	1	1
Mississippi	1	1
North Carolina	1	< 1

[a]Number of communities in the state that are served by combined systems or by combined *and* separate systems.

[b]The percentage figure shows the relation between (1) communities with combined systems (or combined and separate systems) and (2) all communities in the state that have identifiable types of sewer systems. Thus, the 271 Pennsylvania communities that have combined systems (or combined and separate systems) form 32 percent of the communities in the state that have identifiable types of sewer systems.

[c]The following states have no combined systems: Alabama, Louisiana, New Mexico, Oklahoma, South Carolina, Utah, Wyoming, Hawaii.

DIVERSION OF STORM-WATER DRAINAGE

Sources of nutrients and other pollutants, including street-gutter storm runoff, reaching Green Lake in Seattle were studied because aquatic vegetation interfered with recreational use. Recommendations were to discontinue most urban storm-water drainage to the lake and to add 10,000 gallons per day of city water supply to the lake to reduce overenrichment (Sylvester, 1960).

GROUND RECHARGE OF STORM-WATER RUNOFF

On Long Island, New York, the sandy subsoil makes possible an economical means of disposing of storm-water runoff. Ground recharge basins in this area are used by the state for disposing of road drainage and by housing contractors and industry for disposing of storm-water runoff. Many of the

FIGURE 2   Distribution of communities with combined sewer systems. Source: U.S. Public Health Service (1964a).

small basins are underground pits; others are fenced open pits along rural roads. Larger installations are excavated open pits sized to contain runoff from a large storm. One of the excavated open pits includes two dugouts, one serving as a desilting basin and the other as a seepage basin (Brice *et al.*, 1949–1951). This pit receives road drainage.

PARTIAL SEPARATIONS

Toronto, Canada, has started a road sewer program involving construction of new storm sewers to receive street storm-water runoff only. This is expected to divert 30 to 60 percent of the flow from heavily loaded combined sewers to where it is needed and so relieve local flooding and increase sewage-interception efficiency. The unit capital cost for this is about $7,500 per acre instead of $17,000 per acre for a new separate sanitary-sewer system or $21,500 per acre for a new separate storm-sewer system (Dunbar and Henry, 1966).

COMPLETE SEPARATION

The cost for complete eventual separation in metropolitan Washington, D.C., is estimated at $20,000 per acre. Interim control measures include separation of residence and building sewers and increasing interceptor capacities to reduce overflows in sensitive areas (Johnson, 1958).

STORAGE IN COMBINED-OUTLET SEWERS

In the Twelve Town Drainage District, South Oakland County, Michigan, a high weir skimming structure at the outlet of combined sewers serving a 24,500-acre drainage area creates 100 acre-feet of storage in outlet sewers before overflow. This has reduced the number of spills. Oil, grease, and floatables are retained. Spills do not appear aesthetically offensive. The contents retained in the sewer are metered into the Detroit system for treatment. The 54-in. magnetic flow meter automatically limits maximum flow to the purchased capacity. The installation includes automatic flow-gauging and sampling equipment for dry- and wet-weather flows, rainfall gauges, and a laboratory. Analyses include suspended solids, volatile suspended solids, oil and grease, biochemical oxygen demand, dissolved oxygen, pH, chromium, cyanide, copper, and microbiological determinations (C. E. Hubbell, unpublished data, 1966).

STORAGE IN COMBINED-UNDERFLOW TUNNELS

Chicago plans a tunnel 33 to 50 ft in diameter and 200 to 400 ft deep in limestone rock, together with branch tunnels and shafts for storage of excess combined flow, which will be pumped to the sewage-treatment plants later. The tunnels are expected to absorb the flow from a rainfall as large as 2 to 2½ in., which occurs once a year on the average, without discharging any untreated water from the mainstream underflow. The tunnel and surface system could handle the flow from a large storm (the kind that occurs once in a hundred years) without overflow to Lake Michigan. The cost was estimated at some $4,200 per acre, which is 10 percent of an estimated cost of $42,000 per acre for separation and a new sanitary-sewer system for Chicago. An "underflow sewer" is about to be constructed under Lawrence Avenue as part of the system. Boring with a tunnel-boring machine (the "mole" method) is expected to cost $2 million less than the open-cut method and save 2 years of construction time (Pikarsky and Keifer, 1967).

HOLDING BASINS FOR EXTRANEOUS SANITARY SEWER FLOWS

Construction of holding basins has been reported for parts of Johnson County, Kansas, and Kansas City, Missouri. Their purpose is to accommodate extraneous flows in sanitary sewer systems during wet weather. Accommodating these flows is expected to eliminate troubles from sewer backup and overflow. The flows resulted from areaway and foundation drain connections to the sanitary sewers (Weller and Nelson, 1965).

TREATMENT OF COMBINED-SEWER OVERFLOW

Sanitary and combined sewers serving Harper Woods and Grosse Pointe Woods, Michigan, discharge to the Grosse Pointe interceptor and to the Detroit municipal system. Wet-weather combined flows in excess of 8,000 gpm are diverted to a sedimentation-skimming tank, the effluent of which discharges to Milk River and Lake St. Clair during storms causing overflow. The contents of the tank are flushed to the interceptor, when capacity becomes available, for treatment by Detroit (Hirn, 1962).

MONITORING COMBINED-SEWER OVERFLOWS

Cincinnati has almost completed installation of a combined-sewer overflow-monitoring system, parts of which are now in operation. Signaling switches on each of 250 combined-sewer overflow-interceptor locations grouped on 87 party lines signal any overflow from storm or clogging of the interceptor device. An operator calls the political subdivision responsible for maintenance at the interceptor location. A central map lights the location of the overflow. The operator delays his call 10 minutes. If during that time numerous other signals within the area show overflows, rain is inferred and another course of action is taken. Daily printout records enable study of overflow patterns and maintenance operations. The operation should reduce pollution problems throughout the metropolitan Cincinnati area (Caster, 1965).

STORM-WATER TANKS FOR COMBINED SEWAGE

Two thirds of West Berlin, Germany, has separate sewers; the rest has combined sewers. The city has installed covered settling tanks at a number of pump-station locations. Interceptor devices in the combined sewers divert flows 5 or 6 times the volume of the average dry-weather sewage flow to central treatment. The untreated remainder, which formerly overflowed to local streams, now flows into the covered settling tanks. Thus, small overflows are completely contained by the tanks, and large overflows receive sedimentation treatment as they pass through the tanks to local streams. After a storm, the contents of the tanks are flushed back to the sewers to be intercepted and led to central treatment. Before the tanks were built, about 55 percent of the total waste reaching the sewers received treatment. After the tanks were put into operation, 85 percent received treatment by the central plant or the storm-water settling tanks. It was difficult to find space for some of the tanks. One tank was installed below a schoolyard; the offset manholes were outside the yard at a pump station. Two tanks were installed beneath a street. Most of the tanks were installed at pump-station grounds (Cohrs, 1962).

I have given these examples to show the variety of practices that cities have adopted for controlling pollution from sewers carrying storm water or storm water and sewage. Information on plant-nutrient control in storm water and in combined-sewer overflows is scarce.

The discharge of polluting wastes from storm-drainage systems and overflows from combined sewers is a distinct challenge to the ingenuity of municipal officials, consulting engineers, universities, corporations engaged in research and development, and equipment manufacturers. Polluting discharges from combined storm-and-sanitary sewers occur during wet-weather periods when the carrying capacity of the sewers is exceeded because of the large amounts of storm water entering the sewers. The normal, or dry-weather, flow is prevented from overflowing continuously by means of overflow weirs, mechanical regulators, valves, and other devices. They permit overflows to occur when sewer flows reach a predetermined level.

Separation of the storm water from the sanitary sewage can be at least a partial solution to the problem. If the systems are completely separated, the most concentrated waste load can be conveyed to, and treated at, the waste-treatment plant. We have come to recognize in recent years, however, that surface runoff also contains significant amounts of pollutants—sometimes nearly as much as sewage. That is why we believe separation of sanitary wastes is only a partial solution.

Congress had these factors in mind when it authorized grants for demonstrating control of pollution in sewers. The Federal Water Pollution Control Act authorizes

... grants to any State, municipality, or intermunicipal or interstate agency for the purpose of assisting in the development of any project which will demonstrate a new or improved method of controlling the discharge into any waters of untreated or inadequately treated sewage or other wastes from sewers which carry storm water or both storm water and sewage or other wastes ...

The Federal Government can provide up to 75 percent of the estimated reasonable cost of individual research, development, and demonstration projects. The applicant must provide assurances that local funds are or will be available to pay the remainder of the cost. Application for contract support for pertinent research and development projects will also be considered.

The Federal Water Pollution Control Administration, which administers the Federal Water Pollution Control Act, has prepared informational material dealing with technical aspects of its program. The material would be helpful to persons wishing to participate in the task of controlling or abating pollution from storm sewers and combined sewers. It may be obtained from the Office of Research and Development, Federal Water Pollution Control Administration, U.S. Department of the Interior, Washington, D.C.

REFERENCES

Benzie, W. J., and R. J. Courchaine. 1966. Discharges from separate storm sewers and combined sewers. J. Water Pollut. Control Fed. 38:410–421.

Brice, H. D., C. L. Whitaker, and R. M. Sawyer. 1949–1951. A progress report on the disposal of storm water at an experimental seepage basin near Mineola, N.Y. U.S. Geological Survey, Washington, D.C. 34 p.

Camp, T. R. 1963. Water and its impurities. Reinhold Publishing Corp., New York. 355 p.

Caster, A. D. 1965. Monitoring stormwater overflows. J. Water Pollut. Control Fed. 37:1275–1280.

Cohrs, A. 1962. Storm water tanks in the combined sewerage system of Berlin. Gas und Wasserfach 103:947–952.

Dunbar, D. D., and J. G. F. Henry. 1966. Pollution control measures for stormwaters and combined sewer overflows. J. Water Pollut. Control Fed. 38:9–26.

Evans, F. L., E. E. Geldreich, S. R. Weibel, and G. G. Robeck. 1968. Treatment of urban storm-water runoff. J. Water Pollut. Control Fed. 40:129–256.

Gambell, A. W., and D. W. Fisher. 1964. Occurrence of sulfate and nitrate in rainfall. J. Geophys. Res. 69:4203–4210.

Hirn, W. C. 1962. Providing primary treatment for storm sewage overflows. Wastes Eng. 33:450–452, 483.

Johnson, D. F. 1958. Nation's capital enlarges its sewerage system. Civil Eng. 28:428–431.

Junge, C. W. 1963. Air chemistry and radioactivity, p. 289 In Air chemistry and radioactivity (vol. 4 of International Geophysics Series). Academic Press, New York.

Mackenthun, K. M. 1965. Nitrogen and phosphorus in water; an annotated selected bibliography of their biological effects. U.S. Public Health Service (Pub. No. 1305). U.S. Govt. Printing Office, Washington, D.C. 111 p.

Martin, E. J., and L. W. Weinberger. 1966. Eutrophication and Water Pollution, p. 451–469 In Pub. No. 15, Univ. Michigan, Great Lakes Research Division.

National Academy of Sciences, Committee on Pollution. 1966. Waste management and control. NAS-NRL Pub. 1400. National Academy of Sciences–National Research Council, Washington, D.C. 257 p.

Palmer, C. L. 1950. The pollutional effects of storm-water overflows from combined sewers. Sewage Ind. Wastes 22:154–165.

Palmer, C. L. 1963. Feasibility of combined sewer systems. J. Water Pollut. Control Fed. 35:162–167.

Pickard, J. P. 1967. Future growth of major U.S. urban regions. Urban Land, Feb. 1967, p. 3–10.

Pikarsky, M., and C. J. Keifer. 1967. Underflow sewers for Chicago. Civil Eng. 37(May):62–65.

Riis-Carstensen, E. 1955. Improving the efficiency of existing interceptors. Sewage Ind. Wastes 27:1115–1122.

Sawyer, C. N. 1952. Some new aspects of phosphates in relation to lake fertilization. Sewage Ind. Wastes 24:768–776.

Sylvester, R. O. 1960. An engineering and ecological study for the rehabilitation of Green Lake. Univ. Washington, Seattle.

U.S. Department of the Interior, Federal Water Pollution Control Administration. 1966. Storm water runoff from urban areas; selected abstracts of related topics. 98 p.

U.S. Department of the Interior, Federal Water Pollution Control Administration. 1967. Report of Lake Erie enforcement conference technical committee.

U.S. Public Health Service. 1964a. Statistical summary of 1962 inventory, municipal waste facilities in the United States. U.S. Public Health Service Pub. No. 1165. U.S. Govt. Printing Office, Washington, D.C. 41 p.

U.S. Public Health Service. 1964b. Pollutional effects of stormwater and overflows from combined sewer systems. U.S. Public Health Service Pub. No. 1236. U.S. Govt. Printing Office, Washington, D.C. 39 p.

Vaughn, R. D., and G. L. Harlow. 1965. Report on pollution on the Detroit River, Michigan waters of Lake Erie, and their tributaries; summary, conclusions and recommendations. U.S. Public Health Service, Grosse Isle, Michigan. U.S. Govt. Printing Office, Washington, D.C. 59 p.

Weibel, S. R., R. J. Anderson, and R. L. Woodward. 1964. Urban land runoff as a factor in stream pollution. J. Water Pollut. Control Fed. 36:914–924.

Weibel, S. R., R. B. Weidner, J. M. Cohen, and A. G. Christianson. 1966a. Pesticides and other contaminants in rainfall and runoff. J. Amer. Water Works Ass. 58:1075–1084.

Weibel, S. R., R. B. Weidner, A. G. Christianson, and R. J. Anderson. 1966b. Characterization, treatment, and disposal of urban stormwater. Third International Conference on Water Pollution Research, Proc. (Munich, Germany). Water Pollution Control Federation, Washington, D.C.

Weller, L. W., and M. K. Nelson. 1965. Diversion and treatment of extraneous flows in sanitary sewers. J. Water Pollut. Control Fed. 37:343–352.

Wilkinson, R. 1956. The quality of rainfall runoff water from a housing estate. J. Inst. Public Health Engineers (London) April, p. 70–78.

J. W. BIGGAR AND R. B. COREY
University of Wisconsin, Madison

Agricultural Drainage and Eutrophication

Agricultural drainage, like many other aspects of our total water resources, has only recently become of national interest and concern. This recency of concern is true particularly with regard to the quality of the drainage water and to nutrients contained in the drainage. As a result, relatively few studies have been directed toward elucidating the factors that control the amounts of plant nutrients reaching streams and lakes from agricultural sources. However, a number of investigators have studied nutrient losses from soils in conjunction with investigations designed to measure salt movement, to measure efficiency of fertilizer application, or to clarify soil-development processes. Results of many of these investigations can be applied to the problem of eutrophication of lakes.

Although the current emphasis is on man's contributions to water fertilization, we should not lose sight of the fact that all natural waters contain dissolved materials (plant nutrients included) derived from natural processes (Frink, 1967). For example, estimates of annual solute erosion in 11 western river basins ranged from loads of 180 T per mi^2 in the Willamette basin to 4.2 T per mi^2 in the Gila basin, with an average of 58 T per mi^2 (Van Denburgh and Feth, 1965). The average dissolved-solids content ranged from 54 to 1,500 ppm. The Willamette basin averages 50 in. of annual runoff, and the Gila basin averages 0.02 in. Quantities of solutes eroded are highest in areas of abundant precipitation and runoff, whereas concentrations of solute are highest in areas of low precipitation. The greatest suspended-sediment losses, however, occur from areas of intermediate effective precipitation.

In this paper we are interested principally in the sources, quantities, and distribution of agricultural drainage waters and the extent to which these waters contain constituents that contribute to eutrophication. Agricultural drainage water provides the means by which constituents of eutrophication are redistributed, the destination of immediate interest being streams and lakes. Our discussion will be divided into three main sections:

1. Chemical reactions that nutrient elements undergo in soil-water systems. This part of the discussion is limited primarily to nitrogen and phosphorus. Many other substances contribute to the growth of algae, but these two elements are the only ones that have been definitely established as consistent growth-limiting factors in natural waters.

2. Factors influencing the amounts and kinds of drainage from agricultural lands and the associated transport of nutrients.

3. Studies that provide a basis for estimating the amounts of nutrients entering lakes and streams.

NUTRIENTS IN THE SOIL-WATER SYSTEM

Most of the soluble nutrients that get into lakes and streams from rural areas are first dissolved in water and then moved in solution to the waterways. Some nutrients may also be carried to streams and lakes as components of suspended particulate matter and later converted to soluble forms. Therefore, to fully grasp the problem of water fertilization from agricultural lands, we must understand the factors that affect the forms and solubilities of the nutrients and the manner in which the nutrients are transported to the streams and lakes.

FORMS AND AMOUNTS OF NITROGEN AND PHOSPHORUS

Bartholomew and Clarke (1965) describe all the chemical changes that nitrogen undergoes from its origin and distribution in the soil profile to its loss from the profile. The description includes the many biological reactions that take place. In addition, reviews on the loss of nitrogen, including leaching, have appeared at various times (Allison, 1955, 1966). From an agronomic viewpoint, some sources of nitrogen that ultimately contribute to eutrophication are relatively unimportant. Nitrogen derived from rainfall may range between 1 and 19 lb per acre per year at various locations, averaging 7 or 8 lb. This is not enough to produce marked increases in crop yield, but in runoff or percolative waters it becomes a steady and significant supply. The quantities of N that are fixed by plants and microorganisms also contribute to the potential source of available N for redistribution. Hence 200 lb of N per acre per year might be fixed by legume–microorganism interactions in comparison with 10 to 30 lb under grass.

A large percentage of N in soil occurs in organic forms (e.g., in organic matter). In fact, 95 percent or more of the N may be in this form. For example, a grass-covered prairie soil with 10 percent organic matter, a carbon–nitrogen ratio of 10, and a bulk density of 1.2 g/cm^3 in the surface 1 ft would contain 32,000 lb of N (Stout, 1965). Microbial decomposition of the organic matter results in the release of nitrogen in the ammonium form (NH_4^+), a process called ammonification. Under conditions of good aeration and favorable temperatures, different microorganisms oxidize the ammonium first to nitrite (NO_2^-) and then to nitrate (NO_3^-), a process called nitrification. The step from nitrite to nitrate is usually faster than from ammonium to nitrite, so that practically no nitrite accumulates. If the content of ammonia (NH_3) in the system is high, however, nitrite may accumulate, and it is toxic to many organisms. If nitrate is exposed to conditions of poor aeration (reducing conditions), it will be reduced to gaseous nitrogen forms and lost to the atmosphere, a process called denitrification.

Ammonium ions are held on the cation-exchange sites in soils, so the concentration of ammonium in the soil solution is not very high. The nitrate anion, on the other hand, is completely soluble in the soil solution, and it moves with the soil water. Therefore, nitrate is the form of nitrogen most subject to leaching. But this does not exclude the possible movement of certain organic forms, which may be quite soluble.

The phosphorus content of soils ranges from 0.01 to 0.13 percent. Phosphorus occurs in both organic and inorganic forms, and neither form is considered to be very soluble. The proportions of each kind have been found to range from 3 percent organic and 97 percent inorganic to 75 percent organic and 25 percent inorganic.

The inorganic forms of phosphorus are mainly iron and aluminum phosphates in acid soils and calcium phosphates in alkaline soils. All inorganic forms of phosphate in soils are extremely insoluble. Any phosphorus added as fertilizer or released by decomposition of the organic matter is quickly converted to one of these insoluble forms. Because of the extreme insolubility of these phosphates, the over-all concentration of soluble phosphorus in the soil solution of surface soils seldom exceeds 0.2 mg per liter, and concentrations in the range of 0.01 to 0.1 mg per liter are common. Displaced soil solutions contained 0.03 ppm of phosphorus as inorganic orthophosphate (Pierre and Parker, 1927). Other data indicate values of 0.07 to 0.17 ppm. Phosphorus concentrations in the soil solution of subsoil layers are frequently less than 0.01 mg per liter.

SOLUBLE NITROGEN AND PHOSPHORUS
IN SURFACE RUNOFF AND PERCOLATES

The concentrations of nitrogen and phosphorus in surface runoff are considerably different from those in soil percolates. The ammonium and nitrate forms of nitrogen are very soluble. If these materials are present at the surface of the soil at the beginning of a rain, the first rain that falls will dissolve them and carry them into the soil. If the surface runoff occurs later, little soluble nitrogen will be left at the surface to be carried away with the runoff. Therefore, runoff waters usually contain very little soluble inorganic nitrogen. In fact, the nitrate contents of runoff waters are usually lower than the average nitrate content of rainwater. The first rain that falls sweeps most of the nitrate from the air and carries it into the soil. The rain that falls later and runs off has a lower nitrate content.

Though runoff waters in humid areas contain relatively little nitrate, water that percolates through the soil may contain considerable amounts. As stated before, nitrate is completely soluble in the soil solution and moves with it. If the nitrate ions manage to evade absorption by the plant roots as they move downward, they will be present in the drainage waters that move to the lakes and streams by base flow. Thus soil percolates generally contain more nitrate than do surface runoff waters. This nitrate eventually reaches the waterways unless the water emerges in a marsh, where it may be absorbed by the vegetation or reduced to gaseous nitrogen.

The relative concentrations of soluble phosphorus in surface runoff and soil percolates are the reverse of the nitrogen system. Phosphorus applied to the surface of the soil tends to saturate the "fixing" sites at the surface and locally raise the concentration of phosphorus in the soil solution.

This is a near-equilibrium system, and although infiltrating waters will carry the soluble phosphorus downward, more will quickly dissolve to maintain the concentration in solution. Runoff water will contact this surface soil, and the phosphorus concentration in the runoff could conceivably approach the equilibrium concentration. If phosphorus fertilizers were applied to the soil surface, the equilibrium concentration of phosphorus in a thin surface layer could reach 1 mg per liter or more, and the concentration of phosphorus in the runoff water might range up to a few tenths of a milligram per liter. This concentration of phosphorus is speculative, at best, since few data are available pertaining directly to this problem. However, soluble phosphorus concentrations in surface runoff frequently approach or exceed the average concentrations expected in the soil solution, a fact that tends to support this contention.

In the water that percolates through the soil, the soluble phosphorus concentration is usually very low because the phosphorus precipitates in the subsoil. Therefore, most of the soluble phosphorus should reach the waterways via surface runoff. This route contrasts with that of nitrogen, since most of the soluble inorganic nitrogen should reach the waterways by percolation and base flow. These conclusions assume that the soils are not frozen. If the soils were frozen, a relatively large proportion of all soluble nutrients at the soil surface would be carried away in the runoff waters. This is undoubtedly the case during the initial stages of the spring thaw, and is of special significance for nutrients in manure or fertilizers applied on frozen fields.

EFFECTS OF SUSPENDED MATERIAL

The energy associated with the impact of falling raindrops tends to break down aggregates of soil particles at exposed soil surfaces. Runoff waters can then pick up the finer particles and carry them downslope, causing sheet erosion. Much of the suspended material is usually deposited at the base of the slopes or on the terraces or floodplain of a stream, but some is carried into the stream itself. When runoff water is concentrated in channels, its erosive power is increased by its increased velocity, and deep gullies may be formed. Much of the finer material eroded from a gully, mainly subsoil material, may end up in a stream. During periods of high flow, the streams erode their banks and carry some of the eroded material downstream in suspension. Suspended materials, whatever their source, undoubtedly affect the nutrient status of the water. However, no data are available to indicate the magnitude of their effects. Therefore, the following discussion is based mainly on theory.

Nitrogen in suspended particles is present mainly in the organic form. Some of these particles will settle out when the water velocity decreases, to be covered later by other sediments. They do not contribute significantly to the soluble nitrogen supply. Other organic particles may be attacked by microorganisms, with the nitrogen being converted to soluble inorganic forms in the decomposition process. Fresh organic materials are quite readily decomposed by microorganisms, but humified-soil organic matter is quite resistant to decomposition. Thus, the contribution of the suspended organic matter to the soluble nitrogen content will depend on the nature of the organic materials.

Phosphorus in suspended particles is present in both organic and inorganic forms. The organic forms undergo microbial transformations, as does nitrogen. However, the inorganic forms present a more complex system. The phosphorus bonded to iron, aluminum, or calcium in the mineral particles tends to equilibrate with the phosphorus in solution. If the particles come from a surface soil high in phosphorus, they will tend to support a relatively high concentration of phosphorus in solution. If, on the other hand, the particles come from a subsoil low in phosphorus, they will support a low concentration of phosphorus in solution. In fact, if subsoil particles were introduced into a stream containing a moderate or high concentration of soluble phosphorus, they would adsorb phosphorus from the water, thereby lowering the phosphorus concentration in solution. Since much of the sediment in streams during high flow is derived from stream-bank erosion, the phosphorus status of the sediments in the streambeds and stream banks may well be an important factor affecting the concentration of soluble phosphorus in the water during periods of high flow.

The contribution of eroded particulate matter to the nutrition of the algae may be associated more with its effects on the concentrations of soluble nitrogen and phosphorus in the incoming waters than with the total or "extractable" nitrogen and phosphorus in the particles themselves.

OTHER CONSTITUENTS

Constituents found in drainage water in addition to nitrogen and phosphorus are known to be required for adequate lake fertility. Nutrients may be deficient and sometimes are added for greater production. Most, if not all, of these constituents are found in soils and the parent materials from which the soils are derived. Cations such as Ca^{2+}, Mg^{2+}, K^+, and Na^+ are often present in major proportions, and more information on concentrations of these in runoff and percolating waters from particular land sources should be sought. However, elements present in trace amounts are rarely investigated. Table 1 gives a listing, together with the range of concentrations, of trace elements that are of interest in agriculture. It does not indicate the concentrations that might be found in drainage waters.

Goldman (1965) reviewed the requirements of natural phytoplankton for molybdenum, copper, vanadium, cobalt, manganese, zinc, boron, iron, sodium, and calcium. In Castle Lake, a mountain lake in California, sediments

TABLE 1   Range of Total Content of Trace Elements in Soils[a]

Element	Range (ppm)	Element	Range (ppm)
B	2–200	Cr	5–1,000
Co	1–40	V	20–500
Cu	2–100	Ni	5–500
Mn	200–3,000	Pb	2–200
Mo	0.2–5.0	Se	0.1–2.0
Zn	10–300	Li	5–200
As	1–50	Sr	50–1,000

[a]Source: Swaine (1955).

TABLE 2   Elemental Composition of Some Drainage Waters from Arctic Alaska[a]

Drainage Source	Elements (ppb)								
	Fe	Al	Cu	Mn	Pb	Ga	Cd	Co	Zn
I Barrow	35–900	< 2–26	3.1–50	13–75	3.9–41	<0.1– 2.6	<1 –2.7	<0.05– 1.8	4.4– 91
Calcareous	<25–215	8–155	2–13	<6–41	<0.1– 9.1	<0.1–10	<1 –8.1	<0.05– 1.4	3.7–4,000
Arctic	<25–250	41–131	5.3–73	<6–14	2.7–12	<0.1–21	<1 –1.4	<0.05–19	3.6– 250
Lakes	<25– 49	9–490	2.6–5.3	<6–25	<0.1– 6.6	<0.1	<1 –3.3	0.05– .74	12.0– 110
Moraines	<25–480	3– 31	—	<6–64	<0.1– 1.9	<0.1	<0.1	<0.05–24	6.4

[a]Source: Brown et al. (1962).

410

to a depth of 4 m contained (in μg/kg) Mo $>$ 100; Va, 100 to 1,000; Co, 10 to 10^5; and Mn, 10 to 10^6.

Some idea of natural levels of various low-level elements in drainage waters can be obtained from a report by Brown *et al.* (1962) on the mineral composition of some drainage waters from Arctic Alaska. These waters are least likely to be contaminated by man's domestic, industrial, and agricultural activities. Unfortunately, no nitrogen and phosphorus data are given. Elements are grouped in Table 2 according to different geologic materials, lakes, and streams from moraines.

The most useful indication of quantities of these elements resulting from agricultural drainage would be the drainage water. Wiklander and Hallgren (1960) reported that drainage water from drains at a depth of 1.1 m in a clay soil contained (in kilograms per hectare per year) Cu, 0.038; Mn, 0.0; SO_4–S, 2.3; and NO_3–N, 4.3.

Czekalski and Kocialkowski (1963) reported concentrations in drainage water as follows: P, 0.11 to 0.53 mg per liter; Cu, 0.8 to 6.5 μg per liter; Mo, 0.05 to 0.32 μg per liter; Mn, 80 to 800 μg per liter; Ni, 1.35 to 2.5 μg per liter; and Zn, 5.5 to 30 μg per liter.

Further assessment of the sources and nature of many of the trace elements in agricultural drainage water will require additional investigation.

DISPOSITION OF AGRICULTURAL DRAINAGE

As an introduction to the sources of water that may ultimately contribute to agricultural drainage and eventually eutrophication, it is worthwhile to consider briefly some of the components of the hydrologic cycle.

Precipitation from the atmosphere reaching the soil surface is disposed of by (1) surface runoff, (2) groundwater runoff (interflow), (3) deep percolation, (4) storage, and (5) evaporation and transpiration. The first three of these can, and do, contribute to eutrophication by providing pathways of nutrient movement to the lakes and streams. Water that does not run off or evaporate from surface catchments infiltrates the soil. Some of this water percolates into deeper layers, eventually joining the groundwater, and many rivers and lakes receive a major part of their water from groundwater flows. Part of the water that enters the soil drains downslope to reappear at a lower elevation as surface water or seepage. This water may also contribute fertility to the lakes.

Other processes involved in the hydrologic cycle contribute indirectly to changes in the fertility load of the water. Some of the precipitation is intercepted by the plant canopy and returned to the atmosphere by evaporation. In addition, water that is evaporated may return salts to the surface, where they may be lost by runoff and seepage or redistributed in the profile with the next water application. Transpiration reduces percolation losses, thereby modifying nutrient movement patterns.

In certain areas, additional water is applied to the soil as irrigation. In most cases this water is drawn from surface streams, impoundments, lakes, or groundwater, and thus represents a recycling of the water derived from runoff, seepage, and percolation. Since irrigation water is generally applied because precipitation is inadequate or poorly distributed in time and space, irrigation often increases the amounts of nutrients moving to lakes and streams. Examples of this are presented later.

SURFACE RUNOFF

Whenever precipitation occurs more rapidly than it can be absorbed by the soil, it runs off into drainageways or into depressions. Because of the importance of surface runoff to streamflow and erosion, much effort has been expended to predict the quantity of both runoff water and suspended matter, usually without regard for the chemical or physical nature of the soluble or suspended materials (Smith and Wischmeier, 1962).

A number of factors, including storm characteristics, soil properties, topography, and plant cover, appear to influence the amount of runoff derived from a given storm. Consequently, these factors have been included in a number of empirical equations designed to predict runoff characteristics. However, these equations have been developed for particular watersheds and do not apply in general. Recent trends in hydrograph synthesis reflect these problems (Van de Leur *et al.*, 1966).

An empirical equation proposed by Enderlin and Markowitz (1962) ignores storm characteristics other than total rainfall:

$$Q = \frac{(P - I_a)^2}{(P - I_a + S)} \cdot \qquad (1)$$

In this equation, the authors relate Q, the runoff, to P, the rainfall, I_a, the rain that falls before runoff, and S, a parameter that groups soil and vegetative cover characteristics, including surface storage, soil permeability, and antecedent moisture.

Surface runoff from a watershed follows a general pattern. The hydrograph, which is a plot of discharge rate versus time, is characteristically made up of a rising limb, a crest or peak, a recession limb, and base flow. Very little is known about the changes in chemical composition of the discharge resulting from runoff and seepage. But this knowledge might prove useful in predicting nutrient losses from storm characteristics and in assessing the sources of water contributing to the discharge. For these purposes, it may not be necessary to have an equation that predicts the entire shape of the hydrograph, if most of the nutrients in the runoff water are derived from the surface canopy and soil during either the rising limb or recession limb.

Different rainfall rates may produce vastly different runoff percentages, as shown in Figure 1.

Though a watershed may have a large capacity for water storage within the soil, the water-intake rate of the soil may limit the percentage of a rainfall that can enter it. The percentage of a storm entering the soil depends on the extent to which the rate of rainfall exceeds the rate of intake. Once the intake rate is exceeded, stream discharge is more sensitive to storm intensity and duration than to any other factors.

As shown in Figure 1, the percentage of runoff of the Russian River is relatively independent of rainfall. The percentage of runoff of Beargrass Creek, in contrast, is quite sensitive to rainfall. Thus, it is difficult to generalize about runoff from one watershed to another.

Within a watershed, however, it should be possible to combine chemical and meteorological data with measured storm, surface, and vegetative character-istics to form theories of nutrient losses in relation to factors of rainfall disposition.

Rowe and Coleman (1951) reported on rainfall disposition at Bass Lake in a 70-year-old ponderosa pine forest on a 40-percent slope under (1) natural conditions and (2) annual burn of ground cover in the autumn. This study is an example of one in which measurement of nutrient losses could have been usefully combined with rainfall-disposition measurements.

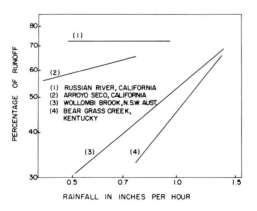

FIGURE 1   Rainfall rates and percentage of runoff for four watersheds. (Crawford, 1966)

Table 3 clearly shows the effect of removing the ground cover each autumn. Relative nutrient losses by runoff would probably be much lower under natural conditions than under annual burn of ground cover. Likewise, nutrient movement that is particularly susceptible to percolation redistribution would occur more readily under normal conditions. Admittedly, the above example is not an agricultural situation. In some ways, nutrient losses from an agricultural area would be easier to estimate because we have a better knowledge of soil characteristics and crop uptake. On the other hand, soil and vegetative conditions are likely to be more diverse under cultivation than under range or forest conditions. Some examples of agricultural runoff and erosion will be given later.

INFILTRATION AND PERCOLATION

It is obvious that the processes of infiltration and percolation are interrelated with runoff and rainfall. Water that is not otherwise lost from the soil may infiltrate and percolate to shallow depths from which it may reappear as groundwater seepage (interflow) at the surface, or it may percolate to perched water tables or deeper underground aquifers. Considerable study has been devoted to infiltration and percolation because replenishment of soil water and groundwater depends on these processes. Likewise, methods have been sought for increasing filtration to reduce runoff and related erosion.

If one understood infiltration and percolation theory and could apply this theory, he could predict not only the distribution of drainage waters but also the expected nutrient removal. These predictions might come from combining water-flux estimates with the mixing factors that determine nutrient distribution within the soil-water system. Various reviews have recently been prepared outlining the extent of our progress in the development of the theory of soil-water movement (Marshall, 1959; Gardner, 1960; Swartzendruber, 1966). One is impressed by the lack of application of these theories to the prediction of water movement in both cultivated soils and natural watersheds. That many natural factors appear to preclude application

TABLE 3   Rainfall Disposition at Bass Lake, 1940–1945[a,b]. [Sample depth, 6 ft; rainfall, 49.6 in.]

Groundcover	Total Evaporation	Interception	Runoff	Estimated Percolation
Natural	22.8	5.9	0.2	26.7
Annual burn in autumn	26.1	5.9	9.3	14.7

[a]Source: Rowe and Coleman (1951).

of the equations is evident from the assumptions upon which they are based. Porosity characteristics of the surface layer, vegetation, initial water content, and shrink–swell characteristics all affect the intake rate.

Not unrelated to the infiltration process is the movement of water within the soil profile. In this case, problems associated with surface conditions are secondary to heterogeneities within the soil materials above the water table. Attempts are being made to use an equation proposed by Childs and Collis-George (1951) to describe water-content changes in the profile for a variety of conditions.

The equation for vertical flow is

$$\frac{\partial \Theta}{\partial t} = \frac{\partial}{\partial z} \left(D \, \frac{\partial \Theta}{\partial z} \right) + \frac{\partial k}{\partial z}, \tag{2}$$

where Θ is the volumetric water content, t is the time, z the depth, D the diffusivity, and k the capillary conductivity. The diffusivity is given by

$$D = -k \, \frac{d\tau}{d\Theta} \, ,$$

where τ is the soil-water pressure. Both D and k are functions of the water content of the soil. The water content may fluctuate through a wide range because of evaporation, transpiration, drainage, and so on. In deeper layers, however, the flow may become steady, and for such conditions, many practical calculations of the flux can be obtained from Darcy's law. In this equation, $Q = -K \, dl/dz$, where Q is the flux, z the vertical distance, l the total potential (soil-water tension and gravity), and K the hydraulic conductivity.

For more restricted conditions, Philip (1957) presents an equation relating cumulative infiltration, I, to time as

$$I = St^{\frac{1}{2}} + At. \tag{3}$$

For very large times, the equation is not applicable, and the infiltration rate approaches the hydraulic conductivity, K, of the profile. S has been called the sorptivity; it is a measure of the initial rate and depends on the initial water content and soil properties, both important factors in relating runoff to rainfall.

Horton (1940) and Free et al. (1940) developed two empirical equations that have been useful and probably will continue to be used for some time. Horton's equation is

$$i = i_c + (i_o - i_c)e^{-kt}, \tag{4}$$

where i, i_c, and i_o are the infiltration rate at time t, steady infiltration rate, and the initial rate, respectively, and k is a soil constant that must be determined for different soils. The equation assumes an adequate supply of precipitation and has been used to estimate runoff. The second equation used in field observations by Free *et al.* (1940) expresses the decrease in the amount of infiltration by

$$i = at^b , \qquad (5)$$

where a and b are constants.

In practice, a and b take on values that vary according to soil conditions. Similarity is evident between (4) and (5) and the more formal Equation 3, which is based on Equation 2. The coefficient b may take on values ranging from 0.2 to 0.70. Ruben and Steinhardt (1963) described more recent attempts to use equations such as Equation 2 in rainfall infiltration problems; in these attempts, computers were used.

From what has been said, it is evident that wide ranges in infiltration rates exist. Musgrave (1955) has grouped soils on the basis of minimum rates as measured with little cover and long periods of precipitation (see Table 4).

Because these rates were established after a long period of precipitation, they might represent the rate of flow or flux through the profile. The flux within the soil profile cannot exceed the infiltration rate; thus, the infiltration rate becomes the upper limit for the percolation rate. Initial infiltration rates are often much greater than those given in Table 4. Short, intensive rains, therefore, frequently produce little runoff, and light, steady rains that do not exceed the long-time intake rates also produce little runoff. Nutrient movement by percolation would most likely occur under these conditions. On the other hand, two storms, one exceeding the intake rate, the other not, but both producing the same total precipitation, will not distribute nutrients the same way, either in the profile or in runoff.

Maximum percolation rates are established while water is being applied to the soil surface. When application ceases, infiltration no longer takes place, but water continues to move in the profile. In general, a decreasing water content at any point in the profile is accompanied by a decreased rate of water movement. Even though less water is moving through a given depth, the nutrient redistribution continues. The rate of water movement may become very slow; if it ceases, nutrient movement over long distances also ceases. Rapid removal of water in the root zone by crops minimizes downward loss of both water and nutrients during certain periods of the year. Below the root zone, the water movement is not complicated by plant withdrawal; hence, water may continue to drain. The nature of nutrient movement throughout the entire range of water fluxes occurring in soils requires additional study.

TABLE 4   Minimum Infiltration Rates of Soils of Different Texture[a]

Soil Material	Infiltration Rate (cm/hr)
Deep sand and well-aggregated soils	0.8–1.1
Sandy loams	0.4–0.8
Clay loams	0.1–0.4
Swelling clay soils	<0.1

[a]Source: Musgrave (1955).

Furthermore, it would be helpful to ascertain the minimum rate of water movement beyond which there is little significant movement of nutrients.

Few field studies report the flux of water moving through soils at various depths and changes in the flux that take place with time. One of the most useful long-time studies of moisture loss from a soil profile was conducted by Edlefsen and Bodman (1941). The investigation over a period of 28 months provided data on the average velocity of water moving downward at various depths for different time intervals in a Yolo silt loam soil covered to prevent evaporation. Values of the rate are given in Table 5.

The final infiltration rate before irrigation was stopped was 882 cm per month, or 29.4 cm per day, which is more than 1 cm per hour, a good sustained rate for a silt loam. When irrigation was stopped, the rate decreased with time at all depths. At a depth of 274 cm, between 590 and 842 days, the average movement was 5 mm of water per month—not a very fast rate. It is not known whether a sustained rate of this magnitude significantly affects nutrient distribution.

Ogata and Richards (1957) presented data on water loss by percolation from a relatively uniform Pachappa sandy loam profile over a period of 50

TABLE 5   Average Water Velocity during Percolation from Yolo Silt Loam Soil[a]

Time Range (days):	0–3	6–14	58–119	590–842
Depth (cm)	Velocity (cm/mo)			
30	46	1.5	0.1	–
137	278	8.8	0.2	0.3
274	448	24.0	0.8	0.5

[a]Source: Edlefsen and Bodman (1941).

days. Even after 50 days, the percolation gave no indication of ceasing, and the water-content changes could be adequately described by an equation of the form

$$W = at^{-b} \tag{6}$$

where W is the water content of the surface layers of soil, t is the time, and a and b are constants. The equation appears similar to the infiltration equation. Results of Nielsen and others (1964) clearly demonstrated that movement of water (and presumably nutrients) can take place in a profile even though there is no significant change in the water content at a particular depth.

In a recent investigation, J. M. Wolfe (unpublished data)* calculated the flux of water through various layers of Panoche clay loam over a period of 28 days. His results, presented in Figure 2, after 24 days show a flux of 0.02 cm per day at 1 ft and 0.14 cm per day at 5 ft. Chloride and dibromo-chloro-propane profiles were still changing perceptibly at these fluxes 16 to 36 days after infiltration ceased. Nitrate data not presented in the report showed responses similar to chloride. At water flux rates of 1.5 mm per day, nutrient distribution in the soil changes measurably, giving some indication of continued water and nutrient redistribution at or below the questionable "field capacity" of this soil.

GROUNDWATER

When nutrients percolate to the groundwater, their subsequent movement to lakes or streams involves the general movement of groundwater. Because many of our water supplies are derived from the groundwater, the velocity and direction of water movement as well as the quantities are investigated when possible. Understanding the mixing between the resident groundwater and replenishment water is important, particularly if the invading water contains dissolved constituents that are undesirable. The same mathematical models and techniques that are used for studying mixing in any porous material are used in studying groundwater. Consequently, the dispersion equation (Day, 1956) is an example of one description. The processes involved are the same, although the magnitude of individual components may vary.

The mixing will be very different in a deep sand aquifer than in fractured limestone. Thus, extensive dilution between source and discharge may occur in the deep sand aquifer, whereas in the fractured limestone the amount of mixing with resident fluid will be small, although the path may be tortuous.

*Movement of Chloride, Nitrate and 1,2-Dibromo-3-Chloro-Propane in Panoche Clay Loam. M.S. thesis, Univ. California (1967).

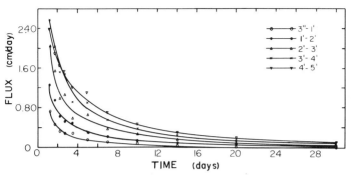

FIGURE 2   The change in the flux of water with time at five depths in Panoche clay loam as measured in the field. (Wolfe, 1967)

Adding to the complexity are layering, lenses, downdips and updips, fine pores, cracks, heterogeneous materials, and adsorbing surfaces.

In a deep aquifer, complete mixing is probably rare. It is not uncommon to have layers of more dense or less dense fluid flowing over and under each other. Therefore, it is not safe to assume that nutrients derived from percolating waters will be diluted by the entire groundwater mass prior to discharge into a lake.

Smith (1965) observed incomplete mixing in explaining nitrate contamination from feedlots and bat caves. Stout *et al.* (1965) observed a nitrate "cap" on the groundwater of Arroyo Grande. Freshwater mounds over salt water are common in Hawaii and Florida. The authors have observed saltwater layers over low salt groundwater in the Imperial Valley, California.

Patterns of contaminated zones of water in the ground have been discussed by Legrand (1965). The velocity of groundwater movement may vary from as little as 1 ft per year to several miles per month. Displacement of all the water from a particular aquifer may take several hundred years, even though parts of the aquifer may be readily conducting water.

On the other hand, Carlson (1964) reported the mean residence time of groundwater recharge in a Wisconsin drainage basin as 45 days, and in a New Jersey basin, 30 days. Continuing investigations of groundwater movement and mixing are needed. Sternau *et al.* (1967) are adopting the radioisotope-tracer method to elucidate the characteristics of aquifers. The possibility of combining nutrient movement studies with aquifer studies has to be explored.

PREDICTIONS FROM CLIMATOLOGICAL DATA

Allison (1965) has suggested that probable leaching losses of nitrogen (and also other nutrients) from soils may be evaluated by comparing measured precipitation with estimates of water loss derived from climatological data during a given period. The idea of using microclimatic studies to predict nutrient losses and to assist in their measurement needs further study. The two graphs in Figure 3 illustrate the use of climatological data for predicting nutrient movement.

At Minneapolis the annual rainfall is about 28 in., and at Jackson, Mississippi, it is approximately 52 in. From November to February or April, the amount of rainfall exceeds the evaporation–transpiration losses at both sites; during the rest of the year this may not be true. Such climatological information coupled with water-storage and runoff data could prove to be very useful in estimating volume of percolate and accompanying nutrient losses.

Various methods involving measurements of runoff and soil water have been used to estimate water use by plants on watersheds (Penman, 1963). With an ever-increasing amount of data on plant use and the development of reasonably accurate equations for predicting water use from climatological

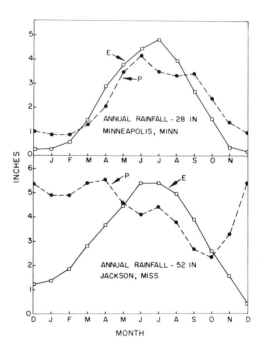

FIGURE 3   The potential evapotranspiration and precipitation during the year at Minneapolis, Minnesota, and Jackson, Mississippi. (Allison, 1965)

measurements, it should now be possible, if we know rainfall and plant use, to reverse this procedure and estimate both interflow (seepage) and deep percolation.

In the Great Plains area, with an average of 20 in. of precipitation, about 18.8 in. is lost by evaporation and transpiration, 1 in. as stream flow (surface and interflow runoff) and 0.2 in. as percolate (Bertrand, 1966). The losses by leaching and runoff would understandably be small under these conditions. In contrast, in the humid Southern Piedmont rainfall is about 50 in. About 11 in. runs off to lakes and streams, and 39 in. is lost by evapotranspiration and percolation. Between 6 and 8.5 in. is needed to produce a corn crop; the remainder is available for storage and deep percolation. If all but 10 in. of the 39 is lost by evapotranspiration, the remaining 21 in. of runoff and percolation is sufficient to reduce significantly the nitrogen content of many soil profiles. As was pointed out in the introduction, larger solute-removal loads occur in regions of higher precipitation even though the concentration in low runoff areas is much higher (Van Denburgh and Feth, 1965).

Allison (1966) concluded that very little nitrogen loss by leaching occurs in regions where precipitation is 50 in. or less during the months of May to October, except in sandy soils or in areas with unusually heavy rainfalls. The reasoning behind this conclusion is that during these months water loss to the atmosphere exceeds the supply; therefore, downward movement is very slow. In many areas fall precipitation reduces the moisture deficit in the profile accumulated during the growing season. Because the soil is dry, absorption of the precipitation is generally greater than during the winter when the soil is frozen or during the spring when the soil is wet. Consequently, both runoff and percolation are frequently greater in the spring than at other times of the year, and this is significant to nutrient movement.

MOVEMENT OF NUTRIENTS IN SOLUBLE AND PARTICULATE FORMS

NUTRIENTS IN PERCOLATE WATERS

Despite rapid advances in the development of water-movement theory, the theory has not been extensively used to describe nutrient movement. Examples, to be given later, indicate that loss of nutrients by leaching is a significant pathway for fertilization of lakes.

Contrary to previous thoughts, water and its dissolved constituents do not move in a one-to-one manner. There is sufficient evidence to indicate that the mixing that occurs between the resident soil solution and invading water, either from precipitation or irrigation, must be considered as a major process affecting the movement and distribution of soluble materials, both in the plant-root zone and beneath it. Hence, the processes of mixing determine the concentration changes that occur as a function of both time and direction of flow. The reasoning behind this conclusion is that (1) dissolved constituents can move independently of the solvent and (2) macroscopic parameters, such as average hydraulic conductivity, cannot describe processes that are responsible for mixing on a microscopic scale.

On a microscopic scale, soils are made up of particles of various shapes and sizes. In like manner, the spaces between the particles are of varying shapes and sizes, the variations giving rise to channels that differ in direction, size, and shape. Through these channels or pores water moves. Sometimes the channels are full of water, with no air phase; sometimes they are only partly full. As a result of the variation in the paths of flow, the water moving through these channels changes direction and rate from point to point. Consequently, when a band of water containing a dissolved constituent is introduced into a system of microscopic channels conducting water at various velocities, it soon becomes spread out, or diffuse (Day, 1956). The length of the diffuse zone increases more or less as the square root of the distance traveled and is affected by the velocity. Mixing, therefore, takes place between the resident solution and the invading solution.

Other interactions between the invading solution and the solid matrix also modify the distribution of the constituent. Ions such as ammonium or phosphate may undergo exchange or adsorption and may be "delayed" in their movement in the direction of fluid flow. Anions such as nitrate or chloride, which are usually not adsorbed, may be repelled by the solid matrix so that the volume available for water flow is different from that through which the negatively charged ion moves. On the other hand, the adsorption of water by the matrix and the accompanying modification of its physical properties, especially viscosity, may reduce the area of effective flow and create a variation of velocity within the cross-sectional areas of a pore.

Not to be overlooked is the process of diffusion of the dissolved constituent that can result from activity gradients within the soil-water system. Although not a spectacular process, diffusion is inevitable, and it is likely to contribute to mixing at flow velocities comparable with those found in many soils and some aquifers. Diffusion cannot be expected to move large quantities of salts long distances in short times, but it can move large quantities of salt short distances in any given time span. Figure 4 gives an example of the spreading or mixing of a band of solution containing NO_3^- and SO_4^{2-} as it moves through the soil. The solution containing the two ions was added to the surface of a moist soil and was followed by additional water without salt. If it can be assumed that the NO_3^- ion is more-or-less free to move and is not removed by organisms, then the diffuse nature of the downward-moving NO_3 band is due entirely to the mixing of the band with

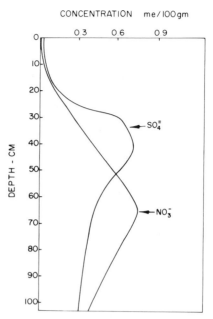

FIGURE 4   The distribution of nitrate and sulfate in a sandy loam soil after application of a slug of irrigation water followed by ion-free water.

water in the column according to the principles previously described. On the other hand, because this soil partially adsorbs sulfate ion, it retards sulfate movement and reduces maximum concentration of the sulfate.

Yaalon (1965), who observed similar effects, proposed that the depth at which maximum concentration occurs can be calculated from the equation

$$\overline{X} = \frac{\text{Pef} \cdot \text{Rm,}}{\text{wa}} \tag{7}$$

where $\overline{X}$ equals depth to maximum concentration (cm), Pef equals effective precipitation (cm), Rm equals relative migration coefficient, and wa equals fractional volume occupied by mobile moisture, considered to be the "available water" or water between field capacity and permanent wilting percentage. R for Cl^- according to Yaalon is near 1.0 and for sulfate about 0.66 for his soil. It is defined as

$$\text{Rm} = \frac{\text{distance traveled by the ion}}{\text{distance traveled by moisture}}.$$

We shall see later that Dyer found departures from 1.0 to be large for both Cl^- and NO_3. It may be anticipated that further displacement would reduce the maximum concentration and distribute the solute more widely with depth.

Day (1956) considered the movement of a band of dissolved constituent through a porous material when the main process of mixing was by mass flow of the water. For the initial conditions of a band with concentration C_o located in a small depth x_o of porous material the mixing is described by the equation

$$C = C_o x_o \, (4 \, Dt)^{-\frac{1}{2}} \exp \left[-(x-vt)^2/4Dt \right], \tag{8}$$

where C equals concentration at depth x; C_o, the initial concentration; D, factor of dispersion; v, the velocity; and t, the time. Day defines an index of dispersion $\beta = 2D/v$.

The maximum concentration or the peak in the distribution curve as is illustrated in Figure 4 for NO_3^- is given by

$$C_p = C_o x_o / \sqrt{2\beta x_p},$$

where C_p is the concentration at the peak at depth x_p.

The equation has been tested in the laboratory, but no adequate field tests have been reported. Day and Forsythe (1957) report values of the dispersion coefficient, which is a measure of the mixing for different materials, as follows: Ottawa sand, 0.006; Monterey sand, 0.006; Ben Lomond sand, 0.080; and Dowex resin, 0.061 cm^2/sec. The parameter could prove to be a very significant one for estimating nutrient distribution from water-movement data.

Although it would be helpful to be able to predict the movement of dissolved constituents based on the water movement, it might prove to be more useful to predict water movement on the basis of the movement patterns of dissolved constituents. Examples of this approach include the use of tracers, which has been discussed in detail elsewhere (Nielsen and Biggar, 1967).

Many models have been proposed as being suitable for describing the redistribution of dissolved constituents. One of these is van der Molen's application of the Gluecauff theory to the leaching of soluble salts. The equation, also adopted by Dyer (1965a), is

$$C = \frac{C_o}{2} \, \text{erfc} \, \frac{P-1}{1.414P} \, (N)^{\frac{1}{2}}, \tag{9}$$

Where C equals the concentration of an ion (e.g., NO_3^-) in the soil solution at depth d, C_o equals initial concentration of ion which is assumed uniform throughout the unleached profile, P equals the ratio of percolating water to total soil solution above depth d, N equals the number of theoretical plates above depth d, and erfc equals the error function complement. Van der Molen estimated the number of theoretical plates in the profile from the center of the transition zone to the surface by the equation

$$N = 2\pi \left[\frac{d}{Bo} - \frac{dB}{dd} \right]_B^2 = Bo/2 , \tag{10}$$

where Bo and B are the nutrient content per unit dry weight of unleached soil and leached soil, respectively, and d is the depth to the center of the transition zone or breakthrough curve. The thickness of the theoretical plates is given by d/N.

For purposes of simplifying the actual process, the soil profile is divided into a number of plates or layers. The concentrations in each plate are considered uniform, and equilibrium is assumed to exist between adsorbed and solution ions or molecules.

Dyer (1965a) collected a Panoche sandy loam soil from the sixth foot of a soil profile in western Fresno County, California, for leaching studies. The area overlying the sample had not been cultivated. The soil at the field-moisture content of 13.4 percent had a NO_3^- concentration of 94 mEq/liter and Cl^- content of 210 mEq/liter, a significant nitrate content under any circumstances. The dry soil was packed into a column and leached with the equivalent of 12.5 in. of a saturated $CaSO_4$ solution.

Figure 5 shows the distribution (redrawn) of Cl and NO_3^- as measured by Dyer after 27 days of drainage. We immediately note the decrease in concentration at the surface, the sigmoid shape of the tail part of the breakthrough curve, and the bulge of increase in concentration above the original ion content given by the dotted line. Although the Cl^- and NO_3^- displacements are similar, they are not the same. Dyer noted that the salt front should be located at 39.9 in. if perfect or "piston" displacement has occurred. In fact, this front was located between 42 and 60 in., with the average occurring at 53.5 in. Consequently, the salt and water did not move in a like manner; mixing did occur, and this mixing was a result of the processes described above.

It is worth noting that the initial condition in this particular case is different from the previous example of NO_3^- and SO_4^{2-}. In the previous example, we noted the spreading of a band of solute as it moved through the profile. In the latter case, the constituent is initially more-or-less uniformly distributed, and the increase in water content, coupled with flow, moves the salt deeper, increasing the concentration.

We note here the possibility of such distributions occurring in profiles not

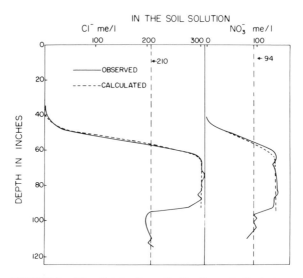

FIGURE 5   The distribution of chloride and nitrate in a
Panoche sandy loam soil after application of 12.5 in. of
CaSO$_4$ water. The vertical dotted lines are the initial salt
distribution, and the calculated line is from Equation 8.
(Dyer, 1965a)

as a result of irrigation but merely as a result of periods of excessive rainfall
(Soubies *et al.*, 1952; Black, 1966). Such distribution could result in a
pulse-type nutrient contribution to surface waters or groundwaters over
periods of years, accounting for the rise and decline of NO_3^- in some well
waters (Feth, 1966). On the other hand, a more-or-less uniform distribution
of NO_3^- may occur in a profile over long times where the input of NO_3^- is
more-or-less constant and the precipitation is uniform (Stout *et al.*, 1965). It
should be recognized that the soil depths below the Dyer profile would tend
to reduce or dilute the slug as it moves down if the soil water contained a
lower NO_3^- or Cl^- content.

Dyer (1965b) also investigated the distribution of Cl^- and NO_3^- in
adjacent soil profiles in the field. One profile received irrigation water but its
counterpart did not. Differences in the two salt profiles presumably reflect
changes brought about by adding irrigation water in excess of rainfall and by
cropping.

In Figure 6, the nitrate profiles for location number II, as referred to by
Dyer (1965b), have been redrawn from that source. The data have been
smoothed, but the essential features of the distribution are retained. The
irrigated area, similar before irrigation to the nonirrigated area, had 4 years of
alfalfa and 4 years of barley. In the nonirrigated area, the concentration of
NO_3^- increased rather sharply in the upper 10 ft and remained fairly constant

with depth below this. In the irrigated area, percolation of excess water leached the upper 20 ft, increased the concentration gradually to a maximum at 30 to 35 ft, and then declined below the original concentration. The displacement proceeded slowly. From Equations 8 and 9, the plate height for the laboratory column was calculated at 1.3 cm, and the field data, at 20 cm. Nonuniformities of displacement, decreasing velocity, degree of water saturation, and method of water application affect the mixing and therefore affect the calculated plate height (Bigger et al., 1966; Miller et al., 1965).

Although the present sections consider only the nature of the displacement, it is noteworthy that nitrate concentrations such as are found in these profiles are very significant to eutrophication. Without further speculation on the various features of the distribution, we can see the general picture clearly: in natural field profiles, even to great depths, redistribution follows the same general pattern.

In a study of the vertical movement of nitrogenous matter in the Grover City–Arroyo Grande Basin, Stout et al. (1964) made some estimates of the nitrogen balance. They had to make certain assumptions concerning consumptive use, percolation, and nitrogen imports and exports. They estimated that the net residual nitrogen would be 74 lb per acre. This residual

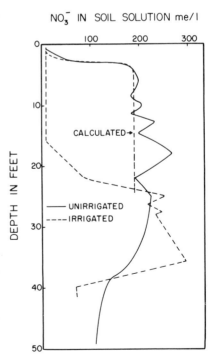

FIGURE 6   The distribution of nitrate in a field soil before and after several seasons of cropping and irrigation. The solid line represents the distribution in an unirrigated profile, and the dotted line represents the distribution under cropping and irrigation nearby. The calculated line is from Equation 8. (Dyer, 1965b)

in water moving to the water table could raise the concentration of N in the groundwater 10.9 mg per liter.

As an example, irrigation wells in the study areas showed NO_3-N concentrations of 19 to 27 mg per liter. Other wells not used for irrigation showed NO_3-N concentrations of 0.1 to 33 mg per liter. Groundwater in a creek bed, which may reflect the concentrations to be found in runoff or interflow water, contained 0.1 mg per liter; other samples of creek water ranged from 0.2 to 0.5 mg per liter. Several soil profiles were sampled, some to a depth of 43.5 ft. The range of NO_3-N in solution was rather wide—from 0.3 to more than 450 mg per liter—and several profiles had concentrations ranging between 23 and 110 mg per liter. Various bands of nitrate were found; some of these were believed to have been present before irrigation began.

Significant increases in nitrates in these groundwaters from domestic and agricultural activities seem certain. Recycling of the water through irrigation apparently concentrates nitrates further. The nature of the water and nutrient movement with respect to soil stratification, however, remains unclear.

Stout *et al.* (1964) speak of pulses of nitrate interspersed with low-nitrate water. They also point out that soil-solution nitrate is characteristic of all fertile soils, with or without additions of fertilizers. Hence, fertile uncropped sandy soils in the area generate about 30 mg per liter of NO_3^--N in soil solutions, a heavily fertilized clay, 40 mg per liter, and a conservatively managed strawberry farm, 20 mg per liter. The shallow wells reflecting groundwater quality are consistent with the soil-profile measurements.

The work of Sylvester and Seabloom (1962) can give us an additional insight into the magnitudes of nutrient concentrations and the differences in concentration when the drainage is from surface sources and when it is from subsurface sources. The study involved the Yakima Basin. Average mean rainfall at Yakima is 7.2 in. With this amount of rainfall, it is necessary to apply about 4.25 acre-feet of irrigation water per acre. A few results are summarized in Table 6.

Surface-drain concentrations reflect some overland flow, subsurface water, concentration by evaporation, and removal by aquatic growth. All the concentrations given are considered indicative of significant nutrient contributions to this river. For an irrigated area, the nitrate values are quite low compared with some others reported. Although phosphorus levels represent insignificant agricultural losses, they are not insignificant relative to eutrophication. What is considered low agriculturally may be high for receiving waters. The constancy of concentrations from subsurface drainage throughout the year is very interesting. Also, the lack of phosphorus reduction by vegetative growth in open ditches is surprising. However, this constancy may involve the balance between N and P required by the aquatic life, and there may be an excess of phosphorus.

TABLE 6   Nitrogen and Phosphorus Content of Subsurface and Surface Drains during Irrigation and Nonirrigation Seasons

Water Source	Nutrient (mg/liter)		
	Nitrate as N	Soluble P	Total P
Irrigation season			
Applied water	0.25	0.07	0.10
Subsurface drain	2.5	0.22	0.28
Surface drain	0.8	0.19	0.27
Nonirrigation season			
Subsurface drain	2.5	0.23	0.26
Surface drain	1.9	0.20	0.26

The authors of this useful study have made the following estimates of pounds of nutrients per acre contributed to the river by return flow:

	Nitrogen (lb)	Soluble Phosphorus (lb)
Irrigation season	32	0.3
Nonirrigation season	36	0.7

These values are higher than many reported in humid regions. One drain, D-12, believed to be a subsurface drain that discharges all year, yielded 33 lb of N and 1.3 lb of P per acre per year.

Water samples taken from shallow wells in the regions investigated yielded average concentrations as follows (data expressed in milligrams per liter):

	NO_3 as N	Soluble P	Total P
Irrigation season	2.3	0.12	0.16
Nonirrigation season	1.6	0.08	0.10

These concentrations of phosphorus are somewhat higher than those usually thought to exist in waters in subsoils.

Doneen (1966) investigated the salt distribution in field profiles at five locations, three that were drained by tile and two nonirrigated virgin areas. The soluble salts in the soil were determined by bringing the soil samples to their saturation percentage, extracting the solution, and analyzing for cations and anions (saturation extract). Each of the locations had a drainage system that allowed measurement of the flow and periodic sampling for salt analysis.

The operators supplied information about cropping histories, fertilizers applied, and quantities of water used.

Drain field DS, installed 9 years previously, appeared to have reached a more-or-less dynamic equilibrium in the surface with the irrigation water. Cotton, rice, barley, and lettuce were grown on this location during 1962–1966. The average application of N amounted to 167 lb per year as fertilizer. Water applied ranged from 7.4 to 12.1 acre-feet per acre per year. The NO_3–N measured in the drain effluents averaged 44.5 mg per liter, or 2 to 4 times the concentration in the saturation extract of the soil-profile samples; the irrigation water contained 1.7 mg per liter.

The second location, with drains more recently installed, was cropped to tomatoes, cotton, wheat, safflower, barley, and lettuce. It had received about 100 lb of N per year since 1964. Water applications ranged from 2.5 to 6.1 acre-feet per acre per year, the water containing 2.1 mg per liter of NO_3–N. The average NO_3–N content of the drainage water as nitrogen was 23 mg per liter. The third location, cropped to safflower and cotton, received N applications of 75 to 131 lb per acre per year and an average yearly water application of 1.7 acre-feet per acre. The average nitrogen content of the drainage water was 37 mg per liter.

Since nitrogen found either in the groundwater or the drainage effluents is considered lost from plant roots, it could eventually find its way to other bodies of water. All the quantities are significant in terms of adding to the nitrogen content of receiving waters.

Comparisons of the nitrate content of nonirrigated virgin soils can be made with those previously mentioned only on the basis of the NO_3^- in the saturation extract or groundwater analysis, since no drainage effluents are available. One of the nonirrigated sites contained NO_3^- levels higher than found in the irrigated areas; on the other site the levels were lower. The implication here is that neither irrigation water nor applied fertilizer is necessary for high concentrations of NO_3^- to be found in the profiles of some soils.

Some estimates were made of the amount of total N in two profiles. Despite the wide differences in cropping practices and soil characteristics, the total N was high enough to cause both profiles to contain more than 21 mg per liter of N as NO_3^-.

Johnson et al. (1965) studied N and P loss in tile-drainage effluents from a number of tile-drainage systems in irrigated areas in the San Joaquin Valley of California. A number of cropping practices were involved in the study, with crops (cotton, alfalfa, rice), fertilizers, and irrigation water applications as variables. Initial analysis of tile effluent in a previously unirrigated, noncropped area showed an N concentration of 1 mg per liter. A tile system that had been cropped to alfalfa and had a low discharge over the period of a year yielded a range of N between 2.0 and 14.3 mg per liter. In systems where high rates of N fertilizer were applied, the concentrations ranged up to 62.4 mg per liter. In the systems reported, the concentrations ranged from 1.8 to 62.4 mg per liter, with a weighted average, based on flow, of 25.1 mg per liter. In the same effluents, the concentration of P ranged from 0.053 to 0.23 mg per liter with a weighted average of 0.079.

Some attempt was made to estimate the pounds of N and P lost in the tail water—water that has reached the end of the field over the surface and is either discharged to drains or recycled.

It is not possible to make anything more than interesting observations from the data because the magnitude of flow is not given. However, one system received 1,263 lb of N in the irrigation water and lost 1,539 lb by tail-water loss. Since the amount of tail water is likely to be less than the water applied, the increase must come about by solution from the soil surface. The loss by tail water represents about 10 percent of the loss by drainage. By contrast, one third less P was lost by tail water than was applied in the water, the comparison with N indicating a net removal relative to the N. In most cases, more P was lost over the surface than was lost by tile effluent.

G. E. Smith (unpublished data)* conducted an extensive survey of NO_3^- contamination of well waters in parts of Missouri. There seems to be some correlation between animal population and NO_3^- content of the water. Animal feedlots storing manure for many years are readily leached, the percolate moving down to groundwater. In Figure 7, Smith's data illustrate the distribution of NO_3^- under two feedlots. The Clinton County lot has been a feedlot for more than 25 years. A nearby drilled well over 100 ft deep contained 15 to 20 mg per liter of NO_3^-–N. Figure 7 illustrates the possibility of a continuous supply of NO_3^- and suggests that NO_3^- is moving into the groundwater even though the maximum in the soil-NO_3^- distribution appears at 6 ft.

The Macon County location was near an old barn lot that had not accommodated livestock for 10 years. A 30-ft well nearby contained 150 to 170 mg per liter of NO_3^-–N. This, too, may indicate that nitrate has moved below the depth sampled. The band of nitrate derived from the feedlot is undoubtedly moving down. Future samplings could reveal the rate at which this is occurring. The nitrate profiles in Figure 7 are strikingly similar to profiles shown in Figures 5 and 6, and could be approximately described by Equation 8. Analysis for nitrogen in other forms, including soluble organic forms, would be useful, particularly where the source could provide other forms.

*Nitrate Problems in Plants and Water Supplies in Missouri. Presented at annual meeting of American Public Health Association (1964).

That rainfall can be quite effective in redistributing nitrate-nitrogen has been demonstrated by Soubies and others (see Black, 1957). Calcium nitrate was applied to the surface of a fallow sandy loam soil. The amount of rainfall involved over 153 days was only 34.7 cm, but the maximum in the peak of N concentration was well below 100 cm. The displacement with each increment of rainfall is evident in Figure 8.

Other examples of nitrogen movement by rainfall reflected in subsequent crop growth have been given by Wetselaar (1962).

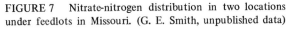

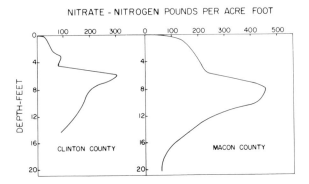

FIGURE 7   Nitrate-nitrogen distribution in two locations under feedlots in Missouri. (G. E. Smith, unpublished data)

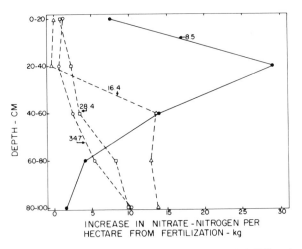

FIGURE 8   Increase in nitrate-nitrogen content at different depths in a fallow sandy loam soil as a result of fertilization with calcium nitrate. The numbers refer to the cumulative rainfall in centimeters corresponding to the nitrate-nitrogen distribution shown. (Black, 1957, p. 206)

NUTRIENTS IN RUNOFF WATER

Adding to the difficulties of predicting amount of runoff water from storms is our inadequate knowledge of the chemical composition of the runoff. In particular, we know little about the nutrient content of the runoff.

Catchment basins have provided data on the total sand, silt, and clay (suspended solids) loads that accompany a particular volume of runoff. The mineralogical characteristics usually are not measured, and organic fractions are rarely considered. Both colloidal and dissolved fractions should be considered.

Lipman and Conybeare (1936) estimated the loss of nutrients from harvested crop areas in the United States for the year 1930. The values are given in Table 7 to establish some basis for comparing present and past estimates as well as for computing relative losses by runoff and leaching. Irrigated areas are omitted. Losses by erosion and leaching are separated by combining nutrient content of soils and the amount of eroded material measured at various locations. Lysimeter and river analyses were used to arrive at the leaching losses. Thus, roughly 5 million T of N were lost by removal of suspended matter (solution and sediment) and 5 million T by leaching in the year 1930. Two million T of P were lost mainly by erosion. It is noticeable that phosphorus loss was attributed to erosion. As might be expected, the method of cropping affected the rate of removal by both erosion and leaching. Under conditions favorable to erosion, more N is lost by this process than by leaching. Neal (1944) estimates that the N in eroded material is 5 times greater than in soil remaining in place. The estimates are based on the fact that organic matter has a low density relative to soil and therefore is more subject to wind and water action.

TABLE 7   Loss of Plant Nutrients from Harvested Crop Areas, United States, 1930[a]

	Nutrient[b]		
	N	P	K
Intertilled crops			
Leaching	17.1	–	39.1
Erosion	48.1	21.0	280.7
Annual crops not intertilled			
Leaching	32.5	–	37.6
Erosion	11.1	4.9	65.0
Biennial and perennial crops			
Leaching	23.0	–	
Erosion	24.2	10.6	141.1

[a]Source: Lipman and Conybeare (1936).
[b]Data are expressed in pounds lost per acre per year.

For a more restricted area, Sawyer (1947), on the basis of analysis of one stream tributary to each lake, estimated the removal rates of N and P given in Table 8. Inorganic N losses were 4.4 to 6.4 lb per acre per year, and organic N losses were 1.6 to 1.8 lb per acre per year. These values are one sixth to one fourth as large as those in Table 7. As for phosphorus, the differences are one fortieth to one fourth as large as those of Lipman.

Weibel *et al.* (1966) conducted a number of investigations, involving both rainfall and runoff from agricultural lands, that yielded information on nutrient removal. One area was a 1.45-acre cultivated field in winter wheat near Coshocton, Ohio. The constituent concentration ranges and means from composite samples from five storm-runoff events are given in Table 9. The area was contour plowed, adequately fertilized, treated to control pests, and planted to an annual rotation of corn, wheat, first-year meadow, and second-year meadow.

TABLE 8   Contribution of Fertilizing Elements in Drainage from Agricultural Lands[a]

| | Fertilizing Elements[b] | | | |
Drainage Area	Inorganic N	Organic N	Inorganic P	Organic P
Tributary to				
Lake Monona	4.4	1.6	0.055	–
Lake Waubesa	4.9	1.8	0.097	0.29
Lake Kegonsa	6.4	1.8	0.097	0.31

[a]Source: Sawyer (1947).
[b]Data are expressed in pounds per acre per year.

TABLE 9   Constituent Concentrations in Rural Land Runoff from a 1.45-Acre Cultivated Field in Winter Wheat near Coshocton, Ohio (March 1964 through February 1965), and in the Rainfall[a]

| | Runoff[b] | | Rainfall[b] |
Constituent	Range	Average	
Suspended solids	5–2,074	313	11.7
N[c]	2.2 –12.7	9	1.17
Inorganic N as N[d]	0.2 – 8.2	5	0.86
PO_4 (total hydrolyzable as P)	0.08– 1.07	0.6	0.03

[a]Source: Weibel *et al.* (1966).
[b]Data are expressed in milligrams per liter.
[c]Average of four forms of N.
[d]$NH_3 + NO_2 + NO_3$.

Table 9 also presents the constituents measured in the rainfall during the same period and presented as average concentrations during this period.

It is of interest to note the rather wide range in the suspended solids and particularly the high concentration in the rainfall. If we assume a threshold concentration for algal growth of 0.3 mg per liter of N and 0.03 mg per liter of P, as proposed by Sawyer (1947), both the rainfall and the runoff water exceed these values significantly. In some cases, the rainfall concentration exceeds the minimum concentration in the runoff, which suggests that overland flow can reduce some of the constituents found in rainfall.

Table 10 presents a comparison, from the same report, of concentration of constituents in runoff from various areas having different agricultural uses. "Improved" in the table refers to such agricultural practices as contour plowing, strip-cropping, adequate fertilization, and pest control; it is in contrast to "prevailing practice." The two fields were in an annual rotation of corn, wheat, first-year meadow, and second-year meadow. "Mixed farm" refers to a meadow, tillage, dairy barn, pasture, woodlot complex totaling 303 acres. A 300-tree apple orchard on a 5-acre tract near Ripley, Ohio, is the remaining treatment.

Table 10 shows a loss of soluble phosphorus content ranging from 0.01 to 0.03 lb per acre per day for the various farm conditions, with the exception of the mixed farm, which has a removal rate of 0.07 lb per acre per day. In some cases the improved farming methods did not reduce losses, a fact that may be linked with the higher fertility obtained under improved farming conditions and the susceptibility of this surface fertility to erosion loss. Again, the contribution from rainfall is reduced after contact with the soil.

TABLE 10   Constituent Loads in Runoff under Improved and Prevailing Agricultural Practices[a]

Agricultural Practice	Constituent[b]		
	Suspended Solids	PO$_4$ as P	Total N
None (rainfall only)	4.5	0.03	0.58
1.5-acre field			
Improved	8.8	0.03	0.81
Prevailing practice	19.5	0.02	0.85
7.5-acre field			
Improved	21.0	0.02	0.19
Prevailing practice	0.94	0.01	0.16
Mixed farm	165	0.07	0.83
Orchard, improved	12.6	0.02	0.48

[a]Source: Weibel et al. (1966).
[b]Data are expressed in pounds of constituent per acre per day of rainfall.

The range for N removal is relatively uniform except for the 7.5-acre farm. The values given in Table 10 reflect the rate per acre per day even though the duration of rainfall varied. The nutrient values given by Weibel *et al.* (1966) appear to lie somewhere between removal rates reported by Lipman and those reported by Sawyer.

Engelbrecht and Morgan (1961) studied land drainage in the Kaskaskia River basin. Farm lands are drained mainly by tile drains placed at a depth of 24 in. The stream was sampled at various times during the period April to September. One sampling station was located so the only source of phosphorus was land drainage. The concentration of ortho plus hydrolyzable P ranged from 0.01 to 0.4 mg per liter, with an average of 0.07 mg per liter, which corresponds to a value of 0.04 lb of P per day per square mile. At other locations, where the stream runoff was greater, the P lost in stream runoff was calculated at a mean of 0.6 lb per mi^2 per day, a value somewhat higher than that proposed by Sawyer.

NUTRIENTS IN ERODED MATERIALS

Mention must be made of suspended-solid losses even though data relating the effect of sediment loads on the nutrient status of receiving waters is not well documented. Suspended solids include both sediments and colloidal material that remains suspended in lakes and rivers.

Numerous reports giving the quantities of sediments removed by water erosion at various locations around the world emphasize the varied conditions that give rise to these losses. Separating nutrients carried off in solution in runoff water and those associated with sediment is difficult and makes identification of the source of nutrient arbitrary. Therefore, total analyses of nutrient losses in sediments do not reflect that which is in solution and that which is adsorbed—a distinction that can be very useful.

Nutrient losses by erosion tend to be selective in the sense that organic matter and clay particles, which in soil are relatively high in nutrients, are more subject to erosion than coarser particles.

That eroded material frequently differs in composition from the original soil is good evidence that erosion losses will bring about nutrient changes in receiving waters. To express this change, an enrichment ratio defined as

$$ER = \frac{\text{concentration of the element in soil material in runoff}}{\text{concentration of the element in soil from which runoff originated}}$$

is often used (Barrows and Kilmer, 1963).

Since organic matter is a source of many nutrients and is readily lost, it is of importance both as a possible nutrient source in streams and as a valuable resource removed from the land. Magnitudes of loss range from 337 lb per acre per year to 1,149 lb per acre per year (Knoblauch *et al.,* 1942). Massey and Jackson (1952) reported organic matter enrichment ratios of 1.15 and 1.24 for Fayette and Almena silt loams, and Neal (1943) gave values of 4.0 for Collington silt loam. The loss of organic matter is not linearly related to the loss of soil, and Massey *et al.* (1953) showed an inverse relation in this respect. As the surface layers of higher organic content are eroded, the percentage would be expected to change.

Nitrogen losses are related to organic-matter losses. Thus nitrogen ER values for Almena silt loam and Fayette silt loam were 1.34 and 1.08, respectively, and these compare with the organic-matter ER values given above (Massey ,and Jackson, 1952; Massey *et al.,* 1953). Neal (1944) gave values much higher at 5.0 and 4.7 for nitrogen and organic matter. Actual magnitudes of N loss are quite variable. Hays *et al.* (1948) reported total N losses of 48 lb per acre per year from a silt loam soil, and Knoblauch *et al.* (1942) gave values of 67 lb per acre per year from a sandy loam on 3.5-percent slope. A cover crop and manure reduced this loss to 19 lb.

Bryant and Slater (1948) found 10.4 lb of dissolved nitrogen per acre removed from a fallow silt loam on a 5-percent slope, whereas less than 1 lb was lost from a fallow silt loam on an 8-percent slope. Duley and Miller (1923) reported annual nutrient removal in runoff and erosion from a Shelby loam with a 3.5-percent slope and no fertilizers added (Table 11). The differences in nutrient removal between cropping systems are apparent. Also apparent is the wide difference in $NO_3^- -N$ losses compared with total N losses. Losses of phosphorus are generally higher than reported elsewhere, the greater losses emphasizing the role of erosion.

Phosphorus losses in eroded material may be much more significant than nitrogen in relation to other avenues of loss. Phosphorus is relatively

TABLE 11   Annual Nitrogen and Phosphorus Content of Runoff and Eroded Material from Shelby Loam, 3.6-Percent Slope[a]

	Nutrient (lb/acre)		
Cropping System	Total N	$NO_3^- -N$	Total P
Not cultivated	99	1.38	48
Spaded 8 in. deep	74	0.56	33
Bluegrass sod	0.6	0.07	0.1
Wheat annually	30	0.32	11
Rotation: corn, wheat, clover	6	0.02	2
Corn annually	40	0.02	8

[a]Source: Duley and Miller (1923).

immobile, tending to reside where it is placed. Since most of the phosphorus is held in an immobile form, it is not leached as nitrate is. Placing phosphorus fertilizer beneath the soil surface avoids immediate loss by erosion but later cultivations may expose it to runoff.

Bedell showed that soil planted to corn lost 4.5 T of soil and 20 lb of phosphorus, 60 percent in the organic form. The ER values for phosphorus range between 1.0 and 3.1. Rogers (1942) reported ER values of 1.3 for total phosphorus from a Dumore silt loam and 3.3 for soluble P as determined by 0.002 NH_2SO_4. Neal (1943, 1944) gave an ER range of 1.5 to 3.1, and Massey *et al.* (1953) noted a range of 1.9 to 2.2.

Massey and Jackson (1952) developed regression equations for enrichment ratios for N, P, K, and organic matter from Almena and Fayette silt loam and Miami loam. These and other equations could prove useful, upon further testing, for estimating losses of particular nutrients. The equations are of the form

$$Y = a + Bx + cz ,$$

where Y = log enrichment ratio
  x = $-$ log tons of solid per acre-inch of runoff
  z = $-$ log tons of total solids lost per acre
  $a, b, c,$ = constants unique for each nutrient.

The order of removal is exchangeable potassium $>$ available phosphorus $>$ organic and ammoniacal nitrogen $>$ organic matter, with the average ER being 19.3, 3.4, 2.7, and 2.1, respectively.

Many other examples of sediment erosion involving nonagricultural conditions can be given. A summary of relative erosion values and cropping practices in the Pacific Northwest is given in Table 12.

These and other examples emphasize the relation between runoff, erosion, crop cover, slope, and soil type. Much of the material that is eroded is probably not measured in catchment areas. Consequently, there is no evaluation of the eroded colloidal reactive phase (including organic matter) in soils. These fractions may be significant in eutrophication of lakes. It would be useful to have measurements that consider all parts of the eroded material and the chemical changes involved.

The fact must not be overlooked that sediments and related materials not only contribute nutrients by releasing them to the receiving waters but also may adsorb or fix nutrients. An eroded colloidal soil, low in phosphorus, might remove measurable amounts of phosphorus from water, provided there is sufficient mixing to bring the sediment into contact with large volumes of water. By the same reasoning, all nutrients that potentially might become

TABLE 12   Relative Erosion According to Cropping Practice in the Pacific Northwest (United States)[a]

Crop or Practice	Relative Erosion
Forest duff	$10^{-3}-1$
Pastures, humid region or irrigated	$10^{-3}-1$
Range or poor pasture	5–10
Grass/legume hayland	5
Lucerne	10
Orchards, vineyards (with cover crops)	20
Wheat, fallow (stubble not burned)	60
Wheat, fallow (stubble burned)	75
Orchards, vineyards (clean tilled)	90
Row crops and fallow	100

[a]Source: Musgrave (1954).

available to lakes and streams from suspended sediments entering the water may not become available because they are deposited and buried under later sediments. From the few examples given, it would appear that total N losses for a variety of conditions are estimated to be between 4 and 50 lb per acre per year by runoff and percolation. This is equivalent to 1.3 to 16 T per mi^2 per year. Feth (1966) reported NO_3-N loads of near zero to 1.8 T per mi^2 per year for several rivers in the United States and 0.45 T for European rivers. From this comparison it is apparent that not all nitrogen lost to streams ends up as nitrate. However, it is also probable that nitrate analyses provide indications of the total nitrogen contributing to eutrophication that are too low. Consequently, if an average figure for NO_3-N must be used, it might safely lie between 1 and 10 lb per acre per year. (Much of this is derived from subsurface flows.)

The problem of estimating phosphorus losses significant to eutrophication is even more difficult. Total phosphorus losses, mostly through erosion, range from near zero to about 20 lb per acre per year. Losses of soluble P plus hydrolyzable P range from 0.04 to 2 lb per acre per year.

For specific situations, values may be higher than these. And, where possible, values appropriate to the specific situation should be used. There may be a tendency to overestimate losses based on small catchment areas and the more intensive storms that produce the most runoff. Projecting short-time yields over large areas on an annual basis may result in an overestimate of the losses. Finally, if a particular *ratio* of nitrogen to phosphorus is the significant factor that determines the acceleration of eutrophication of lakes, then total losses given here should be evaluated on this basis.

AGRICULTURAL DRAINAGE TO LAKE MENDOTA

Corey and Beatty (1965) have evaluated agricultural sources of nutrients to Madison lakes in the Lake Mendota watershed. Precipitation and potential evapotranspiration are about equal, with potential evapotranspiration exceeding precipitation from May through August (Corey, 1967). Consequently, the losses by percolation and runoff would appear to be minimized, compared with the Jackson, Mississippi, example given earlier. Thus, in an average year, about 24 in. are lost by evapotranspiration, 2 in. by runoff, and 4 in. by seepage and percolation.

In the watershed, 73 percent of the land area is cropland; 7 percent, woodland; 8 percent, pasture and other; 5 percent, major wetland; and 7 percent, urban.

If we assume an annual runoff of 2 in., we can form the following estimates of losses on the basis of data provided by Eck *et al.* (1956):

Losses (lb per acre)

Nutrient	Croplands and Pasture	Woodlands	Manured Lands
Soluble inorganic N	0.06	0.03	3.0
Soluble inorganic P	0.04	0.003	1.0

The disposal of manure on frozen lands is recognized as a potentially significant source of nitrogen and phosphorus. For the Mendota watershed, an equivalent cow population of 20,000 represents a major nutrient source, each cow producing 15 T of manure annually, half of which is spread on frozen ground. Data of Midgley and Dunklee (1945) indicate that 3 lb of N and 1 lb of phosphorus were lost from a 10-ton-per-acre application of manure on frozen soil with an 8-percent slope.

Estimates of the amounts and sources of available nutrients in Wisconsin cultivated soils given in Table 13 clearly show the major contribution of manure to these nutrients relative to other sources. Also, the addition of phosphorus to lakes and streams from manure could be significant because of the high levels of these nutrients present in a relatively mobile form subject to direct runoff.

By using these estimates of nutrient losses from agricultural sources, along with estimates of nutrients derived from other sources, we can compile an inventory of nutrients entering Lake Mendota (Table 14).

Table 14 indicates the large contribution of N from groundwater sources, which is related to the presence of nitrate in percolating waters noted previously. In contrast, phosphorus from agricultural sources arises mainly

TABLE 13   Sources and Estimates of Available Nitrogen and Phosphorus Derived from These Sources in Cultivated Wisconsin Soils

Source	Nitrogen (lb/acre)	Phosphorus (lb/acre)
Fertilizer	10	8
Legumes	12	–
Precipitation	8	–
Organic matter decomposition	45	5
Manure	42	12

TABLE 14   Nutrient Source and Percentage of Total Nutrients Entering Lake Mendota

Nutrient Source	Total Nutrients Entering Lake (%)	
	Nitrogen	Phosphorus
Precipitation	17	2
Groundwater	45	2
N fixation	14	–
Runoff		
Rural lands (not manured)	1	12
Rural lands (manured)	8	30
Urban area	5	17
Waste waters (municipal and		
industrial	8	36

from runoff and erosion. It would appear that different methods of manure disposal could significantly reduce the N and P inputs from agricultural drainage. Undoubtedly, much of the nitrate content of groundwaters arises from agricultural lands. But it is difficult to evaluate the effects of management practices because of our limited knowledge of nutrient sources and flow characteristics of percolating waters.

ADDITIONAL INVESTIGATIONS

From the foregoing discussion it may be concluded that a number of investigations would provide immediate and long-term benefits with regard to understanding agricultural contributions to eutrophication. Most useful would be the characterization of the chemical and mineralogical contents of runoff waters over a period of time involving seasonal changes. It is essential to ascertain the changes in these constituents as they progress from the point of origin to their deposition in lakes and streams. These changes, in turn, can be related to soils, crops, rainfall, runoff, and storm characteristics. Continued effort should be made to understand the nature of groundwater movement and the mixing of contaminating zones into these flows. The transport of water in the unsaturated zone between the soil surface and the groundwater, particularly below the root zone, should receive more study. Such studies should involve movement of nutrients, including the various forms that are yet unknown. The relation of these movements to the intermixing with groundwater flows has received very little study. We should seek not only more data but also improved physical models of nutrient losses by runoff, erosion, and percolation so that agricultural drainage contributions to eutrophication of lakes and streams may be more accurately assessed and corrected.

The senior author wishes to acknowledge support by a U.S. Public Health Special Research Fellowship at the University of Wisconsin while on leave from the University of California.

REFERENCES

Allison, F. E. 1955. The enigma of soil nitrogen balance sheets. Advan. Agron. 7:213–250.

Allison, F. E. 1965. Evaluation of incoming and outgoing processes that affect soil nitrogen *In* W. V. Bartholomew and F. E. Clark [ed] Soil nitrogen. Agronomy 10:573–606.

Allison, F. E. 1966. The fate of nitrogen applied to soils. Advan. Agron. 18:219–258.

Barrows, H. L., and V. S. Kilmer. 1963. Plant nutrient lanes from soils by water erosion. Advan. Agron. 15:303–316.

Bartholomew, W. V., and F. E. Clarke. 1965. Soil nitrogen. Agronomy 10:1–615. Amer. Soc. Agron., Madison, Wis.

Bertrand, A. R. 1966. Water conservation through improved practices, p. 207–235 *In* W. H. Pierre, D. Kirkham, J. Pesek, and R. Shaw [ed] Plant environment and efficient water use. Amer. Soc. Agron., Madison, Wis.

Biggar, J. W., D. R. Nielsen, and K. K. Tauji. 1966. Comparison of computed and experimentally measured ion concentrations in soil column effluents. Amer. Soc. Agr. Eng., Trans. 9:784–787.

Black, C. A. 1957. Soil-plant relationships. John Wiley and Sons, Inc., New York. 332 p.

Black, C. A. 1966. Crop yields in relation to water supply and soil fertility, p. 177–206 *In* W. H. Pierre, D. Kirkham, J. Pesek, and R. Shaw [ed] Plant environment and efficient water use. Amer. Soc. Agron., Madison, Wis.

Brown, J., C. L. Grant, F. C. Ugolini, and J. C. F. Tedrow. 1962. Mineral composition of some drainage waters from Arctic Alaska. J. Geophys. Res. 67:2447–2453.

Bryant, J. C., and C. S. Slater. 1948. Runoff water as an agent in the loss of soluble materials from certain soils. Iowa State Coll. J. Sci. 22:269–312.

Carlson, C. W. 1964. Tritium-hydrologic research: Some results of the U.S. Geological Survey research program. Science 143:804–806.

Childs, E. C., and H. Collis-George. 1950. The permeability of porous materials. Roy. Soc. (London), Proc., A. 201:392–405.

Corey, R. B. 1967. Contributions from rural lands, p. 16–28 *In* F. H. Schraufnagel [ed] Execessive water fertilization. Report to the Water Subcommittee, Natural Resources Committee of State Agencies, Madison, Wis.

Corey, R. B., and M. T. Beatty. 1965. Nutrient contributions to Lake Mendota by surface runoff from rural lands. Dep. Soils, Univ. Wisconsin, Madison. 8 p.

Crawford, N. H. 1966. Some observations on rainfall and runoff, p. 343–353 *In* A. V. Kneese and S. C. Smith [ed] Water research. The Johns Hopkins Press, Baltimore, Md.

Czekalski, A., and F. Kocialhowski. 1963. [Spectographic analysis of major nutrients and trace elements in drainage waters.] Roczn. Glebozn., Suppl. 13:238–243.

Day, P. K. 1956. Dispersion of a moving salt-water boundary advancing through saturated sand. Amer. Geophys. Union, Trans. 37:595–601.

Day, P. R., and W. M. Forsythe. 1957. Hydrodynamics dispersion of solutes in the soil moisture stream. Soil Sci. Soc. Amer., Proc. 21:477–480.

Doneen, L. D. 1966. Effect of soil salinity and nitrates on tile drainage in San Joaquin Valley, California. Dep. Water Sci. and Eng. Paper 4002, Univ. California, Davis. 47 p.

Duley, F. L., and M. F. Miller. 1923. Erosion and surface runoff under different soil conditions. Missouri Agr. Exp. Sta. Res. Bull. 63. 50 p.

Dyer, K. L. 1965a. Unsaturated flow phenomena in Panoche sandy clay loam as indicated by leaching of chloride and nitrate ions. Soil Sci. Soc. Amer., Proc. 29:121–126.

Dyer, K. L. 1965b. Interpretations of chloride and nitrate ion distribution patterns in adjacent irrigated and nonirrigated Panoche soils. Soil Sci. Soc. Amer., Proc. 29:170–176.

Edlefsen, H. E., and G. B. Bodman. 1941. Field measurements of water movement through a silt loam soil. J. Amer. Soc. Agron. 33:713–731.

Enderlin, H. C., and E. M. Markowitz. 1962. The classification of the soil and vegetative cover types of California watersheds according to their influence on synthetic hydrographs. *In* National engineering handbook–The hydrology guide (Supplement A, Section IV). Soil Conservation Service, U.S. Dep. Agr. U.S. Govt. Printing Office, Washington, D.C.

Engelbrecht, R. S., and T. T. Morgan. 1961. Land drainage as a source of phosphorus in Illinois surface waters, p. 74–79 *In* Algae and metropolitan wastes. SEC TR W61-3, U.S. Public Health Service. Robert A. Taft Sanitary Engineering Center, Public Health Service, Cincinnati, Ohio.

Feth, J. H. 1966. Nitrogen compounds in natural water–a review. Water Resources Res. 2:41–58.

Free, G. R. E., G. M. Browning, and G. W. Musgrave. 1940. Relative infiltrations and related physical characteristics of certain soils. U.S. Dep. Agr. Tech. Bull. 729. 52 p.

Frink, C. R. 1967. Nutrient budget: Rational analysis of eutrophication in a Connecticut lake. Env. Sci. Technol. 1:425–428.

Gardner, W. R. 1960. Soil water relations in arid and semi-arid conditions, p. 37–62 *In* Arid zone research 15. United Nations Educational, Scientific and Cultural Organization, Paris.

Goldman, C. R. 1965. Micronutrient limiting factors and their detection in natural phytoplankton populations, p. 121–135 *In* C. R. Goldman [ed] Primary productivity in aquatic environments. Mem. Int. Ital. Idrobiol. (Suppl. 18). University of California Press, Berkeley.

Hays, O. E., C. E. Bay, and H. H. Hull. 1948. Increasing production on an eroded loess-derived soil. J. Amer. Soc. Agron. 40:1061–1069.

Horton, R. E. 1940. Approach toward a physical interpretation of infiltration capacity. Soil Sci. Soc. Amer., Proc. 5:399–417.

Johnson, W. R., F. Ittihadich, R. M. Daum, and A. F. Pillsbury. 1965. Nitrogen and phosphorus in tile drain effluent. Soil Sci. Soc. Amer., Proc. 29:287–289.

Knoblauch, H. C., L. Kolodny, and G. D. Brill. 1942. Erosion losses of major plant nutrients and organic matter from Collington sandy loam. Soil Sci. 53:369–378.

Legrand, H. E. 1965. Patterns of contaminated zones of water in the ground. Water Resources Res. 1:83–95.

Lipman, J. G., and A. B. Conybeare. 1936. Preliminary note on the inventory and balance sheet of plant nutrients in the United States. New Jersey Agr. Exp. Sta. Bull. 607. 23 p.

Marshall, T. J. 1959. Relations between water and soil. Tech. Comm. No. 50, Commonwealth Bureau of Soils, Harpenden Commonwealth Agricultural Bureaux. Farnham Royal, Bucks, England. 91 p.

Massey, H. F., and M. L. Jackson. 1952. Selective erosion of soil fertility constituents. Soil Sci. Soc. Amer., Proc. 16:353–356.

Massey, H. F., M. L. Jackson, and O. E. Hayes. 1953. Fertility erosion on two Wisconsin soils. Agron. J. 45:543–547.

Midgley, A. R., and D. E. Dunklee. 1945. Fertility runoff losses from manure spread during the winter. Vermont Agr. Exp. Sta. Bull. 523. 19 p.

Miller, R. T., J. W. Biggar, and D. R. Nielsen. 1965. Chloride displacement in Panoche clay loam in relation to water movement and distribution. Water Resources Res. 1:63–73.

Musgrave, G. W. 1954. Estimating land erosion—sheet erosion. Int. Ass. Sci. Hydrol. Tenth General Assembly (Rome). 1:207; 215.

Musgrave, G. W. 1955. How much of the rain enters the soil, p. 151–159 In U.S. Dep. Agr. Yearbook 1955. U.S. Govt. Printing Office, Washington, D.C.

Neal, O. R. 1943. The influence of soil erosion on fertility losses and on potato yield. Amer. Potato J. 20:57–64.

Neal, O. R. 1944. Removal of nutrients from the soil by crops and erosion. J. Amer. Soc. Agron. 36:601–607.

Nielsen, D. R., J. M. Davidson, J. W. Biggar, and R. T. Miller. 1964. Water movement through Panoche clay loam soil. Hilgarde 35:491–505.

Nielsen, D. R., and J. W. Biggar. 1967. Radioisotope and labelled salts in soil-water movement, p. 61–76 In soil-moisture and irrigation studies. International Atomic Energy Agency, Vienna.

Ogata, G., and L. A. Richards. 1957. Water content changes following irrigation of bare-field soil that is protected from evaporation. Soil Sci. Soc. Amer., Proc. 21:355–356.

Penman, H. L. 1963. Vegetation and hydrology. Tech. Commun. No. 53, Commonwealth Bureau of Soils, Harpenden Commonwealth Agricultural Bureaux. Farnham Royal, Bucks, England. 124 p.

Philip, J. R. 1957. The theory of infiltration. 4. Sorptivity and algebraic infiltration equations. Soil Sci. 84:257–264.

Pierre, W. H., and F. W. Parker. 1927. Soil phosphorus studies. II. The concentration of organic and inorganic phosphorus in the soil solutions and soil extracts and the availability of the organic phosphorus to plants. Soil Sci. 24:119–128.

Rogers, H. T. 1942. Losses of surface-applied phosphate and limestone through runoff from pasture land. Soil Sci. Soc. Amer., Proc. 7:69–76.

Rowe, P. B., and E. A. Coleman. 1951. Disposition of rainfall in two mountain areas of California. U.S. Dep. Agr. Tech. Bull. 1048. 84 p.

Rubin, J., and R. Steinhardt. 1963. Soil water relations during rain infiltration. I. Theory. Soil Sci. Soc. Amer., Proc. 27:246–251.

Sawyer, C. N. 1947. Fertilization of lakes by agricultural and urban drainage. J. New Engl. Water Works Ass. 61:109–127.

Smith, D. W., and W. H. Wischmeier. 1962. Rainfall erosion. Advances Agron. 14:109–148.

Smith, G. E. 1965. Water form–Nitrate problems in water as related to soils, plants, and water. p. 42–52 In Missouri Agr. Exp. Sta. Special Report 55.

Soubies, L., R. Gadet, and P. Maury. 1952. Migration hivernale de l'azote nitroque dans un sol limoneaux de la région Toulousaine. Ann. Agron. 3:365–383.

Sternau, R., J. Schwarz, A. Mercado, Y. Harpaz, A. Nir, and F. Halevy. 1967. Radioisotope tracers in large scale recharge studies of ground water. International atomic energy agency. Symposium on isotopes in hydrology (Vienna), p. 489–505.

Stout, P. R., A. G. Burau, and W. R. Allardice. 1965. A study of the vertical movement of nitrogenous matter from the ground surface to the water table in the vicinity of Grover City and Arroyo Grande–San Luis Obispo County; report to Central Coastal Regional Water Pollution Control Board, San Luis Obispo, California. Dep. Soils and Plant Nutrition, Univ. California, Davis. 51 p.

Swaine, D. J. 1955. The trace element content of soils. Tech. Comm. No. 48, Commonwealth Bureau of Soils, Harpenden Commonwealth Agricultural Bureaux. Farnham Royal, Bucks, England. 157 p.

Sylvester, R. O., and R. W. Seabloom. 1962. A study on the character and significance of irrigation return flows in the Yakima River Basin. University of Washington. 104 p.

Van Denburgh, A. S., and J. H. Feth. 1965. Solute erosion and chloride balance in selected river basins of the western coterminous United States. Water Resources Res. 1:537–541.

Van de Leur, D. A., F. E. Shulze, and F. O. Donnell. 1966. Recent trends in hydrograph synthesis. Proc. Tech. Meeting No. 21, Versl. Meded. Comm. Hydrol. Onderz. T.N.O. 13, The Hague.

Weibel, S. R., R. B. Weidner, J. M. Cohen, and A. G. Christianson. 1966. Pesticides and other contaminants in rainfall and runoff. J. Amer. Water Works Ass. 58:1075–1084.

Wetselaar, R. 1962. Nitrate distribution in tropical soils. III. Downward movement and accumulation of nitrate in the subsoil. Plant and Soil 16:19–31.

Wiklander, L., and G. Hallgren. 1960. Utlakning av naring samnen ur dranerad jord. Grundforbattring 13:71–77.

Yaalon, D. H. 1965. Downward movement and distribution of anions in soil profiles with limited wetting, p. 157–164 In E. G. Hallsworth and D. V. Crawford [ed] Experimental pedology. William Cloves and Sons Ltd., London.

CHARLES F. COOPER
The University of Michigan, Ann Arbor

Nutrient Output from Managed Forests

Forests, typically occupying areas of thin soil or steep topography and supplied with abundant precipitation, are the source of much of the water that reaches streams and lakes in temperate and tropical latitudes. Forest drainage ordinarily carries a minimal concentration of nutrient elements. Nevertheless, its aggregate amount is sufficient to transport large quantities of dissolved and suspended material into streams and lakes each year.

Forests, even those intensively managed by present standards, are less influenced by the activities of man than are most other important runoff sources. Water draining from essentially undisturbed forests sets a base line for water quality; substantial effort will ordinarily be required to reduce nutrient content below this level of natural enrichment. It therefore seems feasible to use forest-runoff data to define geochemical provinces. The observed concentration of nutrients in water at a point of use in such a province can be compared with the appropriate forest base line values as an index of man's additions.

NUTRIENT INPUTS AND OUTPUTS

Over the long run, nutrient outputs must balance inputs. Mineral and organic nutrients available for export enter the forest ecosystem from three sources: geologic weathering of parent rock; meteorological inputs; and biological inputs, including the activities of man. Chemical and physical weathering transforms rock minerals into soluble or transportable form. Dissolved and particulate matter, including organic compounds and mineral ions, are added

446

to forest systems in precipitation, dust, and other aerosols. The biological inputs are principally materials gathered elsewhere and deposited in the forest by animals, including man. Feces of migratory birds and mammals are significant biological inputs in certain waterfowl marshes and in watering areas where large mammals congregate. Other biological inputs are garbage and body wastes produced by humans from food grown outside the forest, and artificial fertilizers used to increase forest production. Fixation of atmospheric nitrogen by soil microorganisms, an important biological input to agricultural croplands, is of relatively little importance in forest ecosystems.

Nutrients can leave the sytem in one of several ways:

- Removal of wood or forage harvested by man or animals
- Release to the atmosphere, a normal part of the nitrogen cycle and an occasional cause of phosphorus and nitrogen loss during fires
- Water transport in solution
- Removal of particulate organic or inorganic matter by wind or water

Over a relatively short span of time there may be accumulation, balance, or net loss of nutrients from the system. Man is generally increasing his inputs of mineral elements as forest management intensifies. Accompanying the increased inputs should be efforts to ensure that they are not carried away in water flowing from the area. Instead, they should be removed as useful products or as wastes deliberately collected for disposition elsewhere. Land managers, acutely conscious of erosion, have seldom considered the hazards of nutrient loss.

HYDROGEOCHEMISTRY OF MATURE HUMID-TEMPERATE FORESTS

Hydrogeochemistry, or chemical hydrology, is the study of those aspects of the hydrologic cycle that relate the chemical character of water to its physical and biological environment. Although most of the basic theory of general hydrology has now been formulated, hydrogeochemistry lags well behind other topics in hydrology (Back and Henshaw, 1965). Nevertheless, enough is known about the hydrogeochemical reactions of mature forests to provide a background for later discussions of the effects of human activities on the chemistry of forest runoff.

The more important elements of the hydrochemical cycle can be summarized as follows (after Back and Henshaw, 1965):

1. Wind blowing over the ocean carries sodium, chloride, sulfate, and

other substances landward and picks up additional terrestrial material as it moves across the continents.

2. As water vapor in the atmosphere condenses, nitrogen, oxygen, and carbon dioxide dissolve in the snow crystals or raindrops and, together with dissolved atmospheric salts, are carried to the ground.

3. Additional carbon dioxide dissolves in water percolating through biologically active soil.

4. Minerals dissolve, releasing many cations and anions.

5. Hydrogen and other ions in percolating water are exchanged for ions adsorbed on the surfaces of soil colloids.

6. Minerals and nitrogen are carried into plants in the transpiration stream, are incorporated into plant tissue, and then are returned to the soil either directly or after consumption by animals.

7. Part of the water returns to the ocean as streamflow or groundwater discharge, carrying dissolved, and perhaps suspended, matter with it.

In the terrestrial system, appreciable quantities of nitrogen, sulfur, and other elements are incorporated in precipitation and in dust and dry fallout between rains. Gorham (1961) has reviewed the atmospheric factors influencing the supply of major ions reaching the earth. Junge and Werby (1958) have summarized the available data on the concentration of chloride, sodium, potassium, calcium, and sulfate in rainwater over the United States; Junge (1958) has reviewed the similar distribution of ammonium and nitrate ions. All these studies show markedly uneven and largely unpredictable spatial distributions of the major ions. Junge's survey (1958) of nitrate and ammonium data showed no correlation with thunderstorm activity, density of industry, population, or agricultural activity. Areas having soils of low pH, capable of absorbing ammonia, correlate with areas having low concentrations of ammonium and nitrate in rainwater. Conversely, regions of alkaline soil tend to be associated with consistently high levels of nitrogen in rainwater.

Few reliable data are available on the phosphorus content of precipitation, but levels are generally low away from major cities. Eight of 19 samples of snow from the Sierra Nevada of California contained too little phosphorus to be detected even by sensitive analysis. The highest concentration in any of the samples was about 40 ppb of elemental phosphorus; the mean was about 10 ppb (Feth et al., 1964). The latter figure accords well with Voigt's observations (1960) in Connecticut and with other published data. Atmospheric phosphorus is, however, more likely than nitrogen to be associated with industrial air pollution.

Table 1 summarizes representative values of the concentration of important ions, excluding phosphorus, in precipitation over eastern North Carolina and Virginia. The actual quantities added to the soil per unit area

TABLE 1   Concentration of Important Ions in Rainwater and Dry Fallout over 1-Year Period in North Carolina and Virginia[a]

Ion	Monthly Average Concentrations (mg/liter)		Yearly Average Concentration (mg/liter)	Average Total Annual Input (kg/ha)
	High	Low		
Cl	1.1	0.1	0.57	6.3
Ca	1.2	0.2	0.65	7.0
Na	1.1	0.3	0.56	5.9
Mg	0.2	0.1	0.14	2.1
K	0.3	0.1	0.11	1.4
SO_4	3.2	1.1	2.18	23.5
NO_3	1.0	0.2	0.62	6.7

[a]From Gambell and Fisher (1966).

were obtained by multiplying the concentrations by the total annual rainfall. This total quantity supplied by precipitation is the basic input to the hydrochemical processes of forest ecosystems.

Nutrient elements circulate actively within terrestrial plant and animal communities. Elements move at different rates and in different quantities among four compartments: the atmosphere, the pool of available soil nutrients, living and dead organic matter, and soil and rock minerals (Bormann and Likens, 1967). Nutrients are taken by plant roots and are incorporated into plant tissue, and thence are returned to the atmosphere or to the soil. Some of this return is in the form of soluble compounds leached from leaves and stems; the ionic composition of rainfall beneath a forest canopy is markedly different from that in the open (Attiwill, 1966; Voigt, 1960). The chemical composition of forest precipitation is not a good indicator of the composition of runoff, however, because of the selective retention of specific ions in the soil and the preferential release of others.

Most of the nutrients taken by the plants return to the soil in leaf and litter fall or in animal products. The total internal circulation has been studied in detail in several types of forest ecosystems (Ovington, 1965; Nye, 1961; Remezov et al., 1964), but these studies have only an indirect bearing on nutrient output from forest ecosystems. Nutrient output in runoff is simply leakage from the internal circulation. There is no clear evidence of correlation between the rate of turnover of nutrients within the forest ecosystem and this leakage. The rate of biological cycling of nutrient elements in tropical forests is at least 3 times that in temperate forests (Bazilevich and Rodin, 1964), but it is not at all certain that the output of nutrients is in the same ratio.

Although leakage may occur from any of the four compartments in which nutrients reside, the soil is much the greatest contributor of elements to runoff. The other parts of the ecosystem, through their influence on soil properties and soil processes, also play a major role in determining the chemistry of forest outflow.

Water entering a forest as precipitation can leave in one of four modes: evaporation and transpiration, overland flow, deep seepage to groundwater, and subsurface flow. Although water beneath a forest sometimes flows over the surface for short distances, as over matted fallen leaves, it usually enters the soil quickly; extensive overland flow is rare in an undisturbed forest. Some of the water that enters the soil moves downward by gravity until it reaches a water table; thence it moves under a hydraulic gradient through saturated aquifers to a stream or lake.

In steep or rolling topography, much of the runoff accompanying rainstorms or rapid snow melt takes place by lateral movement through the soil layers near the surface. This runoff is referred to as subsurface flow or interflow. The highly permeable surface layers characteristic of undisturbed forests allow more rapid lateral than downward percolation of water. Subsurface flow, therefore, is quantitatively more important than either overland flow or deep seepage.

The mode of water flow through the soil has an important influence on the ionic composition of water leaving the forest. Clay and organic colloids are chemically the most active components of a forest soil. One of the basic properties of soil colloids is their ability to exchange ions in solution for those adsorbed on a solid. Clays have high exchange capacities compared with most other minerals because of their large surface area per unit volume and because of their negative electrical charge. To oversimplify a complex situation, cations adsorbed on soil colloids are selectively exchanged for hydrogen ions from the soil water. The hydrogen ions come principally from the solution of carbon dioxide in water and dissociation of the resulting H_2CO_3 molecule into a hydrogen ion and a bicarbonate ion. The dissolved carbon dioxide originates chiefly in the metabolism of microorganisms and plant roots in the soil. Thus only water remaining for some time in interstices between soil particles is likely to pick up an appreciable load of dissolved carbon dioxide and consequently to be effective in leaching mineral ions.

Nutrient concentrations, and sometimes even total output of nutrients, are commonly lower at times of high flow than at low flow (Livingstone, 1963). The high concentration at low flow is due to the long residence time of water in the soil during dry weather, which affords more opportunity for chemical reactions. Furthermore, nutrient levels at a given flow rate are often greater when a stream is falling than when it is rising (Toler, 1965). The lower nutrient level when a stream is rising can be explained by the fact that during periods of stream rise following rain, most water movement occurs as subsurface flow. In addition, some rainwater enters root channels, worm holes, and other crevices and flows through these openings without coming into real contact with the soil. The energy relations of soil moisture, however, prohibit water from entering such openings unless the soil is at or near saturation; thus when streamflow begins to recede, moisture is confined to soil pores, where it comes into intimate contact with active colloids.

It has only recently been fully appreciated that thin films of water moving slowly through unsaturated soil make a major contribution to base flow of streams during dry periods (Hewlett, 1961; Elrick, 1963). Unsaturated flow is particularly important in the relatively steep basins with deep soil that characterize many forest catchments. The precise mechanism of mass flow through unsaturated soil is still not fully understood, but active water films are in close contact with soil particles for long periods. More investigation is needed of the removal and transport mechanisms of ions in unsaturated flow and of the role of this flow in determining the ionic composition of streams.

The complex interaction of soils, vegetation, geology, and topography in determining water chemistry is exemplified by the nitrogen and phosphorus inflow to two small lakes on an island in southwestern Alaska (Dugdale and Dugdale, 1961). One lake had a large, flat drainage basin; the other lake had an appreciably steeper basin. The steeper basin yielded a lower concentration of phosphorus because of faster runoff and smaller opportunity for leaching. The nitrate concentration, however, was greater in the lake fed by the steeper basin. This anomaly was attributed to the abundance of alder alongside the streams of the steeper catchment. This species is known to fix atmospheric nitrogen in the same manner as legumes, bringing more nitrogen into the cycle and making it available for leaching.

SOURCES OF DATA ON FOREST HYDROCHEMISTRY

Three principal methods have been used to measure the contribution of forests and forest soils to chemical content of water. Each of these methods is applicable to the hydrochemistry of undisturbed stands and of forests modified by man.

The first and most widely used method is analysis of samples of water obtained from streams draining representative basins. Most of the large number of analyses of water chemistry reported by government agencies have been made to determine the suitability of the water from a particular stream for some such specific use as irrigation or domestic consumption. For this reason, most such analyses have been made on fairly large rivers draining diverse soils and types of land use, and have emphasized alkali salts rather than the nutrient elements important in eutrophication.

The second method is the detailed study of the input—output relations of a small forested watershed. For instance, data on weathering rates, biogeochemical processes, input, output, and the annual budget of several ions are being accumulated in six small catchments in the northern hardwood forest zone of New England (Bormann and Likens, 1967). Studies of this type can give information obtainable in no other way, but they are so expensive and laborious that wide representation is not possible. However, the U.S. Forest Service is beginning a program to monitor the chemical composition of outflow from barometer watersheds located in diverse geologic and vegetation situations throughout the national forests of the United States.

The third method is the use of lysimeters in which drainage water from a small artificially restricted area is collected. Although widely used in the past for soil studies, lysimeters are losing popularity because of their expense and the frequent lack of accuracy in their results (Allison, 1966). The presence of a lysimiter may change the normal flow regime enough to cause lysimeter waters to be more concentrated than natural waters and to have an ionic composition different from that of natural waters (Gorham, 1961).

IONIC COMPOSITION OF FOREST DRAINAGE

Viro (1953) concluded that "The soils of a river basin govern the quantity and the rocks the quality of the solids in the water." To this generalization we may add that the rock substrate of essentially undisturbed ecosystems largely determines the relative concentrations of metallic ions, whereas the relation of biological and biochemical processes in the soil to precipitation largely governs the yield of anions. Most of the common metallic ions in water are derived from rock weathering. Much of the chloride, nitrate, and sulfate (at least in the absence of abundant sulfide minerals) originates from the atmosphere; the output of these ions is regulated by soil processes. For instance, water at the heads of streams in the Sierra Nevada of California is notably low in chloride, sulfate, and nitrate. Chloride and sulfate, added to the system in precipitation, seem in some cases to be removed from solution by adsorption (Feth et al., 1964).

In well-weathered podzolic soils, the concentration of most alkali metal ions in soil water is effectively buffered and stabilized by cation exchange and the equilibrium between soil colloids and water. In some instances, sharp changes in potassium concentration accompany periods of plant growth and dormancy, suggesting that potassium may be the most sensitive of the major alkali metal ions as an index of annual biological activity (Likens et al., 1967).

Phosphorus is typically low in drainage from forest ecosystems. Some otherwise thorough studies of water chemistry in forest areas simply do not mention this element. In other instances, phosphorus has been sought but not found. Miller (1961) determined the ionic composition of water from 23 small streams draining three rock types in grazed but otherwise unmodified forests of the Sangre de Cristo Range, New Mexico. Spectographic analysis failed to detect phosphorus in any of the samples. Lysimeter studies of the nutrient outflow from oak forests in the forest-steppe zone near Moscow, U.S.S.R., showed that phosphorus was the least mobile of the nutrients studied and that loss of this element beyond the root zone was insignificant (Remezov *et al.*, 1964). In the Douglas fir region of Washington State, Cole and Gessel (1965) determined that the escape of phosphorus beyond the root zone amounted to about .034 kg/ha per year. The phosphate carried by rivers draining diverse forest landscapes in Finland was equivalent to an annual phosphorus loss in the range of 0.17 to 0.27 kg/ha (Viro, 1953). These values are typical of those reported elsewhere in the literature. It is important to note that the concentration of biologically active elements like phosphorus should be determined near the source in hydrochemical studies of forests. Dissolved phosphorus may be removed by aquatic plants and animals to such an extent that it is undetectable in inorganic solution at a downstream point (Livingstone, 1963).

Nitrogen levels are likewise usually low. Feth (1966) has reviewed the sources and status of nitrogen compounds in water without commenting on the output of this element from undisturbed ecosystems. The soil—plant—atmosphere cycle is the dominant process in the nitrogen economy of natural communities; even more than with other elements, nitrogen output in runoff is leakage from an active internal circulation. Intensive study has been devoted to the nitrogen cycle in agricultural crops and to the fate of nitrogen applied to soils (Allison, 1966). But Weetman's literature review (1961) of the nitrogen cycle in temperate forests includes little information on volatilization and less on leaching of this element.

The data that do exist suggest that the output of nitrogen in runoff from actively growing forests is nearly in balance with the input in precipitation. Because of the intensity of the nitrogen cycle within the community, however, this relation cannot be relied on for predictive purposes. Viro (1953) calculated that precipitation supplied twice as much N to the catchments of five Finnish rivers as was lost in the drainage water. Crisp (1966) constructed a balance sheet for the elements Na, K, Ca, P, and N for a stream catchment of 83 ha in Great Britain. The surface was about 80 to 85 percent blanket bog vegetation, the remainder being eroding peat and a little grassland. Outputs of K and N in the stream water were less than the inputs in precipitation, but both elements were also lost by erosion of peat.

Table 2 shows representative data on total quantities of selected elements

TABLE 2　Nutrient Output and Ionic Composition of Drainage from Forested Watersheds in Northern Minnesota[a]

Watershed	Area (ha)	Nutrient Output (g/ha per day)					Ionic Concentration (mg/liter)				
		Ca	NH$_4$	NO$_3$	Organic N	P	Ca	NH$_4$	NO$_3$	Organic N	P
2	2,150	646	0.43	7.3	5.9	0.30	54	0.04	0.50	0.18	0.032
3	2,850	646	0.50	6.0	5.7	0.50	64	0.08	0.91	0.69	0.060
4	958	113	0.31	0.7	2.6	0.23	38	0.06	0.42	0.66	0.047
6	13,000	259	0.43	3.2	5.5	0.37	17	0.03	0.34	0.38	0.023

[a]U.S. Forest Service data.

removed from forested areas in northern Minnesota and the ionic composition of the drainage waters from these areas. All values are for elemental
concentrations (nitrate N and ammonium N, for example) and were
determined from the weighted average of weekly samples collected from
August to November 1966. These data, provided by Michael Barton of the
U.S. Forest Service (personal communication), are among the few that
include analyses for phosphorus and organic nitrogen. The basins are located
in a heavily glaciated area of low relief and poor drainage. The presence of
many small lakes in the stream network undoubtedly influences the water
chemistry. Vegetation is predominantly aspen-birch forest, with extensive
areas of spruce-fir forest and pine forest, and smaller areas of brush and of
swamp forest.

INFLUENCE OF FOREST HARVEST

Forest cover regulates nutrient output both by storing nutrients in standing
vegetation and by influencing the flow of water through the soil profile. It
has been demonstrated repeatedly that removal of forest cover lowers
transpiration and increases runoff; however, only rather severe reduction of
vegetation affects runoff significantly (Goodell, 1966).

During the fall and winter of 1965–1966, the forest biomass of a small
catchment at Hubbard Brook, New Hampshire, was completely leveled.
Regrowth of vegetation has since been prevented by aerial application of
herbicides. This treatment shifted much of the organic material from the
living state to debris subject to breakdown and decay and resulted in
transformation of nutrients from organic form to soluble inorganic form. At
the same time, the herbicidal treatment has blocked uptake of nutrients by
vegetation. Reduction of transpiration allows more water to move through
the profile. Agricultural experiments have shown that these effects account
for leaching of nitrogen fertilizer from fallow soil in substantially greater
quantities than from similar soil under crops (Allison, 1966).

Cutting the hardwood forest ecosystem altered the nitrogen cycle
sufficiently to cause an increase in nitrogen outflow from less than 2 kg/ha to
more than 60. Nitrate concentrations in stream water exceeded established
pollution standards (10 ppm) for more than a year, and algal blooms have
appeared for the first time. Measured outflow is approximately equal to the
normal annual turnover of nitrogen in this ecosystem before clear-cutting.
The striking increase in nitrate outflow is clear evidence of the occurrence of
nitrification in the cutover ecosystem, and of the ability of a normal forest
ecosystem to hold on to accumulated nutrients (Bormann et al., 1968).

Man usually removes harvested products instead of allowing the accumulated biomass to decompose in place, as was done at Hubbard Brook. Regrowth of vegetation is usually sufficient to prevent drastic losses of nutrients after logging. Gessel and Cole (1965) found that in the course of a year, about 0.02 percent of the nitrogen stored in the soil profile percolated beyond the root zone in a Douglas fir plantation on the watershed supplying the city of Seattle, Washington. The corresponding loss the first year after clear-cutting the stand was 0.04 percent. Similar effects were noted for K and Ca. However, substantially more water passed through the soil after clear-cutting, so the actual concentration of N and K in the runoff water was slightly reduced, and that of Ca remained about the same.

WEATHER MODIFICATION

Research aimed at increasing precipitation through cloud seeding and other techniques is under way in the United States and elsewhere. The avowed goal is to increase precipitation reaching critical target areas by 10 to 20 percent. Present indications are that the total quantity of nutrients removed from the forest will be somewhat increased by such a change, but not as much as the increase in runoff. Thus the actual concentration of important elements in forest runoff is apt to be lowered by successful weather modification.

Monitoring of Ca, Mg, K, and Na in streamflow at Hubbard Brook, however, has shown that there is little seasonal variation in cationic concentrations even under drought conditions that drastically diminished the quantity of discharge (Likens et al., 1967). This lack of variation was somewhat unexpected because of the strong inverse relation between ionic concentration and streamflow that has been demonstrated elsewhere. A moderate increase in mean annual precipitation in this area could be expected to alter the total output of nutrients from the basin, but probably not their concentration in the runoff water.

TABLE 3   Output of Mineral Nutrients beneath the Root Zone in Untreated, Fertilized, and Clear-Cut Douglas Fir Forests[a]

	Nutrient (lb/acre)			
Treatment	N	P	K	Ca
Control	0.48	0.03	0.81	3.63
Clear-cut	0.87	0.10	0.95	7.81
Fertilized (200 lb/acre)				
Urea	0.62	0.07	0.85	5.07
Ammonium sulfate	0.94	0.15	2.32	7.59

[a]After Cole and Gessel (1965).

The season of increased outflow, whether a result of forest harvest or of increased precipitation, is of considerable limnological importance, particularly with respect to phosphorus. Even though levels of P in runoff are low, the concentration in the receiving lake may be still lower during the period of active plankton growth. Forest runoff in summer may add P to the lake system, whereas runoff of the same concentration during winter flushes away excess P. Most of the added flow resulting from forest harvest occurs during the growing season in areas where snow accumulation is not important, and during the spring in snow zones.

FOREST FERTILIZATION

Until very recently, artificial fertilization of extensive forest stands was written off as economically unattractive. Changes in markets and in silvicultural practices have altered the picture to the extent that potential gains from aerial application of fertilizers are now larger and more immediate than those from most other forms of intensive silviculture (Swan, 1965). There is likely to be accelerated use of fertilizers in forestry, particularly in high-value stands.

Forest fertilizers properly applied probably are not an important threat to water quality. A major danger is aerial application directly into streams. Not only will precautions have to be taken to avoid adding fertilizer to the water itself, but strict attention must be paid to the riparian zones adjacent to streams. These are often highly productive forest sites that can be expected to respond to fertilization. Fertilizers applied here have only a short distance to move through the soil to reach flowing water. Little or no work is now being done to devise safe and effective methods of applying fertilizer under these conditions.

Fertilization of sites at a distance from streams is less apt to be a problem. Cole and Gessel (1965) found that in the first 10 months after nitrogen fertilization of a Douglas fir plantation in Washington, nutrient losses were small, although levels of N, P, K, and Ca in the leachate were 1.3 to 4 times greater than the low initial levels. The release of P was accelerated even though no P was added in the fertilizer. Under fertilization, about 20 percent of the capital of available phosphorus in the profile was lost in the first 10 months, compared with 3.4 percent in the untreated stand and 7.7 percent in a clear-cut but unfertilized stand (Table 3).

These results were obtained on a gravelly glacial drift with a moderate base-exchange capacity. The situation might be different in the sandy soils of the Lake States, where lack of nutrients frequently so limits tree growth that forest fertilization might prove economically attractive. Leaching may become a problem in such soils because of their low base-exchange capacity. Since these areas are directly tributary to the Great Lakes, with their long residence time for water, great caution is called for in applying fertilizer to forests of this region.

Actually, however, most forest ecosystems show a remarkable capacity for accumulating and retaining mineral ions, a capacity that may be worth exploiting for renovation of treated sewage effluent. In a series of experiments carried out by The Pennsylvania State University, effluent containing an average of about 12 mg/liter of organic and nitrate N and 8.5 mg/liter of P was applied in frequent 1- and 2-in. irrigations to a red pine plantation and a natural hardwood stand. Soil percolate measurements indicated that the surface 12 in. of soil retained from 62 to 85 percent of the N and virtually all of the P (Pennypacker et al., 1967). At the application rates used in this experiment, only 129 acres would be required for continuous disposal of 1 million gal per day of treated effluent, equivalent to the average discharge from a city of 10,000. Of course, application at this rate year after year might eventually saturate the soil exchange complex and perhaps induce undesirable changes in the vegetation. As a practical matter, a disposal field several times the minimum size would probably be appropriate. Different parts could be intensively irrigated each year, allowing biological processes to dispose of accumulated N and P more effectively.

With proper silvicultural techniques, the resulting forest could be highly productive. The regular inputs of fertilizer elements would stimulate growth. The extensive installation of water-distribution facilities would allow irrigation of the entire forest during drought periods in addition to the excessive (and possibly detrimental) irrigation applied to certain portions of the stand each year. Intensive management of intermediate-value land near cities for disposal of sewage effluent concurrently with production of high-quality forest products would appear to offer an important, unexploited public-investment opportunity. This approach may in many instances be superior to disposal through application to agricultural crops or to mere maximum-rate application to wasteland.

EROSION AND FIRE

Removal of elements by erosion is in some degree selective in that organic matter and fine soil particles relatively high in mineral nutrients are more vulnerable to erosion than are the coarser soil fractions (Barrows and Kilmer, 1963). There seem to be no good estimates of the quantity of nutrient elements carried into lakes and streams from forests by erosion. It is clear, however, that many of the mineral ions in streams are adsorbed on suspended and colloidal matter resulting from erosion (Livingstone, 1963). For instance, data from rivers draining the steppes of central Asia show that suspended sediments are high in N, P, K, Ca, and Mn (Kuznetsov et al., 1965). The phosphorus fraction of these sediments is almost entirely in unavailable or insoluble form. Activity of bottom-dwelling organisms and other biological processes in lakes and streams may, over time, liberate significant quantities of adsorbed nutrients from erosion sediments.

It has often been postulated that forest and brush fires, by reducing organically bound elements to soluble form, considerably increase the mineral concentration of streams draining the burned area. Specific evidence on this point, however, is almost wholly lacking. A moderately severe fire in a conifer stand tributary to Sagehen Creek in the Sierra Nevada of California had no specific effect on Ca, Mg, Na, K, or HCO_3 in the stream (Johnson and Needham, 1966). The authors postulated that ash constituents were dissolved by light rainfall and leached into the permeable forest soil before the first snow. Because of the acidic nature of the soil, the dissolved cations were adsorbed on the exchange complex instead of being washed directly into the stream. Of course, torrential rains following a severe fire could move large quantities of soluble ash compounds into flowing streams. Fish kills and other detrimental effects from this source have been noted.

HUMAN WASTES

Forests have long been a favorite site for recreational activities. The attraction is particularly great when the forest is combined with water. As more and more people have the time and money to enjoy vacations near forest lakes, the formerly simple problem of disposing of their wastes becomes increasingly difficult. There are practically no data from which to estimate the magnitude of this problem. It is known that many intensively developed recreational lakes have become greatly enriched through seepage from septic tanks and other waste-disposal systems (Mackenthun and Ingram, 1964). This is in effect a problem of urban sanitary engineering without the concentrations of the urban population to make a solution feasible.

Large numbers of campgrounds, picnic areas, and other semirustic recreational developments are being installed in forest areas by government agencies at all levels. Most of these are carefully planned for sanitary disposal of wastes, but attention has seldom been paid to the nutrient aspects of waste disposal. This appears to be one of the acute problems facing future managers of forests and waters.

PREDICTIVE MODELS OF CHANGES IN WATER CHEMISTRY

"The purpose of chemical geohydrology is to . . . predict changes in the chemical character of water over space and time that result from the natural functioning of a hydrologic system or from artificial disturbances" (Back and Henshaw, 1965). Among the environmental disturbances most likely to alter the chemical character of forest drainage are timber harvest and associated land-use practices, use of commercial fertilizers, recreational use, and planned and inadvertent modification of climate. In many instances enough is known to suggest the approximate effects of environmental changes on water chemistry, but the data are not adequate for quantitative prediction. A truly predictive model will require more information from the field and from the laboratory.

We need many more empirical evaluations of the effects on water quality of management practices likely to be prevalent in the future. These field measurements should be accompanied by more intensive studies of the chemical thermodynamics of water in a forest environment. Only in the last 10 years have the concepts of free-energy and low-temperature thermodynamics been widely used to relate solute concentrations to the stabilities and surface properties of rock and soil minerals; this has brought about a greatly expanded understanding of the chemistry of natural water solutions (Back and Henshaw, 1965).

The great complexity of forest–soil–water systems and, even more, the highly variable nature of their hydrologic and biological inputs make it unlikely that precise deterministic models will soon be developed to predict the effects of environmental modification on water chemistry. This complexity suggests the desirability of combining computer simulation models incorporating stochastic inputs with expanded empirical evaluations and intensified laboratory and field studies of chemical thermodynamics and geochemical processes.

Such a combination of methods has not yet been applied directly to this problem, although enough has been done in related fields to point the way. One approach is through hydrologic simulation of the type used in the Stanford Watershed Model (Crawford and Linsley, 1966). A large number of empirical relations between weather variables and the physical properties of specific stream basins have been combined in such a way that when a previously observed pattern of weather events is introduced into the model, the simulated runoff from the basin closely approximates the observed flow pattern under that particular weather regime. Other input patterns, not previously observed, are then supplied to the model to test how they affect runoff distribution. This scheme could be extended to incorporate data on water chemistry.

A second and related approach is that of compartment or balance studies. In these studies, each element of interest is distributed among several distinct compartments. These compartments might be identified with plant leaves, plant roots, animal bodies, soil solution, exchange surfaces of clay minerals, and others as appropriate. A balance study is an application of the law of conservation of mass to an open system (Jacquez and Mather, 1966). At any time, the rate at which an element enters the system minus the rate at which it leaves must equal the rate of accumulation, positive or negative, in the system. For many systems, the rate of excretion of a particular element is proportional to its concentration or chemical activity in critical compartments and to the concentration of water in contact with those compartments. If inputs are constant, such systems approach a steady state in which output equals input.

Compartment models of this sort have been successfully used in a number of ecological situations, particularly those involving transfer of radionuclides within a closed system (Olson, 1965). A system of differential equations is set up representing the transfer rates between compartments. The equations are solved to determine the steady-state distribution of material among compartments. If the input or the quantity of material in the source compartment is then changed to a new level, the system goes through a transition period of variable length before settling down to a new steady state.

In most forest–soil–water systems, however, the inputs change too rapidly for a true steady state to develop. Jacquez and Mather (1966) have attempted to deal with this situation in a physiological system in which the rate of excretion is directly proportional to the quantity of the element in a compartment and the intake is a random sample from a stationary distribution. They have provided analytical and simulation solutions for excretion from the system under a regime of inputs varying randomly with a specified mean and standard deviation. The inputs may then be altered to a similar distribution about a new mean value, and the pattern of excretion after the change is followed. This model, with its simulation features and its use of stochastic inputs, seems particularly appropriate to nutrient output studies. It has to be extended to account for the observed fact that the rate of transfer of an element from one ecosystem compartment to another is not a simple linear function of the amount in the compartments.

Models are useful for predicting the behavior of a system under various stimuli, for providing insight into the operation of the system, and for aiding in the design of new experiments to improve predictive ability. By using such models and by closely integrating detailed field and laboratory studies of the input–output relations and biogeochemistry of whole ecosystems, as in the study at Hubbard Brook (Bormann and Likens, 1967), we can greatly improve our ability to understand and predict the ecological consequences of man's use of land and water resources.

REFERENCES

Allison, F. E. 1966. The fate of nitrogen applied to soils. Adv. Agron. 18:219–258.

Attiwill, P. M. 1966. The chemical composition of rainwater in relation to cycling of nutrients in mature eucalyptus forest. Plant and Soil 24:390–406.

Back, W., and B. B. Henshaw. 1965. Chemical geohydrology. Adv. Hydroscience 2:50–109.

Barrows, H. L., and V. J. Kilmer. 1963. Plant nutrient losses from soils by water erosion. Adv. Agron. 15:303–316.

Bazilevich, N. I., and L. E. Rodin. 1964. [The biological cycle of nitrogen and ash elements in plant communities of the tropical and subtropical zones.] Bot. Zhur. SSR 49:185–209. Trans. in Forest Abs. 27:357–368.

Bormann, F. H., and G. E. Likens. 1967. Nutrient cycling. Science 155:424–429.

Bormann, F. H., G. E. Likens, D. W. Fisher, and R. S. Pierce. 1968. Nutrient loss accelerated by clear-cutting of a forest ecosystem. Science 159:882–884.

Cole, D. W., and S. P. Gessel. 1965. Movement of elements through a forest soil as influenced by tree removal and fertilizer additions, p. 95–104 In C. T. Youngberg [ed] Forest-soil relationships in North America. Oregon State University Press, Corvallis.

Crawford, N. H., and R. K. Linsley. 1966. Digital simulation in hydrology: Stanford Watershed Model IV. Stanford University, Dep. Civ. Eng. Tech. Rept. 39. 210 pp.

Crisp, D. T. 1966. Input and output of minerals for an area of Pennine moorland: the importance of precipitation, drainage, peat erosion, and animals. J. Appl. Ecol. 3:227–348.

Dugdale, R. C., and V. A. Dugdale. 1961. Sources of phosphorus and nitrogen for lakes on Afognak Island. Limnol. Oceanogr. 6:13–23.

Elrick, D. E., 1963. Unsaturated flow properties of soils. Austral. J. Soil Res. 1:1–8.

Feth, J. H. 1966. Nitrogen compounds in natural waters—a review. Water Resources Res. 2:41–58.

Feth, J. H., C. E. Roberson, and W. L. Polzer. 1964. Sources of mineral constituents in water from granitic rocks, Sierra Nevada, California and Nevada. U.S. Geol. Surv. Water-Supp. Paper 1535I:1–170.

Feth, J. H., S. M. Rogers, and C. E. Roberson. 1964. Chemical composition of snow in the northern Sierra Nevada and other areas. U.S. Geol. Surv. Water-Supp. Paper 1535J:1–39.

Gambell, A. W., and D. W. Fisher. 1966. Chemical composition of rainfall, eastern North Carolina and southeastern Virginia. U.S. Geol. Surv. Water-Supp. Paper 1535K:1–41.

Gessel, S. P., and D. W. Cole. 1965. Influence of removal of forest cover on movement of water and associated elements through soil. J. Amer. Water Works Ass. 57:1301–1319.

Goodell, B. C. 1966. Watershed treatment effects on evapotranspiration, p. 477–482 In W. F. Sopper and H. W. Lull [ed] Forest hydrology. Pergamon Press, Oxford.

Gorham, E. 1961. Factors influencing supply of major ions to inland waters, with special references to the atmosphere. Geol. Soc. Amer. Bull. 72:795–840.

Hewlett, J. D. 1961. Soil moisture as a source of base flow from steep mountain watersheds. U.S. Forest Service, S. E. For. Expt. Sta., Sta. Paper 132.

Jacquez, J. A., and F. J. Mather. 1966. Balance studies on compartmental systems with stochastic inputs. J. Theor. Biol. 11:446–458.

Johnson, C. M., and P. R. Needham. 1966. Ionic composition of Sagehen Creek, California, following an adjacent fire. Ecology 47:636–639.

Junge, C. E. 1958. The distribution of ammonia and nitrate in rain water over the United States. Trans. Amer. Geophys. Union 39:241–248.

Junge, C. E., and R. T. Werby. 1958. The concentration of chloride, sodium, potassium, calcium, and sulfate in rain water over the United States. J. Meteorol. 15:417–425.

Kuznetsov, N. T., O. A. Shelyakina, and I. A. Klyukanova. 1965. [Physico-chemical properties of suspended sediments in the Amu-Dar'ya River delta.] Pochvov. 1965 (5):50–56. Trans. in Sov. Soil Sci. 1965:503–509.

Likens, G. E., F. H. Bormann, N. M. Johnson, and R. S. Pierce. 1967. Calcium, magnesium, potassium, and sodium budgets for a small forested ecosystem. Ecology 48:772–785.

Livingstone, D. E. 1963. Data of geochemistry: chemical composition of rivers and lakes. U.S. Geol. Surv. Prof. Paper 440G:1–64.

Mackenthun, K. M., and W. M. Ingram. 1964. Limnological aspects of recreational lakes. U.S. Pub. Health Serv. Pub. 1167. 176 pp.

Miller, J. P. 1961. Solutes in small streams draining single rock types, Sangre de Cristo Range, New Mexico. U.S. Geol. Surv. Water-Supp. Paper 1535F:1–23.

Nye, P. H. 1961. Organic matter and nutrient cycles under moist tropical forest. Plant and Soil 13:333–346.

Olson, J. S. 1965. Equations for cesium transfer in a Liriodendron forest. Health Phys. 11:1385–1392.

Ovington, J. D. 1965. Organic production, turnover, and mineral cycling in woodlands. Biol. Rev. 40:295–336.

Pennypacker, S. P., W. E. Sopper, and L. T. Kardos. 1967. Renovation of wastewater effluent by irrigation of forest land. J. Water Pollut. Control Fed. 39:285–296.

Remezov, N. P., Ye. M. Samoylova, I. K. Sviridova, and L. G. Bogashova. 1964. [Dynamics and interaction of oak forest and soil.] Pochov. 1964(3):1–13. Trans. in Sov. Soil Sci. 1965:222–232.

Swan, H. D. 1965. Reviewing the scientific use of fertilizers in forestry. J. Forestry 63:501–508.

Toler, L. G. 1965. Relation between chemical quality and water discharge in Spring Creek, southwest Georgia. U.S. Geol. Surv. Prof. Paper 525C:209–213.

Viro, P. J. 1953. Loss of nutrients and the natural nutrient balance of the soil in Finland. Comm. Inst. Forest. Fenn. 42(1):1–45.

Voigt, G. H. 1960. Alteration of the composition of rainwater by trees. Amer. Mid. Natur. 63:321–326.

Weetman, G. F. 1961. The nitrogen cycle in temperate forest stands. Woodland Res. Index Pulp Paper Res. Inst. Canada No. 126. 28 pp.

HUGH F. MULLIGAN
Cornell University, Ithaca, New York

Management of
Aquatic Vascular Plants and Algae

BIOLOGICAL ROLE OF AQUATIC PLANTS IN LAKES AND PONDS

Plants adapted to the aquatic environment include floating and benthic macroscopic plants, phtoplankton, and periphyton. Macroscopic aquatic plants and phytoplankton will be considered in this discussion.

In large, deep bodies of water, the littoral zone occupies a small area, and the benthic plant communities contribute little to the total primary productivity of the aquatic system. The phytoplankton account for most of the primary productivity. The situation is reversed in smaller lakes with extensive littoral zones; here the aquatic macrophytes contribute most of the primary productivity (Stråskraba, 1965).

The role of macroscopic plants in the aquatic ecosystem is not well understood. Shelford (1939) stated,

The rooted vegetation is eaten only to a small extent. We could probably remove all the larger rooted plants and substitute something else of the same form and texture without greatly affecting the conditions of life in the water, that is, as far as the life habits of the animals are concerned.

Conclusions differing from this view have been presented by Moore (1915), Needham and Lloyd (1916), Butcher (1933), Frohme (1938), Berg (1949), MacGaha (1952), and Smirnov (1959), who have shown experimentally that aquatic insects utilize aquatic plants for food and nest-building material. Ruttner (1966) considers two types of aufwuchs communities: (1) annual on living plant tissues; and (2) perennial on inanimate objects. The aufwuchs

464

communities on benthic plants are subjected to metabolic effects exerted by the living plants at the epidermis—water interface (Ruttner, 1966). Foerster and Schlichting (1965) contend that macrophytes are important because of the unique periphyton communities that colonize them.

Benthic macrophytes play a vital role in the aquatic environment. These organisms

- Produce oxygen through photosynthesis.
- Shade and cool the sediments of the littoral zone.
- Slow water movements and provide habitats for sessile benthic organisms.
- Regenerate substances from the sediments to the water.
- Provide surfaces for attachment by bacteria, periphyton, and aquatic insects.
- Serve as food, nest-building material, and sites for egg attachment for aquatic insects and fish.
- Provide nesting sites for fish.
- Protect small fish from predation.
- Convert inorganic material to organic matter.
- Serve as food for game birds and animals.
- Anchor the soil in place by means of their attenuated root systems.

The role of phytoplankton is more easily defined. Most workers agree that phytoplankton are directly involved in the food chain leading to fish production (Swingle, 1947); others raise a question of alternate pathways (Meehan, 1935; Nauwerk, 1963). Phytoplankton

- Oxygenate the water.
- Convert inorganic material to organic matter.
- Serve as food for zooplankton and young fish.
- Sink to the bottom after death, retaining nutrients within the water body.

There appears to be a definite interaction between the growth of benthic aquatic plants and phytoplankton. Kofoid (1903) reported that blooms of phytoplankton did not coincide with large growths of benthic plants. Pond (1905) noted that there was competition for nutrients between the phytoplankton and benthic plants, but only among aquatic vascular plants that have no root systems (*Ceratophyllum demersum*) and certain macroscopic algae (*Characeae*). Prowse (1955) concluded that the presence of macrophytes stimulated the production of the algal cells that are normally in the phytoplankton. Hasler and Jones (1949) suggested that large plants had an antagonistic effect on algae and rotifers. Swingle (1947) carried out experiments showing that blooms of phytoplankton restrict the growth of benthic plants by reducing the amount of incident illumination.

WEEDS IN THE AQUATIC ENVIRONMENT

Although plants are essential parts of the ecosystem, intense blooms of plankton algae and large populations of benthic macrophytes often produce undesirable changes in the aquatic environment. During the day, phyto-plankton blooms supersaturate the surface water with oxygen; at night their respiration causes extreme oxygen deficiencies. This condition of extreme oxygen concentrations does not result directly from the growth of aquatic macrophytes because of the presence of large air-storage tissues (Arber, 1963; Ehlrich, 1966), but is brought about indirectly by restricting water movement. Blooms of certain species of plankton algae cause taste and odor problems in municipal water supplies, clog filters (Moore and Kellerman, 1904), and excrete toxins into the water (Schwimmer and Schwimmer, 1964). Reduced fish production is often the result of excessive filamentous algal growths (Lawrence, 1958) and large benthic plant populations (Surber, 1961).

In addition to the effects upon the other aquatic communities, blooms of algae and large standing crops of macrophytes often interfere with multiple uses of lakes and rivers. Swimming, fishing, boating, and aesthetic enjoyment may be hampered by excessive plant growths.

All aquatic plants must not be considered as weeds and therefore undesirable. Certain aquatic plants are able to "out-compete" others for the same habitat. These plants have an inherently superior competitive ability. The best definition of a weed would be a plant that is a very successful competitor for space and nutrients (King, 1966).

Several aquatic plants have earned the title of aquatic weeds. These plants are not necessarily new to an area, but often build up large populations where they naturally occur. In Australia, property owners can be fined for not taking action against plants declared by the government to be "noxious weeds" (Green, 1966). This list of noxious weeds includes *Myriophyllum brasiliense* (parrot feather) and *Eichornia crassipes* (water hyacinth), two important aquatic weeds.

Man, through the process of cultural eutrophication, has disturbed many natural aquatic systems. These disturbances accelerate changes taking place in the environment, creating situations that favor the development of fewer species of organisms (Mulligan, unpublished).

The following discussion will be limited to aquatic plant management in multiple-use water bodies (e.g., streams, lakes, ponds, and marshes) rather than single-use waters (e.g., irrigation and drainage ditches) where modifications of the aquatic environment may be of lesser consequence.

In the northeastern United States, only a few aquatic plants fit within the category of noxious weeds. *Myriophyllum spicatum* var. *exalbescens* (water milfoil) and *Potamogeton crispus* (curly-leaved pondweed) often form dense masses in ponds and shallow zones of lakes. The latter species was introduced into the eastern part of the United States and has rapidly spread to the West Coast (Ogden, 1943). *Trapa natans* (water chestnut) is abundant in restricted areas of major river systems in this region (Stennis and Stotts, 1966).

Aquatic plants that cause difficulty in other parts of the United States include *Eichornia crassipes* (water hyacinth), *Pistia stratiotes* (watter lettuce), *Alternanthera philoxeroides* (alligator weed), *Heteranthera dubia* (water-stargrass), *Myriophyllum brasiliense* (parrot feather), *M. spicatum* var. *spicatum* (eurasian water milfoil), *Najas guadalupiensis* (southern naiad), *Potamogeton pectinatus* (sago pondweed), *Elodea canadensis* (elodea), and *Phragmites communis* (common weed). Other aquatic plants cause problems in local areas. Although *Potamogeton pectinatus* is the most noxious weed in irrigation and drainage ditches, where it retards water movement (Timmins, 1966), it is the most important duck food plant in the United States (Martin and Uhler, 1939).

Undesirable aquatic plants can be controlled by chemical, mechanical, or biological means.

CHEMICAL CONTROL

Because of the acknowledged success of chemicals in controlling growth of undesirable upland plants and because of the minimum expenditure of time and energy involved in applying them, chemical control is looked upon as a primary method for controlling aquatic plants.

Consider the regulations established by the State of New York on the use of chemicals for the control of aquatic vegetation (New York State Department of Health, 1964). Three chemicals have been authorized to be used at specific rates for the control of aquatic plants in New York waters, as follows:

Copper sulfate for algae control: 0.3 ppm in the upper 6 ft of water surface of bodies of water larger than 2 acres.

Sodium arsenite for submerged weeds: 4.0 ppm As_2O_3 for complete treatment of water masses of 10 acres or smaller, or up to 7.5 ppm for partial treatment of larger areas.

TABLE 1 Weed-Control Chemicals Used in New York State Waters, 1961–1966[a]

Year	CuSO$_4$ Number of Treatments	Amount (lb)	2,4-D Number of Treatments	Amount (lb)	Sodium Arsenite Number of Treatments	Amount (lb)	Silvex Number of Treatments	Amount (gal)
1961	10	1,900	13	11,000	2	70,000	3	40
1962	9	2,900	10	28,000	5	14,000	1	10
1963	11	5,000	12	29,000	0	0	1	40
1964	13	4,700	5	70,000	2	400	0	10
1965	11	2,700	7	48,000	1	100	1	0
1966	12	3,900	5	1,000	0	0	1	100
Total	66	21,100	52	187,000	10	84,500	7	200

Year	Endothal Number of Treatments	Amount (gal)	Endothal + Silvex Number of Treatments	Amount (gal)	Diquat Number of Treatments	Amount (gal)	Fenac Number of Treatments	Amount (gal)	Bodies of Water (number)
1961	1	20	0	0	0	0	0	0	29
1962	0	0	0	0	0	0	1	40	26
1963	0	0	1	80	0	0	0	0	25
1964	2	300	2	110	2	160	0	0	28
1965	0	0	1	20	8	400	0	0	27
1966	1	30	8	430	13	930	0	0	40
Total	4	350	12	640	23	1,490	1	40	175

[a]Compiled from permits issued by the New York State Health Department.

Esters, amines, and salts of 2,4-D (2,4-dichlorophenoxy acetic acid) for emergent plants: Up to 8 lb of active ingredients per acre.

In bodies of water larger than 10 acres, sodium arsenite and 2,4-D may not be used for weed control farther than 200 ft from shore or in water that is more than 6 ft deep.

Other chemicals may be used, but each treatment must be approved on an individual basis by a representative of the New York State Water Resources Commission.

Table 1 shows treatments that were permitted in New York State from 1961 to 1966. This, however, is not a complete listing of treatments of New York State public waters; nonauthorized treatments also occurred. The use of chemicals is nearly impossible to regulate without public cooperation. Chemicals can be purchased in most hardware and garden stores. Although individuals are subject to fines for not complying with state regulations governing the use of chemicals for controlling aquatic plants, some persons apply chemicals without the permission of lawful authorities. To protect the public, the use of chemicals in aquatic plant control must be carefully regulated.

Copper sulfate is the only one of the three chemicals authorized by New York State that has not declined in popularity. Sodium arsenite and 2,4-D are being replaced by newer chemicals that promise better aquatic plant control. As of June 1, 1967, eight permits had been issued for treatments during 1967. These include the following: copper sulfate (2); diquat (2); sodium arsenite (1); mixture of diquat and sodium arsenite (1); mixture of 2,4-D and 2,4,5-T (1); and mixture of endothal and silvex (1).

Table 2 presents instances of multiple treatments of bodies of water. Thirty-six bodies of water in New York State received more than one chemical treatment between 1961 and 1966. Twenty-four of these were treated with a single chemical, nine with two different chemicals, and three with three different chemicals. These data indicate the temporary nature of chemical control on plant populations, customer dissatisfaction with the results of treatments, and the search for more effective chemicals. Repeated treatments with the same chemical usually involved applications of copper sulfate for the control of plankton algae.

The primary considerations in selecting chemicals for controlling aquatic plants are dangers to the applicator, long-range effects from residues in the environment, biological concentration, effects on nontarget organisms, and ability of the chemical to kill the target weeds.

COPPER SULFATE

Copper sulfate is the most commonly used aquatic herbicide. It is used

TABLE 2   Summary, Chemical Treatments in New York State Waters, 1961–1966

Years of Treatment	Bodies of Water Treated
1	57
2	14
3	10
4	4
5	4
6	4
Total	93

primarily for the control of phytoplankton. Extensive tests have been carried out with this chemical since its introduction in 1904. In spite of this extensive testing program, new facts are still being discovered (Riemer and Toth, 1967).

Copper sulfate has a low mammalian toxicity (Woodford and Evans, 1965), is inexpensive, and is effective in controlling a wide range of plankton algae (Lawrence, 1962; Fitzgerald et al., 1952).

This chemical is toxic to fish and fish-food organisms at low concentrations (Lawrence, 1961; Hasler, 1949; Maloney and Palmer, 1956) and has been used as an irritant (Tompkins and Bridges, 1958) and fish killer (Smith, 1935). Copper is precipitated on the gills of fish and causes suffocation (Vernon, 1954).

The toxic effect of copper sulfate on plants is due to inactivation of enzymes and precipitation of proteins by the heavy metal divalent ion Cu^{2+} (Brian, 1964).

The amounts of oxygen, organic matter, and carbonates in the water determine the dosages required for effective plankton algae control and fish toxicity (Moore and Kellerman, 1904, 1905; McBrien and Hassall, 1967). Copper is precipitated from the water in the form of CuO_2H_2 (Ellms, 1905) and is adsorbed on soil particles and organic matter (Riley, 1939). It is persistent in the bottom muds, concentrating in the deepest sediments (MacKenthun, 1952), and can be regenerated from the sediments (Riley, 1939). Although usually assumed to be removed from the water in 24 hours, it has been shown to be persistent in the water for several days (Riemer and Toth, 1967). In order to reach the desired concentration levels in the water, it should be applied in a soluble form (Riemer and Toth, 1967).

SODIUM ARSENITE

Sodium arsenite frequently has been used to control submerged aquatic plants. It is not as popular now as it has been. It has a high mammalian toxicity (Woodford and Evans, 1965), making the application of the chemical very dangerous to an inexperienced applicator. Because of its toxic nature, it is no longer available in Great Britain for use as a herbicide (Woodford and Evans, 1965).

This chemical is toxic to fish at levels slightly above recommended levels (Gilderhus, 1966) and is toxic to important groups of fish-food organisms at and below recommended application levels (Ball and Hooper, 1966). Arsenic is rapidly taken up from the water by the fish and stored in fish tissue (Wiebe *et al.*, 1931). Most of the chemical is removed from the water 1 week after treatment (Surber and Meehan, 1931), having been adsorbed on the bottom muds (Cope, 1965). After the application of NaAs, phosphorus levels in the water increase because of the release of phosphorus from decaying plant tissues and the replacement of phosphates in the sediment by arsenite (Lawrence, 1958).

Sodium arsenite penetrates into the bottom sediments (Ball and Hooper, 1966), but no residual effects on aquatic organisms were noted the year after treatment (B. C. Cowell, unpublished data).* Sodium arsenite inhibits respiration of land plants (Christiansen *et al.*, 1949) and upsets normal mitosis in the roots of *Vicia narbonensis* (Mallah and Dawood, 1956).

POTASSIUM PERMANGANATE

Potassium permanganate has been suggested for use in the control of phytoplankton blooms by Fitzgerald (1966). It is an oxidizing agent that removes algae and other organic matter from the water. It is toxic to fish at low concentrations (Kemp *et al.*, 1966). Data from Fitzgerald (1966) indicate that it does not provide a much wider spectrum of control than copper sulfate. When large concentrations of algae must be removed, larger quantities of potassium permanganate must be used. This increases the danger of toxicity to fish (Kemp *et al.*, 1966).

MIXTURE OF COPPER SULFATE AND SILVER NITRATE

This mixture in a 300:50 weight ratio has been reported to be effective in controlling phytoplankton in lakes in Czechoslovakia (Štěpánek *et al.*, 1960; Kocurová, 1966). Kocurová contends that this mixture is less toxic than copper sulfate alone. However, because silver is more toxic than copper to fish (Doudoroff and Katz, 1953), more work must be done on residue studies.

*Ph.D. thesis, Cornell University (1963).

2,4-D (2,4-DICHLOROPHENOXY ACETIC ACID)

This was the first organic herbicide developed. When applied at low doses, it is selective, killing only broad-leaved plants. At higher doses 2,4-D will kill both monocots and dicots. It is recommended at rates of 0.5 to 1.0 lb per acre for weed control in cereal crops (Woodford and Evans, 1965) and is used at rates of 8 to 40 lb in controlling aquatic plants. This herbicide has been used in liquid and granular forms in treating emergent and submerged weeds. It is recommended that a wetting agent, either kerosene or a household detergent, be applied with the chemical for the control of emergent weeds (Lopinot, 1965).

Many formulations of 2,4-D have been tested. Each has different properties. Lawrence (1962) lists the toxicities to fish and fish-food organisms of 14 formulations of 2,4-D. The esters of 2,4-D are much more effective in killing plants than are the amides (Stennis *et al.*, 1962). The esters are also more toxic to fish and fish-food organisms than the amide formulations (Hughes and Davis, 1963), although Rawls (1965) suggests that the 2,4-D butoxyethanol ester was not toxic to shellfish populations.

Residue studies on upland plants indicate that 2,4-D is photo-oxidizable and is decomposed by soil microorganisms (Tutass, 1966), usually breaking down in the soil in 2 to 4 weeks (Audus, 1964) to humic acids (Tutass, 1966). In 1967, the butoxyethanol ester of 2,4-D was used in TVA lakes for control of *Myriophyllum spicatum* (Smith *et al.*, 1967) and for control of eelgrass in Nova Scotia (Thomas, 1967). Stennis and Stotts (1965, 1966) experimented with 2,4-D in combination with other herbicides to control *Myriophyllum specatum* in the Chesapeake Bay and *Trapa natans* in the Hudson and Mohawk rivers.

The use of 2,4-D in tidal areas has raised a serious problem. No acceptable biossay technique measures 2,4-D residues in shellfish (Rawls, 1965). This residue information is essential before the chemical may be used in locations where potential human food can be contaminated. The addition of 2,4-D to water results in large temporary increases in the heterotrophic populations in the water (Petruk, 1965).

Formulations of 2,4-D are considered auxin-type herbicides because they are similar to natural growth-promotion substances present in the plant. They promote the growth of isolated segments of the plant body, and this growth kills susceptible plants (Brian, 1964).

SILVEX [2-(2,4,5-TRICHLOROPHENOXY)-PROPIONIC ACID]

This herbicide is used for the control of woody trees and has been applied to control emergent and submerged aquatic plants. It is nonselective and slow-acting. The herbicide remains in the water at concentrations of 1 ppm up to 5 weeks after application (Estes, 1966). Different formulations of silvex have different levels of toxicity to fish and fish-food organisms (Surber and Pickering, 1962; Bond et al., 1965). Lawrence (1962) reports toxicity data for six formulations of silvex. The least-toxic form is the potassium salt (Mullison, 1966).

FENAC (2,3,6-TRICHLOROPHENYLACETIC ACID)

Fenac is used both as a nonselective soil sterilant in upland crops and as a control of submerged plants. Since it does not readily break down, it persists in the water for a long time (29 weeks at 50 percent reduction) (Grzenda et al., 1966). It is not toxic to fish and fish-food organisms at low concentrations (Crosby and Tucker, 1966) and has a low mammalian toxicity (Woodford and Evans, 1965).

SIMAZINE [2-CHLORO-4,6-BIS(ETHYLAMINO)-1,3,5 TRIAZINE]

Simazine is a nonselective herbicide used as a soil sterilant. It has a low mammalian toxicity (Woodford and Evans, 1965). It is used to control plankton algae, filamentous algae, and submerged plants. The herbicide is slow-acting and is taken up by the roots of benthic plants (Flanagen, 1966). It kills plants by inhibiting the Hill reaction (Sutton and Bingham, 1967). Simazine is increased in concentrations in the water after repeated applications and is adsorbed and held tightly on muck (Sutton et al., 1966). It is taken up into fish tissues when present in the water (Sutton et al., 1967). Fish toxicity data are incomplete, but in field tests simazine was not toxic to zooplankton and large-mouth bass at concentrations of 2 ppm (Snow, 1963).

ENDOTHAL (3,6 ENDOXOHEXAHYDROPHTHALIC ACID)

Endothal is used as a cotton defoliant and as a herbicide in sugar beets. It is usually nonselective (Woodford and Evans, 1965). This herbicide is also used for controlling submerged aquatic plants. The various formulations exhibit different degrees of fish toxicity (Walker, 1963). The sodium salt has a low acute toxicity to fish (Surber and Pickering, 1962). Yet, Seaman and Thomas (1966) experienced fish toxicities under apparently safe conditions. It has a high mammalian toxicity (Woodford and Evans, 1965). The herbicide is inactivated after a few days in the water (Hiltebran, 1962). It is available in several formulations. Virtually nothing is known about the mode of action of endothal (Brian, 1964). Endothal inhibited the maximum development of

proteolytic activity of intact squash seeds (Ashton *et al.*, 1967). This herbicide is frequently used in combination with silvex for the control of
submerged aquatic plants.

DIQUAT-DIBROMIDE (1,1-ETHYLENE-2,3-DIPYRIDYLIUM DIBROMIDE)

This is a new herbicide that kills submerged plants on contact. It was used
frequently in New York State waters in 1965 and 1966. According to Boon
(1967), the uptake of this herbicide interferes with plant metabolism, the
interference resulting in the production of large quantities of H_2O_2, which
kills the plants. Boon also notes that oxygen and light are essential for
herbicidal activity. Diquat quickly kills zooplankton. When diquat is removed
from the water, conditions are again favorable for zooplankton growth. It has
a low toxicity to fish (Bond *et al.*, 1960) and a low mammalian toxicity
(Woodford and Evans, 1965). Diquat is quick-acting, and in two tests it was
removed from the water 14 to 30 days after treatment (Grzenda *et al.*,
1966). The compound is adsorbed on clay particles on the sediment or in the
water and is inactivated (Boon, 1967). Soil residues of diquat are unavailable
for uptake by plants (Sheets, 1965). Twenty-four weeks after treatment with
different concentrations of diquat, bottom sediments (collected with an
Ekman dredge) contained 10 ppm of diquat for each ppm applied to the
water (Gilderhus, 1967). Diquat breaks down rapidly on upland soil by the
exposure to light and the activity of soil microorganisms (Boon, 1967). The
fate of this chemical in the bottom muds is not known.

SUMMARY

These and other chemicals have been used, and will continue to be used,
although the full implications of their use are not known. At present (1967),
inadequate information is available concerning (1) herbicide residues, (2)
rates of breakdown, and (3) long-term effects on aquatic organisms of
sublethal concentrations of herbicides. Experimental methods for assessing
toxicities to fish and fish-food organisms are not uniform, and many of the
results obtained are of regional value only.

When plants are killed by chemical application, oxygen levels in the water
decline rapidly and the nutrients previously tied up in plant material are
released into the water. Addition of these nutrients often results in the
production of massive plankton blooms. The large standing crop of algae
produces major fluctuations in the oxygen content of the water. Fish must
filter more water to obtain sufficient oxygen for respiration (Reiff, 1966).
Thus a small quantity of toxic chemical residue in waters with low levels of

oxygen could produce the same effect as a much larger quantity of residue under normal conditions of oxygenation.

Chemical treatments offer, at most, temporary relief from the problem of overabundance of aquatic vegetation. Even when the best possible results are obtained, the treatments must be repeated on an annual basis.

The present (1967) aquatic herbicides are nonselective; they usually kill all plants with which they come into contact. It would be much more desirable to use aquatic herbicides that possess a selective toxicity against specific aquatic weeds. This would permit the re-establishment of more desirable plants. The cost of developing such herbicides would be greater than for developing an agricultural herbicide because of the requirements for specialized testing, and in most cases the potential market would not be as great.

Mechanical control and biological control are two alternatives to chemical treatments.

MECHANICAL CONTROL

Mechanical control is very promising. It involves (1) hand-pulling or (2) cutting and harvesting of plants from restricted areas. This subject is covered in detail in a paper by Livermore and Wunderlich in this volume (page 494). At its current stage of development, this is an inefficient, time-consuming process that does not lend itself to the control of algae.

Mechanical control usually provides temporary relief from aquatic weeds. Riemer and MacMillan (1967) have shown that repeated cuttings at 2-week intervals during one summer were sufficient to eliminate a stand of spaddendock (*Nuphar advena*). Esler (1965) estimates that three fourths of the waterchestnut (*Trapa natans*) population must be removed annually in order to arrest the spread of this species. He was able to bring isolated populations under control by mechanical means. Lange (1965), in Texas, has experimented with the mechanical harvester and the collection of the aquatic weed *Myriophyllum spicatum* for use as food. This plant had potential as an additive in poultry feed because of its high xanthophyll, high protein, and low fiber content (Creger *et al.*, 1963; Bailey, 1965). Bailey's work showed that cut weeds produced more growth than uncut ones, and weed species grew faster than the more desirable plants. This was fortunate in this specific case, since *Myriophyllum spicatum* had the highest nutritive value. Gortner (1934) and Pirie (1960) also suggest mechanical collection of aquatic plants for food. Szumiec (1963) indicated that species diversity and biomass are less in mowed areas but that more insects are present.

BIOLOGICAL CONTROL

Biological control has met with only limited success, but it is an avenue of great potential. This approach would include introducing organisms that are inimical to the target organisms and manipulating the existing aquatic environment. In a sense, this kind of control uses biology against biology. Among the organisms that have been utilized in the control of aquatic plants are fish, snails, mammals, and beetles.

Biological control is the only method that could provide a permanent solution to the problem of excessive aquatic vegetation. For example, the introduction of flea beetles (*Agasicles* sp.) into the Savannah River system might provide a permanent control for alligator weed (*Alternanthera philoxeroides*) (Hawkes, 1965). Increasing the fecundity of, and providing better protective measures for, the manatees in southern Florida might enable them to remove large quantities of floating, emergent, and submerged vegetation from the canal system (Allsopp, 1960; Bertram and Bertram, 1962).

Herbivorous fish, *Tilapia* sp., have proven effective for controlling weeds in restricted areas where the fish are able to overwinter and build up populations (Pierce and Yawn, 1965; Lahser, 1967). Native carp control submerged weeds by disturbing roots and muddying the water (King and Hunt, 1967). *Marisa* snails have shown some measure of control over nuisance aquatic vegetation in Florida and Puerto Rico (Seaman and Porterfield, 1964). Perhaps these organisms could be introduced into restricted bodies of water in the Temperate Zone during the late spring. It might be possible to establish populations dense enough to cause a reduction in submerged plant populations. In addition to these studies, which are already in operation, aquatic entomologists are searching for new biological control agents in Central and South America (Anderson, 1965; Bennett, 1966).

It is apparent that these biological control agents will not eradicate the weed species. Populations of the introduced species and the target organism will likely fluctuate initially, with the target organism being severely reduced. Later, both populations should become relatively stable (Odum, 1961).

Biological control can be manifested in another manner. The macroscopic and microscopic aquatic floras are not static. There is rapid succession among the plankton algae (Mulligan, unpublished). Continuous mixing of deep reservoirs could be carried out, preventing the development of strong summer thermoclines and the growth of obnoxious algal communities commonly associated with them.

Fertilization of ponds is another means of reducing the standing crop of submerged macrophytes (Swingle, 1947; Surber, 1961; Clugston, 1963). The fertilizer is taken up by phytoplankton and filamentous algae, which bloom and shade the benthic macrophytes and prevent their development. This plan involves exchanging one problem for another but may be practical under some circumstances. This is a self-perpetuating procedure; if it is not carried out several times during each summer, larger growths of submerged vegetation may develop as a result of the added nutrients. The procedure also increases the possibility of winter kill (Ball and Tanner, 1951).

In the northeastern United States, *Potamogeton crispus* is an early-summer aquatic weed, whereas *Myriophyllum spicatum* var. *exalbescens* reaches abundance in late summer. This suggests there are physiological differences between aquatic plants. Aquatic plants have been shown to respond to small changes in the environment. Unpublished records of the weed beds in Sodus Bay, Lake Ontario, collected by the New York State Conservation Department, indicate that the weed beds change in composition, size, and dates of maximum development from one year to the next. Bernatowicz (1966), working in Poland, demonstrated that eight aquatic plants possessed different degrees of shade-tolerance. Hollingsworth (1966) attributed annual fluctuations in the standing crop of *Heteranthera dubia* in the Mississippi River to changes in water depth and turbidity. Lowering the water levels 6 ft for a period of 21 to 25 days during the winter provided a 90 percent reduction in the acreage infested with *Myriophyllum spicatum* in the Tennessee Valley Authority lakes (Smith *et al.*, 1967).

FIGURE 1   Laboratory building and several of the nearly 100 experimental ponds constructed at Cornell University for the study of aquatic plants.

Cornell University has constructed nearly 100 one-quarter-acre experimental ponds (CRF-1 funds) in which the growth and environmental requirements of aquatic plants are being studied (Figure 1). In this outdoor laboratory, it is possible to establish single variables while maintaining adjacent ponds as controls. Populations of plants and animals are exposed to natural physical and chemical changes. Results obtained from these studies can easily be extrapolated to similar field conditions. After environmental requirements of specific aquatic weeds in these experimental ponds have been determined, it should be possible to predict plant and animal species composition and abundance as a consequence of eutrophication and to suggest modifications in aquatic environments to hinder growth of "weed" species and stimulate growth of more-desirable aquatic plants.

REFERENCES

Allsopp, W. H. 1960. The manatee: ecology and use for weed control. Nature 188:762.

Anderson, W. H. 1965. Search for insects in South America that feed on aquatic weeds. Proc. S. Weed. Conf. 18:586–587.

Arber, A. 1963. Water plants: a study of aquatic angiosperms. Reprint. Hafner Publishing Co., New York. 436 p.

Ashton, F. M., D. Penner, and S. Hoffman. 1967. The influence of endothal, dichlorobenil and bromoxynil on proteolytic activity in the cotyledons of squash seedlings. Abstr. Weed Soc. Amer. 56–57.

Audus, L. J. 1964. Herbicide behavior in the soil, p. 163–206 In L. J. Audus [ed] The Physiology and Biochemistry of Herbicides.

Bailey, T. A. 1965. Commercial possibilities of dehydrated aquatic plants. Proc. S. Weed Conf. 18:543–551.

Ball, R. C., and F. H. Hooper. 1966. Use of [74]As-tagged NaAs in a study of effects of a herbicide on pond ecology. Isotopes in weed research 5M-69/10:149–163. Int. Atom. Energy Agency, Vienna.

Ball, R. C., and H. A. Tanner. 1951. The biological effects of fertilizer on a warm-water lake. Michigan State College. Tech. Bull. 223:32.

Bennett, F. D. 1966. Investigations on the insects attacking the aquatic ferns Salvinia spp. in Trinidad and Northern South America. Proc. S. Weed Conf. 19:497–504.

Berg, C. O. 1949. Limnological relations of insects to plants of the genus Potamogeton. Trans. Amer. Micros. Soc. 68:279–291.

Bernatowicz, S. 1966. The effect of shading on the growth of macrophytes in lakes. Ecologia Polaska-Seria A. Tom XIV Nr31:607–616.

Bertram, G. C., and C. K. Bertram. 1962. Manatees of Guiana. Nature 196:1329.

Bond, C. E., R. H. Lewis, and J. L. Fryer. 1960. Toxicity of various herbicidal material to fishes. Trans. 2d. Seminar on Biological Problems in Water Pollution. U.S. Public Health Service. R.A. Taft Sanitary Engineering Center Tech. Rep. W60-3:96–101.

Bond, C. E., J. D. Fortune, and F. Young. 1965. Results of preliminary bioassays with Kurasol SL and Dicamba. Progr. Fish. Cult. 27:49–51.

Boon, W. R. 1967. The quaternary salts of bipyridyl—a new agricultural tool. Endeavour 26:27–32.

Brian, R. C. 1964. The classification of herbicides and types of toxicity, p. 1–37 *In* L. J. Audus [ed] The Physiology and Biochemistry of Herbicides.

Butcher, R. W. 1933. Studies on the ecology of rivers. J. Ecol. 21:58–61.

Christiansen, G. S., L. J. Kunz, W. D. Bonner, and K. V. Thimann. 1949. The action of growth inhibitors on carbohydrate metabolism in the pea. Plant Physiol. 24:178.

Clugston, J. P. 1963. Lake Apopka: A changing lake and its vegetation. Quart. J. Fla. Acad. Sci. 26:168–174.

Cope, O. B. 1965. Some responses of fresh water fish to herbicides. Proc. S. Weed Conf. 18:439–445.

Creger, C. R., F. M. Farr, E. Castro, and J. R. Cooch. 1963. The pigmenting value of aquatic flowering plants (abstr.), Poultry Sci. 42:1262.

Crosby, D. G., and R. K. Tucker. 1966. Toxicity of aquatic herbicides to *Dalphnia magna*. Science 154:289–291.

Doudoroff, P., and M. Katz. 1953. Critical review of literature on the toxicity of industrial wastes and their components to fish. II. The metals as salts. Sewage and Ind. Wastes 25:802–839.

Ehlrich, S. 1966. Two experiments in the biological clarification of stabilization pond effluents. Hydrobiologia 28:70–80.

Ellms, J. W. 1905. Behavior and uses of copper sulfate in the purification of hard and turbid waters. J. Nav. Eng. Water Works. Ass. 19:496–503.

Esler, H. J. 1965. Control of waterchestnut by machine in Maryland. Proc. Northeast Weed Control Conf. 20:682–687.

Estes, R. D. 1966. Some preliminary observations and findings on the effects of silvex on the aquatic environment. Proc. S. Weed Conf. 19:416–419.

Fitzgerald, G. P. 1966. Use of potassium permanganate for control of problem algae. J. Amer. Works Ass. 58:609–614.

Fitzgerald, G. P., G. C. Gerloff, and F. Skoog. 1952. Studies on chemicals with selective toxicity to blue-green algae. Sewage and Ind. Wastes 24:888–896.

Flanagan, J. H. 1966. Progress report on use of simazine as an algicide and aquatic herbicide. Proc. S. Weed Conf. 19:387–392.

Foerster, J. W., and H. E. Schlichting. 1965. Phyco-periphyton in an oligotrophic lake. Trans. Amer. Micros. Soc. 84:485–502.

Frohme, W. C. 1938. Contribution to the knowledge of the limnological role of higher aquatic plants. Trans. Amer. Micros. Soc. 57:256–268.

Gilderhus, P. A. 1966. Some effects of sublethal concentrations of sodium arsenite on bluegills and the aquatic environment. Trans. Amer. Fisher. Soc. 95:289–296.

Gilderhus, P. A. 1967. Effects of diquat on bluegills and their food organisms. Prof. Fish. Cult. 29:67–74.

Gortner, R. A. 1934. Lake vegetation as a possible source of forage. Science 80 (2054): 531–533.

Green, K. R. 1966. Laws designed to stop the spread of weeds. Agr. Gaz. New S. Wales 77:399–403.

Grzenda, A. R., H. P. Nicholson, and W. S. Cox. 1966. Persistence of four herbicides in the water. Proc. S. Weed Conf. 18:521–524.

Hasler, A. D. 1949. Antibiotic aspects of copper treatment of lakes. Wis. Acad. Sci. Arts and Lett. 39:97–103.

Hasler, A. D., and E. Jones. 1949. Demonstration of the antagonistic action of large aquatic plants on algae and rotifers. Ecology 30:359–364.

Hawkes, R. B. 1965. Domestic phases of program designed to use insects to suppress alligatorweed. Proc. S. Weed. Conf. 18:584–585.

Hiltebran, R. C. 1962. Duration of toxicity of endothal in water. Weeds 10:17–19.

Hollingsworth, E. B. 1966. Waterstargrass as an aquatic weed. Agr. Exper. Sta. Div. of Agr., Univ. Arkansas. Bull. 705. 35 p.

Hughes, J. S., and J. T. Davis. 1963. Variations in toxicity to bluegill sunfish of phenoxy herbicides. Weeds 11:50–53.

Kemp, H. T., R. G. Fuller, and R. S. Davidson. 1966. Potassium permanganate as an algicide. J. Amer. Water Works. Ass. 58:255–263.

King, D. R., and G. S. Hunt. 1967. Effect of carp on vegetation in a Lake Erie marsh. J. Wildlife Mgmt. 31:181–188.

King, L. J. 1966. Weeds of the world: Biological control. Intersci. Publ. Co. Inc., N.Y. 526 p.

Kocurová, E. 1966. The application of the algicide Ca_{350} in the Lubi Reservoir near Trebic. Czechosolovakia Hydrobiologia 18:223–240.

Kofoid, C. A. 1903. The plankton of the Illinois River, 1894-1899, with introductory notes on the hydrography of the Illinois River and its basin. Part I, Quantative investigations and general results. Bull. Ill. State Lab. Nat. Hist. 6:95–629.

Lahser, C. W. 1967. *Tiliapia mossambica* as a fish for aquatic weed control. Prog. Fish Cult. 29:48–50.

Lange, S. R. 1965. The control of aquatic plants by commercial harvesting, processing and marketing. Proc. S. Weed Conf. 18:536–542.

Lawrence, J. M. 1958. Recent investigations on the use of sodium arsenite as an algicide and its effects on fish production in ponds. Southeastern Ass. Game and Fish. Comm. Conf. Proc. 11:281–287.

Lawrence, J. M. 1962. Aquatic herbicide data. U.S.D.A. Agriculture Handbook No. 231. Washington, D.C. 133 p.

Lopinot, A. C. 1965. Aquatic weeds—their identification and methods of control. Ill. Dept. Conserv. Fish. Bull. 4. 52 p.

MacGaha, Y. J. 1952. The limnological relations of insects to certain aquatic flowering plants. Trans. Amer. Micros. Soc. 71:355–381.

Mackenthun, K. M. 1952. The biological effect of copper sulfate on lake ecology. Wis. Acad. Sci., Arts and Lett. Trans. 41:177–187.

Mallah, G. S., and M. M. Dawood. 1956. Cytological studies of the effect of sodium arsenite on *Vicia narbonensis* roots. Alexandria J. Agr. Res. 4:91–105.

Maloney, T. M., and C. M. Palmer. 1956. Toxicity of six chemical compounds to thirty cultures of algae. Water and Sewage Works 103:509–513.

Martin, A. C., and F. M. Uhler. 1939. Food of game ducks in the United States and Canada. U.S.D.A. Tech. Bull. 634:1–157.

McBrien, D. C., and K. A. Hassall. 1967. The effect of toxic doses of copper upon respiration, photosynthesis and growth of *Chlorella vulgaris*. Physiologia Plantarum 20:113–117.

Meehan, O. L. 1935. The dispersal of fertilizing substances in ponds. Trans. Amer. Fish. Soc. 65:184–188.

Moore, E. 1915. The potamogetons in relation to pond culture. Bull. U.S. Bur. Fish. Wash. 33:251–291.

Moore, G. T., and K. F. Kellerman. 1904. A method of destroying or preventing the growth of algae and certain pathogenic bacteria in water supplies. U.S.D.A. Bureau Plant Industries Bull. 64. 44 p.

Moore, G. T., and K. F. Kellerman. 1905. Copper as an algicide and disinfectant in water supplies. U.S.D.A. Bureau of Plant Industries Bull. 76. 55 p.

Mullison, W. R. 1966. Some toxicological aspects of silvex. Proc. S. Weed Conf. 19:420–435.

Nauwerck, A. 1963. Zooplankton and phytoplankton im See Erken. Symb. Bot. Upsal. 17:1–163.

Needham, J. G., and J. T. Lloyd. 1916. The life of inland waters. Comstock Publishing Co., Ithaca, N.Y. 438 p.

New York State Department of Health. 1964. Rules and regulations governing the use of chemicals for the control and elimination of aquatic vegetation. Albany. 14 p.

Odum, E. P. 1961. Fundamentals of ecology. W. B. Saunders Co. Philadelphia. 546 p.

Ogden, E. C. 1943. The broad-leaved species of *Potamogeton* of North America, north of Mexico. Rhodora 45:57–105.

Petruk, G. F. 1965. The effect of herbicides on heterotrophic microorganisms of ponds. U.S. Joint Publ. Res. Ser. Transl. Mikrobiol. U.S.S.R. 33:1018–1021.

Pierce, P. C., and H. M. Yawn. 1965. Six field tests using two species of *Tilapia* for controlling aquatic vegetation. Proc. S. Weed Conf. 18:582–583.

Pirie, N. W. 1960. Water hyacinth: a curse or a crop? Nature 185:116.

Pond, R. H. 1905. The relation of aquatic plants to the substratum. Rep. U.S. Fish. Comm. 19:483–526.

Prowse, G. A. 1955. The role of phytoplankton in studies on productivity. Internat. Ass. Theor. Appl. Limnol. 2:159–163.

Rawls, C. K. 1965. Field test of herbicide toxicity to certain estuarine animals. Ches. Sci. 6:150–161.

Reiff, B. 1966. Factors influencing the testing of chemicals and effluents for toxicity to fish. Inst. Anim. Technol. J. 17:116–121.

Riemer, D. N., and W. W. MacMillan. 1967. Effects of defoliation on spadderdock (*Nuphar advena*). Abstr. Proc. Northeast Weed Conf. 21:556.

Riemer, D. N., and S. J. Toth. 1967. Behavior of copper sulfate in small ponds. Abstract. Proc. Northeast Weed Control Conf. 21:534–540.

Riley, G. A. 1939. Limnological studies in Connecticut. Ecol. Monogr. 9:53–94.

Ruttner, Franz. 1966. Fundamentals of limnology. Univ. Toronto Press, Toronto. 295 p.

Schwimmer, D., and M. Schwimmer. 1964. Algae and medicine, p. 368–413 *In* D. F. Jackson [ed] Algae and Man.

Seaman, D. E., and W. A. Porterfield. 1964. Control of aquatic weeds by the snail *Marisa cornuarietis*. Weeds 12:87–91.

Seaman, D. E., and T. H. Thomas. 1966. Absorption of herbicides by submerged aquatic plants. Proc. Calif. Weed Conf. 18:11–12.

Sheets, T. S. 1965. Herbicides residues in soils. Proc. Calif. Weed Conf. 18:8–10.

Shelford, V. E. 1939. Animal communities in temperate America. Univ. Chicago Press, Chicago. 368 p.

Smirnov, N. N. 1959. Consumption of emergent plants by insects. Internat. Ass. Theor. Appl. Limnol. 14:232–236.

Smith, G. E., T. F. Hall, and R. A. Stanley. 1967. Eurasian watermilfoil in the Tennessee Valley. Weeds 15:95–98.

Smith, N. W. 1935. The use of copper sulfate for eradicating predatory fish populations of a lake. Trans. Amer. Fish. Soc. 65:101–114.

Snow, J. R. 1963. Simazine as an algicide for bass ponds. Prog. Fish. Cult. 25:34–36.

Stennis, J. D., and V. D. Stotts. 1965. Tidal dispersal of herbicides to control eurasian watermilfoil in the Chesapeake Bay. Proc. S. Weed Conf. 18:507–511.

Stennis, J. D., and D. U. Stotts. 1966. Recent tests on waterchestnut control. Proc. Northeast Weed Control Conf. 20:476–479.

Stennis, J. D., D. U. Stotts, and C. R. Gillette. 1962. Observations on distribution and control of eurasian watermilfoil in Chesapeake Bay, 1961. Proc. Northeast Weed Control Conf. 16:442–448.

Štěpánek, M., J. Chapula, L. Mašínová, J. Pokorný, and J. Svec. 1960. Application of algicide CA 350 in reservoirs. Sborník Vysoké Školy chemicko-technologické v Praze. Oddil Fak. Technol. Paliv a Vody 4(2):375–402.

Stráškraba, M. 1965. Contributions to the pond of the littoral regions of pools and ponds. I. Quantitative study of the littoral zooplankton of the rich vegetation of the backwater Labicko. Hydrobiologia 26:421–443.

Surber, E. W. 1961. Improving sport fishing by control of aquatic weeds. U.S. Fish-Wildlife Serv. Circ. 128. 51 p.

Surber, E. W., and O. L. Meehan. 1931. Lethal concentrations of arsenic for certain aquatic organisms. Amer. Fish. Soc. Trans. 61:225–239.

Surber, E. W., and Q. H. Pickering. 1962. Acute toxicity of endothal, diquat, hyamine, dalapon and silvex to fish. Prog. Fish Cult. 24:164–171.

Sutton, D. L., and S. W. Bingham. 1967. Some effects of simazine and diquat on the movement of $C^{14}O_2$ or its metabolites in *Myriophyllum specatum.* Proc. Northeast Weed Control Conf. 21:552–555.

Sutton, D. L., T. D. Evrard, and S. W. Bingham. 1966. The effects of repeated treatments of simazine on certain aquatic plants and residues in water. Proc. Northeast Weed Control Conf. 20:464–468.

Sutton, D. L., T. D. Evrad, and S. W. Bingham. 1967. Analyses for simazine in fish and water samples from treated ponds. Proc. Northeast Weed Control Conf. 21:541.

Swingle, H. S. 1947. Experiments on pond fertilization. Ala. Exper. Sta. Bull. 264. 34 p.

Szumiec, J. 1963. The influence of emergent vegetation and the manner of its mowing on the bottom fauna of fish ponds. Acta Hydrobiol. 5:315–335.

Thomas, M. L. 1967. Experimental control of eelgrass (*Zostera marina*) in oyster growing areas. Proc. Northeast Weed Control Conf. 21:542–549.

Timmins, F. L. 1966. Control of weeds harmful to water uses in the west. J. Waterways and Harbors Div. 92(WW1):47–58.

Tompkins, W. A., and C. Bridges. 1958. The use of copper sulfate to increase fyke-net catches. Prob. Fish. Cult. 20:16–20.

Tuttass, H. 1966. Photodecomposition of 2,4-D. Proc. Calif. Weed Conf. 18:13–14.

Vernon, E. H. 1954. The toxicity of heavy metals to fish with special reference to lead, zinc and copper. Canada Fish Cult. 15:32–37.

Walker, C. R. 1963. Endothal derivatives as aquatic herbicides in fishery habitats. Weeds 11:226–232.

Wiebe, A. H., E. G. Gross, and D. H. Slaughter. 1931. The arsenic content of largemouth black bass (*Micropterus salmoides lacepede*) fingerlings. Trans. Amer. Fish. Soc. 61:150–163.

Woodford, E. K., and S. A. Evans. 1965. Weed Control Handbook. Blackwell Sci. Publ., Oxford. 356 p.

R. T. OGLESBY
University of Washington, Seattle

Effects of Controlled Nutrient Dilution

on the Eutrophication of a Lake

A limnological investigation of Green Lake, Seattle, Washington, was conducted from March through November of 1959 (Sylvester and Anderson, 1960, 1964). The study objectives included defining water quality, ascertaining the reasons for the lake's eutrophic condition, and discovering means of decreasing primary production. Upon finding that Green Lake was naturally eutrophic, with the principal nutrient source being water entering via subsurface seepage, the investigators proposed a unique approach to combating eutrophication: dilution.

The city of Seattle is blessed with an abundant supply of good water that is also inexpensive, since no treatment other than disinfection is required. The investigators recommended that surplus water from this municipal supply be diverted through the lake with the object of diluting the concentrations of nitrogen and phosphorus and thereby limiting the primary production. Money for the project was obtained by public bond issue. Following construction of facilities and some dredging of shallow areas in the lake, the first dilution water was added in May 1962. Subsequently some 2.5×10^7 m^3 (6.5×10^9 gallons) of water from the city supply have been flushed through the lake.

Since January 1965, a limnological study by the author has been in progress, the principal objective being an evaluation of the efficacy of this program in decreasing the undesirable effects of eutrophication (Oglesby and Edmondson, 1966). Sampling and analytical procedures used in the 1959 study have been generally followed in order to provide a valid comparison of data.

483

DESCRIPTION OF GREEN LAKE

The geological history and the influence of man on Green Lake have been previously described (Sylvester and Anderson, 1960, 1964). Morphometric data are summarized and some additional computations are included in Table 1. The shape of the basin is diagrammatically presented in Figure 1. Green Lake was formed following the recession of the ice sheets that constituted the Vashon glaciation 20,000 to 25,000 years ago. In a geological sense, natural enrichment of the lake took place with considerable rapidity; it is estimated that in 1907, when the city of Seattle acquired Green Lake, the original volume of the basin was already two thirds filled with sediment.

It has been estimated that the rate of sedimentation in recent years has been about 0.9 cm per year. This sediment consists mostly of organic material, since only minor storm drainage introduces allochthonous particulates into the lake.

Originally, Green Lake drained to the east through a natural channel. In 1911, however, the water level was lowered about 7 ft, and outflow was diverted over skimming weirs through a series of drains to a large storm sewer. Present inflows and outflows and a hydrologic budget for 1965 have been described elsewhere (G. Pederson, unpublished data).* A hydrologic budget and nutrient budgets were also calculated during the 1959 study (Sylvester and Anderson, 1960, 1964).

Water temperatures in Green Lake range from a summer high of about $22°$ C at the surface ($19°$ C in the deepest part) to a winter low of $0°$ C at the surface. In recent years little or no ice cover has formed. During the summer the lake is generally well mixed, with persistent thermal stratification occurring at about the 20-ft contour. Thus only a very small percentage of

TABLE 1   Morphometric Data for Green Lake, Seattle, Washington

Parameter	Measured or Calculated Value
Surface area (A)	104 ha (256 acres)
Shoreline (L)	4.7 km (2.9 mi)
Average water depth $(\bar{Z})$	3.8 m (12.5 ft)
Maximum depth (Z_m)	8.8 m (29 ft)
Normal water content	4.12×10^6 m^3 (1,088 million gallons)
Development of the shoreline $(D_L = L/2 \sqrt{\Pi A}$	1.30
$\bar{Z}: Z_m$	0.43

*An Annual Carbon Budget for a Eutrophic Lake. M.S. thesis, Univ. Washington (1966).

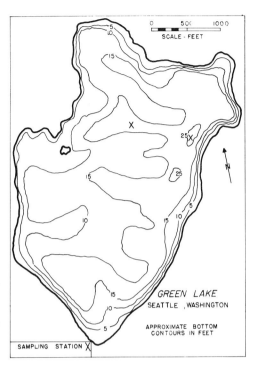

FIGURE 1   Bathymetric map of Green Lake, Seattle, Washington, showing approximate bottom contours in feet and location of principal sampling stations in the lake.

the volume is included in a persistent hypolimnion. Despite a high level of organic production, completely anoxic conditions have never been observed.

Chemically, Green Lake is typical of lowland lakes in western Washington. Some of the more important physical and chemical properties are summarized in Table 2 for the period May–September 1966. Values are derived from samples taken near the middle of the lake at depths of 0, 3, 4, and 8 m. In addition, occasional measurements indicate the following approximate values: total filtrable solids, 60 mg/liter; total hardness, 35 mg/liter as $CaCO_3$; silica as SiO_2, 7 mg/liter; calcium, 8 mg/liter; magnesium, 3 mg/liter; sodium, 1.5 mg/liter; potassium, 2.5 mg/liter; and iron, 0.1 mg/liter. In November 1959 a mean sulfate concentration of 10.3 mg/liter and a chloride level of 4 mg/liter for samples taken near the center of the lake were reported. In September the copper concentration was observed to vary from 0 to 67 μg/liter (Sylvester and Anderson, 1960, 1964). A carbon budget prepared for Green Lake in 1965 indicated that between early July and mid-December the average content of total carbon on an areal basis was 7.55 mg of C/cm^2 (G. Pederson, unpublished data).*

*An Annual Carbon Budget for a Eutrophic Lake. M.S. thesis, Univ. Washington (1966).

TABLE 2   Selected Physical and Chemical Characteristics of Green Lake, May–September 1966

Measurement	Units	Number of Observations	Mean Value	Range of Values
Temperature	°C	25 vertical profiles	–	Surface 14.3–22.5
				25 feet, 12.2–19.5
Conductivity	μmhos/cm at 25°C	86	86.9	59–116
Secchi disc transparency	m	24	3.0	1.1–4.3
Turbidity	mg/liter SiO_2	91	4.4	0–44
Color	color units	86	17.5	5–70
pH	pH units	103	7.7	6.6–9.0
Dissolved oxygen	mg/liter	106	8.6	0.75–11.0
Dissolved oxygen	% saturation	99	89.5	8–122
Total alkalinity	mg/liter as $CaCO_3$	100	32.6	20–50
Nitrate N	μg/liter	89	23.0	0–91
Orthophosphate P	μg/liter	85	8.6	0–34.3
Total phosphate P	μg/liter	74	30.7	3.3–75.0
Chlorophyll a	μg/liter	79	3.89	0.55–23.21

Green Lake has probably been eutrophic for some seven millennia. It exhibits the high level of blue-green alga production typical of such lakes. Genera of algae prominent in the phytoplankton since 1965 include *Dictyosphaerium, Oöcystis, Stauraustrum, Pediastrum, Dinobryon, Eudorina, Melosira, Fragilaria, Asterionella,* and *Anacystis.* The principal cyanophytes have been *Anabaena circinalis, Anabaena constricta,* and *Gleotrichia echinulata.* Since most of Green Lake is relatively shallow, the littoral zone covers a large part of the entire area. Among the aquatic spermatophytes observed, *Elodea* sp. is abundant in the nearshore areas and *Nymphaea* sp. is fairly common. In late summer *Potamogeton* sp. occasionally reaches the surface from depths of up to 15 ft. Aquatic herbicides are periodically used for the control of rooted vegetation.

The zooplankton of Green Lake have received little study. In general, however, it can be stated that several species of cladocerons are seasonally very abundant and that copepods are relatively scarce. The fish population is dominated by rainbow trout (*Salmo gairdneri*). This is an artificial situation maintained by annual stocking and by poisoning the entire fish population every 3 or 4 years. After poisoning, the lake is restocked with trout.

RESULTS OF THE DILUTION PROGRAM

A comparison of data collected in 1965–1967 with those obtained in 1959 indicates that substantial changes in the water quality of Green Lake have followed the addition of dilution water. The net effects of the program on the concentrations of phosphorus and nitrogen are illustrated in Figure 2. Even ignoring the very high total phosphate value reported for March 26, 1959 (291 μg/liter representing average values from samples taken at the surface and at 3 m), it is apparent that generally more phosphate phosphorus was present in 1959 than in 1965 or 1966; this is particularly true for August and September, when maximum growths of nuisance cyanophytes occur.

Variations in nitrate nitrogen between the years under comparison are even more striking. The very high maximum of 470 μg/liter on May 14, 1959, and the maximum occurring during most of July may be particularly noteworthy since these peaks slightly preceded marked increases in chlorophyll concentrations, as illustrated in Figure 3.

Figure 3 also illustrates that with the exception of the months of September and October, the transparency has improved significantly since the addition of dilution water. During the period May through August the mean Secchi disc transparency was 1.3 m in 1959, 3.1 m in 1965, and 3.5 m in 1966. On May 26, 1967 (not shown in Figure 3), a transparency reading of 6.3 m, the highest summer value ever obtained for Green Lake, was recorded.

Because of the limited number of observations, changes in many of the other chemical constituents of Green Lake water cannot be accurately

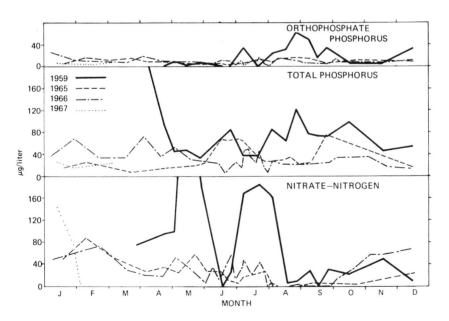

FIGURE 2 Concentration in μg/liter of orthophosphate phosphorus, total phosphorus, and nitrate-nitrogen during 1959, 1965, 1966, and early 1967.

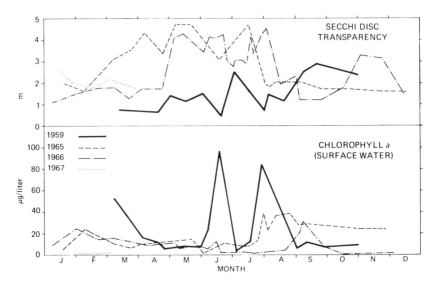

FIGURE 3   Secchi disc transparency in meters and chlorophyll *a* (surface water) in μg/liter for 1959, 1965, 1966, and early 1967.

assessed. The following approximations ot some of these changes between 1959 and 1965–1966 are based on single observations: calcium increased 15 percent, magnesium decreased 66 percent, potassium increased 230 percent, sodium decreased 86 percent, and total hardness decreased 46 percent. Total alkalinity, determined routinely in both studies of Green Lake, decreased 31 percent between 1959 and 1966.

Routine identification and counting of the phytoplankton was carried out in 1959 and in 1965–1966. Changes observed following addition of dilution water included a shifting of dominance between the major taxa and the virtual disappearance of at least one of the primary nuisance forms. *Aphanizomenon holsaticum* (= *flos-aquae*) was an important component of the plankton from mid-June until the end of the study period in 1959 and produced intense nuisance conditions during late July. In 1965 and 1966 this alga was seen only occasionally and then solely in samples taken in September and October. *Anabaena circinalis,* currently the major nuisance form, seems to have increased in relative abundance. *Gleotrichia echinulata,* also a prominent member of the phytoplankton, appears to have occurred somewhat later in 1959 and for a more restricted period than in 1965–1966. However, because of the large (1 to 2-mm diameter) colonies in which this alga occurs, it is often overlooked or its importance is artificially diminished in standard plankton counts. For this reason, special procedures were instituted for its enumeration during 1965–1966 (T. D. Roelofs, unpublished

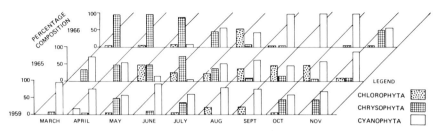

FIGURE 4   Percentage composition of the major phytoplankton taxa summarized by months for 1959, 1965, and 1966.

data).* This may account for the differences reported between the two studies.

The percentage occurrence of the major phytoplanton taxa are shown in Figure 4 with data summarized by months. In 1959 blue-green algae were dominant in every month of the study. In contrast, cyanophytes were the most prominent forms for only 6 months during 1965 and for 5 months in 1966. The general pattern of change seems to have been a marked decrease in the relative importance of planktonic blue-greens during the spring and early summer and, possibly, a relative increase in late summer and autumn.

DISCUSSION

Green Lake exhibited the characteristics typical of a eutrophic body of water to a significantly lower degree in 1965–1966 than in 1959. It could be postulated that such changes were due primarily to the effects of adding large quantities of dilution water or that these changes were occasioned by factors extrinsic in the dilution program. The most obvious possibility included in the latter category would be climatological differences between the years 1959 and 1965–1966.

Solar radiation, temperature, and wind mixing all affect phytoplankton growth, which in turn may regulate the presence of nutrients. Climatological data applicable to Green Lake are summarized in Table 3. With only 3 years of data for comparison, it is difficult to say what, if any, differences are significant, but certainly no major climatological variations are apparent between 1959 and 1965–1966. The increased solar radiation reaching the lake's surface in September and October may be a causative factor in the transparency decrease and rather high level of primary production (Figure 3) observed during this period in 1965–1966.

*The Ecology of Gleotrichia echinulata, a Planktonic, Blue-Green Alga, in Green Lake, Seattle, Washington. M.S. thesis, Univ. Washington (1967).

TABLE 3   Comparison of Climatological Data for Seattle, Solar Radiation as Measured at the University of Washington, and Mean Monthly Surface Temperature for Green Lake[a]

Time of Observation		Solar Radiation[b]	Air Temperature[c]	Surface Temperature[c]	Speed[d]
March	1959	227	8	9	14
	1965	330	9	7	10
	1966	246	7	7	13
April	1959	355	11	13	13
	1965	344	11	11	8
	1966	341	10	11	11
May	1959	470	13	15	13
	1965	442	13	9	10
	1966	485	13	15	11
June	1959	500	17	19	13
	1965	563	17	19	10
	1966	491	15	18	11
July	1959	613	20	22	11
	1965	539	19	21	10
	1966	536	17	20	10
August	1959	459	18	21	13
	1965	400[e]	19	21	10
	1966	457	18	22	8
September	1959	269	15	18	11
	1965	308	16	18	8
	1966	320	17	21	8
October	1959	166	12	14[f]	13
	1965	210	14	8[f]	10
	1966	189	11	14	11
November	1959	–	8	7[f]	13
	1965	92	11	9[f]	11
	1966	76	7	11	10

[a]Climatological data provided by Climatological Data Section, U.S. Weather Bureau, Seattle, Washington.
[b]Gram calories per cm^2 per day.
[c]Degrees Celsius.
[d]Kilometers per hour.
[e]Data available for only 2 weeks.
[f]Based on a single measurement.

If differences in climate between the years under study are ruled out as major factors producing the observed changes, attention can be focused on the mode of action of the added dilution water. The existence of at least three mechanisms may be postulated. First, the addition of water from the city supply may have produced a direct dilution of the principal algal nutrients within the lake. Second, a critical trace element, or elements, may have been diluted to the extent of becoming limiting for algal growth. Third, as the washout rate of the algae more closely approached growth rate, the standing crop was necessarily reduced.

The first of these mechanisms, direct dilution of nitrogen and phosphorus, could be evaluated only if accurate nutrient budgets could be calculated. Although this calculation was attempted in 1959 (Sylvester and Anderson, 1960, 1964), the rate of recycling remained questionable, as in most such studies. Furthermore, a large percentage of the nutrient income was attributed to subsurface seepage, and this seepage had to be determined in the budget by difference. As a further complication, data on the inflow of city water into the lake are available only in the form of monthly summaries. Finally, the nitrogen and phosphorus levels in the dilution water have proved to be highly variable and, in the past, sampling intervals were not close enough together to allow for accurate integration of the data.

Despite these difficulties in determining the precise quantitative effect of adding dilution water on the lake's nitrogen and phosphorus levels, Figure 2 would indicate a significant decrease in both elements. There is indirect evidence to support this conclusion in the case of saline nitrogen. Of the current nuisance species in Green Lake, *Anabaena circinalis* is known to be able to fix nitrogen, and the probability that *Gleotrichia echinulata* also possesses this ability has been demonstrated (T. D. Roelofs, unpublished data).* Conversely, *Aphanizomenon holsaticum* (= *flos-aquae*), which apparently cannot use gaseous nitrogen directly (Williams and Burris, 1952), was virtually absent from the 1965–1966 phytoplankton in contrast to its high level of abundance in 1959.

The possibility that the addition of city water has resulted in the depletion of some trace element has not been extensively investigated. Iron seems to have remained at about the same level in the lake. It is also unlikely that molybdenum has become limiting; nitrogen-fixing species of blue-green algae, such as those that now dominate the summer phytoplankton, require molybdenum in especially large amounts (Bortels, 1938, 1940).

Because Green Lake is a generally well-mixed system, the rate of change of algal numbers or biomass might be expected to fit equations developed for bacteria in continuous, completely mixed culture systems. Thus, the following basic equation from Herbert (1958):

$$\frac{dx}{dt} = Kx - Dx$$

where x = the biomass
K = growth rate constant of the entire phytoplankton community
D = flow per unit volume or dilution rate

*The Ecology of Gleotrichia echinulata, *a Planktonic, Blue-Green Alga, in Green Lake, Seattle, Washington.* M.S. thesis, Univ. Washington (1967).

This equation in its simplest form shows that the dilution rate equals the growth rate under steady-state conditions. In other words, the standing crop at any given time is a function not only of growth but also of washout from the system. This rather obvious fact seems to have been generally ignored in consideration of the relationship of algal growth rates to standing crops in lakes and estuaries. This factor may be even more important in Green Lake than the equation indicates. The water from the city supply enters the lake at the bottom. During the summer it is generally at a lower temperature than the lake. The nuisance blue-green algae tend to concentrate at or near the surface during periods of calm or of relatively light wind. Since water flows out of Green Lake over surface weirs, the net effect should be a displacement of phytoplankton from the lake at a rate greater than that predicted from equations applicable to completely mixed systems. A project to evaluate this phenomenon was proposed for the summer of 1968.

OTHER DILUTION PROGRAMS THAT HAVE BEEN CONSIDERED

A project similar to the one carried out on Green Lake has been proposed to control eutrophication in Lake Bled (Sketelj and Rejic, 1966). This lake, located in the Slovene subalpine region, exhibits a high ratio of epilimnion to hypolimnion, has a rather long flushing time, and has exhibited changes in phytoplankton indicative of developing eutrophication. Following a limnological and engineering study, it was concluded that the only practical means of arresting this trend was to flush large quantities of water from the Radovna River, which is almost katharobic in character, through the lake.

A similar project has been proposed (Sylvester and Oglesby, 1964) for improving the water quality in Moses Lake, a large (2,480 ha), highly eutrophic lake in eastern Washington. In this instance an easily utilized source of dilution water is available from a nearby irrigation canal carrying high-quality water from the Columbia River irrigation system.

GENERAL APPLICATION OF DILUTION AS A MEANS FOR COMBATING EUTROPHICATION

In our present age of water shortages and increasingly degradative use of water, it might seem that few instances would exist where water, adequate in quality and quantity, would be available for dilution programs. In the case of lakes located in or adjacent to metropolitan areas, however, just such a resource of water is almost always available. Every domestic water supply must dispense greater quantities of water through the distribution system than are actually used. Such "waste" water is, in many instances, readily available for the flushing of recreationally or aesthetically important municipal lakes. This availability assumes special significance when we consider that such lakes are frequently eutrophic and have relatively limited volumes and that their waters often have long detention times within the lake.

The example of Moses Lake cited above offers an illustration of how advantage might be taken of water-resources management programs, existing or planned, to improve the quality of water in lakes experiencing or threatened by eutrophication.

If hopes are realized, continued research on Green Lake will provide a kinetic model that scientists and engineers can use in developing similar programs elsewhere. It is also hoped that data collected and analyzed in this study will help to further our understanding of the eutrophication process and, particularly, the relationship of water quality and retention time in the lake to the occurrence and growth of particular phytoplankton populations.

The research reported in this paper was supported by the Federal Water Pollution Control Administration, Grant Number WPD 38-01 (RI) and Contract Number PH-86-66-33.

REFERENCES

Bortels, H. 1938. Entwicklung und Stickstoffbindung bestimmter Mikroorganismen in Abhängigkeit von Spurenelementen und vom Wetter. Ber. deut. Botan. Ges. 56:153–160.

Bortels, H. 1940. Über die Bedeutung des Molybdäns für stickstoffbindende Nostocaceen. Arch. Mikrobiol. 11:156–186.

Herbert, D. 1958. A theoretical analysis of continuous culture systems. p. 21–53 In Continuous culture of microorganisms. S.C.I. Monograph No. 12, Symposium at Univ. College, London, 31 Mar.–1 Apr., 1960.

Oglesby, R. T., and W. T. Edmondson. 1966. Control of eutrophication. J. Water Pollut. Control Fed. 38:1452–1460.

Sketelj, J., and M. Rejic. 1966. Pollutional phases of Lake Bled, p. 345–362 In Advances in water pollution research. Vol. 1, Proc. 2nd Internat. Conf. Water Pollut. Res., Pergamon Press, Ltd., London, England.

Sylvester, R. O., and G. C. Anderson. 1960. An engineering and ecological study for the rehabilitation of Green Lake. A report to the Seattle, Washington, Park Board, 194 p.

Sylvester, R. O., and G. C. Anderson. 1964. A lake's response to its environment. J. San. Eng. Div. Proc. Amer. Soc. Civil Engr. 90, SA1:1–22.

Sylvester, R. O. and R. T. Oglesby. 1964. The Moses Lake water environment. A report to the Moses Lake Irrigation and Rehabilitation District, Moses Lake, Washington, 62 p.

Williams, A. E., and R. H. Burris. 1952. Nitrogen fixation by blue-green algae and their nitrogenous composition. Amer. J. Bot. 39:340–342.

D. F. LIVERMORE
University of Wisconsin, Madison

W. E. WUNDERLICH
U.S. Army Corps of Engineers, New Orleans, Louisiana

Mechanical Removal

of Organic Production from Waterways

The forms of life that constitute the organic production of waterways (lakes, streams, canals, bayous, and so on) are innumerable. A rough, general classification might include the following major divisions:

Very small, even microscopic, plants and animals designated as plankton
Larger animals, such as fish, turtles, and crustaceans
Larger plants, including submersed, emergent, and floating species, chara, and the filamentous algae
Bottom-dwelling organisms (not considered here)

A typical body of water normally contains elements from the first three of these divisions. The extent and distribution of organic production depend on many factors, including the age, location, size, and fertility of the body of water.

What constitutes a desirable condition or state? The answer depends on the uses to be made of the waterway. For example, dense algae "blooms" might be actively promoted in a southern farm pond, whereas similar conditions in a municipal-water-supply reservoir or a recreational lake would be considered disastrous.

The discussion that follows pertains chiefly to situations wherein productivity of a waterway, because it is excessive or consists of undesirable species, would reduce the utility of the waterway and contribute to accelerated eutrophication.

Dense growths of one or more forms of aquatic vegetation create most of these problems. The problems range from relatively minor reductions in the

494

FIGURE 1   Stream in northern Louisiana blocked by water hyacinths (*Eichornia*).

recreational and esthetic quality of the water to almost complete loss of utility for many purposes. For example, in Kariba Lake (1,700 mi^2; 440,000 ha), a man-made impoundment in the southern Rhodesia, large parts of the lake have been covered by an almost impenetrable mat of the floating fern *Salvinia auriculata*. Such problems with floating vegetation in Africa's man-made lakes and other important waterways (e.g., the Nile River) are increasing in importance (M. A. Aziz and T. Wright, unpublished report).*

In waterways of the southern United States, control of the water hyacinth (*Eichornia crassipes*) and alligator weed (*Alternanthera philoxeroides*) has been the objective of a long and continuing struggle. Figure 1 indicates the almost solid appearance of a stream blocked by water hyacinths.

Of increasing concern are problems with algae and submersed aquatic plants. Planktonic algae, under "bloom" conditions, can result in dense accumulations along shorelines and in bays; the subsequent decomposition of these deposits may produce acute nuisance conditions.

*The Spread and Control of Aquatic Weeds on African Man-Made Impoundments. Limnology Seminar, Univ. Wisconsin, 1966.

Bottom-rooted aquatics, often draped with filamentous algae and festooned with collections of trapped algae and debris, may produce particularly objectionable conditions in recreational lakes if significant areas are affected. Such conditions in Lake Mendota (Madison, Wisconsin) in the summer of 1965 led to the formation of a Lake Mendota Problems Committee and to the purchase of machinery for harvesting aquatic plants.

Quantitative relationships between the organic production of a lake (waterway) and its rate of eutrophication are not firmly established. However, it is well known that residues left from the decomposition of dense lake "crops" result in a gradual filling of the lake and contribute to a more eutrophic condition. As a consequence, harvesting of plant and animal produce has often been suggested as a possible means of slowing down eutrophication and improving water conditions (quality). The publication *Limnological Aspects of Recreational Lakes* (Mackenthun *et al.*, 1964) discusses many of the factors that lead to nuisance conditions in lakes. In an unpublished report on water fertilization,* R. B. Corey and others discuss the removal of nutrients by forms of harvesting and compare the harvests with estimated supplies of nutrients. Their report contains statements from many sources pertaining to water fertilization and related problems.

Corey and his colleagues report fish harvests, on a sustained basis, of more than 300 lb per acre of water surface. Sport fishing accounts for probably one third, or less, of the total. This fish harvest represents removal of about 7 lb of nitrogen and 0.6 lb of phosphorus per acre (8 kg of N_2 and 0.673 kg of P per ha).

Another approach to the problem of nutrient removal might be repeated harvesting of large quantities of plankton. Birge and Juday (1922) and E. A. Birge (unpublished pamphlet)† estimated the annual "crop" of plankton in Lake Mendota at about a ton of dry material per acre (2,240 kg per ha) of lake surface, with something like 200 lb per acre (224 kg per ha) present at a given time. During a "bloom," the plankton density may be several times greater than this. Concentrated algae colonies might be removed during a bloom by filtration or screening, but general harvesting of plankton appears difficult because of the minute size of the organisms and their wide horizontal and vertical distribution throughout the water.

Removal of bottom sediments by dredging has been widely used as a means of deepening lakes and rivers. Since bottom deposits may contain residues from the decomposition of many years' organic production, removal of

*R. B. Corey, A. D. Hasler, F. H. Schraufnagel, and T. L. Wirth. 1967. *Excessive Water Fertilization.* Report to Water Subcommittee, Natural Resources Committee of State Agencies, Madison, Wisconsin.

†*Lake Mendota—Origin and History.* 1936. Pamphlet of the Technical Club of Madison, Wisconsin.

these materials may have important implications for nutrient removal besides being a feasible means of reversing the gradual filling process.

In addition, if the water can be sufficiently deepened, rooted aquatics will be largely eliminated from the dredged areas (Sylvester and Anderson, 1964). For dredging to be economically feasible, it is essential that dredged materials not have to be pumped or moved over excessive distances. Hence, zoning, regulation of shoreline developments, "dumping privileges," and careful long-range planning are of utmost importance in areas adjacent to waters that may require dredging in the future. No specific discussion of either the benefits to be derived from removal of bottom sediments or the methods for accomplishing it are included here, because other papers in this volume treat this in considerable detail.

In addition to fish life and plankton, most waterways produce "crops" of larger aquatic plants that represent fairly harvestable material. These submersed, emergent, or floating plants are large enough to be handled by machinery, and many mechanical methods for cutting, harvesting, or destroying them have been used or proposed. While the nutrient removal that could theoretically be achieved by concerted harvesting of these plants does not appear to be large, little work along these lines has actually been reported.

Work by Gerloff and Krombholz (1966) demonstrated a luxury uptake of nitrogen and phosphorus in fertile waters. The amounts they found in six species of plants in Lake Mendota averaged, on a dry basis, 3 percent for nitrogen and 0.47 percent for phosphorus—nearly double the amounts reported by Schuette and Alder (1928). Rickett (1922) estimated the crop density of the larger aquatics in shallow areas of Lake Mendota at 1,800 lb of dry material per acre. Recent productivity studies indicate a standing crop of about 1,600 lb per acre in University Bay of Lake Mendota, with considerable seasonal variation occurring. Bailey (1965) reports a harvest of 5,600 lb dry weight of milfoil (*Myriophyllum exalbescens*) from three successive cuttings about 18 in. below the surface in Caddo Lake. He estimated that five or six cuttings might constitute a total yearly cycle in this lake; thus, an acre actively harvested might yield 6,000 to 9,000 lb of dry material per year. With a dry harvest of 6,000 lb, containing 3 percent nitrogen and 0.47 percent phosphorus, the nitrogen harvest would be 180 lb per acre per year and the phosphorus harvest would be 28 lb. These amounts would represent significant nutrient removal in many lakes. The harvesting of these large amounts of vegetation would tend also to reduce the rate of filling by the residues left from decomposition of organic materials.

A major portion of this paper is devoted to a survey of work that has been done in mechanically removing or destroying larger aquatic plants. The objective of such work usually is the alleviation of a nuisance condition. In a few instances, the objective has been to harvest a crop for its economic value (ocean kelp) or for research purposes. No work has been reported where large-scale harvesting has been employed for nutrient-removal purposes.

The types of machinery developed and the methods used have depended on the objectives to be achieved and the scale of the work to be done. Management objectives in the water differ from those on land. An important reason is the difference in criteria for determining that a plant is a weed. Nearly all plant species that grow abundantly in water are considered to be weeds. But on land, only certain species are weeds, to be suppressed. Other species on land are encouraged to grow luxuriantly and take up the nutrients; then they are harvested as crops. Thus, in the water, management objectives may be to grow essentially no vegetation, neither weeds nor a crop, even though the nutrient supply is abundant.

CONTROL OR HARVESTING PROGRAMS FOR AQUATIC PLANTS

FORMS OF CONTROL

Measures for controlling aquatic vegetation may take many forms, the form depending on the nature of the problem and on the objectives of control.

H. J. Elser (unpublished report)* has classified the categories of control as physical, biological, chemical, and mechanical. Physical controls include such procedures as (1) lowering water levels to allow the roots of plants to dry out and die, and (2) shading out light with black polyethylene sheeting or dye. Biological controls are those dealing with the introduction of diseases, competitor species, or animals or fishes that would eat the vegetation. Experiments are under way with the manatee and herbivorous fishes, snails, and water flea-beetle (Alsopp, 1966; Seaman and Porterfield, 1964; Swingle, 1957).

Chemical versus mechanical control   Chemical nuisance-control measures are so widely applied (C. T. DeWitt, unpublished report†; H. McCarthy, unpublished report‡; Speirs, 1948) that it seems appropriate to compare chemical and mechanical controls before discussing specific mechanical methods and machinery. Further comments on the pros and cons of chemical

Status of Aquatic Weed Problems of Tidewater, Maryland, Spring 1966. Report to Maryland Department of Chesapeake Bay Affairs, Annapolis, Maryland (1966).

†*Aquatic Nuisance Control Procedures in the United States.* River Basin Planning Seminar, Univ. Wisconsin (1966).

‡*Survey Study on Methods of Controlling Aquatic Weeds and Their Effectiveness.* Report for FWD (Four-Wheel Drive) Corp., Clintonville, Wisconsin (1961).

and mechanical controls are to be found in a number of references (e.g., Mackenthun, 1962; Mackenthun *et al.,* 1964; Oglesby and Edmondson, 1966; Sawyer, 1962; Seaman, 1958). An unpublished report by R. B. Corey and others is also pertinent.*

Certain plant growths may be prevented or retarded, or existing growths may be destroyed, by the proper application of suitable herbicides. The effects, however, depend on many factors, including timing of treatment, type of chemical, method of application, and wind currents. Chemical control is often thought to be the easiest, most effective, and least expensive control procedure.

For waterborne algae, chemical control appears to be the only practical method in most instances. Mechanical removal would presumably require filtration of impossibly large volumes of water unless the algae were in extremely heavy concentrations. For larger aquatic plants in relatively open water, either chemical or mechanical methods may be useful. Chemical methods may be used in conjunction with mechanical methods, chemical treatment being reserved for areas where mechanical methods are impossible or impractical.

Chemicals can sometimes be applied in advance to prevent a nuisance problem or to suppress a particular species. Pre-emergence herbicides are examples of chemicals that can be applied for such purposes (R. C. Hiltibran, unpublished report).†

A major disadvantage of chemical control is that a possibly toxic material is added to the water. The short-range effects of this addition can be undesirable, and the longer-range effects may be cumulative or perhaps not fully understood. R. B. Corey and others warn in an unpublished report‡ that when "chemicals are used to kill or prevent the growth of nuisance organisms [in water], complex distortions [of the plant and animal communities] take place and a chain of undesirable situations may occur." They continue:

Some of the chemicals used affect a variety of organisms and their younger stages adversely, and some accumulate in the lake soils. It is difficult to control the influence of a chemical added to a lake or to apply it at the proper time. A great deal more research is needed to produce specific inhibiting agents which do not have undesirable side effects on other organisms and which disintegrate promptly after application. . . . There are as yet no chemicals which are specific enough, in their inhibiting effect, to be used with impunity.

*See footnote on page 496.

†*The Chemical Control of Some Aquatic Weeds.* Illinois Natural History Survey, Section of Aquatic Biology, Urbana, Illinois (1961).

‡See footnote on page 496.

A second major disadvantage of chemical controls is that the plants destroyed remain in the water where they decompose, depleting the oxygen supply and releasing stored nutrients for support of new growths. While the equipment needed for the application of chemicals is relatively inexpensive, the chemicals themselves may prove to be quite expensive, so costs do not necessarily favor chemical methods.

Chemical control measures are used extensively, and a great deal of material has been published on effectiveness, persistence and toxicity, treatment procedure, and so on. The following references are cited as indicative of work being done: Bartch (1954); Burbank (1963); Frye (1963); Hiltibran (1962 and 1963); Lapham (1966a, b); Mackenthun (1950 and 1960); Philippy and Burgess (1966); Smith (1962); Steel and Ewing (1954); Steenis and Elser (1967). Most states regulate the application of herbicides, and many states and the federal government publish bulletins on procedure, equipment, and concentrations. The following references are examples: Harrison (1964); Klussman and Lowman (n.d.); Mackenthun *et al.,* (1964); Meyer (1964); Michigan Department of Conservation (1964); Minnesota Department of Conservation (1966); New York State Department of Health (1964); Surber (1961); Wisconsin Committee on Water Pollution (n.d.).

Mechanical control measures usually have consisted of cutting the nuisance plants loose from their moorings by some means, then collecting the cut parts either immediately, as with aquatic harvesters, or later in a secondary operation. Removal of the cut parts is usually necessary to avoid nuisances caused by the large quantities of drifting and decomposing plants. Currently these harvested plants have no commercial value, and collecting and disposing of them is expensive.

The effects of mechanical cutting on most plant species are not well established. At present it is difficult to specify optimum cutting procedures, seasonal timing, and the like, to accomplish desired objectives. Either chemical treatment or drastic mechanical measures may greatly alter the plant distribution, and may merely substitute one type of problem for another (R. B. Corey and others, unpublished report*; H. J. Elser, personal communication; Hasler and Jones, 1949; Wunderlich, 1966). Success in developing chemicals that are specific for nuisance species should contribute to making the effects of control measures more predictable, and so should research on the effects of cutting on various species.

Although cost and expediency appear to favor chemical methods, mechanical control or harvesting is usually thought to be ecologically more sound. First, it does not introduce foreign substances into the water. Second, it may actually remove nutrient materials from the lake cycle, and should tend to reduce the rate of lake filling by plant residues. Third, if carefully

*See footnote on page 496.

done, it probably does not tend to alter the plant and animal life balances as drastically as chemical treatments do. Another advantage of mechanical methods is that they can provide immediate relief from nuisance conditions.

MACHINERY AND METHODS

Hand methods   The earliest nuisance-control methods were undoubtedly hand-pulling and cutting of the larger aquatic plants. Scythes and sickles for cutting and rakes and forks for removing the cut plants from the water are still widely used to clear rooted plants from small areas. A weed saw consisting of a long steel band with saw teeth on the edges and long handles at each end can be used to cut fairly sizable areas around beaches and docks. Forking and raking following mechanical cutting or for general shoreline cleanup of weeds and debris are still probably the most widely used collection methods.

Powered cutting and uprooting devices   The most widely used and most successful powered cutters for both submersed and emergent plants have been versions of reciprocating mower bars similar to the types used on agricultural machinery. Because of the inherent simplicity of these cutters and the ease with which most aquatic vegetation can be cut, many variations of these units have been built by lake associations, sport clubs, government agencies, and other groups. Several companies* build commercial versions, ranging from small units (4- to 6-ft swath) that can be mounted on the bow of a small boat, to larger units (8- to 15-ft swath) that are mounted on self-propelled steel barges. Figure 2 shows a unique cutter that was developed in New Zealand. It is reported to be widely used there and in England on rivers, canals, and drainage ditches.

Knife-type cutters, which are dragged through the plant beds and cut by a scything action, have been used. These usually take the form of a "V," having sharpened, and sometimes serrated, leading edges. A continuous chain-type cutter carrying sets of moving blades over stationary shearing knives has been used on some recent models of aquatic harvesting machines. An experimental kelp-harvesting unit, developed by the Scottish Seaweed Research Association, made use of rotating blade cutters similar to those in rotary power lawn mowers (Mackensie, 1947). Various types of saws, crushers, and choppers have been used to destroy vegetation, particularly water hyacinths and alligator weeds. Brief descriptions of some of these machines and methods are included in a later section.

As a means of uprooting vegetation, chains and draglines are often pulled along the bottoms of waterways behind powerboats. Another uprooting

*For example, Aquatic Controls Corp., Waukesha, Wisconsin; The Hockney Co., Silver Lake, Wisconsin; Jari Products, Minneapolis, Minnesota.

system is the hydraulic-jet device shown in Figure 3; high-velocity jets of water are used to dislodge the rooted plant. It is claimed that this unit not only uproots the vegetation but also buries heavy debris, glass, and the like. It is especially effective in clearing out swimming areas.

Harvesting systems   Probably the most common harvesting method consists of power-cutting large aquatic plants and collecting the cut parts from locations to which they have drifted. One of its advantages is that relatively simple cutters can be used to cut large areas. Winds or currents may then concentrate the harvest along shorelines or against obstructions where it can be collected in the secondary operation. The practicality of such a two-stage operation depends on the ease of collection. Where shorelines are obstructed and inaccessible from either land or water, the method may be impractical. Heavy concentrations of cut plants usually must be removed fairly promptly to prevent them from becoming a serious nuisance.

The most common collection procedures are probably (1) hand removal by raking or forking onto adjacent shorelines, and (2) hand forking from the water onto barges for transport to convenient shoreline unloading areas. Power-assisted systems have also been widely used. In a shoreline operation, men wading in shallow water can push large quantities of floating plants into the path of a barge-mounted conveyor or weed scoop, which lifts the heavy

FIGURE 2   "U"-shaped reciprocating mower drawn behind boat. (Photograph courtesy of B. Powell, Christchurch, New Zealand.)

FIGURE 3   Hydraulic-jet device for uprooting submerged aquatics. (Photograph courtesy of D. Talbott, Annapolis, Maryland.)

masses of plants from the water and deposits them on the deck of the barge. Rake-type units mounted on the bow of a barge or boat (Figure 4) are sometimes used to push cut plants to shore or to aid in moving them to desired locations in the water where they can be collected by another unit.

Cables, with clips attached to entangle the plants, have been used to collect cut plants. After a cable had collected a load, it was drawn in with a power winch (Domagalla, 1926).

Truck-mounted mobile cranes can be used effectively to remove cut plants from the water at convenient access points where the plants have been concentrated by wind or current action following cutting, for example, in a canal or river behind a "boom" (Miles, 1965).

Other variations of two-stage cutting-and-collection systems have been used, and many more are possible.

Several versions of harvesters are in use that incorporate powered cutting that is followed by immediate collection of the cut plants. Kelp harvesters have been used for many years in coastal waters. Harvesters suitable for use on inland waterways are manufactured by at least two companies in the United States.*

The cutting and collecting systems of these harvesters consist essentially of cutters mounted directly ahead of a porous elevating conveyor. After the cut plants are elevated from the water, they are conveyed to a small storage area at the rear of the harvester barge. Periodically, when this storage area is filled, the plants are transferred to a transport barge. When the transport barge is filled, it travels to an unloading area where the plants are transferred to shore

*Aquatic Controls Corp., Waukesha, Wisconsin, and Grinwald-Thomas Corp., Hartland, Wisconsin.

or loaded onto a truck for disposal. More than one transport barge may be required to ensure continuous operation by the harvester. The number required depends on the distance between the harvesting areas in the lake and the unloading dock.

Many schemes have been proposed for handling the plants after they have been removed from the water. The system described by Crouse (1941) used a baler to compress the plants and bind them together. After they were bound, they could be floated to shore. Another arrangement utilized a hydraulic press on a weed-scooping unit to squeeze from the plants a large part of the entrained water and some fluid. The squeezing greatly reduced the density of the plants, simplifying handling and storage.

One commercial harvester uses nearly the full deck of the harvester barge for weed storage. The main collection conveyor can be inclined upward to the front, and its direction of travel can be reversed for use in unloading.

Figure 5 shows a model of a proposed large-scale harvesting system in which the harvester unit consists only of a cutting and collection system

FIGURE 4   Underwater mower and weed rake. Rake is mounted in front of small mower on bow of pontoon boat. (Photograph courtesy of Jari Products, Minneapolis, Minnesota.)

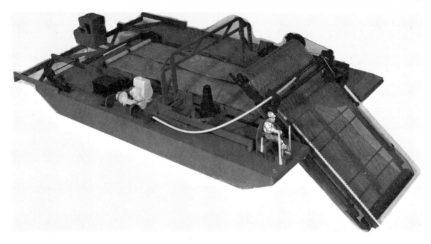

FIGURE 5   Model of proposed large-scale harvester. Cutter and conveyor are mounted on pontoons, which "straddle" shuttle-barge.

pushed along by a transport barge having a powered conveyor on its deck (Livermore, 1954).

Experimental machines have been built to harvest plants from considerable depths in coastal waters (Mackensie, 1947; Jackson, 1957). The basic cutting-and-collection unit consists of a rotary- or mower-type cutter incorporated into a suction head that is mounted on skids at the lower end of a long, flexible suction tube. When this harvester unit is dragged along the ocean bottom, weeds that fall within its cutting path are chopped into short lengths and then drawn through the suction pipe by a pump and discharged onto a drainage conveyor on the ship deck. The chopped weeds are dumped into storage tanks aboard the vessel and later unloaded at port by immersing them in water and pumping them to shore.

EXAMPLES OF CONTROL PROGRAMS USING MECHANICAL METHODS

LOUISIANA

Probably the most extensive and persistent program for controlling nuisance aquatic vegetation has been the U.S. Army Corps of Engineers program for opening southern U.S. waterways clogged by water hyacinths and alligator weeds. The history of this work has been well documented (e.g., Wunderlich, 1938 and 1966).*

*Also by W. E. Wunderlich: *The Role of Machinery in Aquatic Vegetation Work,* presented at 1964 meeting of the Weed Society of America, and *Mechanical Research,* internal report, Aquatic Growth Control Section, U.S. Army Corps of Engineers, New Orleans, Louisiana (1966).

Operations in Louisiana between 1937 and 1950 were conducted with machinery as the only means of control. Thousands of acres of hyacinths and alligator weeds growing in the main waterways were destroyed. Simple experiments showed that when the rhizome of the water hyacinth was bruised, the plant was destroyed. In 1937 a machine was built that could accomplish this on a large scale (Wunderlich, 1938). Figure 6 shows the arrangement of the machine. The vegetation clogging the stream was cut by outrigger saws into a mat 15 ft wide. The mat was picked up by an inclined conveyor operating at 100 ft/minute and fed by gravity through two corrugated rollers operating under heavy pressure. The crushed residue was carried over the side and back into the water by a small high-speed conveyor. This unit, called the "Kenny," proved to be effective against both hyacinths and alligator weeds and was operated regularly from 1937 to 1951. Coverage rate with this unit has been estimated at more than 200 acres (81 ha) per month.

Another type of unit, known as a "sawboat," has also proved effective against these surface plants. The saw units consist of cotton-gin saw blades, with teeth having a pitch of about 1 in., mounted on horizontal spindles and spaced about 1 in. apart. These spindles, driven at 800 to 1,000 rpm and immersed about to their centerlines into the water, are fed forward into the surface vegetation, which they destroy by shredding. Figure 7 shows this unit in operation. The largest of these units can shred a swath 40 ft wide at a forward speed of about 2 mph. A number of smaller units of this type have also been used extensively.

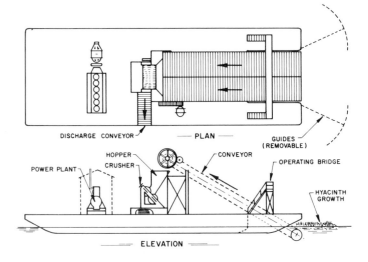

FIGURE 6 Diagram of the "Kenny" plant for destroying water hyacinths. (Drawing courtesy of U.S. Army Corps of Engineers, New Orleans District.)

FIGURE 7   "Sawboat"-type unit for destroying water hyacinths shown in action. (Photograph courtesy of U.S. Army Corps of Engineers, New Orleans District.)

One other type of unit that was widely used in Louisiana against the hyacinth consisted merely of an elevating conveyor and a side-discharge conveyor. The elevating conveyor lifted the plants from the water and the side-discharge conveyor deposited them ashore, where they soon died.

While mechanical units proved effective against hyacinth and alligator weeds in relatively open areas, they could not reach many areas back in the swamps, and chemical spraying, primarily with 2,4-D (V. T. Lapham, personal communication; unpublished report by V. T. Lapham, F. Myers, and T. Dutton*; unpublished report by W. E. Wunderlich†), is the principal method of control in these "source" areas. However, it has been demonstrated clearly that both alligator weeds and hyacinths can be destroyed by damaging the plants mechanically.

Experimental work in 1963—1964 showed that alligator weeds fed through

*Hyacinth Control Section Report, 10th biennial report of Louisiana Wildlife and Fisheries Commission, New Orleans, Louisiana.

†Mechanical Research, internal report, Aquatic Growth Control Section, U.S. Army Corps of Engineers, New Orleans, Louisiana (1966).

an Asplundh chipper only once had a regrowth rate of about 5.0 percent when returned to the water. When the vegetation was fed through the chipper a second time, it was reduced to sludge and there was no regrowth (W. E. Wunderlich, unpublished report).* Tests on some submersed vegetation gave similar results. Much better results could be expected with equipment specifically adapted for this work.

None of the methods used against these nuisance plants was aimed at harvesting the plants for nutrient removal or other reasons; as is the case when chemical control measures are used, destruction of the plant growths was the sole objective. Such "destroy only" measures have not been widely applied to rooted submersed vegetation. Cutting only, with no follow-up collection except by residents along their own shorelines, is widely used. In these situations, floating cut vegetation is a nuisance, and much of it may reroot when it finally settles to the bottom. Since collection is a major problem of materials handling and disposal, it might be possible to accelerate coverage by feeding plants through machines that would damage them in such a way that they would sink rapidly to the bottom and would not reproduce from cuttings. No nutrients would be removed by such a system, but neither would chemicals be added to the water to destroy the plants.

CHESAPEAKE BAY AND TRIBUTARIES

The information in this section is taken almost entirely from data published by H. J. Elser (1966a, 1966b), unpublished papers and reports by him,† and personal communications received from him.

Mechanical control measures against water chestnut (*Trapa natans* L.) in the Potomac River and its tributaries were initiated by the Corps of Engineers in the 1920's. After about 10 years of annual cutting with machines of the mower-bar type, the infestation was reduced to a low level, although the Corps still sends men into the field to handpull what few plants they can find (41 plants in 1965). In 1939 Wunderlich used a combination device consisting of a submerged cutter and bank of rotary cutters and saws to destroy large areas of water chestnuts in the Potomac River. Rosettes of plants were cut below the surface by a forward cutter bar and fed into a partially submerged trough in which closely spaced gin-saw blades revolved against a backing bar. Refuse was returned to the water; it sank rapidly.

Mechanical Research, internal report, Aquatic Growth Control Section, U.S. Army Corps of Engineers, New Orleans, Louisiana (1966).

† *Use of Aquatic Weed Cutters in Maryland,* 1964; presented at 1965 meeting of Aquatic Weed Control Society. *The Formation of Sea-Lettuce Problems, Mayrland, 1965;* internal report, Maryland Department of Chesapeake Bay Affairs, Annapolis, Maryland. *Status of Aquatic Weed Problems of Tidewater, Maryland, Spring 1966;* internal report, Maryland Department of Chesapeake Bay Affairs, Annapolis, Maryland.

In 1964, the Maryland Department of Chesapeake Bay Affairs purchased an aquatic harvester that has been used against both water chestnut and milfoil. Chemical control measures are not widely used in many of these areas, because only 2,4-D can be used in low concentrations, and permits even for this will not be issued for work closer than 1 mile to areas producing clams or oysters. The harvester proved to be effective against water chestnut in the Sassafras River where the plant concentrations were sufficient to warrant its use. The most effective cutting depth was found to be about 8 in. below the surface. At this depth, some cut rosettes escaped the collection conveyor, and these were picked up by hand from a skiff.

The disposal scheme that was used is worthy of note. Square frames of wood two-by-fours were assembled, and snow fencing was nailed around the outside so that the staves extended 3 ft into the water and 1 ft above. The cut plants were dumped into this "pit" in large masses, which dried on top and rotted on the bottom. It is reported that the pits could absorb a surprisingly large amount of plant material and that the mass was reduced to a small part of its original volume in a day or two.

In 1964 the harvester was used against milfoil (*Myriophyllum*) in the Chesapeake Bay area. Estimates of the area infested by milfoil in the bay and its tributaries run as high as 200,000 acres (P. F. Springer, G. F. Beaven, and V. D. Stotts, unpublished paper*; Steenis *et al.,* 1962). Thus a single harvester could cover only a minute fraction of this area in a season. According to Elser (1966a and b), the machine was effective in cutting and collecting plants at depths to 4½ ft and at rates up to about 1 acre per hour. Salt-water corrosion of the machinery caused many maintenance problems, greatly reducing the amount of time the machine could be used. In addition, work around pilings and piers is necessarily slow. Under good operating conditions in a reasonably heavy growth area, it was estimated that the cost of harvesting would be about $30 per acre.

Regrowth occurred after cutting, but at widely varying rates. Some cut areas remained open to boating all summer; others closed again in about 2 weeks. Redhead-grass (*Potamogeton perfoliatus*) was sometimes found to replace milfoil following cutting of the milfoil.

WINTER PARK, FLORIDA

The city of Winter Park, Florida, has operated an aquatic harvester on its lakes since the fall of 1963. According to Blanchard (1966), herbicides had not proved wholly effective against Naiad (Najas) and Elodea (Anacharis), and eel grass (Vallisneria) had "proved to be quite resistant to most

*Euransian Watermilfoil—a Rapidly Spreading Pest Plant in Eastern Waters; presented at Northeast Wildlife Conference, Halifax, Nova Scotia (1961).

herbicides." He reports that harvesting rates have been rather slow for a number of reasons:

The difficulty in reaching shorelines makes it necessary for the transport barge to travel long distances to dumping areas.

The harvester is required to do an unusual amount of maneuvering near shorelines and around piers and water pipes.

Harvester operators must cope with fairly frequent mechanical problems.

The combination of equipment-moving time, travel time on the lakes, and servicing for normal maintenance and breakdowns has resulted in a utilization factor of about 65 percent. Operating costs were estimated at $35.56 per acre. Blanchard reports that weeds in deeper water are slower to renew growth after cutting than those near shore and that an effective combination may be use of the harvester away from shore and use of the more-promising herbicides in inaccessible shoreline areas.

JAMESTOWN, NEW YORK

For the past several years, aquatic harvesting machinery has been used for alleviating nuisance conditions caused by submersed aquatics in Chautauqua Lake, the primary problem species being milfoil and water weed (*Elodea canadensis*). Two harvesters and several transport barges are used. H. F. Meese* estimates the cost of operation at $50 to $70 per acre for harvesting and states that mechanical harvesting is difficult and expensive in Chautauqua Lake. He believes that the cost of using chemical controls would be less than $40 per acre, although water clarity might be reduced. He reports that the stalks of perennial plants become quite rugged after cutting and are a nuisance in swimming areas. Apparently no papers or reports have been written on mechanical harvesting operations in Chautauqua Lake.

NEW YORK CITY

The Bureau of Water Supply, New York City, operated aquatic harvesting machinery in its Muscoot reservoir during 1964, 1965, and 1966. In 1965 it removed an estimated 1,351 tons of weeds in 76 working days (total coverage of perhaps 40 acres). A. Groopman (personal communication) reports the plants are primarily milfoil and coontail, with some Elodea. Addition of a second transport barge, purchased in late 1965, "definitely increased productivity of the operation." Shallow water, frequent underwater

*Personal communication. Mr. Meese is head of the Intersciences Branch, Cornell Aeronautical Laboratory, Cheektowaga, New York.

obstructions, and difficult access prevent an efficient operation in the reservoir. Difficulties in procuring spare parts also slowed operations.

WISCONSIN

A number of communities in Wisconsin utilize mechanical harvesting or control measures alone and in combination with chemical controls. The city of Madison has been using both methods for more than 40 years. In July 1965, it acquired a harvester and transport barge for use in cutting and removing submersed vegetation. In 1966, 267 tons of wet plants were removed from the Madison lakes with this machinery. The city has a number of barge-mounted cutters that are used to supplement the harvester operations. In 1966, cleanup teams removed 1,087 tons of weeds and debris from shorelines (T. W. Burnham, unpublished data).* This activity was in addition to weed removal by harvester crews. As with most lakes where access is limited, disposing of the plants is a major problem.

Other communities on Lake Mendota usually have their own crews remove weeds on the shoreline. Occasionally they have the work done by crews from Madison. The village of Shorewood Hills has acquired a barge-mounted cutter and a transport barge mounted on pontoons and driven by an outboard engine (H. S. Roth, unpublished report).† In 1966, weeds were cut during periods of on-shore winds and were handforked from the water onto two chain-link "baskets," 8 ft by 10 ft, laid across the transport barge. When the baskets were filled, the barge was moved alongside the dock. A tractor with a front-end loader was used to move the basketfuls of weeds from the barge into a waiting truck. Roth concludes that the program was a success in 1966, although improvements in handling can reduce costs. Labor costs were only $468 for cutting and collecting from an estimated 40 acres of lake.

Operations on Pewaukee Lake also have been extensive for many years. Since this is a highly developed lake, and both chemical and mechanical methods have been widely used for shoreline work, sometimes simultaneously, little specific information is available for the mechanical-removal work. The fact, however, that the Pewaukee Lake Sanitary District recently acquired a harvester speaks well for the success of mechanical methods.

The village of Rib-Lake acquired an aquatic harvester in 1963 and has successfully operated it on their lake since that time (E. Juse, personal communication). The lake covers 324 acres and has a maximum depth of about 9 ft. Weed-removal work has been necessary on about one third of the

*Weed Harvesting—1966. Report to Department of Public Health, Madison, Wisconsin (1966).

†Annual Report on Lake Weed Cutting. Village of Shorewood Hills, Wisconsin (1967).

area. Costs, exclusive of depreciation, for 1964–1966 are interesting: 1964, $2,154; 1965, $1,825; 1966, $259. Weed growth in 1966 was minor, which accounts for the low cost for the control program. Whether this minor growth was a result of heavy harvesting in 1964 and 1965 or of weather conditions in 1966 is not known.

A number of other communities in Wisconsin have acquired harvesting units and others are considering doing so. Among those that have harvesters are Oconomowoc and Rice Lake. Many other communities use cutters. The Mid-Lake Improvement Association (Minoqua) uses a commercial reciprocating cutter. The cut weeds blow ashore and are removed by persons living along the shore (O. Wergen, personal communication). Assistance is available for property owners who are unable to handle the accumulations. The community shifted to mechanical methods after using arsenic compounds unsuccessfully for many years.

Sturgeon Bay uses a large, "V"-shaped, scythe-type cutter, which is mounted on the bow of a pontoon barge. The device, driven through the water at 5 to 8 mph, is said to cut 2 to 2½ acres per mile of travel. Large areas can be cut quickly, but the usual cleanup problems occur along shorelines. In general, cleanup is left to residents. Weeds are raked onto the beach and, after they have dried, are burned. The operation is considered a success, although complaints have been made about the accumulations of weeds along the shorelines. The village carried out collection work in 1966.

RECENT RESEARCH

Many research projects are in progress that are related, directly or indirectly, to removal or control of aquatic vegetation by mechanical means.

Studies in plant ecology are being conducted to establish more firmly the relations between aquatic plants and their environments, the effects of various forms of control measures on the plants that are cut or treated, and the types of plant successions that follow removal or control measures. C. J. Antonie (unpublished data)* has handcut various rooted plants in Fish Lake, Wisconsin, for 2 years, studying the response of the plants. In general, several deep cuttings seem to reduce drastically the plant populations that appear the following season. Antonie reports, however, that Chara appears to be moving into the areas vacated by the vascular plants. C. T. Lind (unpublished data)† has classified the large aquatic plant population that was present in University

---

*Preliminary Observations on the Effect of Cutting Aquatic Plants. Univ. Wisconsin, Madison (1966).
†The Submerged Vegetation of University Bay. M.S. thesis, Univ. Wisconsin, Madison (1967).

Bay of Lake Mendota during 1966, and has estimated the standing crop of large aquatics that was present at various times during the growing season. Not surprisingly, *Myriophyllum* is the dominant species.

Since about 1960, several fundamental studies have been made of the relations between nutrient supplies available in the water and the content of various elements in the plants (e.g., Gerloff and Krombholz, 1966). The relations between various nutrients in the water and algae growth have also been studied (Gerloff and Skoog, 1957). Few reports have been made on the effects that mechanical methods of control or removal have on submersed aquatics when these methods are applied on a large scale. Work in controlling plants in irrigation and drainage ditches and in canals has been reported, but most of this work seems to be oriented toward management rather than research (V. F. Bruns, unpublished data*; Hodgson *et al.,* 1962; Stephens *et al.,* 1963; Timmons, 1960; Timmons and Hodgson, 1962).

Some small-scale harvesting has been done to obtain samples of plants so that their usefulness for various purposes could be determined. Usually, such harvesting was done in the hope of finding uses for the plants that would help to alleviate the disposal problem associated with mechanical removal. Apparently there has been no report on work in which nutrient removal was a major objective.

Extensive studies at Caddo Lake (Texas) have given fairly encouraging results regarding the possibility of using dehydrated plants as animal feeds. Bailey (1965) evaluates this work from the point of view of a processor interested in a new feed product. He also gives interesting data on regrowth rates, observing that milfoil regrowth almost crowded out other species where first cuttings involved more than one species. Unfortunately, the xanthophyll content of the plants did not stay high enough throughout the growing season for the plants to be commercially exploited. Cregar *et al.* (1963) reported on poultry-feeding experiments carried on at Texas A & M University in connection with the Caddo Lake studies. H. F. Stern (unpublished report)† and Lange (1965) gave additional data and background on this rather extensive project, which was sponsored by the Area Redevelopment Administration of the U.S. Department of Commerce. Additional reports may be forthcoming from Texas A & M on their stock-feeding experiments with the Caddo Lake plants. An apparent difficulty in commercial development of products from lake plants is that management for recreational use of the lake could well be different from management for optimum harvesting.

Submersed Aquatic Weed Control in Irrigation Systems. Presented at Washington State Weed Conference, 1965.

†*Feasibility Study Report on Harvesting and Processing of Aquatic Plants from Caddo Lake.* Report to Administrator, Area Redevelopment Administration, U.S. Dept. Commerce, Washington, D.C. (1964).

Analyses of milfoil and other aquatic vegetation in dehydrated form usually indicate a protein content comparable with that of alfalfa hay (Anderson *et al.,* 1965; Bailey, 1965; Hasler, 1963). Tests on three steers at the University of Maryland indicated that the nutritive value of milfoil was about the same as high-grade hay. Pelleted milfoil was eagerly eaten by the animals.

H. J. Elser* reports tests showing that milfoil is promising as a mulch-fertilizer. The NPK analyses for milfoil are about twice as high as for dried cow manure but, of course, much lower than for commercial fertilizers. The high water content of the plants "as received" would probably make harvesting them uneconomical in a commercial sense. But the nutrient and fertilizer value are considerable, and it appears that the plants might be converted into products for which there would be a modest demand. The demand might at least be sufficient to facilitate disposing of harvested plants.

Van Breedveld (1966) reports some very promising results on the use of seagrass as a mulch-fertilizer in growing tomatoes and strawberries in experimental plots in St. Petersburg, Florida. Beds fertilized with seagrass produced larger yields of both fruits than did beds treated with a compost or a commercial fertilizer (6-6-4).

Research on methods and machinery for harvesting or controlling larger aquatic vegetation by mechanical means is apparently quite limited. Manufacturers and users are modifying and improving existing harvesters and cutters, but no really large-scale efforts are being made. Where a crop of economic value is available, special machinery and techniques have been more highly developed—for example, those developed by the reed-growing and reed-processing industry in the Danube River Delta area in Rumania (Rudesco *et al.,* 1965). Methods for removing algae from municipal water supplies have received considerable development as needs have arisen, and microstrainers of large capacity are in use (Berry, 1961). The Rex centrifugal brush-type filter appears to show promise for algae removal. T. Rudolph and T. Dewitt (unpublished report)† state that this type of system may combine the features of centrifugation and filtration to give a high degree of self-cleaning and reduced power requirements. Their paper includes a literature survey on algae-removal methods.

Status of Aquatic Weed Problems of Tidewater, Maryland, Spring 1966. Report to Maryland Department of Chesapeake Bay Affairs, Annapolis, Maryland (1966).

†*Design of a Bench Scale Apparatus for Investigating the Separation of Solids from Liquids Using the Centrifugal Filtration Principle.* Water Resources Center, Univ. Wisconsin, Madison (1966).

SUGGESTIONS FOR FURTHER RESEARCH AND STUDY

Certainly the most fundamental areas in which a great deal of additional information is required are those dealing with plant ecology in relation to control measures of all types, nutrient supplies and utilization, and the over-all ecology of a given waterway, including tributary marshes, as it affects water quality. Studies in these areas will determine to a large extent the most fruitful direction for development of improved methods. The interrelations between plant species are often a matter of speculation. In 1967, A. Groopman (personal communication) reported that algae counts in the Muscoot reservoir appeared to be reduced downstream from areas in which mechanical harvesting was being done. H. J. Elser (personal communication) reports that, in one instance, *Cladophora* grew explosively in 1966 after successful chemical treatment of milfoil beds at the top of Middle River, an inlet of the Chesapeake Bay, and became a much worse pest than the milfoil had been. Would the same result have occurred if the milfoil had been mechanically controlled or harvested?

Although combinations of chemical controls in restricted areas combined with mechanical methods in open areas are used on a limited scale, much additional information is needed to arrive at optimum combinations. Selective herbicides might be used to inhibit undesirable species, and mechanical methods might be used to harvest desirable species. The purpose of harvesting might be to prevent the favored plants from becoming a nuisance, to remove nutrients, or to obtain a crop. A more intensive effort should be made to find uses for harvested plants.

Although mechanical methods have been used widely for many years, little information about their effects on plants is to be found in the literature or in government records. Since studies have indicated that cutting or harvesting can have marked effects on plant populations and distributions, it would seem desirable to require permits for extensive cutting and to require that records be maintained showing the areas cut, the machinery used, and the effects on plants in the area. (The state of Minnesota does require a permit for extensive cutting.) That is, where either chemical or mechanical methods are followed, some scientific evaluation should be made of the problem and of the effects of control procedures.

The practicality of including aerial photographs in records should be considered. Such photographs might be a means of spotting weed beds and discovering inroads of eutrophication. J. P. Scherz (unpublished data)* is one of several workers who have obtained interesting results with aerial

Aerial Photographic Techniques in Pollution Detection. M.S. thesis, Univ. Wisconsin, Madison (1967).

photography. By studying photographs of water, one can detect conditions that are undetectable from surface observations.

Further productivity studies might be carried out to determine maximum harvests for different species and the amounts of nutrients that can be obtained from them. Bailey (1965) indicates that the standing crop may represent only a small part of the potential harvest.

Attention should be given to improving mechanical harvesting and control systems through research and development. We have equipment that can rapidly cut large, open areas, but we lack satisfactory means of collecting the cut plants. Mechanical harvesters, although slower, can cover large areas, both cutting and collecting the plants. Larger and more efficient units undoubtedly can be built. The real bottleneck in harvesting systems is transportation of harvested plants to remote disposal areas. This transportation usually involves transfer from the harvester to a transport barge, travel to an unloading area, transfer to a truck, and delivery to a suitable disposal area.

By drawing on experience gained in mechanically destroying water hyacinths, we might devise machines that could destroy plants over large areas and return the residues to the water, where they would sink and decompose. The method would not remove nutrients—a disadvantage. But neither would it add chemicals to the water. Since transportation and handling would be largely eliminated, the costs of such a method might well be less than for chemical control. Systems might be developed for depositing the plant residues in certain locations in a lake, from which they could be pumped periodically.

Another procedure that might eliminate much of the transportation problem is the "bottomless-pit" technique reported by Elser for water-chestnut work. Experiments would have to determine how effective this scheme might be for other plants. Enclosed bins or tanks might be placed in the lake to receive large volumes of cut plants and to provide for their decomposition. These tanks could be pumped out periodically. The concentrated nutrients might make an excellent fertilizer.

If nutrient removal is no object, presses or rolls to reduce the volume and water content of the cut plants would seem appropriate. This reduction of bulk would greatly reduce the handling and disposal problems inherent in present equipment, although it would add minor complications and increase power requirements for the harvester.

Shoreline cleanup work is another area where improved mechanical-handling systems are needed. Even in lakes where weeds are not cut, large volumes of weeds, dead fish, algae concentrations, and other debris are often driven onto the shore, into the shallow water, and around docks and piers. Cleaning up such materials is a continuous task. Efficient machinery to supplement or replace hand labor is badly needed.

CONCLUSIONS

While modern machinery is quite effective for removing organic lake materials (primarily, rooted, submersed vegetation), truly large-scale, economical systems do not exist. It seems apparent that with today's technology much more efficient systems could be built—systems that might operate in large areas at a small part of the cost of chemical controls. However, recent successes with herbicides in many areas have made most large manufacturers reluctant to move into the mechanical-harvester field. More fundamental data are needed from ecologists before optimum management techniques can be properly evaluated. Mackenthun *et al.* (1964) state the case for mechanical harvesting very well, concluding that "ultimately, harvesting techniques that are effective, feasible, and financially practical must be perfected."

The authors wish to express appreciation to the many persons who generously supplied reference material and personal comments and evaluations essential to the preparation of this paper. They represent a wide variety of groups and agencies concerned with different aspects of eutrophication.

REFERENCES

Alsopp, W. H. L. 1966. The manatee: Ecology and use for weed control. Nature 188:762.

Anderson, R. R., G. B. Russell, and R. D. Rappleye. 1965. Mineral composition of Eurasian water milfoil, *Myriophyllum spicatum* L. Chesapeake Science 6:68–72.

Bailey, T. A. 1965. Commercial possibilities of dehydrated aquatic plants. Southern Weed Conf., Proc. 18:543–551.

Bartch, A. F. 1954. Practical methods for control of algae and water weeds. U.S. Public Health Service. Public Health Reports 69:749–757.

Berry, A. E. 1961. Removal of algae by microstrainers. J. Amer. Water Works Ass. 53:1503–1508.

Birge, E. A., and C. Juday. 1922. The inland lakes of Wisconsin: The plankton. I. Its quantity and chemical composition. Wis. Geol. Nat. Hist. Surv. Bull. 64.

Blanchard, J. L. 1966. Aquatic weed harvester operational report. Southern Weed Conf., Proc. 18:477–479.

Burbank, J. H. 1963. An evaluation of the aquatic pest plant control program at Reelfoot Lake. J. Tennessee Acad. Sci. 38:42–48.

Cregar, C. R., F. M. Farr, E. Castro, and J. R. Couch. 1963. The pigmenting value of aquatic flowering plants. Poultry Sci. 42:1262.

Crouse, W. A. 1941. An automatic weeding machine. Water Works Eng. 94:1054.

Domagalla, B. P. 1926. Treatment of algae and weeds in lakes at Madison, Wisconsin. Eng. News Record Dec. 9:3–7.

Elser, H. J. 1966a. Control of water chestnut by machine in Maryland, 1964–1965. Northeastern Weed Control Conf., Proc. 20:682–687.

Elser, H. J. 1966b. How Maryland uses "manatee" to cut waterchestnut. Weeds and Turf. November:10–13.

Frye, J. 1963. The milfoil battle of Glebe Creek. National Fisherman & Maine Coast Fisherman. March:33.

Gerloff, G. C., and F. Skoog. 1957. Nitrogen as a limiting factor for the growth of *Microcystis aeruginosa* in southern Wisconsin lakes. Ecology 38:556–561.

Gerloff, G. C., and P. H. Krombholz. 1966. Tissue analysis as a measure of nutrient availability for the growth of angiosperm aquatic plants. Limnol. Oceanogr. 11:529–537.

Harrison, D. S. 1964. Aquatic weed control. Circular 219A. Univ. Florida, Gainesville. 16 p.

Hasler, A. D. 1963. Wisconsin, 1940–1961, p. 55–94 *In* D. G. Frey [ed] Limnology in North America. Univ. Wisconsin Press, Madison.

Hasler, A. D., and E. Jones. 1949. Demonstration of the antagonistic action of large aquatic plants on algae and rotifers. Ecology 30:359–364.

Hiltibran, R. C. 1962. Duration of toxicity of endothal in water. Weeds 10:17–19.

Hiltibran, R. C. 1963. Effect of endothal on aquatic plants. Weeds 11:256–257.

Hodgson, J. M., V. F. Bruns, F. L. Timmons, W. O. Lee, L. W. Weldon, and R. R. Yeo. 1962. Control of certain ditchbank weeds on irrigation systems. U.S. Dep. Agr. Production Res. Rep. 60. 64 p.

Jackson, P. 1957. Harvesting machinery for brown sublittoral seaweeds. The Engineer (London) 203:400–402, 439–441.

Klussmann, W. G., and F. G. Lowman. [n.d.]. Common aquatic plants; identification, control. Bulletin B-1018. Texas A & M Univ., College Station, Texas. 16 p.

Lange, S. R. 1965. The control of aquatic plants by commercial harvesting, processing, and marketing. Southern Weed Conf., Proc. 18:536–542.

Lapham, V. T. 1966a. Control of alligatorweed with picloram. Southern Weed Conf., Proc. 19:409–413.

Lapham, V. T. 1966b. The effectiveness of some dimethylsulfoxide herbicide combinations. Southern Weed Conf., Proc. 19:438–442.

Livermore, D. F. 1954. Harvesting underwater weeds. Water Works Eng. 107:118–121.

Mackensie, W. 1947. Seaweed harvesting methods. The Engineer (London) 184:337, 373, 387.

Mackenthun, K. M. 1950. Aquatic weed control with sodium arsenite. Sewage Industr. Wastes 22:1062–1066.

Mackenthun, K. M. 1960. What you should know about algae control. Public Works 91:114–117.

Mackenthun, K. M. 1962. A review of algae, lake weeds, and nutrients. J. Water Pollut. Control Fed. 34:1077–1085.

Mackenthun, K. M., W. M. Ingram, and R. Porges. 1964. Limnological aspects of recreational lakes. U.S. Public Health Service Publ. 1167. U.S. Govt. Printing Office, Washington, D.C. 176 p.

Meyer, F. A. 1964. Aquatic plant control. California Dep. Fish and Game, Inland Fisheries Admin. Rep. 64-2. 32 p.

Michigan Department of Conservation. 1964. Aquatic weeds and their control in Michigan. Revised ed. Michigan Dep. Conserv., Lansing. 31 p.

Miles, W. D. 1965. A history of Deeping Fen and Pode Hole pumping station. Deeping Fen, Spalding and Pinchbeck Internal Drainage Board, Spalding, England. 60 p.

Minnesota Department of Conservation. 1966. Control of aquatic vegetation. Minnesota Dep. Conserv. Inform. Leaflet 6. St. Paul. 4 p.

New York State Department of Health. 1964. Rules and regulations governing the uses of chemicals for the control and elimination of aquatic vegetation. New York State Dep. Health, Albany. 14 p.

Oglesby, R. T., and W. T. Edmondson. 1966. Control of eutrophication. J. Water Pollut. Control Fed. 38:1452–1460.

Philippy, C. L., and J. E. Burgess. 1966. Some observations on the use of concentrated sulfuric acid for control of *Elodea*. Southern Weed Conf., Proc. 19:480–490.

Rickett, H. W. 1922. A quantitative study of the larger aquatic plants of Lake Mendota. Wisconsin Acad. Sci., Arts Letters, Trans. 20:501–522.

Rudesco, L., C. Riculezco, and I. P. Chivu. 1965. The monograph on reed from the Danube delta. Academy of the Socialist Republic of Rumania. 542 p.

Sawyer, C. N. 1962. Causes, effects, and control of aquatic growths. J. Water Pollut. Control Fed. 34:279–288.

Schuette, H. A., and H. Alder. 1928. Notes on the chemical composition of some of the larger plants of Lake Mendota. Wisconsin Acad. Sci., Arts, Letters, Trans. 23:249.

Schuette, H. A., and H. Alder. 1929. Notes on the chemical composition of some of the larger plants of Lake Mendota. Wisconsin Acad. Sci., Arts, Letters, Trans. 24:135.

Seaman, D. E. 1958. Aquatic weed control. Soil and Crop Sci. Soc. of Florida, Proc. 18:210–214.

Seaman, D. E., and W. A. Porterfield. 1964. Control of aquatic weeds by the snail *Marisa cornuarietis*. Weeds 12:87–89.

Smith, G. E. 1962. Eurasian watermilfoil in TVA reservoirs. Southern Weed Conf., Proc. 15:258–264.

Speirs, J. M. 1948. Summary of literature on aquatic weed control. Canadian Fish Culturist 3:20–32.

Steel, E. W., and B. B. Ewing. 1954. Controlling water milfoil in a Texas reservoir. Public Works 85(10):89–90.

Steenis, J. H., and H. J. Elser. 1967. Waterchestnut control with mixtures of 2,4-D and dicamba. Northeastern Weed Control Conf., Proc. 21:550–551.

Steenis, J. H., V. D. Stotts, and C. R. Gillette. 1962. Observations on distribution and control of Eurasian watermilfoil in Chesapeake Bay, 1961. Northeastern Weed Control Conf., Proc. 16:442–448.

Stephens, J. C., R. D. Blackburn, D. E. Seaman, and L. W. Weldon. 1963. Flow retardance by channel weeds and their control. Amer. Soc. Civil Eng. (Irrigation and Drainage Div.), Proc. 2:31–53.

Surber, E. W. 1961. Improving sport fishing by control of aquatic weeds. U.S. Dep. Interior Circular 128. 47 p.

Swingle, H. S. 1957. Control of pond weeds by the use of herbivorous fishes. Southern Weed Conf., Proc. 10:11–17.

Sylvester, R. O., and G. C. Anderson. 1964. A lake's response to its environment. Amer. Soc. Civil Eng. (San. Eng. Div.), Proc. 1:1–22.

Timmons, F. L. 1960. Weed control in western irrigation and drainage systems. ARS 34-14. Agr. Res. Serv., U.S. Dep. Agr. 22 p.

Timmons, F. L., and R. H. Hodgson. 1962. The status and prospects in aquatic weed control. Western Weed Control Conf., Proc. 25:59–64.

Van Breedveld, J. F. 1966. Preliminary study of seagrass as a potential source of fertilizer. Special Scientific Report No. 9. Marine Laboratory, Florida Board of Conservation, St. Petersburg. 23 p.

Wisconsin Committee on Water Pollution. [n.d.]. Manual of policy and practice for aquatic nuisance control. Wisconsin Comm. Water Pollut., Madison. 5 p.

Wunderlich, W. E. 1938. Mechanical hyacinth destruction. Military Eng. 30:5–10.

Wunderlich, W. E. 1966. Aquatic vegetation control operations in Louisiana. Southern Weed Conf., Proc. 19:450–452.

J. H. BEUSCHER
University of Wisconsin, Madison

Shoreland Corridor Regulations
to Protect Lakes

This paper focuses on eutrophic inland lakes and on the protection of these lakes through legal measures regulating shoreland uses.

THE NATURAL AND CULTURAL SETTINGS

In nature these lakes have a life-span: they are young, or middle-aged, or well along toward ultimate death by suffocation. As man begins to use the land near such lakes, he may dramatically speed up the aging process. By increasing the supply of nutrients in the lake, he may cause excessive blooms of algae and rooted aquatic plants. Grown in moderate number, these aquatic plants may be beneficial. But man may bring "cultural" eutrophication; his increased contributions of nitrogen and phosphorus compounds and other materials may cause odoriferous and unsightly conditions. Dense blankets of surface growth may radically reduce the amount of light and oxygen available to subsurface species.

Chief sources of these excess nutrients are domestic and industrial wastes, urban drainage, and runoff from fertilized fields and lawns.

Legal controls that attempt to reduce or eliminate these contributions must, unfortunately, be premised on assumptions, not on solid knowledge. Typically, we do not know the level of nutrient contribution in nature for a given lake, much less man's added contribution. In general, there is a lack of data consistently collected over a long time that showed the assumed acceleration due to man's presence. Posed is the question, "Should we wait

520

with a program of shoreland-use controls until at least major gaps in knowledge have been filled?"

I want to return to this question in the next section of the paper, but here I must point out that accelerated aging is not the *only* reason for a program of shoreland regulation. Man's presence near inland lakes has brought other problems.

In an affluent society, use-pressures on shorelands have greatly increased. People by the thousands seek waterside living. By the tens of thousands they seek water-based recreation. Markets for shorelands are active; land prices are booming. But on many lakes the market has proved to be a poor regulator. Desire for maximum profits has induced division into lots that are too small, with resulting overcrowding and inadequate setbacks from the lake edge. Unimaginative site layouts have ringed lakes with structures crowded next to one another in a tight circle all around the lake. Professional site planning for lakeshore developments is almost nonexistent. Congestion often equals or exceeds that of urban neighborhoods. When the easily developed land along the immediate shore is all sold, an entrepreneur may fill the marshes to create more small lots and more congestion. Coincidentally he may destroy a fish spawning ground as well as habitat for other wildlife. Uses get mixed. Most of the structures are used for seasonal or, increasingly, for permanent residences. But dance halls, boat liveries, motels, youth camps, and other commercial uses intrude. Once the land immediately adjacent to the water is used up, similar small-lot, mixed-use development occurs in rings behind the initial ring.

Typically, there is a private sewage-disposal system for each of these crowding structures. Many septic-tank systems are installed in areas that cannot properly absorb effluents because of high groundwater or bedrock, "tight" soils, or excessive slopes. Wastes from these systems pollute wells and frequently overflow or ooze to the surface. Quite apart from lake enrichment, a direct and serious human health hazard exists.

In the process of this development, ugly boathouses and other structures intrude into the lake, shore cover is removed, banks are bulldozed and leveled into the lake, artificial lagoons are dredged in nearby marshes and then connected to the lake to provide more "waterside" development. Natural beauty and man-made amenities are destroyed. The very values that attracted men to the lake in the first place are sacrificed.

There is now abroad in this country a growing and deep-rooted concern about the quality of the environment in which men live and in which they seek recreation. There is abroad a concern that resource management decisions are being made by default—decisions that may be producing a maze of environmental back alleys from which even our vaunted technology may not be able to rescue us. Accompanying this concern is an increased sensitivity to outdoor amenities, the quality of the landscape, and scenic views. People with little or no background in the sciences are demanding action. They want our lakes and our lake shorelands protected from the abuses I have listed. Some are property owners whose lakeshore property values are threatened by these abuses. Thousands more, though not shore owners, are members of the public who use the lake and claim a "public right" to use a relatively clean lake in a relatively undisturbed setting.

In Wisconsin we have felt a groundswell of pressure to do something to protect our shorelands and thereby protect our lakes. In the impressive Water Resources Act passed in June 1966, the Wisconsin legislature responded by adding some important sections on shoreland protection to a comprehensive law on water-pollution abatement.

I want to describe this shoreland law and what is being done to implement it. But first let me make two general points and then generally describe the legal setting for a program of shoreland management and regulation.

The first point is this: There is no sign that normal market forces will solve our shoreland problems; rather, they seem to accentuate them.

The second point is that patient educational efforts and programs of voluntary action are vital; no shoreland program can succeed without them. But for those who use shorelands wisely and with sensitivity to be protected from those who do not, legal controls are required.

THE LEGAL SETTING

As background for an understanding of shoreland regulation generally, and Wisconsin's new program specifically, I want now to do the following:

- Take a lawyer's look at a lake, at its bed, at its surrounding shorelands.
- Make some comments about public interest in private property and about the importance of showing by factual studies and analyses the need for the shoreland regulatory program.

A LAWYER'S LOOK AT A LAKE

A lawyer sees a lake as an area of open water surrounded by land that is divided into numerous tracts to which attach the private-property rights of the respective owners. In short, he notes that shoreland, like other land, has been allocated through the institution of private property. He will, in many places, also observe that because the land is touched by the lake, special privileges of access and use accrue to the shoreland owners. In most places, members of the public do not have the right to cross this private land to get to the water. Increasingly, however, the state, by becoming the owner of a shoreland parcel, makes the lake available to the public.

A lawyer also wonders about "ownership" of the bed of the lake. In some places, for so-called nonnavigable lakes, bed ownership will be in the shoreland proprietors, with each owning his pie-shaped piece of the bottom. In other places the state is said to own the bed of the lake. One lake is then called "private" and the other "public."

But a lawyer will know that even where the lake is "private," the state has a reserved power reasonably to regulate both the use of the shoreland and the use of the water itself.

Probably, however, a lawyer will find it easier to justify really intensive shoreland or water-use regulation where the lake is conceived by the applicable law to be public. In other words, the intensity of regulation permitted for particular shorelands may depend on whether local law treats the lake as private or as public. Where, as in Wisconsin, virtually all natural lakes are legally public, a lawyer will explain the resulting regulatory situation in property-law terms. He will say that the state holds the lake "in trust" for the people and that the public as "beneficiaries" of this trust have "public rights" to use the lake for navigation, recreational boating, swimming, hunting, fishing, or just to enjoy the natural scenery. It is easy to conceive of this "public trust" and of these "public rights" as burdening privately owned shoreland so as to protect against uses that would interfere with the exercise of public rights. Thus, for example, Wisconsin's shoreland-protection law expressly declares that it was passed "to aid in the fulfillment of the state's role of trustee of its navigable waters." This probably means that Wisconsin can go further in regulating shoreland uses than can a state where the lake is conceived to be private. Or, put another way, some states may have to purchase shoreland owners' use privileges, whereas in Wisconsin such privileges could be limited by regulation, which would not require payment.

PUBLIC INTEREST IN PRIVATE PROPERTY

The shoreland owner, whether on a private or on a public lake, has "private property" in his land. He has *rights* to keep others off, *powers* to dispose of that which he owns, and *privileges* to use his land in a wide variety of ways. Valid land-use regulations reduce the number of use privileges available to him. If as a practical matter the use regulations reduce the owner's privileges to zero, so that he can make no use at all of his land, then almost always these regulations are invalid. This is considered a "taking" of the owner's private property, and compensation must be paid.

Nevertheless, in appropriate cases, public regulations validly may, within a substantial latitude, actually deprive private owners of many, or even of most, of their privileges of use. It is frequently erroneously assumed that the institution of private property in this country has as its sole function the preservation of the status quo. Much in our history demonstrates, on the contrary, that it has been a malleable instrument for dynamic growth in a free-enterprise system. Our local, state, and federal legislatures and courts are constantly remolding and reshaping "private property." So-called "vested" property rights frequently have been pushed aside in the interests of present and long-term returns in social function.

But suppose a regulation takes some but not all of our owners' privileges of use? How many privileges of use can be taken before the brink of invalidity is reached? When is a regulation "reasonable" and therefore valid? When is it "unreasonable" and therefore invalid? The regulation may, for example, limit the owner to residential use and forbid commercial or industrial use. This will be upheld, assuming that studies demonstrate sound reasons for the limitation. But suppose that shoreland control over privately owned wetland forbids filling, draining, building, or any other kinds of development. Is this conservancy control valid?

Such questions must be approached in terms of a particular tract of land on a particular lake in a specific environmental setting. Some of the variables to be considered are as follows:

What uses can the owner, as a practical matter, actually make of his land, and what kind of return can he expect from these uses? What is the impact of the regulation upon the market value of the land?

Is the tract so situated that its development will directly and substantially damage others, as in the case where wetlands are an essential storage area for flood waters?

Can the financial burdens imposed on the owner by the regulation be justified by a clear factual demonstration of (1) public need for the regulation and (2) the responsiveness of the regulation to this need?

This last item indicates the importance of facts and of analysis of their implications. A good lawyer defending a shoreland regulation against constitutional attack is much more interested in a solid record of research findings, measurements, and impact analysis for the particular lake or for similar lakes than he is in black-letter quotations from constitutions or court cases.

There is, then, a vital need for close teamwork between a limnologist, a soils specialist, a biologist, a geologist, a fish and game specialist, an engineer, a land-use planner, and a lawyer in the development of a program of shoreland regulations. In preparing his pioneering County Shoreland Protection Ordinance, Douglas Yanggen, of the University of Wisconsin's Extension Division, recently demonstrated how to use such a team of specialists.

But, as indicated earlier, specialists will sometimes have to say, "In the present state of knowledge, we just cannot be sure." Or, "Reliable data are lacking from which to draw a conclusive inference." The law is quite accustomed to dealing with a state of imperfect knowledge. To wait for perfect knowledge may be to wait forever, or at least until it is too late for effective action. A few years ago the marketing of tens of thousands of pounds of cranberries was forbidden by the U.S. government, to the consternation and grave disruption of the cranberry industry. There was no firm scientific proof that the herbicide used by the growers, in the quantities involved, would actually induce cancer in humans. But research had gone far enough to establish an "indicator of danger," and so the drastic control was imposed and was undoubtedly valid.

The same kind of approach has to be taken for shoreland controls. We have to identify "indicators of danger" and then act on the basis of imperfect knowledge. But our programs have to be flexible enough to enable us to make substantive adjustments in regulations if later research proves that the indicator was wrong.

WISCONSIN'S SHORELAND PROTECTION PROGRAM

Wisconsin's shoreland-enabling legislation is a part of a much larger legislative packet addressed to problems of water quality and water regulation in general. Before we had this statute, "water pollution control" had come to mean, in this state as in others, action at the state level to abate the discharge of wastes into streams (or into lakes that happened to be parts of stream systems) by municipalities, industries, and institutions.

Except for this abatement of waste discharges, regulation of shorelands on Wisconsin's 8,000 lakes was left to local units of government through their zoning, subdivision control, and sanitation control powers. This was a substantial dispersion of power among the 1,908 local governmental units: 72 counties, 1,270 towns, 184 cities, and 382 villages. As you might guess, results in achieving protection of lake shoreland were extremely spotty. There was some effective local action, some ineffective action, and for thousands of lakes, no action at all.

The shoreland-protection part of Wisconsin's 1966 Water Resources Act (Chapter 614, Laws 1965–1966) applies to stream shorelands as well as lake shorelands. Here I am focusing on lakes, but I should at least indicate that corridors 300 ft wide on each side of every navigable stream (or the width of the floodplain, if it is greater) were established. Corridors around natural lakes were set back 1,000 ft from the normal high-water mark of the lake. The program of control that I am about to describe applies not only to lake-shoreland corridors but also to stream corridors, including floodplains.

For the sake of simplicity, the rest of what I have to say about Wisconsin's shoreland program will apply to lake shorelands in counties. I am leaving out the special problems of lands in cities or villages.

Counties are given a mandate to have shoreland regulations in effective operation by January 1, 1968. If a county fails, the state Department of Natural Resources, Division of Resource Development, "shall adopt" such regulations, which are then to be locally administered under state supervision. The Division of Resource Development is instructed to give technical assistance to the counties and to establish standards and criteria for effective shoreland regulations.

Personnel of the Division of Resource Development have been active in working with Wisconsin's lake counties. As I have mentioned, Mr. Yanggen, who had technical assistance representing many disciplines, prepared a shoreland ordinance that can be used as a starting point for county action. The ordinance is unique in that in one legislative package it includes the following:

- Zoning controls affecting, for example, setbacks from the lake, lot size, and use
- Subdivision controls regulating the creation and platting of lots for sale or building development
- Sanitation controls applicable to private sewage-disposal systems
- Tree-cutting restrictions for trees along the lake shore
- Controls on filling, grading, and creating artificial waterways

Lands within the 1,000-ft corridor shown on the U.S. Geological Survey map to be wetlands are placed in a conservancy zone; filling, dredging, and most kinds of building are forbidden. In appropriate cases, the harshness of this as it applies to some wetland owners may be alleviated through the conditional-use permit.

The nonwetland areas within the corridor are zoned for recreational–residential uses, with some commercial uses permitted subject to specified standards through conditional-use permits.

Admittedly, the ordinance is a primitive first attempt to get at sets of complicated problems on thousands of lakes. Although the attempt is premised on technical advice, much work needs to be done in specific localities on specific lakes. A major impediment to progress is a shortage of personnel to handle technical aspects of the work. Moreover, as ordinances are adopted, administrators will need some technical know-how and background, presently lacking.

A copy of this innovating ordinance and the explanatory manual that accompanies it may be obtained from Mr. Douglas Yanggen, Extension Land

Use Planning Specialist, Department of Agricultural Economics, University of Wisconsin, Madison, Wisconsin 53706.

LOOKING AHEAD

A program such as the one I have just sketched has implicit in it demands on the future that, if they are not met, will spell failure.

1. There must be established close liaison between the research scientist and those administering the shoreland program. The program has to keep abreast of new knowledge. This is not just a one-way street. The administrators will be feeding back policy-oriented questions and issues to challenge the researcher.

2. Additional technically competent staff qualified to work on the special problems of shoreland uses are sorely needed. Federal and state technical personnel now in our counties may have to be freed from some of their regular duties for the time being to help get the shoreland program under way.

3. We must give more attention to a carefully worked-out program integrating shoreland regulations with water-use regulations, as such. This is especially true of boating, pier building, water level, and control of noxious aquatic growth.

4. We have to continue to develop an effective, coordinated working relationship between state-agency technicians and local officials and administrators. Shoreland ordinances are complicated pieces of legislation. Local officials must understand them and be provided with the reasons why private owners are being restricted. These officials in turn have vital educational functions to perform in explaining the ordinance and its purpose to local people. Here we have a challenge to create a working partnership, in the best tradition of our democracy, between state and local officials.

5. We have to organize resource technical teams to move from lake to lake, adjusting the program to the special needs of the lake resource. Each shore is unique; it merits special study and planning attention. And these land-use plans oriented to shore corridors have to be carefully fitted and meshed into larger county or regional plans.

6. It may be that as we move ahead we will have to consider easement purchases, or tax subsidies or exemptions, or other means of compensating those owners whose lands are placed in conservancy, nondevelopment zones.

7. Particularly for larger lakes that receive stream inflows, the shoreland program must be viewed as a part of, and fitted into, the state's total water-quality program.

8. The special and difficult problems of pollutants from urban hard surfaces and from fertilized fields and lawns have barely been touched. They need attention.

VI.

CONTRIBUTIONS

TO SCIENCE

FROM THE STUDY

OF EUTROPHICATION

F. E. J. FRY
University of Toronto, Ontario, Canada

Some Possible Physiological Stresses
Induced by Eutrophication

Because of our general lack of understanding of physiological stresses induced by eutrophication, only one aspect will be discussed: stress occasioned by the reduction in the oxygen content of the water. I believe that this is the most pressing source of stress for fishes in a eutrophic lake and that almost all other stresses are incidental to, or aggravated by, that primary one. There are some notable exceptions. Specific toxins are associated with certain algal blooms, for example, and there are cases of too much oxygen. Where there is a heavy growth of rooted aquatics, the water may become oversaturated with oxygen. Incidental to this oversaturation, a very high pH value may occur, which can be fatal to fish.

There are two types of fluctuation of oxygen content: long-term, such as takes place in the hypolimnion and under snow-covered ice; and short-term, which takes place in the phototrophic zone. Although the latter usually occurs daily, an oxygen deficit may be cumulative over several days during cloudy weather.

In stating that the reduction of oxygen is the primary cause of stress, I recognize, of course, that other harmful chemical changes take place concomitantly. In general, however, such harmful effects are aggravated by the lack of oxygen, rather than the reverse. Figure 1 shows the various combinations of oxygen and carbon dioxide that asphyxiate the brook trout or restrict its activity. Since the consumption of 1 mg of oxygen under aerobic conditions produces approximately 1.5 mg of carbon dioxide, the consumption of, say, 15 mg of oxygen will be required even in the least-buffered water before the level of carbon dioxide begins to approach a borderline condition for asphyxiation, even for such a sensitive species as the brook trout. A lesser stress— for example, that which reduces activity to half its normal value (half scope) — would be brought about by a lesser increase in carbon dioxide, but it would also be brought about by a lesser decrease in oxygen. Oxygen, therefore, still has the primary role.

If fish had specific anaerobic mechanisms, there would be no such stress from oxygen as is indicated in Figure 1. We have little information about the degree to which fish can be anaerobic, but there are two remarkable cases in the literature. Bläzka (1958) reported that the crucian carp, *Carassius carassius*, can live for some months in water low in oxygen if the temperature is low. Mathur (1967) reported that one of the Indian carps, *Rasbora daniconius*, can survive a low oxygen content at tropical temperatures. The goldfish can maintain a respiratory quotient of 2 when held in water of low oxygen content (Kutty, 1968a). Thus, some fish are facultative anaerobes to a considerable extent. However, it is noteworthy that the major adaptation of fish in chronically anoxic environments is to turn to air breathing. The rainbow trout, *Salmo gairdneri*, the only other species for which the respiratory quotient has been thoroughly investigated, is extremely aerobic, as is shown in Figure 2.

Figure 3 illustrates how changes in oxygen content operate on the respiration of fish. The two boundary lines in the graph show the two limits

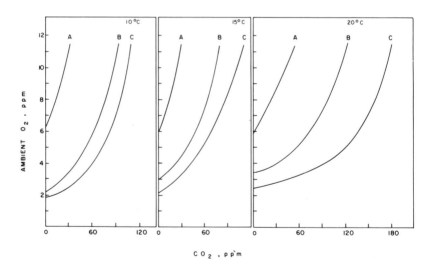

FIGURE 1   Various limiting combinations of oxygen and carbon dioxide for the eastern brook trout, *Salvelinus fontinalis*. A = half scope. B = no excess activity. C = asphyxial. "Half scope" is the ability to extract half as much oxygen for activity as can be removed at air saturation. "No excess activity" represents a level of oxygen consumption that just meets the values for standard metabolism as estimated by Job (1955). Modified from Basu (1959).

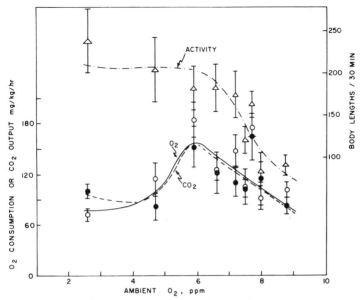

FIGURE 2   Relation of spontaneous activity and metabolism to oxygen concentration in the rainbow trout acclimated to oxygen at air saturation. Temperature $15°C$. From Kutty (1968a).

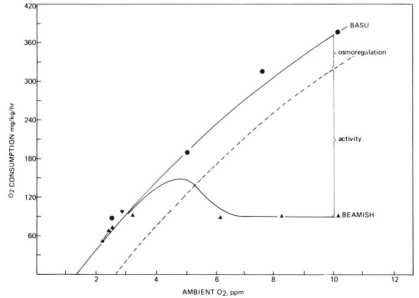

FIGURE 3   Effect of oxygen concentration on the limits of metabolism for the eastern brook trout (Beamish, 1964). The broken line is a hypothetical one to suggest the cost of osmoregulation; it is based on Rao's (1968) data for the rainbow trout.

to the rate of oxygen consumption. The upper line indicates the maximum rate at which oxygen can be consumed; the lower shows the rate for a resting fish. Over the range of concentration where the upper line in Figure 3 falls with decreasing oxygen content, oxygen operates as a limiting factor. It can be considered that the stress of a restricted oxygen supply becomes increasingly exerted from the point where the limiting effect is first apparent (the incipient limiting level). The fish responds to the stress by increased ventilation for a given rate of oxygen consumption and, at a certain level not clearly defined for any species, by increased locomotory activity. The latter response is illustrated in Figure 2. Both responses may expose the fish to further stresses.

The locomotory response to low oxygen, which may be termed the oxykinetic response, may be expected to move the organism from regions of lower to those of higher oxygen concentration even when other conditions may be unfavorable. Figure 4 illustrates the summer migration of the cisco, *Coregonus artedi*, in Lake Nipissing, Ontario. The cisco retreats to the hypolimnion in early summer, presumably because of high temperatures in the epilimnion. In late summer, as the oxygen content of the lower waters is depleted, these fish move upwards from the bottom and are finally forced through the thermocline before the fall turnover. In extreme circumstances such response to oxygen depletion undoubtedly will lead to mass mortalities. The more general response is for such fish to be narrowly restricted in their vertical range until sufficient autumn cooling has occurred.

As would be expected, the major exchange of dissolved materials and the major exchange of water take place through the gills. Two consequences arise from this circumstance. When a fish must increase the rate of irrigation of the gills to compensate for a lowered oxygen concentration, it essentially increases its exposure to any toxic substance that might also enter through the gills. Lloyd (1961) discusses this problem and gives several examples. For instance, the toxic effect of phenols and several metals is considerably greater at 30 percent air saturation of oxygen than it is at air saturation.

One of the important consequences of increased irrigation may be an increased need for osmotic regulation. The only firm data yet available for the cost of osmoregulation is for the rainbow trout (Rao, 1968). Near air saturation, the cost of osmoregulation in this species in moderately hard fresh water (Toronto tap water) at all metabolic rates is approximately 20 percent of the metabolic rate at any given level of metabolism. One of the important consequences of increased irrigation may be that the cost of osmotic regulation will pre-empt an increasingly greater fraction of the total metabolism available to the animal as the ambient oxygen concentration decreases. Such an effect is suggested, in a spirit of pure speculation, in Figure 3. It is assumed in Figure 3 that the cost of osmoregulation is the same in the

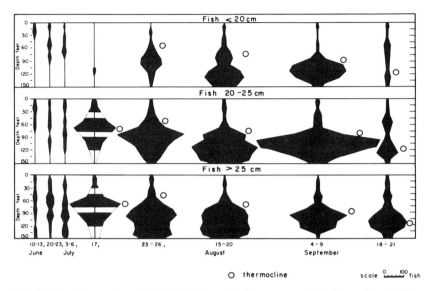

FIGURE 4   Seasonal vertical distribution of *Coregonus artedi*. From Fry (1937).

brook trout as in the rainbow trout and that the rate of irrigation is maximum and constant along the active respiration line, the uppermost line in the figure. Under these assumptions, the cost of osmoregulation for maximum activity at any level of ambient oxygen would be the same, and in the figure it is represented by the difference between the active respiration line and the broken line that lies below it. The metabolism available for maximum activity at a given ambient oxygen, then, is represented by the difference between the broken line and the lower line representing the metabolism of a motionless fish. At approximately 5 ppm of O_2 (at 15°C), the broken line and the line for standard metabolism intersect, so that if no saving feature intervened, activity would be expected to cease at an ambient oxygen not far below that point. The graph, however, is not intended to be quantitative; it is intended merely to indicate the possible order of magnitude of the osmotic effect due to increasing ventilation. However, since there are no comparative studies on the cost of osmoregulation, we have no idea whether the cost of osmoregulation as determined for the rainbow trout is typical for freshwater fish. It is well, therefore, to reiterate that what is said above is pure hypothesis.

Kutty (1968b) has demonstrated that when the swimming speed of a fish is limited by the oxygen content of the water, as was shown, for example, by Davis *et al.* (1963), the effort is limited through some form of nervous control rather than through muscular exhaustion. A fish forced to swim at a certain speed will immediately cease doing so if the oxygen content is dropped to a certain level. It will start to swim again as soon as the oxygen is raised above that level. Thus, voluntary activity is adjusted to the available oxygen supply. However, certain long-term demands can be made on oxygen consumption, for example, those of digestion and assimilation of food. Stewart *et al.* (1967) have demonstrated this phenomenon very clearly for the largemouth bass, *Micropterus salmoides*. Therefore, in eutrophic conditions where there is a large daily range of oxygen concentration, the fish may not be as well off as the mean oxygen concentration would suggest, even though the daily minimum may be well above the lethal level.

Finally, it is possible that the concentration of phytoplankton in the water, when the blooms are dense, may mechanically hinder the uptake of oxygen by fish in the manner demonstrated for pulp fines by MacLeod and Smith (1966). Such an effect, however, remains to be demonstrated for phytoplankton.

REFERENCES

Basu, S. P. 1959. Active respiration of fish in relation to ambient concentrations of oxygen and carbon dioxide. J. Fish. Res. Bd. Canada 32:408–420.

Beamish, F. W. H. 1964. Respiration of fishes with special emphasis on standard oxygen consumption. III. Influence of oxygen. Can. J. Zool. 42:355–366.

Blažka, P. 1958. The anaerobic metabolism of fish. Physiol. Zool. 31:117–128.

Davis, G. E., J. Foster, C. E. Warren, and P. Doudoroff. 1963. The influence of oxygen concentration on the swimming performance of juvenile Pacific salmon at various temperatures. Trans. Amer. Fish. Soc. 92:111–124.

Fry, F. E. J. 1937. The summer migration of the cisco, *Leucichthys artedi* (Le Sueur) in Lake Nipissing, Ontario. Univ. Toronto Stud. Biol. Ser. 44. 91 p.

Job, S. V. 1955. The oxygen consumption of *Salvelinus fontinalis*. Univ. Toronto Stud. Biol. Ser. 61. 39 p.

Kutty, M. N. 1968a. Respiratory quotients in goldfish and rainbow trout. J. Fish. Res. Bd. Canada 25:1689–1728.

Kutty, M. N. 1968b. Influence of ambient oxygen on the swimming performance of goldfish and rainbow trout. Can. J. Zool. 46:647–653.

Lloyd, R. 1961. Effect of dissolved oxygen concentrations on the toxicity of several poisons to rainbow trout (*Salmo gairdnerii* Richardson). J. Exp. Biol. 38:447–455.

MacLeod, J. C., and L. L. Smith. 1966. Effect of pulpwood fibre on oxygen consumption and swimming endurances of the fathead minnow, *Pimephales promelas*. Trans. Amer. Fish. Soc. 95:71–84.

Mathur, G. B. 1967. Anaerobic respiration in a cyprinoid fish, *Rasbora daniconius* (Ham). Nature 214:318–319.

Rao, G. M. M. 1968. Oxygen consumption of rainbow trout (*Salmo gairdnerii*) in relation to activity and salinity. Can. J. Zool. 46:781–786.

Stewart, N. E., D. L. Shumway, and P. Doudoroff. 1967. Influence of oxygen concentration on the growth of juvenile largemouth bass. J. Fish. Res. Bd. Canada 24:475–494.

G. C. GERLOFF
University of Wisconsin, Madison

Evaluating Nutrient Supplies
for the Growth of Aquatic Plants
in Natural Waters

During the last 20 to 30 years, much progress has been made in understanding the detailed nutritional requirements of algae. This progress is indicated by the various review articles on the subject (Krauss, 1958; Provasoli, 1958; Lewin and Guillard, 1963; Hutner and Provasoli, 1964; Fogg, 1965). Although there has been little work on the nutrition of angiosperm aquatic plants, the few studies carried out suggest that approaches and techniques useful with other plants will be applicable to these species (Bourn, 1932; Gerloff and Krombholz, 1966). With time and effort, progress can be made comparable with that in studies of algae nutrition.

Relatively little progress has been made in another aspect of aquatic organism nutrition: developing simple, reliable techniques for evaluating nutrient supplies to aquatic plants (both algal and higher plants) growing in natural environments. This problem is important for several reasons. First, the occurrence of organisms in specific lakes and streams and the periodicity in the blooming of certain algae undoubtedly are influenced by nutrient availability. Second, the ability to evaluate the nutritional status of a body of water by its capacity to support growth of various organisms can be extremely valuable to investigators concerned with controlling nuisance growths of aquatic organisms. These growths often are accentuated by the inorganic nutrients originating in pollution wastes. Sanitary chemists and engineers interested in developing procedures for removal of nutrients from pollution sources (Lea *et al.*, 1954) ask biologists, with justification, to specify which of the 13 essential mineral elements must be removed from pollution sources to reduce nuisance situations effectively. The answers to these questions are difficult because of the lack of procedures permitting ready assessment of nutrient supplies for the growth of aquatic plants.

This discussion will consider some of the general problems associated with (1) evaluating the availability of nutrient elements in natural waters for plant growth and (2) determining which elements might have become limiting for growth. One procedure showing considerable promise will be described in detail. Emphasis will be given to the point that research in this area has much in common with, and perhaps much to gain from, research on comparable problems in terrestrial environments. For more than 100 years, agriculturists and horticulturists have given attention to techniques for diagnosing the nutrient status of soils. During this period, hundreds of investigators and thousands of papers have been oriented toward this problem.

The eventual application of data on the nutrient status of the environment will probably be different in the two situations. In agriculture this information is the basis for increased plant growth through fertilization with the elements likely to limit plant growth; in pollution studies, this knowledge could be the basis for reducing nuisance plant growths by removal of the most likely limiting factors. Nevertheless, progress in either area might well be applicable in the other.

EVALUATING NUTRIENT AVAILABILITY
IN TERRESTRIAL AND AQUATIC ENVIRONMENTS

In comparing the evaluation of nutrient availability in terrestrial and aquatic environments, it is apparent that many of the problems of obtaining representative samples from a specific environment and of interpreting analytical data on the samples are common to the two environments. In both, for example, total concentrations of nutrient elements must be separated into forms available for plant growth and forms not available for plant growth. Also, because of element interactions, not only the absolute concentration of an element but also the concentration relative to other elements may have to be considered in the interpretation of an analysis. Because of slow but continued fixation of nitrogen or the release of elements from bottom muds in aquatic environments and from rocks and minerals in soils, the concentration of an element in any single sample may not accurately reflect availability over a period of time; a low concentration of an element is not necessarily indicative of an inadequate supply.

Some of the problems of sampling and interpretation are encountered to a much greater extent in one environment or the other. Penetration of roots to considerable depths makes it difficult to obtain soil samples representing the substrate from which land plants primarily obtain nutrients. Obtaining representative samples is even more of a problem in aquatic environments because of the high degree of independent movement of the water and organisms. There is no assurance that the water in which organisms are observed represents the substrate from which the nutrients necessary for growth were obtained. Algae may absorb nutrients from, and make most of their growth in, layers or regions of a lake quite unlike the surface layers in which they eventually appear. Many higher aquatic plants are fixed in position, as are terrestrial species, but they absorb nutrients from both the bottom muds in which they are rooted and the waters surrounding shoot portions.

PROCEDURES FOR ASSESSING NUTRIENT AVAILABILITY

I will not attempt a detailed review of the results obtained with, and the advantages and disadvantages of, the several procedures considered for evaluating nutrient supplies and detecting limiting factors for aquatic plant growth. Some aspects of these procedures have been considered by Fogg (1965) and Wetzel (1964a, 1965b). A brief comparison will be made of different types of techniques as applied to both aquatic environments and soils. These are divided into four categories.

FIELD PLOTS

The treatment of field plots with essential elements applied at different rates and in various combinations to establish optimum fertilizer practices dates back at least to the mid-nineteenth century. This simple procedure has been tremendously effective throughout the world in improving agricultural production through fertilizer application. Unfortunately, it is not readily applicable to aquatic environments because of the difficulty of isolating and separating bodies of water comparable with field plots. A recent effort (Goldman, 1962) to isolate water columns in a lake by means of large polyethylene cylinders might be considered an adaptation of this approach. The field-plot technique may also have some potential in demonstrating limiting factors for aquatic plant growth in rare situations in which a series of small ponds or lakes in close proximity are available for experiments.

BIOASSAYS

Through the years, many types of greenhouse and laboratory bioassays have been developed for evaluating soil fertility. These include greenhouse tests with crop plants grown in small pots of soil treated with essential elements in various amounts and combinations, tests based on the intensive uptake of nutrients by large numbers of seedlings in small volumes of soil, and even tests based on the growth of fungi and algae on soil to indicate specific mineral element deficiencies (Bould and Hewitt, 1963). Bioassays are not the most commonly used soil tests at this time.

Bioassays of different types have also been employed for many years in attempts to evaluate nutrient supplies in natural waters. These have involved enrichment cultures in both the field and the laboratory. In some cases, natural populations have been employed as assay organisms (Goldman, 1960 a and b, 1964); in other cases, pure cultures of algae have been used (Potash, 1956; Gerloff and Skoog, 1957; Lund, 1964). Probably the most sophisticated refinement of the bioassay is the *in situ* enrichment culture of natural populations, in which ^{14}C uptake is used as a measure of photosynthetic and biological activity. This technique has found widespread application in the measurement of primary productivity and possible growth-limiting factors, for example, in the reports of Vollenweider and Nauwerck (1961), Goldman (1964), Goldman and Carter (1965), and Wetzel (1964a, 1965b). It has been adapted for measurements both on phytoplankton and larger angiosperm species (Wetzel, 1964b, 1965a).

Bioassays have provided valuable information. Their use on large aquatic plants, as mentioned, is particularly interesting because the bulk of the plant would not have to be isolated but could remain in contact with the natural environment. In general, however, this approach is of limited value because of problems associated with evaluating the degree to which isolated water samples can accurately reflect nutrient availability in a large body of water over a period of time. As shown in Table 1, even in its simplest form a bioassay can be of some help in dividing essential elements into two groups: those likely to limit plant growth and those unlikely to limit plant growth in a particular body of water.

Table 1 presents data on the growth of the blue-green alga *Microcystis aeruginosa* in autoclaved samples of Lake Mendota water to which essential elements had been added singly and in various combinations. It is apparent that the addition to the lake water of only nitrate, phosphate, and iron in treatment *H* resulted in as much growth as the addition of all the essential elements. In fact, treatment *H* produced slightly more growth than was present in the synthetic culture medium. This supports the suggestion that this approach can focus attention on elements that are potentially growth-limiting.

A comparison of the results of treatments *E, F,* and *G,* in which nitrogen, phosphorus, and iron were in turn limiting growth, suggests that nitrogen was relatively less abundant in this particular sample of water than was

TABLE 1   Growth of *Microcystis aeruginosa* in Surface Water from Lake Mendota with and without Additions of Essential Mineral Elements

Treatment	*Microcystis* Cells per Cubic Millimeter[a]	Percent of Growth with N, P, and Fe Added
A Autoclaved lake water	300	1.1
B NO_3	1,750	6.5
C PO_4	325	1.2
D Fe	325	1.2
E PO_4 and Fe	175	0.7
F NO_3 and Fe	1,950	7.2
G NO_3 and PO_4	6,050	22.3
H NO_3, PO_4, and Fe	27,100	100.0
I All essential elements	21,025	77.6
J Synthetic nutrient solution	17,550	—

[a]The number of cells added as inoculum has been subtracted from the cell counts.

phosphorus or iron, and was in fact the primary limiting factor. There was sufficient nitrogen for only about 1 percent of maximum growth, phosphorus for 7 percent, and iron for 22 percent. As already indicated, sampling problems make this observation on the limiting role of nitrogen of relatively little value.

CHEMICAL TESTS

The analysis each year of tens of thousands of samples in soil-testing laboratories as a basis for soil-management recommendations indicates the refinement and widespread use of chemical tests. Successful application of these procedures usually has required the development of chemical extraction procedures (employing dilute acids and buffered salt solutions) that approximate the capacities of crop plants to remove essential elements from the soil. These procedures have been particularly useful in diagnosing deficiencies of the major essential elements.

The possibility of assessing the nutrient status of lakes and streams by analyzing water samples, and thereby measuring concentrations of specific elements, has great appeal. It would be highly desirable if we could state with reasonable confidence that a concentration of phosphorus or other growth-limiting element below a specific concentration would not result in nuisance growths. Furthermore, distinguishing between available and nonavailable forms of an element is probably less difficult in aquatic environments than in soils.

Many attempts have been made to correlate changes in concentrations of essential elements in bodies of water with fluctuations in the growth of algae and other aquatic organisms. These attempts have been successful to some extent. For example, they have been successful in establishing a relation between dissolved silica and diatom growth (Lund, 1964). However, there are obvious problems having to do with obtaining representative water samples, analyzing for the low concentrations of many elements in water, detecting a capacity for continued replacement of the supply of an element in the water, and considering the capacities of different species for absorbing nutrients.

The results of chemical analyses are, for the most part, complex, and it is difficult to determine from them whether the supply of a specific element is limiting plant growth. Undoubtedly, extensive collections and analyses of samples of water from various depths of a lake throughout the year can indicate the general fertility of that body of water. It is doubtful that this approach can be simplified to the point that the availability of specific elements in aquatic environments can be estimated with the efficiency associated with chemical soil tests.

TISSUE ANALYSIS

In recent years there has been rapid development and application of tissue or leaf analysis for the diagnosis of the nutritional status of crop plants growing in soil. Tissue analysis is based on the fact that the concentration of an element in an organism varies over a wide range in response to the concentration in the environment, and that over part of this range, yield of the organism is related to tissue content of the element. The theory, terminology, and details of the tissue-analysis technique are discussed in various review papers (Lundegårdh, 1951; Ulrich, 1952; Bould et al., 1960; Smith, 1962; Chapman, 1966).

The essential relations on which tissue analysis is based are shown in Figure 1. The nearly vertical part of the curve is termed the "Deficient Zone"; here, plant yield is increasing markedly, but the concentration in the organism is changing very little. In the horizontal part of the curve, organism content of the element is increasing, but yield is not; this is the "Adequate Zone," or, more commonly, the "Zone of Luxury Consumption." The "Transition Zone" is that part of the curve between the zones of deficiency and adequacy. Successful application of tissue analysis depends on establishing for a species the critical concentration for each element of interest. According to Ulrich (1952), the critical concentration is that content which is just inadequate for maximum growth. It is approximately the breaking point of the curve.

Plant content of an element below the critical concentration indicates that growth is being limited by the supply of that element. More growth would

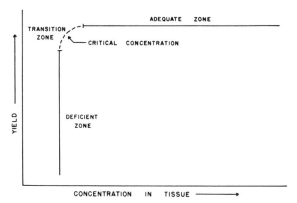

FIGURE 1 Schematic diagram of the relationship between plant growth and the concentration of an essential element in specific plant parts of a definite physiological age. (From Ulrich, 1961.)

result if greater amounts of the limiting nutrient could be absorbed. Plants will absorb an element in concentrations far above the critical level, but this luxury consumption is not associated with increases in yield.

Further details of the development of the tissue-analysis technique as a basis for recommendations on fertilizer application for crop production are considered in hundreds of papers. The technique seems more promising as a means of evaluating nutrient supplies in natural waters than the other possibilities I have discussed. With it, sampling problems could be greatly reduced, because the concentration of an element in an organism serves as an integrated expression of the availability of that element in the various microenvironments to which an organism may have been exposed during growth.

The remainder of this presentation will be concerned with summarizing the preliminary results obtained in attempting to use tissue analysis for assessing nutrient supplies in Wisconsin lakes. The results with *Microcystis aeruginosa* are described in greater detail in Gerloff and Skoog (1954, 1957), and the results with angiosperm plants, in Gerloff and Krombholz (1966).

TISSUE ANALYSIS IN EVALUATING NUTRIENT SUPPLIES FOR ANGIOSPERM AQUATIC PLANTS

ESTABLISHING CRITICAL CONCENTRATIONS

With terrestrial plants, the critical concentrations of elements can be established either in greenhouse cultures or with variously fertilized field plots; for aquatic plants, laboratory cultures are necessary.

Culture of angiosperm aquatic plants required the development of a satisfactory nutrient medium and the establishment of algae-free cultures. Of various nutrient media tested, the one described in Table 2 proved quite satisfactory. This is a modification of Hoagland's culture solution, one of the most commonly used media in studies on crop plant nutrition (Hoagland and Snyder, 1933). It contains only one fifth the concentrations of the major essential elements present in the original medium, but the full-strength concentrations of the trace elements. Iron was added as an Fe-EDTA complex. All media were autoclaved, and the cultures were handled with aseptic technique. Air enriched with CO_2 (approximately 0.5 percent) was bubbled into the flasks at a rate that maintained a pH of 6.0 to 6.5. The over-all appearance of cultures of *Vallisneria americana* is shown in Figure 2.

The most satisfactory technique for obtaining algae-free plants was to immerse the seeds in a dilute Clorox solution (Gerloff and Krombholz, 1966). Seed coats were then cut with a sterile razor blade. Following this, the seeds sprouted readily and gave no indication of prolonged dormancy. Four species, *Ceratophyllum demursum*, *Vallisneria americana*, *Heteranthera dubia*, and *Elodea occidentalis*, made excellent growth under the conditions described. The medium did not appear to be optimal for the culture of *Najas flexilis* and *Zannichelia palustris*.

The procedure for establishing the critical levels of nitrogen and

TABLE 2   Composition of a Modified Hoagland's Solution Used for the Culture of Angiosperm Aquatic Plants

Salt Used	0.5 M Stock Solution in 1 Liter of Final Solution (ml)	Concentration in Final Solution (ppm)
KNO_3	2.0	N -42
		K -47
$Ca(NO_3)_2 \cdot 4H_2O$	2.0	Ca -40
		P $- 6.2$
$MgSO_4 \cdot 7H_2O$	0.8	S -12.8
		Mg $- 9.6$
KH_2PO_4	0.4	
KCl*	$-$	Cl $- 1.77$
H_3BO_3*	$-$	B $- 0.27$
$MnSO_4 \cdot H_2O$*	$-$	Mn$- 0.27$
$ZnSO_4 \cdot 7H_2O$*	$-$	Zn $- 0.13$
$CuSO_4 \cdot 5H_2O$*	$-$	Cu $- 0.03$
$(NH_4)_6Mo_7O_{24} \cdot 4H_2O$*	$-$	Mo$- 0.01$
Fe $\cdot$ EDTA	$-$	Fe $- 0.40$

*Trace-element stock solutions were prepared at 1,000 times the concentration of the final solution. One ml of each stock solution was added to the final culture medium.

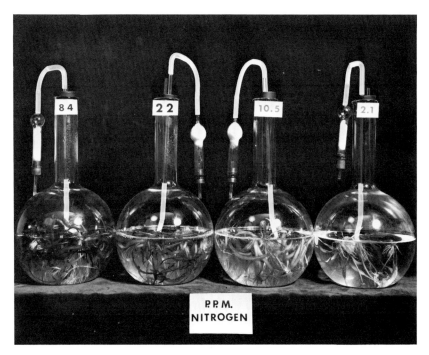

FIGURE 2 The response of *Vallisneria americana* to various levels of nitrogen in algae-free cultures in a modified Hoagland's solution.

phosphorus in a particular species also is indicated in Figure 2. The organism was grown in a series of replicated cultures similar in all respects except for the levels of nitrogen or phosphorus. Flasks representing four of the nine levels of nitrogen included in the nitrogen experiment are shown in this photograph. A marked difference is apparent in the amount of growth at the highest and lowest concentrations of nitrogen. As growth developed, the supply of nitrogen was progressively exhausted at the lowest external concentrations, and concentrations within the plants represented the zone of deficiency. Plants were harvested when ranges of growth and of nitrogen concentration in the tissues were represented.

After dry weight of samples of the plants from each flask had been determined, and after the samples had been analyzed for total nitrogen, the data were plotted to give the curve shown in Figure 3. It is apparent that the nitrogen content varied over a considerable range, from 0.78 to 4.28 percent, but over a considerable part of that range, from 1.3 to 4.8 percent, the yield of *Vallisneria* was at a constant maximum. The critical concentration for nitrogen in this species, therefore, was considered to be approximately 1.3 percent. Tissue contents above this value represented luxury consumption.

The results of a comparable experiment to establish the critical phosphorus concentration for *Vallisneria* are shown in Figure 4. The phosphorus content of the plant tissue varied from 0.10 to 0.70 percent with much of this range representing luxury consumption. The critical concentration was approximately 0.13 percent.

ANALYSES OF PLANTS COLLECTED FROM LAKES

Application of the above data to the evaluation of nitrogen and phosphorus supplies in a specific lake or stream would require collection of samples of a

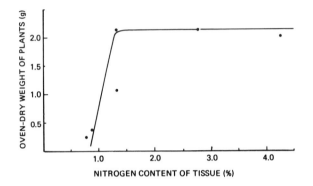

FIGURE 3 Relationship between the oven-dry weight and the nitrogen content of *Vallisneria americana* when cultured in nutrient media containing various concentrations of nitrogen.

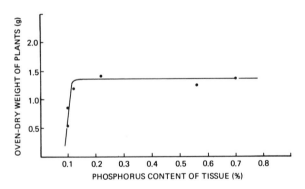

FIGURE 4 Relationship between the oven-dry weight and the phosphorus content of *Vallisneria americana* when cultured in nutrient media containing various concentrations of phosphorus.

species from that body of water, analysis of the samples for nitrogen and phosphorus, and comparison of the data with the tissue content—yield curves. Tissue contents above the critical concentration for an element would indicate adequate supplies of the element; values below the critical level would suggest a growth-limiting role.

To facilitate these comparisons, the critical concentrations and the range of nitrogen and phosphorus in each of the several angiosperm aquatic plants studied are presented in Table 3. *Heteranthera* and *Elodea* showed the same wide range of nitrogen and phosphorus content observed in *Vallisneria*. Surprisingly, approximately the same critical concentrations were established for all three organisms, namely, 1.3 percent nitrogen and 0.13 percent phosphorus. These, of course, must be considered approximate. Only repeated and independent trials will indicate their reliability. Data from earlier work to establish the critical nitrogen and phosphorus concentrations in the blue-green alga *Microcystis aeruginosa* have been included (Gerloff and Skoog, 1954). The critical phosphorus concentration was approximately the same as in higher aquatic plants; the nitrogen value was very much higher.

Samples of higher aquatic plants were collected from nine Wisconsin lakes at intervals during the growing season and were analyzed for total nitrogen and phosphorus. The procedures used are described in detail in an earlier publication (Gerloff and Krombholz, 1966). The results in Table 4 are for plant samples from Lake Mendota. It is readily apparent that no nitrogen or phosphorus value was below the critical concentration or even close to it. At the first sampling date early in the summer, plant growth is not as abundant as later, and there would usually be relatively little pressure on nitrogen and phosphorus supplies. This observation is supported by the high values for total nitrogen and phosphorus at the June sampling. By the July and August samplings, plant growth was very heavy, which would be expected to result in the lowest nitrogen and phosphorus values of the year. Nevertheless, the concentrations of both elements, and particularly phosphorus, were still far in

TABLE 3   Critical Concentration and Range of Content of Nitrogen and Phosphorus in Three Angiosperm Aquatic Plants and in a Blue-Green Alga

Plant Species	Nitrogen (%)		Phosphorus (%)	
	Critical Concentration	Range of Content	Critical Concentration	Range of Content
Vallisneria americana	1.3	0.78−4.28	0.13	0.10−0.70
Heteranthera dubia	1.3	0.71−3.88	0.12	0.11−0.65
Elodea occidentalis	1.4	0.77−4.68	0.14	0.14−0.53
Microcystis aeruginosa	4.0	3.16−7.43	0.12	0.11−0.46

TABLE 4   Total Nitrogen and Phosphorus Contents of Samples of Higher Aquatic Plants Collected at Intervals during Growing Season from Lake Mendota, 1964 (%)

Plant Species	Dates Sampled							
	June 29		July 22		August 18		September 14	
	N	P	N	P	N	P	N	P
Ceratophyllum demersum	4.43	0.75	2.11	0.51	2.17	0.56	3.41	0.71
Heteranthera dubia	3.79	0.55	3.24	0.69	2.32	0.51	2.67	0.58
Myriophyllum spp.	2.72	0.41	2.42	0.35	2.63	0.35	2.77	0.41
Potamogeton richardsonii	–	–	3.73	0.45	3.24	0.32	2.59	0.23
Potamogeton zosteriformis	3.65	0.42	3.70	0.35	3.65	0.59	3.37	0.44
Vallisneria americana	3.85	0.42	2.88	0.42	2.34	0.43	1.98	0.37

excess of the critical concentrations. This result reflects the recognized high fertility of Lake Mendota.

The data in Table 5 compare the nitrogen and phosphorus contents in samples of four plant species collected from two lakes, Mendota in southern Wisconsin and Big Kitten in the northern part of the state. Two of the species (*Vallisneria* and *Heteranthera*) are those for which critical levels were determined in the laboratory. Neither the nitrogen nor the phosphorus content was close to the critical concentration in these plants. However, there was a marked difference in the phosphorus contents of samples from the two lakes. The Mendota samples were much higher. The same pattern of relatively high nitrogen and low phosphorus is apparent in the *Ceratophyllum* and *Potamogeton* samples from Big Kitten Lake. In fact, the phosphorus concentration of 0.14 percent in the July sample of *Potamogeton* suggests that phosphorus supplies were placed under sufficient stress to approach a growth-limiting level.

A primary point of the data in Table 5 is that the analyses indicate a marked difference in nutrient availability in the two lakes. Both lakes seem to contain adequate supplies of nitrogen for plant growth; Big Kitten Lake, however, contains a much smaller supply of phosphorus than Lake Mendota does. The phosphorus contents of Lake Mendota samples were, in general, double those of the Big Kitten Lake collections.

In contrast to the last two tables, the data in Table 6 are from the collection and analysis of one species, *Elodea,* from four lakes. The high nitrogen values for Lake Escanaba samples throughout the season, more than double the critical concentration, suggest it is a lake relatively high in nitrogen. The phosphorus content of the Escanaba samples decreased much

TABLE 5 Comparison of Nitrogen and Phosphorus Contents of Samples of Three Species of Angiosperm Aquatic Plants Collected from Mendota and Big Kitten Lakes during Summer Months, 1964 (%)

Plant Species	Lake Sampled	June 24–29		July 16–22		August 6–18		September 2–14	
		N	P	N	P	N	P	N	P
Vallisneria americana	Mendota	3.85	0.42	2.88	0.42	2.34	0.43	1.98	0.37
	Big Kitten	2.89	0.37	2.27	0.27	2.48	0.35	2.20	0.28
Heteranthera dubia	Mendota	3.79	0.55	3.24	0.69	2.32	0.51	2.67	0.58
	Big Kitten	2.76	0.27	2.58	0.24	2.45	0.33	1.95	0.26
Ceratophyllum demersum	Mendota	4.43	0.75	2.11	0.51	2.17	0.56	3.41	0.71
	Big Kitten	2.28	0.19	2.69	0.33	2.06	0.28	1.82	0.21
Potamogeton richardsonii	Mendota	—	—	3.73	0.45	3.24	0.32	2.59	0.32
	Big Kitten	2.12	0.16	2.35	0.14	1.95	0.20	2.14	0.20

TABLE 6   Nitrogen and Phosphorus Contents of Samples of *Elodea* spp. Collected from Several Wisconsin Lakes during Summer Months, 1964 (%)

Lake Sampled	June 24–25		July 15–17		August 5–7		September 1–3	
	N	P	N	P	N	P	N	P
Escanaba	3.28	0.54	3.13	0.28	3.13	0.22	3.02	0.30
Trout	3.02	0.30	1.78	0.19	2.57	0.33	2.11	0.12
Big Kitten	2.24	0.30	1.88	0.17	1.96	0.18	2.08	0.23
Nebish	–	–	2.10	0.14	2.86	0.24	2.56	0.15

more during the season than did nitrogen, although it did not approach the critical concentration. Samples from Trout, Big Kitten, and Nebish lakes contained much lower concentrations of both elements than did Lake Escanaba samples. This is readily apparent in the July collections. The samples were particularly low in phosphorus, and in one collection from Trout Lake (September) and two from Nebish (July and September) they approached or were below the 0.13-percent critical concentration.

Of the lakes considered in Table 6, Nebish seemed most likely to be deficient in phosphorus. Table 7 presents the results of analyses of four species collected from this lake throughout the season. The relatively low concentrations of nitrogen and phosphorus in comparison with the Lake Mendota samples considered earlier are quite obvious. None of the nitrogen values was below the critical concentration, although in the August sample of *Lobelia*, the nitrogen content of 1.48 percent approached this value. However, in 5 of the 15 samples, the phosphorus content was at or below the average critical concentration established in the laboratory. Therefore, the analyses indicate Lake Nebish to be relatively infertile, with phosphorus supplies more likely than nitrogen to limit higher aquatic plant growth.

TABLE 7   Total Nitrogen and Phosphorus Contents of Samples of Higher Aquatic Plants Collected at Intervals during Growing Season from Lake Nebish, 1964 (%)

Plant Species	Dates Sampled							
	June 24		July 15		August 5		September 1	
	N	P	N	P	N	P	N	P
Elodea spp.	–	–	2.10	0.14	2.86	0.24	2.56	0.15
Eriocaulon septangulare	2.16	0.18	2.26	0.12	2.09	0.11	2.03	0.10
Lobelia dortmania	1.78	0.13	1.89	0.16	1.48	0.10	1.95	0.23
Potamogeton epihydrus	2.79	0.33	3.19	0.30	2.38	0.19	2.84	0.30

Data on the other lakes of the nine sampled will not be presented. However, in summary it can be stated that in none of the 189 samples analyzed was the nitrogen content below the critical concentration. In contrast, the phosphorus concentration in 24 samples was 0.15 percent or less, and in nine samples it was 0.13 percent or less. With the exception of Lake Mendota, this again focuses attention on phosphorus as the more likely growth-limiting essential element.

The data in Table 8 are from earlier work employing the tissue-analysis technique to evaluate nitrogen and phosphorus supplies in southern Wisconsin lakes for the growth of *Microcystis aeruginosa* (Gerloff and Skoog, 1957). In the columns labeled "Original," both the nitrogen and phosphorus values are well above the critical concentrations of 4.0 percent nitrogen and 0.12 percent phosphorus. This would usually be interpreted to indicate that neither element had become limiting for growth of the algae. However, it was noted that there was considerable difference in the amount of sheath material associated with the algae cultured in the laboratory and those collected from the field. Differences in the ratio of protoplasmic to nonprotoplasmic constituents in various samples of an organism contribute to errors in all applications of the tissue-analysis technique. The abundance of the sheath in blue-green algae makes the problem particularly acute with these organisms.

A procedure was developed that seemed to permit comparisons of nitrogen and phosphorus contents on the basis of the same carbohydrate content. This procedure is the basis for the values indicated in the columns labeled "Corrected." The corrected values show that the phosphorus contents, except for the Lake Nagawicka sample, were far above the critical concentration of 0.12 percent, whereas the nitrogen contents in all cases were proportionally

TABLE 8   Comparisons of Nitrogen and Phosphorus Contents of *Microcystis aeruginosa* Collected from Lake Blooms with Critical Nitrogen and Phosphorus Levels before and after Correction for Variations in Carbohydrate Content

Lake Sampled	N Content (%)		P Content (%)	
	Original	Corrected	Original	Corrected
Spaulding	7.61	5.02	0.62	0.41
Delavan	6.97	4.16	0.33	0.20
North	6.28	3.91	0.67	0.42
Nagawicka	5.43	3.88	0.22	0.16
Monona	6.76	4.40	0.44	0.26
Monona	6.15	4.15	0.41	0.28
Nagawicka	6.57	4.20	–	–
Waubesa	7.10	4.90	–	–
Waubesa	6.63	4.43	–	–

closer to the 4.0-percent critical content and in two samples were actually slightly below this value. This suggested that nitrogen in general was more likely to limit *Microcystis* growth in these lakes than phosphorus was. The contrasting conclusions with the alga and the angiosperm aquatic plants is not surprising in view of the very high critical nitrogen concentration in the alga.

CONCLUSIONS

At this time, no procedure for evaluating nutrient supplies and detecting growth-limiting factors in lakes and streams seems to have been developed to the point of general reliability and usefulness. This lack is particularly apparent in comparison with the progress made on similar problems in soil-fertility evaluation. The primary point of this presentation is that for the immediate future the tissue-analysis procedure gives promise of being very useful in studies of aquatic environments. In a single analytical value, it provides for an integrated expression of the availabilities of an element in the various parts of a lake or stream from which an organism absorbed nutrients during growth. Many of the difficult sampling problems associated with the continual movement and shifting of the aquatic environment thus can be minimized.

Enthusiasm over tissue analysis should not be interpreted as lack of appreciation for other techniques. For example, *in situ* enrichment cultures using ^{14}C have given interesting results in recent applications to limiting-factor analysis (Goldman and Carter, 1965; Wetzel, 1966). That procedure does seem somewhat more complex and intricate than tissue analysis. It must be recognized that no procedure is, or will be, entirely free from problems and difficulties. Judging from experience with terrestrial environments, it is also likely that no single technique will be the most satisfactory under all conditions; rather, each will be appropriate in particular situations.

Any indication that the tissue-analysis technique is at an advanced stage of development for use in aquatic environments would be completely erroneous. The work thus far has only suggested its applicability. Among the problems in need of immediate study are the following:

1. Determining the most appropriate plant parts to sample and analyze.

The analysis of entire plants in the present study is probably not the most satisfactory procedure. In agricultural applications, there has been a trend toward the analysis of index tissues, that is, plant parts that are more reliable indicators of the availability of specific elements than are entire plants. It is generally recognized that plants or plant parts to be analyzed should be of approximately the same physiological age.

2. Determining whether analysis for the total content of each element or for an extracted fraction more reliably correlates with yield.

The data presented here involved analyses for total nitrogen and phosphorus. Again, in agricultural experience, analyses for extracted fractions—NO_3–N, for example (Ulrich, 1969)—often have been found more reliable than total contents.

3. Determining the critical concentrations for various species.

It would be highly desirable if the critical concentrations for the species within a broad taxonomic group were similar. Although this was true for the several angiosperm species considered in this study, it cannot be generally accepted without more study. It seems particularly important to determine whether the critical concentration of an element in the dominant species in lakes of low fertility is lower than in the species characteristic of eutrophic waters.

4. Determining whether certain species are particularly reliable indicators of water fertility, that is, whether they would serve as indicator plants.

5. Determining the most reliable procedures for collecting plant samples in a particular body of water.

6. Determining the critical concentrations of essential elements, other than nitrogen and phosphorus, in several of the most important aquatic species.

This would be in accord with the degree of development of this technique in agricultural and horticultural applications. For example, Chapman (1961, 1966) lists for citrus not only the critical concentration but also the "low range," the "satisfactory range," the "high range," and even the "excess range" for 12 of the essential elements. Extension of tissue analysis to include all essential elements has been facilitated by improved analytical instrumentation, for example, the development of direct-reading, computer-programmed emission spectrometers, which permit very rapid and low-cost analyses of plant samples for most of the essential elements.

One feature of applying the tissue-analysis technique to aquatic plants that is markedly different from using it on crop plants is the difficulty of verifying with field data that tissue concentrations of an element below a critical level do actually correlate with a growth-limiting role of that element and with decreased plant yield. This correlation is readily verified with crop plants but undoubtedly will remain difficult with aquatic plants. Of some value may be measurements and general observations of the effects of adding supplemental quantities of a limiting factor to bodies of water indicated by tissue analyses to be low in that nutrient. Rapid bioassays with ^{14}C *in situ* enrichment cultures might also be used to verify a suggested growth-limiting level of an element. Because of evidence that nitrogen or phosphorus is the most likely limiting factor for the growth of plants in at least some of the lakes studied, this investigation was limited to those two elements. In other lakes, other elements might well be primary limiting factors. This has already been suggested in certain situations for the trace elements molybdenum and iron (Goldman, 1960, 1964; Wetzel, 1966).

The various problems mentioned in this discussion do not seem a reasonable basis for hesitancy in proceeding with attempts to develop and apply tissue analysis to aquatic organisms. Through many years of research, comparable problems have been overcome in applying the technique in terrestrial environments. Similar effort should result in marked progress in the difficult area of the nutrition of plants as they occur in natural waters.

REFERENCES

Bould, C., E. G. Bradfield, and G. M. Clarke. 1960. Leaf analysis as a guide to the nutrition of fruit crops. I. General principles, sampling techniques, and analytical methods. J. Sci. Food Agr. 11:229–242.

Bould, C., and E. J. Hewitt. 1963. Mineral nutrition of plants in soils and in culture media, p. 15–133 *In* F. C. Steward [ed] Plant physiology. II. Inorganic nutrition of plants. Academic Press, New York.

Bourn, W. S. 1932. Ecological and physiological studies on certain aquatic angiosperms. Contrib. Boyce Thompson Inst. Plant Res. 4:425–496.

Chapman, H. D. 1961. The status of present criteria for the diagnosis of nutrient conditions in citurs, p. 75–106 *In* W. Reuther [ed] Plant analysis and fertilizer problems (Pub. 8). Amer. Inst. Biol. Sci., Washington, D.C.

Chapman, H. D. 1966. Diagnostic criteria for plants and soils. Univ. California, Div. Agr. Sci. 793 p.

Fogg, G. E. 1965. Algal cultures and phytoplankton ecology. Univ. Wisconsin Press, Madison. 126 p.

Gerloff, G. C., and F. Skoog. 1954. Cell contents of nitrogen and phosphorus as a measure of their availability for growth of *Microcystis aeruginosa.* Ecology 35:348–353.

Gerloff, G. C., and F. Skoog. 1957. Nitrogen as a limiting factor for the growth of *Microcystis aeruginosa* in southern Wisconsin lakes. Ecology 38:556–561.

Gerloff, G. C., and P. H. Krombholz. 1966. Tissue analysis as a measure of nutrient availability for the growth of angiosperm aquatic plants. Limnol. Oceanogr. 11:529–537.

Goldman, C. R. 1960a. Primary productivity and limiting factors in three lakes of the Alaska Peninsula. Ecol. Monogr. 30:207–230.

Goldman, C. R. 1960b. Molybdenum as a factor limiting primary productivity in Castle Lake, California. Science 132:1016–1017.

Goldman, C. R. 1962. A method of studying nutrient limiting factors *in situ* in water columns isolated by polyethylene film. Limnol. Oceanogr. 7:99–101.

Goldman, C. R. 1964. Primary productivity and micro-nutrient limiting factors in some North American and New Zealand lakes. Verh. Intern. Verein. Limnol. 15:365–374.

Goldman, C. R., and R. C. Carter. 1965. An investigation by rapid carbon-14 bioassay of factors affecting the cultural eutrophication of Lake Tahoe. J. Water Pollut. Control Fed. 37:1044–1059.

Hoagland, D. R., and W. C. Snyder. 1933. Nutrition of strawberry plants under controlled conditions: (a) Effects of deficiencies of boron and certain other elements; (b) Susceptibility to injury from sodium salts. Amer. Soc. Horti. Sci., Proc. 30:288–294.

Hutner, S. H., and L. Provasoli. 1964. Nutrition of algae. Ann. Rev. Plant Physiol. 15:37–56.

Krauss, R. W. 1958. Physiology of the fresh water algae. Ann. Rev. Plant Physiol. 9:207–244.

Lea, W. L., G. A. Rohlich, and W. J. Katz. 1954. Removal of phosphates from treated sewage. Sewage Ind. Wastes 26:261–275.

Lewin, J. C., and R. R. L. Guillard. 1963. Diatoms. Ann. Rev. Microbiol. 17:373–414.

Lund, J. W. G. 1964. Primary productivity and periodicity of phytoplankton. Verh. Intern. Verein. Limnol. 15:37–56.

Lundegårdh, H. 1951. Leaf analysis. Hilger and Watts, London. 176 p.

Potash, M. 1956. A biological test for determining the potential productivity of water. Ecology 37:631–639.

Provasoli, L. 1958. Nutrition and ecology of protozoa and algae. Ann. Rev. Microbiol. 12:279–308.

Smith, P. F. 1962. Mineral analysis of plant tissues. Ann. Rev. Plant Physiol. 13:81–108.

Ulrich, A. 1952. Physiological bases for assessing the nutritional requirements of plants. Ann. Rev. Plant Physiol. 3:207–228.

Ulrich, A. 1961. Plant analysis in sugar beet nutrition, p. 190–211 In W. Reuther [ed] Plant analysis and fertilizer problems (Pub. 8). Amer. Inst. Biol. Sci., Washington, D.C.

Vollenweider, R. A., and A. Nauwerck. 1961. Some observations on the C^{14} method for measuring primary production. Verh. Intern. Verein. Limnol. 14:134–139.

Wetzel, R. G. 1964a. A comparative study of the primary productivity of higher aquatic plants, periphyton, and phytoplankton in a large, shallow lake. Internat. Rev. Ges. Hydrobiol. 49:1–61.

Wetzel, R. G. 1964b. Primary producitivity of aquatic macrophytes. Verh. Intern. Verein. Limnol. 15:426–436.

Wetzel, R. G. 1965a. Techniques and problems of primary productivity measurements in higher aquatic plants and periphyton. Mem. Ist. Ital. Idrobiol., Suppl. 18:249–267.

Wetzel, R. G. 1965b. Nutritional aspects of algal productivity in marl lakes with particular reference to enrichment bioassays and their interpretation. Mem. Ist. Ital. Idrobiol., Suppl. 18:137–157.

Wetzel, R. G. 1966. Productivity and nutrient relationships in marl lakes of northern Indiana. Verh. Intern. Verein. Limnol. 16:321–332.

GEORGE W. SAUNDERS, JR.
The University of Michigan, Ann Arbor

Some Aspects of Feeding in Zooplankton

In 1963 Nauwerck (1963) published his extensive study of the relationships between zooplankton and phytoplankton in Lake Erken. He calculated that *Eudiaptomus graciloides* constituted at least 60 percent of the annual herbivore production and that the phytoplankton could not be of primary importance as a direct source of food for the zooplankton in the lake. He indicated that other food sources must be considered and suggested that bacteria and detritus must be presumed to be the most important. He thus focused again on a major problem plaguing both marine and freshwater biologists.

Detritus has normally been considered as relatively non-nutritious (Conover, 1964), but work of marine investigators (Baylor and Sutcliffe, 1963; Riley, 1963; Riley, Wangersky, and Hemert, 1964) has indicated a mechanism for producing a detritus that might be nutritious. This mechanism is flocculation of dissolved organic substrates at the surfaces of bubbles that are formed in sea water.

Certain workers (Hillbricht-Ilkowska *et al.,* 1966; Mauilova, 1958; Rodina, 1958) have emphasized the importance of bacteria as a food for zooplankton. However, little is known quantitatively about this aspect of zooplankton feeding. The matter of zooplankton feeding has been reviewed extensively (Conover, 1964; Edmondson, 1957; Jorgensen, 1966) and has been updated in part by some recent statements (Edmondson, 1966). Therefore I will concern myself with more recent work and certain unpublished work concerning the feeding and nutrition of zooplankton.

THE TRACER KINETICS OF ZOOPLANKTON FEEDING

It is possible to view the feeding of zooplankton as one stage in an irreversible catenary reaction (Figure 1). If a radioisotopic tracer is added to this system as A, it will be transported through the reaction sequence according to first-order reaction kinetics, regardless of the actual order of the nonlabeled kinetics, provided the isotope has been added in truly tracer amounts. The time-dependent flow of specific activity will have the form indicated in Figure 2. It is possible to describe the relation between food source and zooplankton by the equation (Robertson, 1957)

$$C_z \frac{dS_z}{dt} = V_{bz}(S_b - S_z),$$

where C_z = quantity of zooplankton
S_b = specific activity of the food source
S_z = specific activity of the zooplankton
V_{bz} = rate of transfer of the food substance from B to Z

The above equation assumes that growth of the zooplankton is zero. A growth term can be included, but this inclusion requires that the instantaneous growth rate dC_z/dt be known. Since the duration of the experiments discussed below is short, growth of the zooplankton may be neglected within the limits of the experiment without introducing serious error. The above equation is transformed to

$$V_{bz} = \frac{C_z \, dS_z/dt}{(S_b - S_z)}.$$

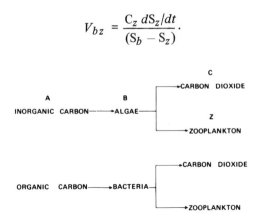

FIGURE 1   Schematic diagram of tracer flow in algal- and bacterial-zooplankton feeding systems.

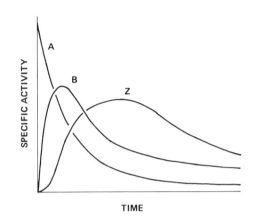

FIGURE 2   Generalized form of tracer carbon distribution in the first three compartments of an irreversible catenary reaction system.

Since dS_z/dt is not linear, the slope of the specific activity curve is difficult to determine accurately. The integral form of the equation is therefore easier to use (Zilversmit *et al.*, 1943) with the integrals being determined graphically.

$$V_{bz} = \frac{C_z \displaystyle\int_{t_1}^{t_2} dS_z}{\displaystyle\int_{t_1}^{t_2} S_b\,dt - \int_{t_1}^{t_2} S_z\,dt} \ .$$

The second term of the denominator is usually very small, compared with the first term, and can be neglected. However, if the zooplankton were grazing the food source to a very great extent, this second term would have to be considered. The theory also requires that the isotope mix completely and instantaneously with the food source and the zooplankton.

The basic procedure is to add radioactive tracer to separate carboys of lake water and to follow the flow of this tracer in the food source and in the zooplankton with time. The scheme indicated above assumes that leakage from bacteria to phytoplankton or the reverse is minimal during the experiment. Therefore the experiments should be conducted for relatively short periods of time.

ASSIMILATION OF BACTERIA BY ZOOPLANKTON

The transfer rate V_{bz} is an estimate of the amount of food source assimilated by the zooplankton per unit time before any losses. If the average weight of a bacterium is known, the number of bacteria assimilated per individual per hour can be calculated. I have assumed that the average size of a bacterium is 1 X 0.5 μ. In aerobic waters this estimate is probably too large. Tullander (Nauwerck, 1963, page 79) indicated that the majority of aquatic bacteria consists of forms less than 0.5 μ. My own experience agrees with this. Therefore the above assumption is conservative since it tends to overestimate the assimilation of bacteria when biomass is the basis of consideration.

The results of several field experiments at different times of the year are given in Table 1. The results are not as precise as indicated by the kinetic formula, since the design of our earlier experiments was not as refined as more recent work. However, this will not change the conclusions. The data show that there is a general relaticnship between size of organism and capacity to assimilate bacteria. But this relationship breaks down. Although the adult copepods and *Daphnia* are of about the same size range, the copepods are much less able to obtain bacteria from the water than the cladocera. Undoubtedly, this difference is related to differences in the structure of their collecting apparatus. The values in Table 1 are not corrected for losses of carbon-14 upon death of the organisms nor for self-absorption by the organisms themselves. I have estimates of these correction factors only for *Daphnia* and *Cyclops bicuspidatus*. Application of such correction factors would only serve to magnify the differences between the larger and smaller organisms.

The assimilation values depend on temperature, size of the herbivores, concentration of bacteria in the water, and probably a few additional factors. Schindler (1967) has demonstrated that temperature, food concentration, animal weight, food-energy content, reproductive state, and quality of food all significantly affect assimilation rate in *Daphnia magna*. I wish merely to establish that the data are consistent, both among different species populations of zooplankton and among size categories within a specific population. Therefore it is possible in certain natural populations of zooplankton, to discriminate the differences between feeding behavior and the nutritional aspects of feeding.

FILTERING-RATE MEASUREMENT
UNDER NATURAL CONDITIONS

There is a body of data concerning filtering rates of both marine and freshwater zooplankton (Jorgensen, 1966). Most estimates of filtering rate have been made in the laboratory and cover a wide range of values, apparently because of differences in experimental design and problems attending the use of the soft beta-emitting carbon-14. Two attempts (Nauwerck, 1959; Richman, 1964) to estimate filtering rates of zooplankton

TABLE 1 Assimilation of Bacteria by Zooplankton in Frains Lake, Michigan, at Different Times of the Year[a]

Zooplankton	Jan. 9, 1961	Apr. 18, 1961	Feb. 15, 1964	Apr. 12, 1964	June 29, 1964	July 20, 1964	Sept. 5, 1964	Jan. 30, 1965
Daphnia	—	13,100	20,800	—	35,300	30,100	49,400[b]	—
							32,600[c]	
							25,200[d]	
Ceriodaphnia	—	—	—	—	—	19,100[b]	33,600[b]	—
						5,300[d]	23,500[d]	
Bosminia	15,900	4,600	9,180	—	—	—	—	—
Adult								
Copepod	5,136	1,600	6,800	5,200	4,250	1,500	5,300	—
Copepodites	798	650	6,810[b]	3,800	250	520	315	—
			1,320[d]					
Nauplius	396	160	264	76	133	—	50	—
Tropocyclops	325	325	—	—	320	250	525	—
Keratella	1,060	650	525	—	130	—	—	—
Conochiloides	—	650	6,810	280	—	—	—	—
Codonella	84	—	—	—	—	—	—	116

[a]Data expressed in number of bacteria assimilated per hour.
[b]Large.
[c]Medium.
[d]Small.

in situ by using carbon-14 may yield underestimates because self-absorption was neglected in one case, and perhaps because exposure times were too long.

Rigler (1961) and McMahon and Rigler (1963), in studying the laboratory feeding behavior of *Daphnia magna* Strauss, developed a technique of labeling foods with carrier-free phosphorous-32. These food cells were fed for short periods to the *Daphnia,* and the filtering and feeding rates were estimated. The filtering rates that Rigler obtained were much higher than those generally obtained by other workers, except Ryther (1954).

Burns (1966) and Burns and Rigler (1967) applied this technique in estimating filtering rates in natural water. In laboratory populations of *Daphnia rosea,* Burns has confirmed the general relationship between feeding rate and food density that had been obtained for *D. magna.* She has also demonstrated that below incipient limiting concentration of yeasts, filtering rate depends on temperature and size of the organism. *Daphnia rosea* has a maximum filtering rate at 20°C. The filtering rate at 5°C is approximately one half the rate at 20°C, and the rate decreases above 20°C. The maximum filtering rate of *D. rosea* in lake water was about 20 ml per individual per day and occurred in late May. In June the filtering rate of natural populations dropped sharply and remained low through early November, when the experiments were terminated. *Daphnia galeata mendotae* displayed the same low filtering rates during the summer and fall. Because measurements were not made before July, it is not known if there was a decrease in filtering rate corresponding to that of *D. rosea.* The summer filtering rates of both species were one fourth to one third of the theoretical rates, calculated from measurements performed in the laboratory and corrected to field conditions. The sharp decrease in filtering rates of *D. rosea* was associated with the development of a bloom of *Anabaena.* The alga may have inhibited *Daphnia* feeding, mechanically or through chemical secretions.

The points I wish to make are these: (1) It is possible to estimate filtering rates with precision under natural conditions. (2) It is not possible to extrapolate, with reliability, filtering rates measured in the laboratory to feeding behavior under natural conditions.

ASSIMILATION OF PHYTOPLANKTON, BACTERIA, AND ORGANIC DETRITUS BY NATURAL POPULATIONS OF ZOOPLANKTON

Since it was possible to estimate bacterial assimilation by zooplankton, the next step was to evaluate this bacterial assimilation relative to assimilation of algae. Simultaneous kinetic analyses of algal and bacterial assimilation by zooplankton were conducted. The results are presented in Table 2. In no instance did the bacteria contribute more than 40 percent of the carbon assimilated. The highest percentages for bacteria occurred during the summer,

TABLE 2   Percentage of Total Carbon Assimilation by Zooplankton Feeding under Near-Natural Conditions in Frains Lake, Michigan, by Date

Zooplankton	Algae(%)	Bacteria(%)
September 5, 1964		
Daphnia		
1.0–1.2 mm	70	30
0.8–0.9 mm	82	18
0.4–0.6 mm	64	36
Ceriodaphnia		
0.5–0.7 mm	81	19
0.3–0.5 mm	67	33
Cyclops bicuspidatus (females)	93	7
Copepodites	92	8
Nauplii	98	2
Tropocyclops prasinus	87	13
January 30, 1965		
Daphnia		
0.5 m deep	93	7
1 m deep	86	14
Bosmina		
0.5 m deep	81	19
1 m deep	92	8
Cyclops bicuspidatus		
0.5 m deep		
Females	97	3
Males	97	3
1 m deep		
Females	98	2
Males	98	2
Copepodites		
0.5 m deep	98	2
1 m deep	97	3
Nauplii		
0.5 m deep	99	1
1 m deep	100	0
Keratella		
0.5 m deep	93	7
1 m deep	98	2
April 29, 1966		
Daphnia	63	37
Cyclops bicuspidatus (females)	71	29

when generation time might be more rapid, and on April 29, 1966, when the bacterial biomass was exceptionally high and the forms were larger than usual. There is also a tendency for fewer bacteria to be assimilated by the copepod *Cyclops bicuspidatus* than by the cladocera, another indication that the collecting apparatus of *Cyclops* is not adapted for collecting very small particles.

Recently we attempted to apply Rigler's method for kinetic analysis of assimilation and filtering rate to natural mixed populations of *Daphnia parvula* and *Daphnia ambigua*. Algal and bacterial assimilation were studied kinetically in separate carboys. Assimilation of detritus necessarily had to be handled in another manner. We also attempted to evaluate algal assimilation independently, with a method analogous to the detritus method.

Labled detritus was manufactured by exposing lake water in a third carboy to sodium bicarbonate–carbon-14 for 3 days while the carboy was suspended in the lake. The carboy was then autoclaved, reaerated aseptically, and resuspended overnight in the lake. On the day of the experiment, zooplankton from an equivalent volume of lake water were removed with a number 125 nylon plankton gauze and added to the carboy.

As an independent check of algal assimilation, the water from which the zooplankton were removed was placed in a fourth carboy. The phytoplankton were labeled with sodium bicarbonate–carbon-14 for 8 hours. The carboy was then placed in a black bag and zooplankton from an equivalent volume of lake water were added to this carboy. It was assumed that leakage of labeled photosynthate was minimal. During the 8-hour dark period, the loss of radioactivity from the phytoplankton did not exceed 25 percent.

The time course of carbon-14 uptake in the *Daphnia* was followed for 8 hours in each carboy. All carboys were suspended in the lake during the experiment. Organic detritus in the lake water was determined by weighing seston and subtracting live weights and inorganic ash. Assimilation of detritus was calculated by prorating uptake of labeled material to that of the lake detritus, on the assumption that the two were qualitatively similar. Assimilation of detritus and of algae in the black-bag experiment was estimated as the slope of the uptake curves at time zero. Ingestion rate was calculated from filtering rates, on the assumption that there was no selectivity in feeding.

The results of these experiments are presented in Table 3. Only gravid *Daphnia* are examined in this table. The results indicate that bacteria may provide 10 percent or less carbon and thus are not of importance in these experiments. The supporting experiment on algal assimilation estimates about 25 percent low, relative to the kinetic analysis. The assimilation rate is determined by taking the slope of the uptake at time zero. This is difficult to

TABLE 3 Percentage of Dry Organic Matter Assimilated by Gravid *Daphnia* per Day

Date	Algae I[a]	Detritus	Bacteria	Algae II[b]	I–II (%)
1967					
Jan. 21	44.0	55.4	0.7	–	–
Apr. 27	43.8	56.2	–	33.6	23.3
May 3	42.1	52.8	5.1	29.4	30.2
May 9	32.4	59.3	8.3	31.9	1.5
May 16	49.4	42.4	8.3	38.3	22.5
May 23	54.0	31.1	14.9	34.7	35.8
May 30	41.6	53.7	4.8	28.8	30.8

[a]Kinetic method.
[b]Black-bag method.

do with accuracy and the value may be somewhat low. If the algae secrete organics during labeling, and these organics are in turn assimilated by bacteria, the specific activity of the algae in this experiment would be overestimated, causing an underestimate of assimilation rate. Therefore I consider that the data are at least consistent and that the kinetic analysis provides a reasonable estimate of true algal assimilation by the cladocera.

Knowing the filtering rates of the *Daphnia* and the assimilation rates as estimated, and assuming no selectivity in filtering, one can calculate assimilation efficiency, which I define as assimilation divided by ingestion. Assimilation efficiencies are presented in Table 4. The data are corrected for loss of carbon-14 by lysis on death of the *Daphnia* and for self-absorption of carbon-14. The algae are the most assimilable food component, apparently being easily digested. Both detritus and bacteria are refractory to digestion.

Detritus may be as useful as the phytoplankton as a source of energy because it is present in much greater quantity in the lake water. The ratio of the dry weight of organic detritus to phytoplankton is given in Table 5. It might be argued that the manufactured labeled detritus was highly nutritious

TABLE 4 Assimilation Efficiency in Gravid *Daphnia* in Percentage of Food Source Ingested

Date	Algae	Detritus	Bacteria
1967			
Apr. 27	47.0	12.7	–
May 3	74.2	13.9	51.8
May 9	67.4	8.4	22.6
May 16	70.0	3.9	21.1
May 23	51.9	11.1	15.5
May 30	88.1	4.1	13.5

TABLE 5   Ratio of Dry Weight of Organic Detritus to Phytoplankton

Date	Ratio
1967	
Jan. 21	5.7
Apr. 27	5.8
May 3	5.5
May 9	13.8
May 16	14.2
May 23	2.7
May 30	11.6

since it could have been derived from newly synthesized protoplasm that disrupted during autoclaving but was easily digestible. However, the detritus carboy after autoclaving contained about the same amount of detritus, per unit volume, as the lake water.

Presumably some part of the phytoplankton turned over in the 3 days it was exposed in the detritus carboy, and probably the more easily assimilable matter was dissolved during sterilization, leaving the more refractory labeled particles. Microscopic examination of manufactured detritus and organic detritus in the lake water did not reveal significant differences in structure or size of particles. Therefore I believe that the manufactured detritus is more nearly representative of the lake-water detritus than it is of easily digestible photosynthate. However, this point does have to be tested.

DISSOLVED ORGANIC MATTER AS A SOURCE OF DETRITUS BY FLOCCULATION

In a carefully conducted experiment, C. Comstock (unpublished data) tested the hypothesis that organic detritus can be formed in filtered lake water. The water was filtered through Whatman No. 12 and HA Millipore ® filters, in sequence. One half the filtered water was bubbled for 3 hours in a column with a sintered glass plate. The other half was not bubbled. Single newborn Daphnia magna were placed in each of 20 containers replicated for bubbled and unbubbled filtered lake water. The mean lengths of life of the Daphnia and growth as mean body length (exclusive of head and spine) were determined. The Daphnia were transferred three times a day into fresh medium in clean containers. This transfer prevented bacteria from developing in the medium. Water was collected from three sources, all in Cheboygan County, Michigan: Douglas Lake, a moderately eutrophic lake; Lancaster Lake, a highly stained eutrophic lake; and Nigger Creek, a clear cold spring-fed stream, where it crosses Crump Road. Five experiments were performed, and the results are presented in Table 6.

TABLE 6   Lengths of Life and Body Lengths of *Daphnia magna* Grown in Filtered Lake Water

Water Source	Mean Length of Life (hr)		Mean Body Length (mm)	
	Unbubbled	Bubbled	Unbubbled	Bubbled
Douglas Lake	72	63	0.72	0.70
Douglas Lake	103	111	0.83	0.79
Douglas Lake	192	232	0.83	0.90
Lancaster Lake	105	102	0.72	0.75
Nigger Creek	100	99	0.74	0.74

In no experiment did the mean body lengths or mean lengths of life of *Daphnia* raised in bubbled water and of those raised in unbubbled water differ significantly (99.9 percent confidence level). In only one experiment did the *Daphnia* molt more than once. No particles could be detected in bubbled water with a compound or dissecting microscope, nor could particles or aggregates of particles be observed after bubbled water was allowed to stand for 24 to 48 hours. Miss Comstock concluded that if particles were formed in her experiments, they were too small to be seen with a light microscope and either were not nutritionally valuable to *Daphnia* or were present in quantities too small to make a difference in the growth and survival of the *Daphnia* raised in the two treatments of filtered water. Since her experiments covered a spectrum of water types, it is highly unlikely that the proposed mechanism for generating organic detritus is a general phenomenon in freshwaters.

DISCUSSION

There is no need to argue the fact that phytoplankton constitutes a primary source of food for zooplankton. However, it is not clear in the sea or in inland waters (Jorgensen, 1966; Nauwerck, 1963) that phytoplankton can in fact supply all the energy requirements of zooplankton. The discrepancies between requirement and supply may arise from two sources. First, estimates of energy requirements as calculated from laboratory data may be too high because of experimental design, handling of the organisms, or use of organisms that have been raised in heavily fed aquaria. Second, there may be a supplementary source of energy that fulfills the discrepancy but has not been considered in the experimental analysis. The supplementary source may be detritus, bacteria, dissolved organic matter, or any combination of the three. The evidence available to date indicates that dissolved organic matter is not an important source of food for zooplankton, either directly or indirectly, through flocculation as detritus.

The assimilation efficiency of the phytoplankton by the *Daphnia* in Frains Lake is high. Conover (1964) categorized assimilation values into two groups: assimilation greater than 60 percent, and assimilation generally less than 60 percent. Assimilation values in the literature actually range from 6.6 to 99 percent. Conover indicates that the differences may be real and attributable to the variability of natural material but that it is equally likely that the variability is related to the different techniques used to measure assimilation. There is very good evidence, however, that algae with heavy cell walls can pass through the intestines of zooplankton and remain viable. Thus it must be expected that assimilation efficiencies will vary widely, the efficiency depending on the composition of the algal food source.

The nannoplankton in the experiments reported here were dominated by phytoflagellates of the genera *Rhodomonas* and *Cryptomonas*. It is probable that these forms are highly digestible. The net phytoplankton, which are probably of limited availability as food, did not exceed 10 percent of the standing biomass of the phytoplankton. The experiment of September 5, 1964, showed a lower relative contribution of phytoplankton to assimilation in the zooplankton. At this time, 86 percent of the phytoplankton biomass consisted of large filamentous and colonial blue-green algae, most of which were *Anabaena planctonica*. The nannoplankton were made up mostly of algae with heavy walls, such as *Peridinium* sp., *Scenedesmus* spp., *Crucigenia* sp., *Oocystis* sp., and *Tetraedron* sp. This may explain in part the lesser contribution of phytoplankton to total assimilation on this date, since these forms are not readily digestible in comparison to the thin-walled phytoflagellates.

Naumann (1921) stated that detritus and bacteria in lakes must be the most important food sources. Pennak (1955) argued that detritus must be the most important food source in the Colorado lakes he studied. Jorgensen (1966) provided data for marine waters suggesting that neither phytoplankton nor bacteria together can normally provide the energy requirements of the filter feeders. Nauwerck (1962) showed that *Eudiaptomus gracilis* can survive long periods on nonalgal food, and he also showed (Nauwerck, 1963) that *Eudiaptomus graciloides* cannot satisfy its energy requirements with phytoplankton during its major pulse in Lake Erken. There was enough organic detritus present to make up the discrepancy.

The mere presence of large amounts of detritus in a volume of water is not sufficient reason to argue for its significance as a food source for zooplankton. The detritus not only must be present, it must be assimilable. Whether it is assimilable will depend on where it was generated, how old it is, and how much of it has been decomposed into refractory material.

The detritus in Frains Lake, Michigan, is probably planktogenic, particularly in the winter, spring, and summer. In the autumn more detritus may be generated from the littoral zone when higher aquatic plants decay.

I have presented arguments both that detritus is of primary importance as a food source for *Daphnia* and that it is refractory. Its importance as a source of food can stem only from its very high concentration relative to the phytoplankton and the bacteria. Obviously the validity of these results depends on how well the detritus generated in our carboy mimics the detritus that occurs in the lake. This question is susceptible to testing, although this has not yet been done. Whether these results can be extrapolated to other bodies of water is problematical.

Wright (1959) tabulated detritus/seston ratios for lakes of various sizes. The ratios vary inversely with the size of the lakes, the variation implying that the seston in larger lakes is planktogenic and proportionately much less concentrated than that in small lakes. Therefore organic detritus would be less important quantitatively as a food source in large lakes. Lake Erken is much larger than Frains Lake; yet, except for the large August pulse of dinoflagellates, the detritus volume during the summer was 2 to 10 times the algal volume. The detritus occurring during the summer appeared to be mostly of organic origin. The detritus in the discontinuity layer in July consisted of large amounts of copepod fecal pellets; in August it was primarily the remains of blue-green algae. My results support Nauwerck's conclusions regarding detritus as a possible food source for zooplankton, and suggest that this aspect of zooplankton nutrition should be investigated more critically.

Several workers have emphasized the importance of bacteria as food for zooplankton. However, there have been no critical experiments testing the assimilation efficiency of bacteria by zooplankton, with the possible exception of Monakov and Sorokin (1961). The evidence for statements emphasizing the importance of bacteria as food is derived from differential cell counts between vessels containing zooplankton and vessels without zooplankton. No attempt to compare the utilization of bacteria and algae has been made. Very often no information is given concerning the number of zooplankters used, or even the species employed.

The fact that there is a difference between experiment and control may indicate that bacteria have been eaten, but it does not demonstrate importance in terms of the energy requirements of the zooplankton. The feeding rate on bacteria by *Daphnia* in Frains Lake ranged from 6×10^6 to 17×10^6 bacteria per individual per day. These values are consonant with those of Russian workers (Manuilova, 1958; Egorova, 1954; Monakov and Sorokin, 1961). McMahon and Rigler (1963) observed a maximum feeding rate of 134×10^6 bacteria per individual per day in 3-mm *Daphnia magna* at

20°C when fed on *E. coli* as the sole source of food. The gravid *Daphnia* in our study range from 0.8 to 1.2 mm long and would have a maximum feeding rate on *E. coli* much less than this.

The short generation times and rapid turnover times of bacteria are sometimes used to indicate the probable importance of bacteria as food. However, turnover time is a measure of metabolic rate, not necessarily of bacterial production. The generation times of bacteria in natural waters are not exceptionally short. The generation times of bacteria in Russian impoundments range from 9 to 120 hours (Krasheninnikov, 1960; Salmanov, 1960, 1964). Very short generation times (2 to 6 hours) have been found in ponds subjected to fertilization (Ivanov, 1954, 1955; Rodina, 1958). The maximum net production of phytoplankton in Lake Erken is about 150 percent of standing crop per day (Nauwerck, 1963). This indicates that for short periods the generation time of the phytoplankton may roughly equal that of the bacteria. The growth of bacteria in natural water does not normally outstrip phytoplankton growth so greatly that the bacteria may continuously dominate the food supply of the zooplankton. On the contrary, it is only when phytoplankton die, releasing large quantities of nutrient substrate for bacterial growth, that the bacteria may temporarily increase their role as an energy source for zooplankton.

The assimilation efficiencies for bacteria in the *Daphnia* in Frains Lake are relatively low. The only other experimental evidence that bacteria are not assimilated as well as algae is that of Monakov and Sorokin (1961). They showed that *Hydrogenomonas flava* is assimilated about one half as well as *Chlorococcum* sp. *Chlorococcum* has a heavy cell wall. Thin-walled phytoflagellates should be more easily digested and assimilated and exhibit an even greater assimilation efficiency than the *Chlorococcum.*

My data compare with those of other workers in several ways. Estimates of zooplankton respiration taken from the literature, if we assume a respiration quotient of 1, range from 3 percent to 45 percent of body carbon per day. The average is about 20 percent. If the respiration quotient is more nearly 0.8, this value would be reduced proportionally. The estimates of assimilation in Frains Lake *Daphnia* range from about 10 to 25 percent of body carbon per day.

Schindler (1967) fed *Daphnia magna* with carbon-14-labeled *Chlorella* and also with the sediments from carbon-14-labeled mixed algae after the *Daphnia* had reingested and defecated the material for several weeks. The *Chlorella* represents a high-calorie diet and the sediments a low-calorie "detritus" diet. When fed the high-calorie diet (5.2 cal/mg), assimilation amounted to 67 percent of body weight per day. When fed the low-calorie diet (2 cal/mg), assimilation amounted to about 26 percent of body weight per day. Schindler also found that the energy content of seston in a Minnesota Lake neve exceeded 4.0 cal/mg of dry weight, and it fell as low as 0.8 cal/mg of dry weight in late winter.

These data suggest two very important effects. First, assimilation values obtained by feeding zooplankton algal cultures will be very different from, and will be greater than, the values obtained when fed the "stuff" occurring in lakes. Therefore it is not possible to extrapolate the laboratory experiments to the natural system with confidence. Second, low-calorie detritus in large amounts will maintain a zooplankton population for a period of time after some catastrophic collapse of the phytoplankton food source, but, acting as a diluent, it may also dampen the assimilation of high-quality algae, particularly when the seston is present in quantities approaching incipient limiting concentration for feeding. This dampening would tend to suppress the growth rate of an exploding zooplankton population and perhaps permit the phytoplankton community to adjust to a high rate of grazing. Thus the detritus may act to buffer the dynamics and energetics of the phytoplankton–zooplankton feeding system.

I would like to suggest a very general scheme for understanding zooplankton feeding in lakes. The food system may be viewed as a buffered system. In lakes with very little organic detritus, the zooplankton depend primarily on the phytoplankton. The detritus and bacteria provide an important food only during the terminal phases of major phytoplankton pulses. Such a food system is lightly buffered. In lakes where organic detritus is relatively high in concentration, the detritus may always occupy a dominant role in zooplankton feeding along with the phytoplankton. Such a system is highly buffered, and any sharp change in one food source is compensated by the other. Normally, the bacteria do not occupy an important role in the system. Their major role involves the processes of mineralization.

The scheme can be taken one step further. Hillbricht-Ilkowska et al. (1966) stated that bacteria seem to be limiting to zooplankton production in those eutrophic lakes they have studied. Russian workers have also emphasized the importance of bacteria as a food for zooplankton. Z. Gliwicz (personal communication) believes that the zooplankton production is dependent on bacteria in eutrophic lakes and more dependent on phytoplankton in lakes of lesser productivity. I would consider the bacteria important as a special case. One example is that of Sorokin (1957), who found a correlation between the distribution of *Daphnia* and the chemosynthetic methane-oxidizing bacteria in Rybinsk Reservoir. Bacteria in the so-called plate in the microaerobic interphase in very productive lakes and at the chemocline in meromictic lakes may develop tremendous densities and provide a very concentrated food source.

A third possible case is more general. We know that in Frains Lake there is a sharp increase in dissolved carbohydrate in mid-June when the blue-green algae develop. There is also a tenfold increase in bacteria, which is maintained throughout the summer. We can now invoke the mechanism suggested by Burns (1966). The blue-green algae inhibit the filtering rate of the zooplankton, either mechanically or chemically. They also provide a large base of organic nutrient upon which the bacteria may increase tremendously. The large shift in the relative proportions of biomass would then permit the bacteria to dominate such a zooplankton feeding system.

The large body of information available on zooplankton feeding has been very confusing. Discrepancies have often been attributed to experimental error or improper experimental design. I would like to suggest that there is perhaps a much greater fundamental unity to this information than we have suspected. The discrepancies may be due to studying only parts of systems or to analyzing very different systems. Thus one would obtain quite different answers. The problem of zooplankton feeding is a very complex one, but its complexity may offer some advantage. I think it suggests that it may be easier and more effective to analyze zooplankton feeding systems in nature than it will be to synthesize such systems in the laboratory. It is the former that we should attempt to do.

This work was supported by Michigan Memorial Phoenix Project Grants No. 192 and 267 and by a grant from the Louis W. and Maud Hill Family Foundation. I would like to thank Dr. D. C. Chandler for his support and encouragement and L. S. Chertkov, D. Brubaker, M. S. Misch, J. C. Roth, and T. A. Storch for help with the experiments.

REFERENCES

Baylor, E. R., and W. H. Sutcliffe, Jr. 1963. Dissolved matter in seawater as a source of particulate food. Limnol. Oceanogr. 8:369–371.

Burns, C. W. 1966. The feeding behavior of *Daphnia* under natural conditions. Ph.D. Thesis, Univ. of Toronto. 89 p.

Burns, C. W., and F. H. Rigler. 1967. Comparison of filtering rates of *Daphnia* in lakewater and in suspension of yeast. Limnol. Oceanogr. 12:492–502.

Conover, R. J. 1964. Food relations and nutrition of zooplankton. Symp. on Exp. Mar. Ecology, Univ. Rhode Island, Occas. Pub. 2:81–91.

Edmondson, W. T. 1957. Trophic relations of the zooplankton. Trans. Amer. Microscop. Soc., 76:225–245.

Edmondson, W. T. 1966. Marine Biology III. N.Y. Acad. Sci., New York. 357 p.

Egorova, A. A. 1954. Opyt primeneniya zhestoi rastitel'nosti v kachestve zelenogo udobreniya v rybkhozakh del'ty reki volgi (Experiments using higher plants as green fertilizer in fish ponds of the Volga delta). Trud. Instit. Mikrobiol. 3:201–211.

Hillbricht-Ilkowska, A., Gliwicz, Z., and Spodniewska, J. 1966. Zooplankton production and trophic dependencies in the pelagic zone of two Masurian lakes. Verh. Int. Verein. Limnol. 16:432–440.

Ivanov, M. V. 1954. Opredelenie vremeni generatzi vodnykh bakterii v rybkhoze del'ty reki volgi (Determination of the generation time of aquatic bacteria in fish ponds of the Volga River delta). Trud. Mikrobiol. 3:213–220.

Ivanov, M. V. 1955. Metod opredeleniya produktzii bakterial'noi biomassy v vodoeme (A method for determining the production of bacterial biomass in natural waters). Mikrobiology 24:79–89.

Jorgensen, C. B. 1966. Biology of suspension feeding. Pergamon, N.Y. 313 p.

Krasheninnikov, S. A. 1960. Mikrobiologicheskaya Kharakteristika Gor'kovskogo vodokhranilishcha vo vtorii god ego sushchestvovaniya (The microbiological characteristics of Gorkovsky Reservoir in the second year of its inundation). Trud. Inst. Biol. Vodokhran. 3:9–20.

Manuilova, E. F. 1958. The question of the role of bacterial numbers in the development of cladocera in natural conditions. Dokl. Biol. Sci. 120:438–441.

McMahon, J. W., and F. H. Rigler. 1963. Feeding rate of *Daphnia magna* Strauss in different foods labelled with radioactive phosphorous. Limnol. Oceanogr. 10:105–113.

Monakov, A. V., and Yu. I. Sorokin. 1961. Kolichestvenny dannye o pitanii *Daphnia* (Quantitative data on the feeding of *Daphnia*). Trud. Inst. Biol. Vodokhr. 4(7):251–261.

Naumann, E. 1921. Spezielle Untersuchungen über die Ernährungsbiologie des tierischen Limnoplanktons. I. Über die Technik des Nährungserwerbs bei den Cladoceren und ihre Bedeutung für die Biologie der Gewässertypen. Lunds Univ. Arsskr., N. F. 17:3–26.

Nauwerck, A. 1959. Zur Bestimmung der Filtrierrate limnischer Planktontiere. Arch. f. Hydrobiol. Suppl. 25:83–101.

Nauwerck, A. 1962. Nicht-algische Ernährung bei *Eudiaptomus gracilis* Sars. Arch. f. Hydrobiol. Suppl. 25:393–400.

Nauwerck, A. 1963. Die Beziehungen zwischen Zooplankton und Phytoplankton im See Erken. Symb. Bot. Upsal. 17:1–163.

Pennak, R. W. 1955. Comparative limnology of eight Colorado mountain lakes. Univ. Colorado Stud., Ser. Biol. 2:1–75.

Richman, S. 1964. Energy transformation studies on *Diaptomus oregonensis*. Verh. Int. Verein. Limnol. 15:654–659.

Rigler, F. H. 1961. The relation between concentration of food and feeding rate of *Daphnia magna* Strauss. Can. J. Zool. 39:857–868.

Riley, G. A. 1963. Organic aggregates in seawater and the dynamics of their formation and utilization. Limnol. Oceanogr. 8:372–381.

Riley, G. A., P. J. Wangersky, and D. van Hemert. 1964. Organic aggregates in tropical and subtropical surface waters of the North Atlantic Ocean. Limnol. Oceanogr. 9:546–550.

Rodina, A. G. 1958. Mikroorganizmy i povyshenie ryboproduktivnosti prudov (Microorganisms and increase of fish production in ponds). A. N. Moscow. 171 p.

Robertson, J. S. 1957. Theory and use of tracers in determining transfer rates in biological systems. Physiol. Rev. 37:133–154.

Ryther, J. H. 1954. Inhibitory effects of phytoplankton upon feeding of *Daphnia magna* with reference to growth, reproduction, and survival. Ecology 35:522–533.

Salmanov, M. A. 1960. Mikrobiologicheskie protzessy v Mingechaurskom vodokhranilishche. (Microbiological processes in Mingechaursky Reservoir). Trud. Inst. Biol. Vodokhr. 3:21–35.

Salmanov, M. A. 1964. Vremya generatzii bakterii i ikh vyedanie zooplanktonom v Kuibyshevskom vodokhranilishche (The generation time of bacteria and bacterial consumption by zooplankton in Kuibyshevsky Reservoir). Zool. Zhurn. 43:809–814.

Schindler, D. W. 1968. Feeding, assimilation, and respiration rates of *Daphnia magna* under various environmental conditions and their relation to production estimates. J. Anim. Ecol. 37:369–385.

Sorokin, Yu. I. 1957. Rol'khemosinteza v produktzii organicheskogo veshchestva v vodoemakh (The role of chemosynthesis in the production of organic matter in water bodies). Mikrobiology. 26:736–744.

Wright, J. C. 1959. Limnology of Canyon Ferry Reservoir. II. Phytoplankton standing crop and primary production. Limnol. Oceanogr. 4:235–245.

Zilversmit, D. B., C. Entenman, and M. C. Fishler. 1943. On the calculation of "turnover time" and "turnover rate" from experiments involving the use of labelling agents. J. Gen. Physiol. 26:325–331.

LUIGI PROVASOLI

Haskins Laboratories, New York

Algal Nutrition

and Eutrophication

Present major sources of eutrophication are sewage, variously treated or not, and runoff from cultivated lands. Since these sources are allochthonous, the conservationist approach to eutrophication centers on restoring pristine conditions by eliminating these external sources of nutrients. Thus the effluents of sewage-disposal plants and the runoffs are diverted downstream from the eutrophic lakes whenever possible, or the treated sewage is further depleted of nutrients by algal tertiary treatment, or the phosphorus is stripped by chemical precipitation.

The approach of the agriculturist, conscious of the ever-increasing need for proteins, is to avoid waste and to employ these sources of nutrients as a regulated fertilization to increase fish production. (See the paper by Larkin and Northcote, page 256.)

Implicit in both solutions to eutrophication is the belief that eutrophication is simply an enrichment with nutrients, especially N and P, the two most important stimulants (aside from photosynthesis) of all algal growth. However, the undesirable feature of eutrophication is not so much the production of more algae as it is the change in type of algae, which results in obnoxious blooms. This change in flora, generally blooms of blue-green algae or growth of macrophytes, upsets the base of the normal food chain and results in the creation of new food chains that lead to far less valuable species of fishes or, more often, to the accumulation of algae because of reduced grazing—or no grazing at all—by the herbivores. Rotting of algal masses, followed by depletion of O_2, may cause fish kills. The main problem, then, is determining what discriminatory or preferential factors govern the growth of algae—beneficial algae (diatoms), blue-green algae, or algae from any other group.

To correct the obnoxious effects of eutrophication, particularly when we want to produce beneficial food chains at will, we need detailed knowledge of the needs, tolerances, and idiosyncrasies of single phytoplankton species and of their interactions with other species. Unfortunately, very few freshwater algae of ecological importance have been grown in bacteria-free culture. And we are unable to compound selective nutrient media for algae as we can for bacteria.

Let us review some selected nutritional aspects of algae and try to find possible group-specific idiosyncrasies and, possibly, the preferential discriminatory factors.

NITROGEN AND PHOSPHORUS

Nitrate and phosphate have been the main concerns of conservationists because runoff waters and treated sewage are so rich in these two nutrients. Since both are indispensable, except for the nitrogen fixers, their increased availability should benefit all algae equally and should result in more growth of the normal flora. This increase in growth generally happens *in vitro* with unialgal bacteria-free cultures. In nature, however, species are never alone, and increase in these nutrients sometimes seems to induce a change in flora. This change is probably due to the tolerances of the various algae and their need for certain levels of N and P, the differential efficiencies of algae in concentrating the nutrients, and the competitive relations of different species of algae.

Oligotrophic species can trap P from extremely low concentrations; some of them seem to be inhibited by concentrations of PO_4-P above 10 to 20 μg per liter. *Dinobryon divergens,* however, which in many lakes blooms only in periods of low P (below 10 μg per liter), was reported to be blooming in a phosphorus concentration of 250 μg per liter by Lefèvre and Farrugia (1958) and was cultured even in a concentration of 1 mg of PO_4-P per liter (Lund, 1965, page 244). One wonders whether this anomaly may not be due to different physiological races.

Various algae differ widely in their ability to concentrate phosphorus, from 140,000 $\times$ in *Volvox* and 200,000 $\times$ in *Pandorina* to 800,000 $\times$ in *Spirogyra;* this ability might give quite an edge to fast-reproducing species.

Green algae are favored by high concentrations of nutrients, such as those in the Great South Bay (Long Island, New York), which receive the effluents of duck farms; in fish ponds of temperate climates that are treated with large quantities of the common soil fertilizers (N, P, K), and in sewage ponds. Is this increased growth of green algae due to need of high N and P concentrations? To tolerance? Or to favorable action of other factors introduced with N and P?

An interesting observation was made in fish ponds in Israel. These ponds were generally fertilized once a month to sustain algal growth. In many cases, the fertilization resulted in vigorous growth of beneficial green algae during the first 15 days but was followed by a drastic drop in green algae and a rise of abundant blue-green algae in the subsequent 15 days. This rise in population of blue-greens may lead to their permanent growth in the ponds and may make it necessary to clean the ponds and reinoculate them with green algae. A possible explanation for this replacement of green algae with blue-greens is that when the nitrogen level falls too low for the needs of the green algae, the nitrogen-fixing blue-greens take over. A simple system for avoiding this preponderance of blue-greens is to fertilize weekly, keeping the monthly amount of fertilizer the same as it was when the ponds were fertilized only once a month. This frequent fertilization sustains growth of the green algae, which keeps the blue-greens in check.

SODIUM AND POTASSIUM

Blue-green algae are an index of eutrophication. But by the time they begin to grow, inorganic nitrogen is often scarce, and P is sometimes almost undetectable, though sometimes it is high. This paradox hardly supports the hypothesis that N and P are the only important components leading to the change in flora induced by eutrophication. Other factors must be at work.

Ten years ago, as I discussed with Dr. Edmondson the events leading to the eutrophication of Lake Washington, it became evident to me that two changes had occurred: urbanization with increased emission of sewage, and a slight increase in salinity due to the intrusion of seawater when a canal was opened to the sea. The latter led me to consider the possible effect of increased minerals, mainly Na, and to recollect that the blue-green algae have an absolute need for Na as well as for K—a pattern apparently not shared by any of the other freshwater algal groups. In a review on the ecology of algae (Provasoli, 1958), I advanced the possibility that these monovalent ions might, with other factors, be responsible for tipping the balance in favor of the blue-greens. Strong circumstantial evidence seems to favor this hypothesis, but no direct proof is available.

Allen (1952), following up previous observations, found that *Synechococcus cedorum*, two *Chroococcus* species, and one *Oscillatoria* needed Na for growth; that several isolates of *Nostoc* required appreciable concentrations of K; and that 20 other blue-green isolates could satisfy their needs with either K or Na. *Anabaena cylindrica* (Allen and Arnon, 1955), *A. variabilis, Anacystis nidulans, Nostoc muscorum* G (Kratz and Myers, 1955), and a *Chroococcus* (Emerson and Lewis, 1942) need both Na and K for logarithmic growth. *Microcystis aeruginosa* most likely needs both monovalents (Gerloff *et al.*, 1952), but *Oscillatoria rubescens* needs only K (Staub, 1961). The essential need for one of these two cations is well established only for *A. cylindrica* and Na (Allen and Arnon, 1955); the other workers employed CP chemicals without further purification. The requirement of *A. cylindrica* for Na is specific (K, Li, Rb, and Cs could not substitute for Na), and 5 ppm or more of Na was sufficient for optimal growth; larger amounts apparently are not harmful (Allen and Arnon, 1955). *Oscillatoria rubescens* needs 10 ppm of K for optimal growth (Staub, 1961). Maximal growth of *Anabaena variabilis* and *Anacystis nidulans* is supported by 40 ppm of Na + 5 ppm of K (Kratz and Myers, 1955).

Eutrophication by soil runoff introduces some Na, often Ca, and especially K, which is a component of most agricultural fertilizers. The relative abundance of Na and K is reversed in treated sewage effluents, where Na is a major addition (Oswald, 1961; Na = 72 ppm, and K = 13 ppm). While eutrophication rarely increases the total solids content to a level excluding the oligotrophic planktonic forms, over a long period it may notably change the mineral balance, especially in shallow small lakes or in lakes having a long retention period.

The data assembled by Beeton (1965) on the Great Lakes are very interesting. During the last 50 years the total solid content remained almost constant in Lakes Superior and Huron, increased 20 ppm in Lake Michigan, and increased 50 ppm in Lakes Ontario and Erie. The main mineral changes during the eutrophication of Lakes Ontario and Erie have been increases in sulfates, chlorides, Ca, Na, and K (see Figure 1). Sulfate also increased in Lake Michigan and in Lake Huron (but much less); chloride increased only slightly, and Ca, Na, and K remained constant. Therefore, the changes, typical only of Lakes Erie and Ontario, are increases in Cl (from 8 to 20 ppm), Ca (from 31 to 40 ppm), and Na + K (from 6 to 12 ppm). Lakes Michigan and Huron are still oligotrophic. In Lake Michigan, however, *Bosmina longirostris* has replaced *B. coregoni,* which was still predominant in 1927. The same change of species happened in Lake Zurich (Minder, 1938) and in Lake Washington (Edmondson *et al.,* 1956) and is considered one of the first signs of eutrophication.

The biological changes in Lake Erie are marked: increased total phytoplankton production; increased intensity and duration of the spring and fall phytoplankton maxima; shift of dominant phytoplankton of the spring pulse from *Asterionella* to *Melosira,* and, in the fall pulse, replacement of *Synedra* by *Melosira, Fragilaria,* and blue-green algae (Davis, 1964); drastic decrease in lake herring, whitefish, blue pike, and walleye; and now dominance of yellow perch, smelts, and freshwater drum, the quantity of the catch remaining constant despite heightened phytoplankton production.

TABLE 1 Ionic Composition (mg/liter)

Lake	pH	K^+	Na^+	Ca^{2+}	Mg^{2+}	HCO_3	SO_4^{2-}	Cl^-	Total PO_4	NO_3^-
Windermere (south basin)[a]	—	0.6	3.6	6.3	0.5	8.6	7.5	6.3	0.002 (as P)	0.3 (as N)
Goose Lake[b]	7.9	3.0	3.5	42.0	11.6	130.2	13.0	4.7	0.05–0.3	0.2
Sylvan Lake (basin C)[b]	9.1	3.4	50.0	61.3	19.5	134.1	40.5	51.5	1.5–3.0	0.2
Lilla Ullevifjärden (1964)[c]	—	3.6	14.0	39.0	7.9	129.0	38.0	15.9	0.018	—
Ekoln (1964)[c]	—	4.1	13.0	47.3	7.0	140.0	40.2	15.8	0.119	—

[a]From Lund (1959).
[b]From Wetzel (1966a).
[c]From Ahl and Willén (1965).

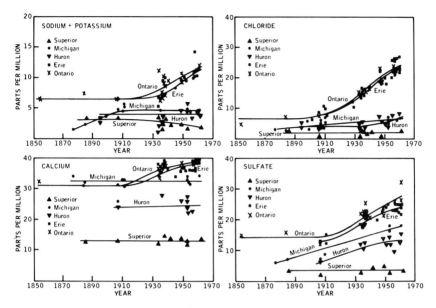

FIGURE 1   Changes in the concentrations of sodium-plus-potassium, calcium, chloride, and sulfate in each of the Great Lakes. (Redrawn from Beeton, 1965.)

Similar increases in Na and K have been recorded for other lakes that have become eutrophic from treated sewage effluents. Wetzel (1966b) studied the productivity of several marl lakes in northern Indiana, concentrating on two of them: Goose Lake, a typical representative of the oligotrophic type of the region, and the hypereutrophic Sylvan Lake, which receives the effluent of a sewage-disposal plant. The typical mineral divergence of Sylvan Lake from the average of the other marl lakes is a pronounced increase of Na and Cl (respectively, from an average of 4.4 to 47 ppm and from 4.0 to 55 ppm) and a minor increase of K (from an average of 1.85 ppm to 3.4 ppm); the Ca, Mg, and SO_4 values for Sylvan Lake are within the range of other lakes (Table 1). Nitrate is similar in both lakes (0.2 ppm), but total PO_4 in Goose Lake is 0.1 to 0.3 ppm, and in Sylvan Lake it is 0.5 to 5 ppm. The annual mean primary productivity is 729 mg of C/m^2 per day for Goose Lake and 1,564 mg of C/m^2 per day for Sylvan Lake; the dominant species of Goose Lake apparently are *Ceratium hirundinella* and *Oscillatoria agardhii*, and of Sylvan Lake, *Anabaena flos-aquae* and *Anacystis*. (Data on the volume or number of cells are lacking in these interesting publications.)

In most cases, Na, K, and PO_4 increased simultaneously, making it difficult to discriminate between their respective roles. The only case I found in which only the level of P varies is the one given by Ahl and Willén (1965) for two parts of the Lake Mälaren system: Lilla Ullevifjarden, which is unpolluted, and Ekoln, which is polluted by sewage water from the town of Uppsala and by leaching of agricultural fertilizers.

The chemistry of these lakes is almost a perfect test case. Lilla Ullevifjarden and Ekoln have almost identical composition of major elements. They have calcareous waters similar in composition to those of Goose Lake except that their content of Na, Cl, and SO_4 is much higher (Table 1). The combined concentrations of K and Na, while not as high as in Sylvan Lake, should be sufficient for optimal growth of blue-green algae according to the laboratory data mentioned above. This is so; even the phytoplankton of the unpolluted Lilla Ullevifjarden is rich in blue-greens: *Aphanizomenon* sp. +, *Oscillatoria* spp. 86,000 cm/liter, *Rhodomonas minuta* 230,000 cells/liter, Peridineae +. As the total content of P increases gradually in the chain of communicating basins from the southern Lilla Ullevifjarden to the northern Ekoln, blue-green algal composition changes gradually: *Aphanizomenon* decreases and *Oscillatoria* increases (150,000 cm/liter) in the first lake above Lilla Ullevifjarden; in the second above (Ryssviken), *Aphanizomenon* disappears, *Oscillatoria* decreases (25,000 cm/liter), and *Microcystis* predominates (1.2 $\times$ 10^6); in Ekoln *Microcystis flos-aquae* becomes the dominant blue-green (3 $\times$ 10^6) and *Rhodomonas minuta* increases tenfold. This change seems to indicate that although freshwater blue-green algae share with the oligotrophic algae the advantage of efficiently utilizing low P and N (even *no* N for the nitrogen-fixing species), they are favored by, need, and can withstand high concentrations of Na, K, Cl, and SO_4, which may put the other species at a disadvantage. An extreme case was found by Iwai (1962) in fertilizing Japanese fish ponds: *Microcystis aeruginosa* kept increasing in growth up to 1.5 g of K/liter. Qualitiative differences in algal composition appeared also in Russian fish ponds enriched differentially with Na, K, and P (Braginskii *in* Lund, 1965, page 254). This circumstantial evidence seems to warrant more attention to the role of Na, K, and Mo (needed for nitrogen fixation) in pollution problems and a reconsideration of their concentrations in eutrophic water. Undoubtedly, more data than the few mentioned here are available.

TRACE METALS

The indispensability of silicon for diatoms is well documented (see reviews of Lewin and Guillard, 1963, and Lund, 1965). But nothing is known about the indispensability of silicon for the Chrysophyceae (which have silicified scales and spores) and for the species of Xanthophyceae that have silicified spores.

Silicon deficiency has often been considered the limiting factor for the growth and persistence of diatom blooms. However, there is no experimental evidence that adding silicon would prolong or restore a declining diatom bloom.*

The need for iron, manganese, boron, molybdenum, zinc, cobalt, copper, vanadium, and other elements has been well documented for a few single algae (reviews of Pirson, 1955, and Wiessner, 1962). We may infer that all algae may need them, but we do not know whether certain species, genera, or algal groups have peculiarly high requirements for any of them. Similarly, few reliable analyses of trace metals are available for freshwater and seawater. Fertilization with manganese and molybdenum increased algal productivity (as ^{14}C uptake) in many lakes (Goldman, 1960, 1964), as did additions to waters of chelators that solubilize trace metals (Schelske et al., 1962; Wetzel, 1966a). Incidentally, the peptides produced by blue-green algae are good solubilizers of tricalcium phosphate (Fogg and Westlake, 1955). Unfortunately, the published accounts of these very interesting experiments do not say which algal species were stimulated by the additions.

Sensitive new techniques (e.g., atomic absorption spectrometry) permit accurate analysis of the fluctuation of most of these elements. Interesting correlations with phytoplankton growth should be forthcoming. The need for these data cannot be overemphasized.

VITAMINS

Most algae studied (147 out of 204 species) require vitamins. Because the requirement for vitamins is absolute and can be satisfied only by the vitamin or its physiological equivalent, the specificity of this requirement is perhaps the greatest differential found in algal nutrition and might be responsible, with other factors, for changes of flora during eutrophication. Vitamin requirements of freshwater and marine organisms are listed by Provasoli (1958), Lewin (1961), Droop (1962), and Provasoli and Droop *in* Oppen-

---

*The addition of silicon alone will probably restore diatom growth only temporarily, and a more complete fertilization may be needed, because other nutrients, especially the nitrates and phosphates, are likely to be close to deficiency levels. Such experiments might prove extremely valuable in eutrophic waters, despite the apparent paradox of suggesting fertilization of already rich waters.

If successful, a sustained high level of diatom growth might check significantly, or even replace, the obnoxious blue-green algal blooms; the result of weekly fertilization of the Israeli fish ponds indicates that this might indeed happen.

The advantages of such change in the flora of eutrophic lakes are obvious: progressive accumulation of organics in the bottom will be checked because the diatoms, being part of the normal food chain, will be eaten by the herbivores, resulting in high production of fish. Even if the fish thus obtained are not the most appetizing, they can easily be fished and transformed into valuable proteins. Their periodic removal would tend to equalize the budget of the lake again and would stem further eutrophication.

TABLE 2 Summary of Vitamin Requirements[a]

	No Vitamins	Vitamins	B_{12}	Thiamine	B_{12} + Thiamine	Biotin	Ratio B_{12}/Thiamine	Major Requirement
Algal Groups								
DO NOT REQUIRE BIOTIN								
Blue-green algae	9	7	7	0	0	—	7/0	B_{12}
Chlorophytes	24	49	16	13	20	—	36/33	none
Diatoms	22	24	18	3	3	—	21/6	B_{12}
Cryptomonads	0	11	2	2	7	—	9/9	none
REQUIRE BIOTIN								
Euglenids	0	11	—	1	9	1	–	none
Chrysomonads	1	27	1	9	14	5	17/24[b]	Thiamine
Dinoflagellates	1	18	13	0	4	5	17/5[c]	B_{12}

[a]For vitamin requirements of single algal species, see Provasoli (1958); Lewin (1961); Droop (1962); Provasoli, remarks and table in Oppenheimer (1966), p. 94–97; and Hutner and Provasoli (1964).

[b]Not in table are: two *Synura* requiring B_{12} + biotin; and *Ochromonas* requiring thiamine + biotin.

[c]Not in table: one *Gymnodinium* requiring thiamine + biotin.

heimer (1966); vitamin requirements of marine algae are listed by Provasoli (1963).

Only three vitamins are required by unicellular photosynthetic algae: B_{12}, thiamine, and biotin. Of these, vitamin B_{12} and thiamine are needed by most auxotrophs (107 and 78, respectively); biotin is needed only by a few species of the chrysomonads and dinoflagellates.

The data indicate that there is no correlation between need for vitamins and source of energy and no correlation between need for vitamins and type of environment. Photoautotrophs and photosynthetic heterotrophs may or may not need vitamins (Provasoli, 1961), and marine and freshwater algae from oligotrophic or polluted environments may or may not be auxotrophs.

Increasingly clear trends seem to differentiate the various algal groups, regardless of their salinity needs (Table 2). The algae may be divided into two main groups on the basis of their need for biotin or lack of it.

The algal groups indifferent to biotin can be subdivided into two groups, as follows:

1. Partially auxotrophic (blue-greens, greens, and diatoms). The ratio autotrophic–auxotrophic varies in these groups and may not be significant; there is a paucity of data.

2. Totally or predominantly auxotrophic (cryptomonads).

All algal groups requiring biotin are predominantly auxotrophic but differ clearly in the incidence of species requiring either B_{12} or thiamine. That the need for a vitamin is predominant is indicated by the number of species requiring single vitamins (columns 3 and 4, Table 2) and by the ratio B_{12}/thiamine. Thus the chrysomonads are considered to have a predominant need for thiamine because 9 out of 10 species requiring only one vitamin need thiamine and because the sum of all the species requiring two or more vitamins shows a preponderance of species requiring thiamine over those requiring B_{12}. (Twenty-seven species require vitamins: 24 of these require thiamine; 17 require B_{12}.) The opposite is true for the dinoflagellates, in which the need for B_{12} strongly predominates.

The type of vitamin needed, the proportion of autotrophs and auxotrophs, and the need for predominant vitamins may serve as elements in an emerging taxonomic key that would characterize and differentiate the algal groups physiologically.

These physiological differences in the algal groups offer hope that it may be possible to find the causes of algal successions in waters.

Since the seasonal succession of algae occurs in all environments, its causes should be independent of the parameters that differentiate environments (e.g., total solids, ionic balance, and pH); it should be related to the seasonal

TABLE 3   Species Composition of the Phytoplankton in the Sargasso Sea off Bermuda during the Spring of 1958 (cells/liter)[a]

Species	Number	Species	Number	Species	Number
March 25		April 15		May 19	
Coccolithus huxleyi	6,400	*Rhizosolenia stolterfothii*	6,500	*Coccolithus huxleyi*	917
Bacteriastrum delicatulum	300	*Bacteriastrum delicatulum*	5,800	*Syracosphaera mediterranea*	612
Lauderia borealis?	120	*Cerataulina bergonii*	3,000	*Syracosphaera heimii*	374
Syracosphaera mediterranea	80	*Leptocylindrus danicus*	2,300	*Rhizosolenia alata*	220
Stephanopyxis palmeriana	40	*Bacteriastrum hyalinum*	1,800	*Nitzschia seriata*	102
Oxytoxum variabile	20	*Eucampia zoodiacus*	1,500	*Syracosphaera pulchra*	102
Syracosphaera heimii	20	*Thalassiosira rotula*	1,200	*Oxytoxum variabile*	102
Rhizosolenia stolterfothii	20	*Nitzschia delicatissima*	1,000	*Katodinium rotundatum*	102
		Chaetoceros laciniosus	1,000	*Rhabdosphaera hispida*	68
		Chaetoceros decipiens	1,000	*Rhizosolenia calcar-avis*	34
		Nitzschia seriata	700	*Calciosolenia granii*	34
		Thalassiothrix mediterranea	600	*Porella perforata*	34
		Thalassiothris frauenfeldtii	300	*Discosphaera tubifer*	34
		Dinophysis ovum	300	*Braarudosphaera bigelowii*	34
		Coccolithus huxleyi	200	*Ceratium teres*	34
		Chaetoceros afinis	200		
		Nitzschia closterium v. recta	200		
		Chaetoceros didymus	200		
		Rhizosolenia calcar-avis	100		

[a]From Hulburt et al. (1960).

phenomena to which the algal groups react specifically (e.g., variation in temperature, light, and nutrients). Of the many nutrients varying seasonally, silicon and perhaps Na and K seem, so far, to affect certain algal groups differentially; too little is known about the trace metals; N and P do not seem to be group differentials. The vitamins, because of the emerging algal-group patterns and the specificity of the requirement, seem to affect several algal groups differentially and therefore may be important in algal succession.

This seems to be so. Because of the lack of reports on the yearly fluctuations of vitamins and of the phytoplankton in freshwaters, a marine example will be used: the excellent work on the Sargasso Sea off Bermuda of the team of Ryther, Guillard, Hulburt, and Menzel. The annual cycle of the phytoplankton is typified by a bloom of diatoms in April—a very small one compared with the diatom bloom in temperate waters. The diatoms are present throughout the year, but at extremely low levels. Among the flagellates, *Coccolithus huxleyi* is the dominant species the year around (Table 3).

The Sargasso Sea off Bermuda has some interesting features: temperature is almost constant, light intensity varies seasonally, N and P are scarce, and the concentration of iron is extremely small and varies little seasonally and with depth (Menzel and Spaeth, 1962a). The quantity of B_{12} in the upper 50 m fluctuates from undetectable to 0.03 ng/liter* from May to November, then starts to rise to a peak in March at 0.07 ng/liter (Figure 2). For the assay, Menzel and Spaeth (1962b) used the diatom *Cyclotella nana*, which was isolated from the Sargasso Sea and was developed as a bioassay organism by Ryther and Guillard (1962). The diatom bloom occurs when B_{12} is high; at the end of the diatom bloom, B_{12} drops to 0.01 ng/liter. The scarcity of B_{12} during the year parallels the paucity and endemicity of the diatoms. The probability that the diatom bloom depends on the B_{12} levels is heightened by *in-vitro* data; most of the diatoms isolated from the Sargasso Sea, including some bloom species, require vitamin B_{12} (Guillard and Ryther, 1962; Guillard and Cassie, 1963).

Reasons for the endemicity and predominance of *Coccolithus huxleyi*, except during the diatom bloom, remained a mystery until it was found that the Sargasso Sea strain and an English strain of *C. huxleyi* need thiamine and not B_{12} (Pintner and Provasoli, 1960, and R. R. L. Guillard, private communication). Field and *in-vitro* data fit beautifully together in showing correlations between fluctuations of B_{12} and growth of diatoms requiring B_{12} and in showing that the small permanent population of *C. huxleyi* probably depends on a low but sustained level of thiamine. Menzel and Spaeth (1962b) do not think that B_{12} limits primary production in the

*ng = nanogram.

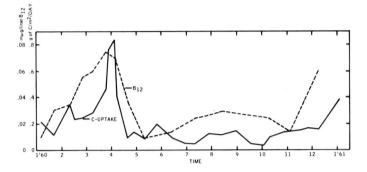

FIGURE 2   Relation between primary production and the mean concentration of vitamin B_{12} in the euphotic zone. (Redrawn from Menzel and Spaeth, 1962b.)

Sargasso Sea but that its fluctuations may exert an important effect on the composition of the phytoplankton.

A previous study by Vishniac and Riley (1961) on the fluctuations of vitamin B_{12} in the richer coastal waters of Long Island Sound shows a similar winter peak of B_{12} (16 ng/liter), the occurrence in the early spring of a large bloom of *Thalassiosira* and *Skeletonema* (both need B_{12}), and the corresponding drop of B_{12} to 4 ng/liter. Termination of the diatom bloom in this case seems to be due to lack of nitrogen, not of B_{12}; however, later chemostatic experiments showed that levels of $B_{12} < 8$ ng/liter reduce the growth rate of Skeletonema by 50 percent or more (E. Adair Wood, unpublished data).* Because B_{12} was found in coastal waters at levels much higher than the levels of B_{12} needed by different algae *in vitro*, Droop suggested that vitamin B_{12} is not limiting and may not be ecologically important (Droop, 1957, 1961, 1966). This opinion has not been widely accepted.† Unfortunately, it was expressed far too early and may have discouraged field and *in-vitro* work of the type so fruitfully pursued in the Sargasso Sea.

The main objections are that Droop's opinion is based on the needs of only one species of marine algae (which may not be typical), that *in-vitro* quantitative data may not be meaningful under natural conditions, and that the need for B_{12} in chemostatic experiments is higher than in stagnant cultures.

---

*Ph.D. thesis, Bingham Oceanographic Laboratory, Yale University, New Haven, Conn. (1962).

†See Provasoli (1963a), the interesting discussion (pages 98–112) and the summation of Lund (pages 227–249) *in* Oppenheimer (1966), and Guillard and Cassie's (1963) data on the number of molecules of B_{12} needed to produce 1 μg^3 of several diatoms.

The lack of data on freshwater species, especially the ecologically important ones, is unfortunate. Eutrophication and pollution, great sources of vitamins and other growth factors, create problems in freshwater bodies that are much more severe and immediate than corresponding problems in the communicating oceans; this is so because of the limited size of the freshwater bodies and their higher exposure to man's impingement. Digested sludge contains 1.5 to 1.7 μg of B_{12} per ml, 19 to 22 μg of B_{12} per g of dry solids (Neujahr, 1955, 1960), and 10 μg of biotin per g of dry solids (Neujahr and Hartwig, 1961). For comparison, liver, considered the richest source of antipernicious anemia factor, contains 0.7 to 0.9 μg of B_{12} per g of dry solids, and the quantity needed by most algae for maximum growth is 0.01 μg or less per liter. Bacteria are the main producers of vitamins: 70 to 80 percent of the soil bacteria in the rhizosphere produce B_{12} (Lockheed and Thexton, 1951), and the enormous growth of bacteria in sewage decomposition is also responsible for the high production of vitamins.

Any natural environment or condition favoring bacterial growth is likely to be vitamin rich—for example, bottom muds, suspended particulate matter, and marshes and underwater vegetation where growth alternates with cycles of decomposition.* Random isolation of bacteria from these locales yielded a good number (20 to 40 percent of cobalamin producers (Ericson and Lewis, 1953; Starr et al., 1957; Burkholder and Burkholder, 1956, 1958). However, only 20 percent of these bacteria produce true vitamin B_{12}; the majority produce other cobalamins. These analogs differ from true B_{12} in their lack of, or variation in, the nucleotide in the side chain of the cobalamin molecule. The specificity of microbes and algae is determined by the nature of the nucleotide and falls into three groups (Droop et al., 1959; Provasoli, 1963a; Guillard, 1968). Most algae have a narrow specificity and can employ only true B_{12}; however, more than half of the marine diatoms tested are able to employ all the analogs besides B_{12}. This versatility may give to these diatoms an ecological advantage over the algae restricted by a narrow specificity.

Since the ratio B_{12}/analogs varies widely in nature, the specificity of the algae for different cobalamins is another important differential. Because of

*Differential enrichment done by adding to natural waters N, P, and trace metals alone and in combinations proved extremely useful in assessing the nutritional properties of the waters, especially in detecting the limiting factors (Goldman, 1960, 1964; Wetzel, 1966a; Schelske et al., 1962; Ryther and Guillard, 1959; Menzel and Ryther, 1961; Menzel et al., 1963). Unfortunately, these nonaseptic conditions cannot be employed to detect whether the phytoplankton of the sample is stimulated by vitamins or whether the vitamins are limiting. It is well known that bacteria find favorable conditions for growth when samples of water are confined in a container, even a sterile one, because surfaces absorb nutrients and offer attachment to bacteria. In turn, bacterial growth will alter the vitamin content widely.

this complex situation, measurements of cobalamins in waters, to be meaningful, should be done with at least two bioassay organisms: *Escherichia coli* to assess "total cobalamins" and *Ochromonas malhamensis* for true B_{12}.

Further caution is dictated by the finding of J. C. and R. A. Lewin (1960) that even different strains of the *same* diatom species may require different vitamins or none at all. Species of the same genus often differ. For example, *Fragilaria capucina* does not need vitamins (Provasoli, 1958) but *F. crotonensis* needs B_{12} (Vollenweider and Saraceni, 1964). Therefore, to avoid gross errors, data on vitamin content of waters must be correlated with laboratory determination of the needs in vitamins obtained *exclusively* on species isolated from the same water.

The effects of vitamins, which may be of great ecological importance, are varied. So far we have considered the vitamins only as indispensable for the requiring species. As mentioned, B_{12} concentrations also affect the growth rate, but detailed studies on the ecologically important species are lacking. This effect alone would offset Droop's opinion. The abundance of vitamins in rich environments has the important effect of eliminating the disadvantage (especially in poor environments) of the vitamin-requirer in its competition with the nonrequirer.

Vitamin B_{12} apparently affects auxospore formation. Guillard and Cassie (1963) found that *Chaetoceros* sp. (clone BBsm) never formed resting spores in vitamin-deficient media, although it produces spores when other nutrients become deficient. They comment that a minimum B_{12} level is needed for spore formation and that this minimum is consistent with the idea that certain estuarine species are excluded from the open sea because their life cycles are not adjusted to the cycle of nutrient supply and depletion in deep waters. Conversely, R. Holmes (personal communication) found that lack of B_{12} induces in *Ditylum brightwelli* formation of auxospores; these auxospores germinate if B_{12} is added. This might explain some important ecological happening such as the reinoculation from year to year.

One wonders also whether these vitamin-needing species are the only consumers and whether this consumption may not speed the rate of growth and increase yield of the nonrequirers by sparing needed synthesis. Vitamin B_{12} was found to be a far more efficient source of cobalt than inorganic cobalt for non B_{12}-requiring blue-green algae (Holm-Hansen *et al.*, 1954), but we do not know whether B_{12} is consumed by these blue-greens in lakes as a source of cobalt. Finally, seaweeds (e.g., *Polysiphonia* sp. and *Dasya* sp.), which lose their normal morphology in the absence of the normal microflora and in synthetic media, acquire different forms (cellular or structural), the form depending on which vitamin is added, even though normal morphology cannot be re-established by a "complete" mixture of vitamins (Provasoli, 1963b).

CHEMICAL SYMBIOSIS

Vitamins are only one of the many chemical interrelationships between species. The overwhelming source of vitamins in nature is microbial, but the algae themselves can excrete surplus vitamins, and this ability may underlie associations and various degrees of intimacy. The diatom *Lichmophora hyalina* normally grows as dense tufts on the pluricellular alga *Spongomorpha arcta*. Lewin (1958), in finding that *Lichmophora* requires B_{12}, wondered if this need might be a reason for its colonization of *Spongomorpha*. He found that young thalli of *Spongomorpha* contain 0.5 μg of B_{12} per g. It is also well known that the algae of lichens often produce abundant thiamine, which is needed by the fungal partner (Zehnder, 1949).

Another interesting example is the need of *Ulva lactuca* and *Monostroma oxyspermum* for morphogenetic substances. When these closely related seaweeds are deprived of their usual microflora, they completely lose the leafy morphology of the thallus and become, respectively, a colony of uniseriate branching filaments (*Ulva*) and an assemblage of loose cells, some round, some composed of 1 to 3 round cells attached to a disproportionately long rhizoidal cell (*Monostroma*). Normal morphology of *Monostroma* can be restored by the supernatant of cultures of two marine bacteria (out of 40 clones obtained by plating various seaweeds), aseptic supernatants of *Sphacelaria* (a brown seaweed) and of several red seaweeds (*Antithamnion, Bangia, Dasya,* and *Polysiphonia*), and by the brown exudate of *Fucus* (Provasoli and Pintner, 1964). The latter, according to Craigie and McLachlan (1964), is a flavanol or catechin-type tannin. Tannins of higher plants are inactive for *Monostroma*. None of the supernatants active for *Monostroma* restore normal morphology of *Ulva*. Because a number of phenolic compounds are known to be active in green plants, supposedly through their ability to stimulate or inhibit indolacetic acid oxidase, we tried a number of simple phenols known to be precursors of tannins, lignins, and anthocyanins as well as the phenolics active on indolacetic acid oxidase. These compounds were tried on *Monostroma* and *Ulva*. All of them were inactive for *Monostroma*, but a few restored normal morphology for *Ulva*, at least in the early stages of growth, up to 2-cm length (Provasoli and Pintner, 1966). Among the compounds active for *Ulva*, especially in combination, are phenylalanine; 2,4-dihydroxybenzoic, 3,5-dibromosalicylic, 3,5-dibromo-*p*-hydroxybenzoic, homovanillic, and ferulic acids; *p*-hydroxybenzaldehyde and syringylaldehyde; vanillyl, dihydroconiferyl, dihydrocoumaryl, and dihydrosinapyl alcohol; and the brominated compound of *Polysiphonia lanosa* (Hodgkin *et al.,* 1966). Interestingly, similar compounds are found when sea muds (Degens *et. al.,* 1963) or the organic-colored river waters (Christman and Ghassemi, 1966) and humic acid (Steelink, 1963) are chemically degraded, and when humic acids are degraded by *Penicillium* (Mathur and Paul, 1966). Soil and mud extracts are well-known panaceas for solving the problems in cultivating algae. Most of this activity can be attributed to vitamins and chelated trace-metal mixtures; evidently, soil extracts and the like are still holding some interesting secrets.

The preliminary work on *Monstroma* is an example of chemical symbiosis between seaweeds, bacteria, and other seaweeds living in the same level of the littoral zone. Antagonisms among algae (Lefèvre, 1964) may also play an important role in the predominance of species and the succession of forms and had to be mentioned briefly to sketch what needs unraveling before we can manage our environment a little more wisely.

This research was aided in part by contract Nonr 4062 with the Office of Naval Research and research grant GB-4860 of the National Science Foundation.

REFERENCES

Ahl, T., and T. Willén. 1965. Mälarundersökningen–en presentation. Svenst Naturvetenskap. 301–316.

Allen, M. B. 1952. The cultivation of Myxophyceae. Arch. Mikrobiol. 17:34–53.

Allen, M. B., and D. I. Arnon. 1955. Studies on nitrogen-fixing blue-green algae. II. The sodium requirement of *Anabaena cylindrica*. Physiol. Plant. 8:653–660.

Beeton, A. M. 1965. Eutrophication of the St. Lawrence Great Lakes. Limnol. Oceanogr. 10:240–254.

Burkholder, P. R., and L. M. Burkholder. 1956. Vitamin B_{12} in suspended solids and marsh muds collected along the coast of Georgia. Limnol. Oceanogr. 1:202–208.

Burkholder, P. R., and L. M. Burkholder. 1958. Studies on B vitamins in relation to the productivity of Bahia Fosforescente, Puerto Rico. Bull. Mar. Sci, Gulf and Caribbean 8:201–223.

Christman, R. F., and M. Ghassemi. 1966. Chemical nature of organic color in water. J. Amer. Water Works Ass. 58:723–741.

Craigie, J. S., and J. McLachlan. 1964. Excretion of colored ultraviolet-absorbing substances by marine algae. Can. J. Bot. 42:23–33.

Davis, C. C. 1964. Evidence of the eutrophication of Lake Erie from phytoplankton records. Limnol. Oceanogr. 9:275–283.

Degens, E. T., K. O. Emery, and J. H. Reuter. 1963. Organic materials in recent and ancient sediments. III. Biochemical compounds in San Diego Trough, Calif. N. Jb. Geol. Paläont. Mh. 5:231–248.

Droop, M. R. 1957. Vitamin B_{12} in marine ecology. Nature 180:1041–1042.

Droop, M. R. 1961. Vitamin B_{12} in marine ecology: the response of *Monochrysis lutheri*. J. Mar. Biol. Ass. U.K. 41:69–76.

Droop, M. R. 1962. Organic micronutrients, p. 141–159 *In* R. A. Lewin [ed] Physiology and biochemistry of algae. Academic Press, New York.

Droop, M. R. 1966. Vitamin B_{12} and marine ecology. III. An experiment with a chemostat. J. Mar. Biol. Ass. U.K. 46:659–671.

Droop, M. R., J. J. A. McLaughlin, I. J. Pintner, and L. Provasoli. 1959. Specificity of some protophytes toward vitamin B_{12}-like compounds. Preprints Intern. Oceanogr. Congr. AAAS, 916–918.

Edmondson, W. T., G. C. Anderson, and D. R. Peterson. 1956. Artificial eutrophication of Lake Washington. Limnol. Oceanogr. 1:47–53.

Emerson, R., and C. M. Lewis. 1942. The photosynthetic efficiency of phycocyanin in *Chroococcus* and the problem of carotinoid participation in photosynthesis. J. Gen. Physiol. 25:579.

Ericson, L. E., and L. Lewis. 1953. On the occurrence of vitamin B_{12} factors in marine algae. *Arkiv Kemi* 6:427–442.

Fogg, G. E., and D. F. Westlake. 1955. The importance of extracellular products of algae in freshwater. Proc. Internat. Ass. Theor. Appl. Limnol. 12:219–232.

Gerloff, G. C., G. P. Fitzgerald, and F. Skoog. 1952. The mineral nutrition of *Microcystis aeruginosa*. Amer. J. Bot. 39:26–32.

Goldman, C. R. 1960. Primary productivity and limiting factors in three lakes of the Alaska Peninsula. Ecol. Monogr. 30:207–230.

Goldman, C. R. 1964. Primary productivity and micro-nutrients limiting factors in some North American and New Zealand lakes. Verh. Internat. Verein. Limnol. 15:365–374.

Guillard, R. R. L. 1968. B_{12} specificity of marine centric diatoms. J. Phycol. 4:59–64.

Guillard, R. R. L., and V. Cassie. 1963. Minimum cyanocobalamin requirements of some marine centric diatoms. Limnol. Oceanogr. 8:161–165.

Guillard, R. R. L., and J. H. Ryther. 1962. Studies of marine planktonic diatoms. I. *Cyclotella nana* Hustedt, and *Detonula confervacea* (Cleve) Gran. Can. J. Microbiol. 8:229–239.

Hodgkin, J. H., J. S. Craigie, and A. G. McInnes. 1966. A novel brominated phenolic derivative from *Polysiphonia Ianosa*. Can. J. Chem. 44:74.

Holm-Hansen, O. G., G. C. Gerloff, and F. Skoog. 1954. Cobalt as an essential element for blue-green algae. Physiol. Plant. 7:665–675.

Hulburt, E. M., J. H. Ryther, and R. R. L. Guillard. 1960. The phytoplankton of the Sargasso Sea off Bermuda. J. Cons. Intern. Explor. Mer. 25:115–128.

Hutner, S. H., and L. Provasoli. 1964. Nutrition of algae. Ann. Rev. Plant Physiol. 15:37–56.

Iwai, T. 1962. Ecological studies on the phytoplankton of the brackish water ponds. J. Fac. Fish. Univ. Mie-Tsu 5:412–506.

Kratz, W. A., and J. Myers. 1955. Nutrition and growth of several blue-green algae. Amer. J. Bot. 42:282–287.

Lefèvre, M. 1964. Extracellular products of algae, p. 337–367 *In* D. F. Jackson [ed] Algae and man. Plenum Press, New York.

Lefèvre, M., and G. Farrugia. 1958. De l'influence, sur les algues d'eau douce, des produits de la décomposition spontanée des substances organiques d'origine animale et végétale. Hydrobiologia 10:49–65.

Lewin, J. C., and R. R. L. Guillard. 1963. Diatoms. Ann. Rev. Microbiol. 17:373–414.

Lewin, J. C., and R. A. Lewin. 1960. Auxotrophy and heterotrophy in marine diatoms. Can. J. Microbiol. 6:127–134.

Lewin, R. A. 1958. Vitamin-benzonoi de algoi. Sciencaj studoj, Copenhagen 187–192.

Lewin, R. A. 1961. Phytoflagellates and algae, p. 401–147 *In* W. Ruhland [ed] Handbuch der Pflanzenphysiologie, Vol. XIV. Springer-Verlag, Berlin.

Lochhead, A. G., and R. H. Thexton. 1951. Vitamin B_{12} as a factor for soil bacteria. Nature 167:1934–1935.

Lund, J. W. G. 1959. Biological tests on the fertility of an English reservoir water (Stocks Reservoir, Bowland Forest). J. Instn. Wat. Engrs. 13:527–549.

Lund, J. W. G. 1965. The ecology of the freshwater phytoplankton. Biol. Rev. 40:231–293.

Mathur, S. P., and E. A. Paul. 1966. A microbiological approach to the problem of soil humic acid structures. Nature 212:646–647.

Menzel, D. W., E. M. Hulburt, and J. H. Ryther. 1963. The effects of enriching Sargasso Sea water on the production and species composition of the phytoplankton. Deep Sea Res. 10:209–219.

Menzel, D. W., and J. H. Ryther. 1961. Nutrients limiting the production of phytoplankton in the Sargasso Sea, with special reference to iron. Deep Sea Res. 7:276–281.

Menzel, D. W., and J. P. Spaeth. 1962a. Occurrence of iron in the Sargasso Sea off Bermuda. Limnol. Oceanogr. 7:155–158.

Menzel, D. W., and J. P. Spaeth. 1962b. Occurrence of vitam B_{12} in the Sargasso Sea. Limnol. Oceanogr. 6:151–154.

Minder, L. 1938. Der Zurichsee als Eutrophierungs-phänomen. Geol. Meere Binnen-gewässer 2:284–299.

Neujahr, H. Y. 1955. On vitamins in sewage sludge. II. Formation of vitamin B_{12}, folic acid, and folinic acid factors in municipal sludge. Acta Chem. Scand. 9:622–630.

Neujahr, H. Y. 1950. On vitamin B_{12} factors in sewage sludge. Almquist Wiksells, Uppsala. 58 p.

Neujahr, H. Y., and J. Hartwig. 1961. On the occurrence of biotin in different fraction of municipal sewage. Acta Chem. Scand. 15:954–955.

Oppenheimer, C. H. 1966. [ed] Marine biology II, Proc. 2nd Internat. Interdisciplinary Conf. N.Y. Acad. Sci. Interdisciplinary Communications Program. 369 p.

Oswald, W. J. 1961. Metropolitan wastes and algal nutrition. Trans. seminar algae and metropolitan wastes. R. A. Taft Sanit. Engin. Center Rep. W 61-3:88–95.

Pintner, I. J., and L. Provasoli. 1960. Nutritional characteristics of some chrysomonads, p. 114–121 In C. H. Oppenheimer [ed] Symposium of marine microbiology. Charles C Thomas, Springfield, Ill.

Pirson, A. 1955. Functional aspects in mineral nutrition of green plants. Ann. Rev. Plant Physiol. 6:71–114.

Provasoli, L. 1958. Nutrition and ecology of protozoa and algae. Ann. Rev. Microbiol. 12:279–308.

Provasoli, L. 1958. Growth factors in unicellular marine algae, p. 385–403 In A. A. Buzzati-Traverso [ed] Perspectives in marine biology. Univ. Calif. Press.

Provasoli, L. 1961. Micronutrients and heterotrophy as possible factors in bloom production in natural waters. Trans. Seminar Algae and Metropolitan Wastes R. A. Taft Sanit. Engr. Center Rep. W 61-3:48–56.

Provasoli, L. 1963a. Organic regulation of phytoplankton fertility, p. 165–219 In M. N. Hill [ed] The sea, 2. Interscience, New York.

Provasoli, L. 1963b. Growing marine seaweeds, p. 9–17 In Proc. 4th Internat. Seaweed Symp. Pergamon Press, New York.

Provasoli, L., and I. J. Pintner. 1964. Symbiotic relationships between microorganisms and seaweeds. Amer. J. Bot. 51:681.

Provasoli, L., and I. J. Pintner. 1966. The effect of phenolic compounds on the morphology of Ulva, p. 23 In 11th Pacific Sci. Congr. Proc. Abst. Papers in Fish. 7.

Ryther, J. H., and R. R. L. Guillard. 1959. Enrichment experiments as a means of studying nutrients limiting to phytoplankton production. Deep Sea Res. 6:65–69.

Ryther, J. H., and R. R. L. Guillard. 1962. Studies of marine planktonic diatoms. II. Use of *Cyclotella nana* Husted for assay for vitamin B_{12} in seawater. Can. J. Microbiol. 8:437–445.

Schelske, C. L., F. F. Hooper, and E. J. Haertl. 1962. Responses of a marl lake to chelated iron. Ecology 43:646–653.

Starr, T. J., M. E. Jones, and D. Martinez. 1957. The production of vitamin B_{12} active substances by marine bacteria. Limnol. Oceanogr. 2:114–119.

Staub, R. 1961. Ernährungsphysiologisch–autökologische Untersuchungen an der planktischen Blaualge *Oscillatoria rubescens* DC. Schweiz, Z. Hydrobiol. 23:82–198.

Steelink, C. 1963. What is a humic acid? J. Chem. Educ. 40:379–384.

Vishniac, H. S., and G. A. Riley. 1961. Cobalamin and thiamine in Long Island Sound: patterns of distribution and ecological significance. Limnol. Oceanogr. 6:36–41.

Vollenweider, R. A., and C. Saraceni. 1964. La richiesta di vitamina B_{12} in *Fragilaria crotonensis* Kitton (Bacillariophyceae, Pennales). Mem. Ist. Ital. Idrobiol. Marco De Marchi 17:223–230.

Wetzel, R. G. 1966a. Productivity and nutrient relationships in marl lakes of northern Indiana. Verh. Internat. Verein. Limnol. 16:321–332.

Wetzel, R. G. 1966b. Variations in productivity of Goose and hypereutrophic Sylvan lakes, Indiana. Invest. Indiana Lakes and Streams 7:147–184.

Wiessner, W. 1962. Inorganic micronutrients, p. 267–286 *In* R. A. Lewin [ed] Physiology and biochemistry of algae. Academic Press, New York.

Zehnder, A. 1949. Über den Einfluss von Wuchsstoffen auf Flechtenbildner. Ber. Schweiz. Botan. Ges. 59:201–267.

DAVID G. FREY
Indiana University, Bloomington

Evidence for Eutrophication
from Remains of Organisms in Sediments

Every participant in this symposium has had to decide what limits to place on his subject within the over-all context of the symposium. As a paleo-limnologist I had the problem of deciding how far back in time to go—early middle postglacial? interglacials? Tertiary?—because eutrophication in the sense of natural change accompanying increased nutrient input has certainly gone on whenever, and probably virtually wherever, lakes have existed. I finally decided to become essentially a neolimnologist and to restrict my attention as much as possible to the relatively recent changes in lakes that have occurred through man's burgeoning population and his use and abuse of the lakes and their watersheds. The question I wish to ask is: what can paleolimnology contribute to our understanding of the causes, consequences, and rates of change in lakes brought about by cultural influences? More specifically, what can the organismic remains in sediments contribute?

The answer will have to be based more on hope than on accomplishment because, although there are a substantial number of paleolimnologists in the world, few have given sustained study to the most recent changes in lake evolution. Much of the work reported on here is only very recently published. Much of it is in the process of being published. Some is not yet in the final manuscript stage. I am indebted to various persons for making their data and conclusions available in advance of publication.

Limnologists (in the sense of neolimnologists) and paleolimnologists have the same objective: determining process and rate and the interactions between the various components of the biotic system of lakes. But there are vast differences in the resources available to them for study and interpretation. The paleolimnologist might be thought of as a historian and the

594

limnologist as a news analyst or stock market analyst plotting trends and making predictions.

The limnologist is working with the dynamic living system. The information available to him concerning the system is limited only by the present sophistication of methods and instrumentation and by his capabilities and endurance. His methods for gathering information generally yield data from discrete moments in time, except for some modern techniques for continuously recording certain variables. And his information is not repeatable, unless he sets aside samples of this time series for later attention, or unless the processes he studies are more-or-less cyclic on a daily or seasonal basis. Changes between measurements must be interpolated and the rates calculated from these changes.

The paleolimnologist, on the other hand, is working with a static derivative of the dynamic system. At least he hopes it is relatively static with respect to diagenesis or other processes that can alter the record. Because of the continuity of the sedimentary record, the paleolimnologist can return at will to any particular time in the past for rechecking his observations and measurements or for getting new information. The amount of information available to the paleolimnologist is much smaller than that available to the limnologist, because many processes and organisms leave no traces in the sediments. However, this incomplete record is all that is available in those many instances where cultural eutrophication is an accomplished fact and where there are no previous data to document what the lake was like in precultural times.

There is really no need to apologize for the amount of information in sediments; it is tremendous, although still largely unappreciated. Some lines of investigation are already quite highly developed and are yielding exciting results. Others are barely perceived, much less explored. To a considerable extent, it is not yet fashionable to study recent sediments. However, even the relatively few studies that have been conducted make it clear that paleolimnology will have a real impact on our eventual over-all understanding of eutrophication and its effects on lake ecosystems.

"Remains of organisms" can be interpreted to mean both (1) the biochemical substances produced by organisms or resulting from their degradation and (2) the morphological fragments that can be identified with the organisms that produced them. I shall not devote much time to the biochemical fossils, since these constitute a large and complex field by themselves. Principal efforts, to date, have been concerned with sedimentary chlorophyll degradation products, carotenoids, amino acids, sugars, and so on (Vallentyne, 1957, 1963; Swain, 1965). Though not all authorities agree to the possibility, chlorophyll degradation products may eventually be useful in inferring past levels of production (Megard, 1967). And myxoxanthin, myxoxanthophyll, oscillaxanthin, and other carotenoids may be useful for demonstrating the occurrence and abundance of groups of organisms—blue-green algae, for example—that either leave no morphological remains in sediments or leave remains difficult to identify specifically (Züllig, 1961; Brown and Colman, 1963).

The studies of biochemical fossils and the general chemistry and mineralogy of sediments have a big advantage over the study of morphological remains: with the newer sophisticated methods and techniques, only small quantities of sediment are needed for analysis. Hence, a close-interval stratigraphy can be constructed easily from a core only an inch or two in diameter. In contrast, morphological fossils have to be freed from their enclosing sedimentary matrix by some suitable means of concentration before they can be studied under a microscope by a person specifically trained in their identification. The size of sample needed to get significant numbers of fossils is determined by the abundance of the particular organisms in the sediments.

Probably all groups of organisms, even the bacteria (Bradley, 1963), leave some morphological remains in sediments, although sometimes only under highly specialized conditions of deposition. All the groups of aquatic animals certainly leave remains (Frey, 1964), and the algae probably do also, as Korde (1966) has indicated. The difficulty in using morphological remains in helping to interpret the response of lakes to man's influence is that the time span involved is very short, particularly in North America; hence, the thickness of the sediments produced during this interval is measured in centimeters rather than meters. To be useful, organisms must have a high resolution; that is, they must be abundant enough for the construction of a stratigraphy in which the sampling intervals are in millimeters or, at most, centimeters. Few types of microfossils are this abundant.

Because of their abundance and taxonomic diversity, the most important morphological remains in freshwater sediments are pollen and spores (chiefly of terrestrial plants), diatom frustules, and various fragments of the Cladocera. These groups have contributed most thus far to sediment studies related to man-induced eutrophication. However, other groups of organisms, such as the ostracods, midges, and neorhabdocoeles, cannot be neglected. Some of these have already contributed important information concerning longer time sequences of sediments, and they may well prove to be important for the interpretation of more recent changes in lakes.

POLLEN AND SPORES

Pollen diagrams are useful to the paleolimnologist for providing a relative chronology of the terrestrial vegetation of the region. Changes discernible in the development of the lake can then be related to changes in the pollen diagram, and inferences can be made as to the cause of these changes. For example, the Längsee in southern Austria, which is now a meromictic lake, apparently began meromixis when agricultural man moved into the region roughly 2,000 years ago and began clearing the land for his crops (Frey, 1955). The increase in slope wash that resulted produced clay turbidity currents that apparently triggered the onset of meromixis. A similar response of other small lakes to early agriculture has been inferred for northern Germany by Helmut Müller (in correspondence).

Other changes in pollen diagrams from Europe indicate that early and medieval man made extensive modifications of watersheds, which had repercussions on their contained lakes. Two examples of such studies are those of Goulden (1962) on Esthwaite Water in England and of Hutchinson (Cowgill and Hutchinson, 1964; Hutchinson, this volume, page 17) on Lago di Monterosi in Italy. We realize more and more from such studies how sensitively a lake responds to changes in its watershed. Although these responses are eutrophication, too, our main concern is with the changes in lakes brought about by the big increase in human population in recent decades.

Perhaps North America has a distinct advantage over most other parts of the world in such studies because western agricultural man entered the picture comparatively recently. He brought with him not only cereals and other domestic plants from the Old World but also the common agricultural weeds of these regions. These weeds flourished and began contributing pollen grains to accumulating lake sediments. As agriculture developed, native weed species also flourished. As a result, in the lakes of the United States, particularly those of the Midwest, the base of the culture horizon is indicated by the appearance and rapid buildup to high relative levels of the pollen of various agricultural weeds, chief among which is *Ambrosia*. The appearance of these pollen grains provides a very nice stratigraphic horizon, which can be dated from historical records showing when man moved into the region and began modifying its ecology. Sufferers from ragweed hay fever and asthma may derive some consolation from knowing that the cause of their allergy is destined to play an important part in the interpretation of recent lake changes.

The rest of this report is concerned with the diatoms and Cladocera. The information available from each sample of sediment is a list of the species present and the number of remains of each species. Numbers are commonly expressed as percentages of the total population, and the totals are related to volume of sediment, or weight of organic matter, or some other parameter. Percentage distributions are not completely satisfactory because of the constraint imposed by the method, and numbers related to volume or weight do not take into account varying rates of sedimentation. The ideal way of expressing changes in population composition and abundance—and the only feasible way to compare one lake with another—is to report everything as numbers of organisms sedimented per square centimeter per year. Where the sediments are varved, where certain stratigraphic horizons can be dated from historical records, or where the rate of sedimentation can be determined from a series of radiocarbon dates, such rates of accumulation of microfossils can be calculated.

Interpretation of the results depends, at least in part, on knowing the ecological requirements of the species and the communities of which they are a part. The activities of past generations of limnologists notwithstanding, such information is woefully lacking, and hence much attention must be paid to studying the living organisms in present-day situations. "Indicator organisms" do not give much promise. A better approach is community analysis, and with this approach significant advances are being made as various persons apply information theory (species diversity, redundancy, equitability, and the like) to the populations they are studying. Such studies can lead to a refinement of the models that have been devised for interpreting community response to changing ecological conditions. The advantage of lake sediments in this regard is that community stability and adjustment can be studied over periods of time measured in hundreds or even thousands of years.

DIATOMS

The diatoms illustrate as well as any other group of organisms one of the central questions of paleolimnology: how faithfully do the remains recovered from the sediments reflect the composition of the producing populations? Dr. Lund's paper in this volume (see page 306) calls attention to the silica cycle of a lake. By "silica cycle" we mean the great reduction that takes place in the silica content of the epilimnion as the silica becomes transferred into deep water and to the surface of the sediments through the sinking of dead and senescent diatoms. This reduction begins with the spring pulse of diatoms. By the following spring the silica content of the surface waters has recovered, which suggests that a considerable portion of the sedimented diatom frustules becomes redissolved. This dissolution has been confirmed by Tessenow (1966), who found that the frustules sedimented in the littoral and upper profundal zones dissolved at a much higher rate than those sedimented in the profundal zone at depths greater than 25 m. *Daphnia* and midges fragment the frustules in their feeding activities and hence aid the process of dissolution.

In a review of diatoms in sediments, Juse (1966) proposed the following grouping of diatoms according to the degree to which their frustules dissolve in the water: (1) species with delicate frustules that dissolve rapidly and almost never are recovered from the sediments (e.g., *Fragilaria crotonensis*); (2) species with frustules that dissolve in part, so that what is recovered from the sediments is not completely representative of the producing population (e.g., *Asterionella formosa* and some planktonic species of *Synedra* and *Fragilaria*); and (3) species whose frustules undergo almost no dissolution. Juse reported that 80 to 90 percent of freshwater diatoms belong in group 3, whereas about half of the marine species belong in groups 1 and 2. She also pointed out, basing her observation on work by Proshkina-Lavrenko, that in some lakes diatoms in the sediments dissolve gradually, with only occasional frustules being found at depths greater than 50 cm.

Such reports suggest that little could be concluded from the study of diatom frustules in sediments. However, one species that Juse lists as almost never being recovered from lake sediments, *Fragilaria crotonensis*, has been recovered abundantly and is one of the important indicators of eutrophication. Furthermore, in the Zürichsee, massive blooms of diatoms have formed laminae of frustules up to several millimeters thick over the lake bottom in deep water. Although there certainly is dissolution of frustules at the surface of sediments, the amount of dissolution varies with depth in the lake and probably from one lake to another. Nevertheless, the diatom assemblages in offshore cores of sediments show progressive changes that are interpretable in terms of the known history of the lakes.

Two of the more significant studies concerning the use of diatoms in interpreting recent cultural changes in lakes are on the Zürichsee in Switzerland and Lake Washington in the United States.

THE ZÜRICHSEE

Since 1895, the deepwater sediments in the Zürichsee have been deposited in discrete annual layers (Nipkow, 1920). Each layer consists of a calcareous member deriving from the period of summer stratification and a darker organic member deriving from the rest of the year. After his initial study, in which he demonstrated the annual nature of these layers, Nipkow (1927) carried out a detailed year-by-year study on the diatoms as well as on the other organismal remains he could recognize, including *Phacotus, Difflugia, Ceratium* and *Dinobryon* cysts, *Staurastrum,* and the cladocerans *Bosmina* and *Daphnia.* He was so enthusiastic about this sedimentary record that he referred to it as a "unique hydrobiological museum."

He carried out a study in two parts, one of which does not contribute much to the present topic, though it is interesting. In this part of the study, Nipkow recorded changes in size-frequency distributions of the various species from year to year. He found a progressive decrease in mean size over a period of several years and then a sudden increase in size, which he associated with the production of auxospores. Auxospores could be found in the sediments. Nipkow observed the same phenomenon in the Baldeggersee in Switzerland, which also has laminated sediments.

The significant part of the study for our purposes concerns his semiquantitative results (Figure 1). Nipkow fractionated each annual layer into four parts, from each of which he removed roughly 1 mm^3 of sediment. The subjective abundance of each species was scored according to four categories he established. The height of the bars in the figure represents this subjective abundance.

Nipkow thought of a lake's response to an increased supply of nutrients as occurring in two stages: first, an increase in abundance of the species already present in the lake; and second, the invasion and establishment of new species. Some invasions of the second type in the Zürichsee were almost explosive, such as the large initial blooms of *Tabellaria fenestrata* in 1896 and of *Oscillatoria rubescens* in 1898. Nipkow was able to show that a number of these first occurrences of spectacular blooms of species already established occurred in the same year as, or immediately following, historical disturbances at the shoreline (slumping or landslides). These disturbances, according to his viewpoint, upset the equilibrium accompanying the slow rate of accumulation of nutrients and triggered a new combination of environmental conditions, enabling a new species to take over irruptively. Although this relationship seems to be valid for the Zürichsee, it may not be valid for

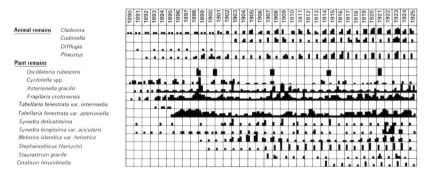

FIGURE 1   Fluctuations in abundance of various organisms, chiefly diatoms, in the stratified sediments of the Zürichsee at a water depth of 140 m. Each annual layer was fractionated into four parts, corresponding to the seasons. The height of each individual bar represents abundance according to the four subjective categories—scarce, scattered, numerous, very abundant. The bars for *Oscillatoria rubescens* denote those years in which blooms of this species were particularly heavy in the lake. The species leaves no morphological remains in sediments. (After Nipkow, 1927).

other Swiss lakes reviewed by Hasler (1947), in which the first appearance of *Oscillatoria* was likewise explosive.

The bloom of *Tabellaria* in 1896 was associated with the first marked decrease in oxygen content of the hypolimnion. The bloom deposited a layer of *Tabellaria* frustules almost 5 mm thick over the lake bottom. Blooms of other species in subsequent years also produced distinct biological laminae in the sediments. From 1908 onward, *Stephanodiscus Hantzschii*, which occurs only during the spring, often produced a brownish-yellow lamina of its frustules 1 to 2 mm thick in the spring portion of the annual sediment layers. *Synedra* had exceptional winter–spring maxima in the years 1922–1924, each of which resulted in a layer of frustules 4 to 5 mm thick. These are examples of how much sediment can be produced by single massive blooms of diatoms.

Nipkow attempted to relate abundance of diatoms to precipitation and temperature in the various years. Not only was there no relation, but years having approximately the same precipitation and temperature could have vastly different productions of diatoms. He concluded, therefore, that the chief control of production from one year to another was the culturally induced irregularities in nutrient input, which masked any effects caused by climatic variation.

Although there have been big shifts in the diatom community of the lake, with *Cyclotella* largely dropping out (this occurs even more markedly in the Baldeggersee) and *Tabellaria* and *Fragilaria* becoming almost continuously abundant, the big troublemaker has been *Oscillatoria rubescens*, which at times has colored the water of the Limmat River in Zürich a deep red. If one had only the sediments available and no other limnological data from the lake itself, would he be able to diagnose the occurrence of this species and the years in which it was especially abundant? We do not know, but I hope that someone will eventually analyze a core from the Zürichsee for oscillaxanthin and related carotenoids. Because *Oscillatoria* leaves no morphological remains in the sediments, the only hope for detecting the occurrence of this alga in lake sediments is from some specific stable biochemicals. Incidentally, E. A. Thomas (personal communication) has reported that *Oscillatoria* has now disappeared from the Zürichsee, at least temporarily. The process began about 3 years ago, and for the past year there have been no filaments in water samples. The cause of the decline is not known, although a virus is suggested.

LAKE WASHINGTON

The very interesting situation in Lake Washington has been investigated by W. T. Edmondson and his co-workers since the early 1950's and is becoming well known through their various papers and reports. For the following material

I am indebted to John Stockner, who furnished me in advance of publication a copy of the manuscript written by him and Woodruff Benson (Stockner and Benson, 1967).

The history of the Lake Washington situation is well documented. From Seattle's early beginnings around 1860 until the early 1900's, the city grew rapidly in population and was undoubtedly contributing a variety of substances to the lake. Between 1910 and 1920, 30 sewage outfalls were contributing raw sewage to the lake. As a result of complaints, primary treatment was begun in 1924. Because this did not alleviate the situation satisfactorily, in 1926 the city began diverting the primary effluent from the lake into Puget Sound. From 1936 to 1941 the lake received only storm-sewer overflow from the city. Further expansion of the city resulted in the construction of new secondary sewage-treatment plants, which began putting their effluent into Lake Washington in 1941. The number of plants kept increasing, the most recent one being constructed in 1959. Now, as Edmondson documented in his report (see page 124), all effluents are being completely diverted from the lake into Puget Sound.

Stockner and Benson were interested in determining if there was any detectable response of the diatoms in the lake to this sequence of enrichment, reduction, and enrichment, as indicated by the frustules recoverable from the sediments. From a core 30 cm in diameter, they extracted their subsamples at closely spaced intervals and made their counts. Except for some changes at about 80 cm, which are related to a clay turbidite layer in this particular core but not generally present throughout the lake, the percentage composition of the various species was really quite constant up to a depth of about 30 cm, above which extensive changes occurred (Figure 2). These changes consisted largely of reciprocal fluctuations in abundance of *Melosira italica* and *Fragilaria crotonensis*, two of the most abundant diatoms in the sediments. A statistical comparison of the rank order of abundance of the seven dominant species above and below the 30-cm level showed that in all instances except one (*Fragilaria construens*) the differences were significant, generally at the 1-percent level. Thus, really major changes in the diatom community of the lake occurred in these top sediments. Calculations of species diversity and redundancy showed much the same pattern, although with less immediate response above the 30-cm level.

Since the response of the diatoms to changes in the lake seemed to occur according to major taxonomic groups, the authors grouped all the centric diatoms together (*Melosira, Stephanodiscus, Cyclotella,* and *Rhizosolenia*) and all species of the tribe Araphidineae together (*Tabellaria, Fragilaria, Synedra,* and *Asterionella*). These two groups of diatoms constituted about 90 percent of those recovered. The remainder were pennate diatoms that occur primarily in the littoral zone as epiphytes.

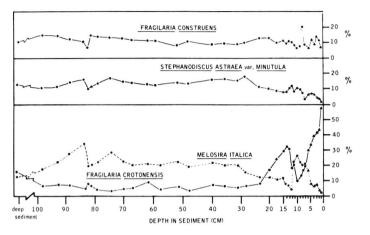

FIGURE 2   Changes in percentage representation of four dominant species of diatoms in the uppermost meter of sediments of Lake Washington. The fluctuations around 80 cm are associated with a clay turbidite layer. The fluctuations above 30 cm are related to recent historical changes in the lake. (From Stockner and Benson, 1967.)

From these data a clear pattern begins to emerge (Figure 3). During the early history of the lake, as represented by this core, there was a typological equilibrium in which the centric diatoms were by far the dominant ones. Only the clay horizon at about 84 cm shows any marked squiggles in the line. A gradual increase in araphidine diatoms then follows to a peak at 15 cm, a sharp decrease to 8 cm, and then a marked expansion at lesser depths to a peak of almost 80 percent at the surface. This performance seems to parallel the known history of enrichment, reduction, enrichment.

What was needed was some kind of time scale for this stratigraphy. Lake Washington does not have the varved sediments that the Zürichsee does (few lakes do), nor could the authors find any good time horizons in the sediments, except a volcanic-ash layer at about 6 m dating from the Mount Mazama eruption 6,700 years ago, and a sand lens presumably dating from 1916 when the lake level was lowered about 3 m. On the basis of these horizons and some other reasoning that seems logical, the authors estimated that the average annual sedimentation rate was about 2.5 mm over the top 30 cm, and that it perhaps increased from about 1.5 mm per year at 30 cm to about 3.5 mm at the surface. With this rough time scale, they re-examined their data to see how the changes in the diatoms agreed with the recorded sanitary history of Seattle. The results show surprisingly good agreement (Figure 4), suggesting that by studying the relative abundance of sedimentary diatoms in these two major groups we can gain considerable insight into the changes in lakes that accompany enrichment.

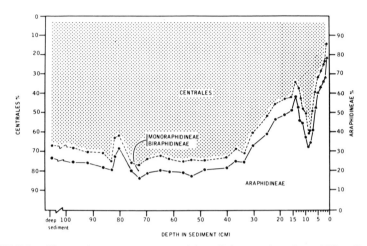

FIGURE 3   Changes in percentage composition of the centric and araphidine diatoms in the uppermost meter of sediments of Lake Washington. The changes around 80 cm are associated with a clay turbidite layer. The changes from 30 cm to the surface reflect the changing nutrient income of the lake derived from sewage plant effluent. (From Stockner and Benson, 1967.)

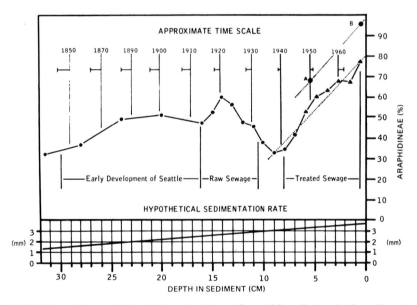

FIGURE 4   Changes in the relative abundance of araphidine diatoms in the sediments of Lake Washington related to the developmental history of the region. The time scale is based on a reasonable average sedimentation rate of 2.5 mm per year over the top 30 cm, increasing linearly from 1.5 mm at 30 cm to 3.5 mm at the surface. (After Stockner and Benson, 1967.)

Other workers have also considered centric diatoms to be more characteristic of oligotrophic waters, and pennate diatoms more characteristic of eutrophic waters. Various workers have established diatom ratios in an attempt to obtain some objective expression of this shift, although they have tended to use all pennate diatoms instead of just the araphidines, as Stockner and Benson have done. In Lake Washington, the Zürichsee, and the Baldeggersee, it is the centrics that go out and the araphidines that come in as enrichment proceeds. Stockner and Benson feel that it may be much more reasonable to use "indicator groups" of diatoms as they have done in the Lake Washington study rather than look for "indicator species."

CLADOCERA

The most abundant animal remains in sediments are the heads and shells of various Cladocera. They have been known to occur in lake sediments for several decades, but only in recent years have attempts been made to use them in reconstructing lake histories. Part of this reluctance was based on the difficulty of their identification. But now these identification problems are solved for Western Europe and can be solved without too much additional effort for other specific localities. From the analysis of cladoceran remains in lake sediments, one obtains for each sample a list of species present and the number of remains of each species, the same as in an analysis of diatoms. The usual procedure in the few studies that have been published is to convert the numbers to percentages and then to plot these against depth, as in a pollen diagram. This enables one to follow the changes in relative abundance of each species from one level to the next.

An example of such a diagram is given in Figure 5, which is from an important study by Goulden (1966b) on the Aguada Santa Ana Vieja in Guatemala. To some extent Goulden's study is significant because of the interpretation of the paleoecology of this site in conjunction with studies on the chemistry of the sediments by Cowgill and Hutchinson (1966). To a greater extent it is significant because it is the first published study on sedimentary Cladocera that attempts some of the more sophisticated analyses now available in community ecology. From this diagram Goulden points out three major zones or communities of Cladocera and speculates as to their environmental controls, including a marked peak in burn agriculture at about 16 cm. However, disturbance of this region by man was relatively minor. There is a progressive increase in the number of species toward the surface (not all species are shown in the diagram) and also a progressive increase in abundance, even though the samples were only semiquantitative. Consequently, the Shannon-Weaver species-diversity index increases toward the

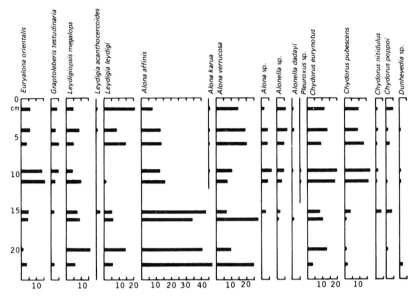

FIGURE 5   Percentage composition of chydorid Cladocera in the sediments of the Aguada de Santa Ana Vieja in Guatemala. Each horizontal line represents a sample from a different depth in the core. For each species the change in length of the bars from level to level reflects the changing relative importance of this species in the total community. (After Goulden, 1966b.)

surface, as does the equitability component. The latter is 1.0 or very close to that at the top four levels.

Goulden compared the abundance of each rank-ordered species at each level with the abundance predicted by the MacArthur Type I distribution (Figure 6). The fit is generally close, with departures explainable either on the basis of one species being much more abundant than expected and the others less abundant or on the basis of two or more congeric species sharing a resource but with some interactions between them. The cladoceran population, at least of this lake, fits a predictable, rigid pattern, which seems to be adequately described by the Type I distribution for nonoverlapping niches. However, it is not at all certain that the niches really are nonoverlapping for these organisms, even though they are claimed to be by the hypothesis. These new techniques enable us to analyze population responses of Cladocera in much greater detail than was formerly possible. Where the distribution of numbers among the species approximates the predicted values, the population is considered to be in balance. Where it does not, then something has disturbed the balance, and the population has not yet had time to re-establish balance. This would be the case in the rapid changes accompanying cultural enrichment.

Another study by Goulden (1966a) also deserves mention. This is likewise concerned with a lake in Guatemala—Petenxil. The sediment cores from this lake are more than 2 m long and almost 4,000 years old, and there is a record of almost continuous activity by man in the immediate vicinity of the lake. Hence, one would expect to find a less good fit with the MacArthur Type I distribution. This expectation is valid, as shown in Figure 7. The equitability component and the species-diversity index parallel each other quite closely, with the equitability always being less than one, denoting an "unbalanced" population, according to the hypothesis.

Because a number of radiocarbon dates were obtained for these cores, it was possible to construct a curve for rate of sedimentation and from this to translate the numbers of organisms into absolute rates of accumulation, namely, numbers of remains per square centimeter per year. This is the ultimate in studies on paleolimnology, because it eliminates any differential diluting effect of the sedimentary matrix, and for individual species it gets away from the constraint imposed by percentages. Goulden did not calculate absolute rates for individual species in the paper cited, and no one has yet done so for Cladocera, although this has been done for pollen. Lacking such

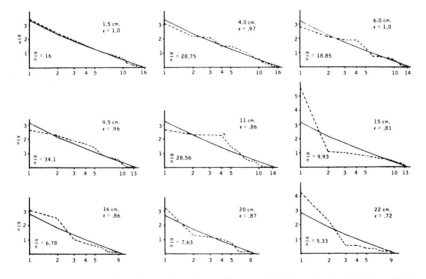

FIGURE 6 Chydorid Cladocera from the sediments of the Aguada de Santa Ana Vieja in Guatemala. In each diagram the species have been rank-ordered, and their actual abundance (broken line) is compared with their predicted abundance (solid line) on the basis of MacArthur's Type I distribution. A further calculation of the equitability component indicates that the communities present at the four uppermost levels correspond almost precisely to the Type I hypothesis for nonoverlapping niches. The chydorid community at these levels was stable and well adjusted. (From Goulden, 1966b).

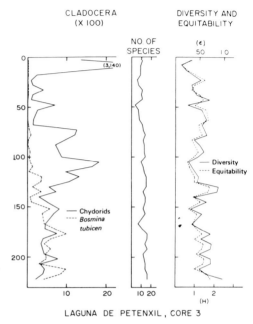

FIGURE 7   Laguna de Petenxil, Guatemala: absolute abundance of Cladocera as number of remains sedimented per square centimeter per year, number of species recovered at each level, and diversity index and equitability. (After Goulden, 1966a.)

absolute rates, the best one can do is to calculate the number of remains per unit weight of sediment or of organic matter. Figure 8 shows that the various curves parallel one another in general course, but understandably, the scales on the abscissa are different. Absolute rates of sedimentation are the only reasonable means of comparing one lake with another.

A number of other studies in the literature, or soon to be in the literature, consider the entire late-glacial—postglacial history of North Temperate lakes. But there have not yet been studies that have been concerned specifically with the changes in lakes occurring during the past few decades under man's influence, or on the response of the Cladocera to these changes. Shifts in species composition and abundance will undoubtedly be found, but they will be difficult to interpret because we know so little about the ecology of the Cladocera.

One change that can be expected is mentioned by Brooks (see page 236). *Chydorus sphaericus*, which is normally a substrate species in the littoral zone, moves offshore into the plankton when blooms of blue-green algae develop. The cladoceran apparently uses these blooms as a substrate, but what the relation is we do not know. By occupying the entire euphotic zone it is able to increase greatly in abundance and hence contribute more remains to the sediments.

One would expect that under such conditions *C. sphaericus* would become very abundant in the sediments and that the other species would become less abundant or even be eliminated. In such comparisons as these, absolute rates of accumulation are needed to prevent the other species from being overwhelmed by this one superabundant species. The only indication we have from sediments that this change in absolute abundance actually happens is in Lake Sebasticook, Maine. There *C. sphaericus* increased in abundance from about 8 percent at a depth of 6 to 8 inches to 67 percent at the surface, with the greatest increase in the uppermost inch (Mackenthun, 1966). A further suggestion that this occurs is in the finding that *Chydorus sphaericus* is the most abundant single species of Cladocera in the surficial sediments of a number of productive lakes: the four lakes at Madison, Wisconsin (Frey, 1960); 14 lakes in northern Indiana and a considerable number in Denmark (Whiteside and Harmsworth, 1967); and a series of lakes along the Mississippi River (DeCosta, 1962).

One would expect that as a lake becomes more productive, either through natural changes or through artificial enrichment, the cladoceran population would build up and contribute more remains to the sediments. In other words, one would expect the quantity of cladoceran remains in the sediments to bear a positive relation to productivity, at least up to some maximum value. The first suggestion that this expectation might be valid was obtained from the four lakes at Madison. The total quantities of cladoceran remains in the surficial offshore sediments of these lakes have the same rank order as the

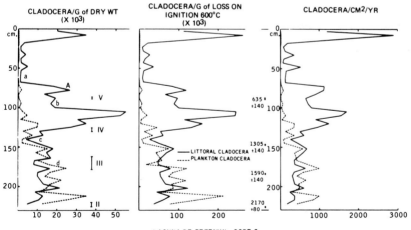

LAGUNA DE PETENXIL, CORE 3

FIGURE 8   Laguna de Petenxil, Guatemala: changing abundance of Cladocera calculated in relation to dry weight of sediments, to loss of ignition, or as numbers sedimented per square centimeter per year. Roman numerals indicate agricultural episodes. Radiocarbon dates provide a measure of the changing rates of sedimentation. (From Goulden, 1966a.)

organic content of the water, the mean standing crop of phytoplankton, and the organic content of the sediments (Frey, 1960), suggesting that all these variables are responding in concert. Unfortunately, data on primary productivity for the Madison lakes are not yet available.

The great changes in absolute rates of deposition in Petenxil (Figure 8) suggest changes in productivity. In other words, as the productivity of a body of water increases, the number of remains of Cladocera arriving at the sediments also increases. The same seems to be indicated by Nipkow's rather crude subjective results on the Zürichsee (Figure 1), based only on the analysis of *Daphnia* and *Bosmina*. In this volume (see page 65), Straškraba demonstrates a strong positive relation between kjeldahl nitrogen in the zooplankton and in the water. These are all indications, although not conclusive ones, that such a positive relation exists and that the quantities of remains in the sediments when properly interpreted can be used as an indication of past productivity.

On the other hand, Harmsworth and Whiteside (1968) could find no positive relation between the quantities of *Bosmina* and chydorids in the surficial sediments and the annual phytoplankton productivity of 19 lakes in Denmark and 14 in northern Indiana for which radiocarbon data are available. Taken at face value, these results would seem to indicate that there is no relation at all between the amount of energy being fixed in a lake and the size of the cladoceran population that is dependent on this energy. It is regrettable that the authors had no basis for calculating absolute rates of accumulation of cladoceran remains or for estimating how much primary productivity is accomplished by the macrophytes and periphyton in the littoral zone.

These conflicting indications may mean that (1) in a single lake the absolute changes in abundance from one level to another may well reflect changes in productivity, and (2) the changes in abundance between lakes in a linear drainage system may reflect differences in productivity where sedimentation rates might be expected to be comparable, but that (3) lakes in different drainages are not so easily compared, either because of overriding local influences affecting productivity or community stability or because of vastly different rates of sediment accumulation.

However, there do seem to be interpretable relations between community structure of the Cladocera and primary productivity of the phytoplankton (Whiteside and Harmsworth, 1967). In the Danish lakes mentioned above there was a significant ($P < 0.01$) negative regression between species-diversity index and annual phytoplankton productivity, whereas in the Indiana lakes, the correlation was not significant at the 5-percent level, although the trend was the same. In both series of lakes there is a highly significant positive relation between species-diversity index and transparency. A possible interpretation of these results is that the lower the productivity, the more transparent the water, and hence the greater the areal extent of the zone inhabitable by rooted aquatic plants. Since these plants constitute the main substrate of the chydorid Cladocera, intuitively we would expect a greater habitat diversity under these circumstances, with a greater number of species of chydorids and a community balance among them. As the transparency decreases (from increased phytoplankton), the zone of rooted aquatics becomes restricted and the habitat diversity declines. Hence, the number of species of the Cladocera would be expected to decrease. Coupled with this decrease would be a big increase in numbers of certain species successful under these circumstances, such as *Chydorus sphaericus*, which would help to reduce the species-diversity index. These results, along with those of Goulden on the Guatemalan lakes, suggest that community analysis of the cladoceran remains will give us much insight into the response of lakes to rapidly increasing rates of productivity.

A number of instances are known from core studies and also from long-term studies on existing lakes where *Bosmina coregoni* has been replaced by *B. longirostris* (reviewed by Frey, 1964). This replacement has been interpreted as evidence of increasing productivity, *B. longirostris* generally being considered characteristic of more productive situations, and *B. coregoni* of less productive situations. However, again we need to know much more about the biology and ecological requirements of these two species or superspecies. Because *B. coregoni* is usually considerably larger than the other species, Brooks (see page 236) has suggested that the species substitution may be a response not so much to increasing productivity as to predation pressure by fishes, which tends to reduce the mean size of the zooplankton, regardless of species.

Cladocera can survive unfavorable conditions (such as those associated with winter) either in embryonic diapause (the so-called resting or ephippial eggs) or as an active adult (the "motile" diapause of Brooks, 1965). If the suggestion of Stross and Hill (1968) is correct that the embryonic diapause is more an adaptation for survival in low oxygen tensions, such as might result under the ice in winter, rather than low temperature or short photoperiod, then as a lake undergoes eutrophication, populations of *Daphnia* that formerly were active over winter should swing into the production of ephippial eggs as the alternate means of survival. If this occurs, then one should find increasing numbers of *Daphnia* ephippia in the sediments above the base of the culture horizon.

This paper is Contribution No. 802 from the Department of Zoology, Indiana University, Bloomington, Indiana.

REFERENCES

Bradley, W. H. 1963. Unmineralized fossil bacteria. Science 141:919–921.

Brooks, J. L. 1965. Predation and relative helmet size in cyclomorphic *Daphnia*. Nat. Acad. Sci., Proc. 53:119–126.

Brown, S., and B. Colman. 1963. Oscillaxanthin in lake sediments. Limnol. Oceanogr. 8:352–353.

Cowgill, U. M., and G. E. Hutchinson. 1964. Cultural eutrophication of Lago di Monterosi during Roman antiquity. Verh. Internat. Verein. Limnol. 15:644–645.

Cowgill, U. M., and G. E. Hutchinson. 1966. La Aguada de Santa Ana Vieja: the history of a pond in Guatemala. Arch. Hydrobiol. 62:335–372.

DeCosta, J. J. 1962. Latitudinal distribution of chydorid Cladocera in the Mississippi Valley, based on their remains in surficial lake sediments. Invest. Indiana Lakes and Streams 6:65–101.

Frey, D. G. 1955. Längsee: a history of meromixis. Mem. Ist. Ital. Idrobiol., Suppl. 8:141–164.

Frey, D. G. 1960. The ecological significance of cladoceran remains in lake sediments. Ecology 41:684–699.

Frey, D. G. 1964. Remains of animals in Quaternary lake and bog sediments and their interpretation. Arch. Hydrobiol., Suppl. Ergebnisse der Limnologie 2:i–ii, 1–114.

Goulden, C. E. 1962. The history of the cladoceran fauna of Esthwaite Water (England) and its limnological significance. Arch. Hydrobiol. 60:1–52.

Goulden, C. E. 1966a. The animal microfossils. *In* The history of Laguna de Petenxil. Mem. Connecticut Acad. Arts Sci. 17:84–120.

Goulden, C. E. 1966b. La Aguada de Santa Ana Vieja: an interpretative study of the cladoceran microfossils. Arch. Hydrobiol. 62:373–404.

Harmsworth, R. V., and M. C. Whiteside. 1968. Relation of cladoceran remains in lake sediments to primary productivity in lakes. Ecology 49:998–1000.

Hasler, A. D. 1947. Eutrophication of lakes by domestic drainage. Ecology 28:383–395.

Juse, A. 1966. Diatomeen in Sedimenten. Arch. Hydrobiol., Suppl. Ergebnisse der Limnologie 4:i–ii, 1–32.

Korde, N. V. 1966. Algenreste in Seensedimenten. Zur Entwicklungsgeschichte der Seen und umliegenden Landschaften. Arch. Hydrobiol., Suppl. Ergebnisse der Limnologie 3:i–ii, 1–38.

Mackenthun, K. M. 1966. Fertilization and algae in Lake Sebasticook, Maine. Technical Services Program, Federal Water Pollution Control Administration, U.S. Govt. Department of the Interior, U.S. Govt. Printing Office, Washington, D.C. 124 p.

Megard, R. O. 1967. Limnology, primary productivity, and carbonate sedimentation of Minnesota lakes. Limnol. Res. Center, Univ. Minnesota, Interim Rep. 1. 69 p.

Nipkow, F. 1920. Vorläufige Mitteilungen über Untersuchungen des Schlammabsatzes im Zürichsee. Schweiz. Z. Hydrol. 1:100–122.

Nipkow, F. 1927. Über das Verhalten der Skelette planktischer Kieselalgen im geschichteten Tiefenschlamm des Zürich- und Baldeggersees. Schweiz. Z. Hydrol. 4:71–120.

Stockner, J. G., and W. W. Benson. 1967. The succession of diatom assemblages in the recent sediments of Lake Washington. Limnol. Oceanogr. 12:513–532.

Stross, R. G., and J. C. Hill. 1968. Photoperiod control of winter diapause in the fresh-water crustacean, *Daphnia*. Biol. Bull. 134:176–198.

Swain, F. M. 1965. Geochemistry of some Quaternary lake sediments of North America, p. 765–781 *In* H. E. Wright, Jr., and D. G. Frey [ed] The Quaternary of the United States. Princeton Univ. Press, Princeton, N.J.

Tessenow, Uwe. 1966. Untersuchungen über den Kieselsäuregehalt der Binnengewässer. Arch. Hydrobiol., Suppl. 32 Heft 1:1–136.

Vallentyne, J. R. 1957. The molecular nature of organic matter in lakes and oceans, with lesser reference to sewage and terrestrial soils. J. Fish. Res. Bd. Canada 14:33–82.

Vallentyne, J. R. 1963. Geochemistry of carbohydrates, p. 456–502 *In* I. A. Breger [ed] Organic geochemistry. Pergamon Press, London.

Whiteside, M. C., and R. V. Harmsworth. 1967. Species diversity in chydorid (Cladocera) communities. Ecology 48:664–667.

Züllig, Hans. 1961. Die Bestimmung von Myxoxanthophyll in Bohrprofilen zum Nachweis vergangener Blaualgenentfaltungen. Verh. Internat. Verein. Limnol. 14:263–270.

RICHARD J. BENOIT

General Dynamics, Electric Boat Division, Groton, Connecticut

Geochemistry of Eutrophication

The state of fertility of surface waters is largely, if not primarily, a matter of their chemical composition. The chemical composition of waters is poised by their origin, their history, and the nature of substances that they are in contact with and that man empties into them.

Geochemists divide the earth into realms, basing the division on the physical state of each realm. Thus, we have:

The lithosphere—the solid crust
The atmosphere—the gaseous envelope
The hydrosphere—the waters
The biosphere—the sum total of living things

These realms interact according to laws that are not completely understood but that are, in some cases, quite straightforward.

Man, as a component of the biosphere, has a dominant influence on many rivers and lakes. His ecology (i.e., human ecology) must be studied and, if possible, regulated somewhat for prudent management of the hydrosphere. The transport and use of natural products (e.g., phosphate rock and nitrogenous fertilizer materials), the disposal of man's body wastes, and the disposal of domestic and industrial waste materials are proper objects of study for geochemists, ecologists, and social scientists. The proper role of scientists of all kinds is to provide, for man's welfare, an understanding of the relevant phenomena in order to make possible a rational, equitable, and advantageous basis for managing the environment.

The composition of seawater is believed to have been constant over millions of years, which suggests that the marine hydrosphere is in equilibrium with the atmosphere, the lithosphere, the land-based hydrosphere, and, equally important, the biosphere. That is not to say that seawaters are not locally variable. But the magnitude of their variability is far less than the magnitude of changes occurring as water moves from the sky to the sea—from the air to the earth's crust; over and through the crust to the headwaters of a river; down the river, finally, to an estuary joining the sea, or to the closed basin of a saline lake.

There is a sameness in the land-based hydrosphere, however, that we tend to overlook. Livingstone (1963) and Rodhe (1949) pointed out that large rivers tend to resemble one another in the composition of the water in the downstream parts. Rodhe attributed this uniformity to the buffering action of ion-exchange reactions with the suspended load or with the soil. But Livingstone held that the integration of the chemical composition of tributaries is sufficient alone to account for the uniformity. According to Livingstone, there is no need to invoke sorption reactions to explain the general chemical uniformity of the water of large rivers. Table 1 gives average contents of the river waters of the six continents and gives world averages weighted for land area and runoff, according to the scheme recommended by the Council of the International Association for Scientific Hydrology in 1957. Dr. I. A. E. Bayly has pointed out that the concept of the standard composition of fresh waters has little meaning or validity in Australia. The details of his facts and reasoning are to be found in two chapters of a recent treatise on Australian inland waters (Weatherly, 1967). The chapters are "Chemical Composition," by W. D. Williams, and "The General Biological Classification of Aquatic Environments with Special Reference to Those of Australia," by I. A. E. Bayly.

Livingstone also pointed out a very strong inverse relationship between flow and total dissolved solids as measured by conductivity. The curves he showed for the Saline River at Russel, Kansas, are nearly perfect mirror images. There is some evidence (Hutchinson, 1957, Table 69) that the ratios of the various ions remain roughly constant over a wide range of flow conditions. Phosphorus (or phosphate) is not included in Table 1, nor is it treated at all by Livingstone in his monographic chapter in the *Data of Geochemistry*. Clearly we must know about variability of phosphorus under varying flow if we are to understand the fertility of lakes and rivers.

When we consider the difficulty of estimating the phosphorus content of the average river water for the world or for the various continents, we can readily understand why Livingstone chose to neglect that element. Clarke (1924), in the old edition of the *Data of Geochemistry*, gave 70 ppb of phosphorus as the average value for world rivers, which Hutchinson (1957) stated to be not inconsistent with the general picture from limnology.

TABLE 1   Composition of River Waters of the World (ppm)[a]

Location	HCO$_3$	SO$_4$	Cl	NO$_3$	Ca	Mg	Na	K	Fe	SiO$_2$	Sum
North America	68	20	8	1	21	5	9	1.4	0.16	9	142
South America	31	4.8	4.9	0.7	7.2	1.5	4	2	1.4	11.9	69
Europe	95	24	6.9	3.7	31.1	5.6	5.4[b]	1.7	0.8	7.5	182
Asia	79	8.4	8.7	0.7	18.4	5.6	5.5[b]	3.8[b]	0.01	11.7	142
Africa	43	13.5	12.1	0.8	12.5	3.8	11	–	1.3	23.2	121
Australia	31.6	2.6	10	0.05	3.9	2.7	2.9	1.4	0.3	3.9	59
World	58.4	11.2	7.8	1	15	4.1	6.3	2.3	0.67	13.1	120
Anions[c]	0.958	0.233	0.220	0.017	–	–	–	–	–	–	1.428
Cations[c]	–	–	–	–	0.750	0.342	0.274	0.059	–	–	1.425

[a]Source: Livingstone (1963).
[b]The data for Asian rivers were based on analyses that give only the sum of Na + K; Livingstone gave 9.3 for the sum, and we have partitioned the sum to make the world averages as shown.
[c]Millequivalents of strongly ionized components.

Hutchinson (1957) gives a full discussion of the climatic, geographic, and geochemical basis of variation in the phosphorus content of lakes. A number of more recent papers of interest have been abstracted and reviewed by Mackenthun (1965). The effect of cultural enrichment on phosphorus levels in lakes and streams has been noted many times, but the data of Englebrecht and Morgan (1959) are of special interest. In eight unpolluted lakes in Illinois they found a mean of 15.5 ppb of phosphorus (orthophosphate) and a mean of 35.5 ppb of phosphorus (total hydrolyzable). For 27 samples from culturally enriched streams of the Illinois River basin, the corresponding means were 180 ppb and 280 ppb. We recently conducted a survey of 23 Connecticut lakes included in the earlier work of Deevey (1940). Our work is not yet published, but a comparison here of the total phosphorus data might be useful.

The consistent increase shown in Table 2 over the 25-year period cannot be attributed to geochemical processes other than those involving human ecology.

EQUILIBRIUM MODELS OF THE COMPOSITION OF
RIVER AND LAKE WATER

An important question is whether the average composition of river water, or the water of any given body of freshwater, is consistent with the known facts concerning the solubility of the lithosphere materials. Kramer (1964) and Sutherland *et al.* (1966) have provided an equilibrium model for the composition of Great Lakes water based on solubility (or activity) equilibria for common lithosphere materials containing the various elements under consideration: calcite, dolomite, kaolinite, Gibbsite, Na- and K-feldspars, and an atmospheric carbon dioxide concentration corresponding to $P_{CO_2} = 3.5 \times 10^{-4}$ atm. The model gives a reasonably satisfactory agreement with the actual compositions observed. But what is more important is the application of the model to other water bodies in the future. Since the model is entirely explicit and rational, departures from what it predicts must be challenged on the basis of the assumptions used in the model or on the basis that solubility conditions are not in equilibrium in the place, or at the time, the water body was sampled.

TABLE 2   Comparison of Total Phosphorus Levels in Connecticut Lakes (ppb)

Reference	Eastern Highland	Western Highland	Central Lowland
Deevey (1940)	10.8	13	20
Benoit (unpublished)	16.7	24.3	24.7

TABLE 4 Kramer Freshwater Model Compared with Actual Concentrations

Lake	pH	Alkalinity (ppm)	Calcium (ppm)	Magnesium (ppm)	Potassium (ppm)	Sodium (ppm)	Phosphorus (ppb)	Fluoride (ppm)	Silica (ppm)
Superior	8.48	102	21	6.9	1.6	11	3.8	0.23	5.6
actual	7.4	46	12	2.8	0.6	1.1	5	0.15	2.1
modified	–	41	8.4	2.8	0.6	4.4	–	–	–
Michigan	8.44	93	25	6.8	1.7	12	3.2	0.18	5.6
actual	8.0	113	32	10	0.9	3.4	13	0.1	3.1
Huron	8.45	96	24	6.9	1.7	11	5.2	0.43	4.6
actual	8.1	82	23	6.3	1.0	2.3	10	–	2.3
Erie	8.39	83	32	6.8	1.9	13	8	0.4	5.6
actual	8.1	97	37	8.0	1.4	10	61	0.1	1.5
Ontario	8.37	80	35	6.8	2.0	14	6.0	0.35	5.6
actual	8.1	96	39	8.2	1.8	9.5	75	0.2	3.0

For some elements, at least, the observed concentrations may be maintained in equilibrium by sorption phenomena rather than solubility equilibria. And although some presumed sorption phenomena can just as readily be explained on the basis of solubility equilibria (Stumm, 1962), sorption processes cannot be overlooked as one basis for departures from solubility equilibrium models. Hsu (1965) discussed the mechanism of phosphate fixation in acid soils and concluded that the concentration of phosphate in solution is determined by the sum of the activities of all forms of aluminum and iron present. Sorption of phosphate on amorphous aluminum hydroxides and iron oxides, rather than the solubility of crystalline phosphate compounds, seems to dominate in acid soils. According to Hsu, the absorption process is chemical rather than physical, however; and Hsu discussed the effects of factors such as pH and concentrations on the process. In neutral and basic soils, the iron phosphate and aluminum phosphate interactions are replaced by calcium phosphate. Solubility of crystalline species probably dominates over sorption phenomena in neutral and basic soils.

A full discussion of the quantitative aspects of the Kramer model is beyond the scope of this paper. But, qualitatively, the model is quite straightforward. The scope and basis of the model are given in Table 3, in which the ionic or molecular species, together with the controlling reactions, are listed. Using recently published values of the solubility equilibrium constants for the reactions listed, Kramer calculated the predicted level of all ions and substances included in the model. Table 4 gives the computed values compared with actual contents of ions in the five Great Lakes. Agreement is reasonably good for all lakes except Lake Superior. In that case, Kramer modified the model to account for the ratio of precipitation on the lake itself (distilled water) to runoff from land (saturated according to the model); with this change, agreement with actual principal ionic contents is good.

TABLE 3   Scope and Basis of Kramer (1964) Freshwater Model

Substance or Ion	Relationship
Ca^{2+}	Calcite $= Ca^{2+} + CO_3^{2-}$
Mg^{2+}	$Mg^{2+} + $ H-illite $= 2\,H^+ + $ Mg-illite
Na^+	$Na^+ + $ K-feldspar $= K^+ + $ Na-feldspar
K^+	Kaolinite + quartz + $K^+ = $ K-feldspar + H^+ + water
Alkalinity	$P_{CO_2} = 3.5 \times 10^{-4}$ atm
Cl^-	Value assumed
SO_4^{2-}	Value assumed; in limit $Sr(CO_3, SO_4) = Sr^{2+} + CO_3^{2-} + SO_4^{2-}$
H_4SiO_4	Quartz + water $= H_4SiO_4$
PO_4^{3-}	$Ca_{10}(PO_4)_6(OH)_2 = 10\,Ca^{2+} + 6\,PO_4^{3-} + 2\,OH^-$
F^-	Dissociation of: $Ca_{9.9}(PO_4)_{5.65}(CO_3)_{.35}F_{2.16}$
H^+	Sum positive charges = Sum negative charges

In a later paper, Sutherland *et al.* (1966) elaborated further on the model and gave stability diagrams for the principal silicate materials of the lithosphere. Stumm and Leckie (1967) gave stability diagrams for sodium and calcium silicates in relation to the weathering of igneous rocks but stressed the sensitivity of the stability diagrams to the dissociation formulas and values of the constants chosen (Figure 1). They also pointed out that the carbon dioxide content of soils, because of microbial respiration, may be a hundred times greater than that of the atmosphere, which obviously affects the solubility of rocks. It might be added that the deep waters of stratified lakes have far more carbon dioxide in solution than the amount demanded by equilibrium with air.

According to Kramer (1964), there is no natural control on chloride and sulfate in lake waters, except that the amounts in rainwater set the minima. The addition of chloride and sulfate by man (principally as the sodium salts) affects the solubility equilibria in known ways, according to the model, principally through the absorption of sodium by clays and feldspar, causing a shift in pH and alkalinity that affects the carbonate dissociation.

In eutrophic lakes the sediments are a very important source of phosphate to the water, especially in unstratified lakes. Livingstone and Boykin (1962) studied the variation in phosphate content in a series of mud samples from the sedimentary column in Linsley Pond, Branford, Connecticut. They found that phosphate was highest in the deeper-lying samples, which formed during the juvenile (oligotrophic) stage in the lake's history. They pointed out that the most important factors governing the release of phosphate from the mud are (1) the exchange capacity of the mud and (2) the total ionic activity of the water. Release of phosphate is obviously inversely related to the former and directly related to the latter. Thus, phosphate would tend to be released from muds formed under a regime of low total-dissolved salts, but bathed by water of higher ionic strength. The discharge of physiologically inert brines of common salt or sodium sulfate could thus promote eutrophication.

RELATIONSHIP BETWEEN PRINCIPAL IONS AND ALGAL GROWTH

Wetzel (1966) compared two Indiana lakes—Goose Lake and the hyper-eutrophic Sylvan Lake. The principal ionic contents of the lakes are shown in Table 5. The differences in Na, Cl, and SO_4 are striking. There is good evidence that sodium in uncontaminated lakes limits growth of blue-green algae. Goldman (1960a, 1960b) and Goldman and Wetzel (1963) have shown that other principal ions can limit productivity, as discussed later. Wetzel cited the Kratz and Myers (1955) report that the threshold level of sodium

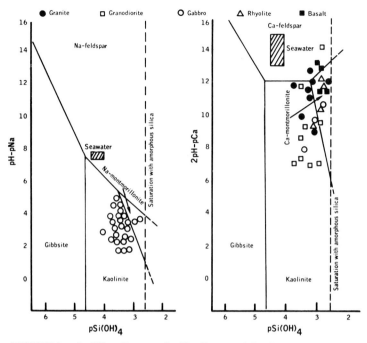

FIGURE 1   Stability diagrams for Na-silicate and Ca-silicate (Stumm and Leckie, 1967).

for near-maximal growth in laboratory cultures of three blue-green species was 4.0 ppm, a level characteristic of many Indiana lakes. Furthermore, Kratz and Myers reported that 40 ppm of Na supported maximal growth with no evidence that higher levels were inhibitory. Provasoli (1958) originally suggested the important role of sodium in blooms of blue-green algae.

TABLE 5   Principal Ionic Contents of Two Indiana Lakes[a]

Lake	Ionic Content (ppm)						
	Ca	Mg	K	Na	Cl	SO$_4$	PO$_4$ (total)
Sylvan Lake (unstratified)	53	19.5	3.4	47	60	53	1.6
Goose Lake (stratified)	43	12	3	72	5	13	0.14[b]

[a]Source: Wetzel (1966).
[b]A single bottom-water value of 2.3 mg per liter is omitted.

Synthetic detergents, which may contain as much as 50 percent by weight of sodium phosphates as "builders," are indicated as important sources of sodium (and phosphate) from urban waste waters. Rock salt used for melting ice and snow on highways may be locally important. Mixing of seawater in the upper reaches of estuaries could also be important.

Goldman (1960a, 1960b) and Goldman and Wetzel (1963), using the ^{14}C method for estimating primary productivity, studied the effect of adding various mineral supplements to natural plankton populations suspended in bottles at various depths in the lakes from which they were derived. In three lakes of the Alaskan peninsula, magnesium, nitrate, and phosphate, singly and in combination, stimulated $^{14}CO_2$ fixation. The stimulating effect of magnesium was somewhat less at stations close to the major tributary of Brooks Lake, indicating a more abundant supply of the limiting element near the tributary. The magnesium deficiency was interpreted by Goldman as being not merely an absolute deficiency but rather as being based on unfavorable ratios of the various cations as first demonstrated by Provasoli et al. (1954). A comparison of Brooks Lake, in Alaska, with the average for world rivers is given in Table 6.

Goldman compared Brooks Lake with world rivers according to Clarke (1924). Livingstone's (1963) estimate differs markedly from the older estimate, but the relative impoverishment of Mg in Brooks Lake is obvious.

It would be unwise to press the relationship of cation geochemistry to productivity in Brooks Lake too far, but the need for more data on more lakes is evident. It is still difficult to say how generally the relative abundances of major cations influence productivity. In spite of the few cases cited here, the importance of major cations to productivity would seem to be limited in lakes and rivers with abundant phosphate and fixed nitrogen. But there may well be specific floral components that are affected; for example, blue-green algae may be limited in their growth by low sodium. What may be just as important to the nuisance value of a blue-green algae bloom are

TABLE 6   Comparison of Equivalents of Principal Cations

Water Body	Ca^{2+}	Mg^{2+}	Na$^+$	K$^+$	Na + K/Ca + Mg	Na + K/Mg	Ca/Mg
	Principal Cation Equivalents (Percent)						
Brooks Lake	63.9	12.7	20.3	3.1	0.3	1.8	5.0
World Rivers[a]	52.7	24.1	19.3	3.9	0.3	1.0	2.2
World Rivers[b]	63.5	17.4	15.7	3.4	0.24	1.1	3.6

[a]Source: Livingstone (1963).
[b]Source: Clarke (1924).

specific effects of cations on habit, that is, whether the algae is colonial or solitary and whether sheath material is produced in abundance or not at all.

Other examples from the work of Goldman include the finding (Goldman and Wetzel, 1963) that sulfate and sulfate plus nitrate stimulated $^{14}CO_2$ fixation in Clear Lake, California. Phosphate was negative, but no measured levels of phosphate in the lake were reported. As far as principal cations are concerned, Clear Lake water contained 17 ppm of Ca, 17 ppm of Mg, 17 ppm of Na, and 1 ppm of K. The sodium level is especially interesting, since Goldman reported a bloom of *Aphanizomenon* (filamentous blue-green) following a September rainfall. Goldman (1960a) reported that K, SO_4, and Mo all stimulated $^{14}CO_2$ fixation in Castle Lake, California. The significance of the stimulation by Mo is discussed later.

TRACE ELEMENTS AND PRODUCTIVITY

The minor elements known to be required in trace amounts for growth of algae include Fe, Mn, Zn, Cu, Co (by some algae), Mo, and B. Although growth-limiting levels of all those elements can be demonstrated readily in laboratory cultures of algae, only very rarely has one of them been shown to limit productivity of waters in nature. In some cases (e.g., Fe, Mn, Zn, and probably Cu and B), the natural abundance of the element is high in relation to the limiting concentration. Soil deficiencies of Co, relative to the levels found in fodder crops grown on the soils and the needs of ruminants for Co in the diet, are common and worldwide in occurrence (Benoit, 1957). There is no known case, however, of cobalt deficiency in aquatic habitats. Cobalt is discussed in detail later.

IRON

The content of iron in ionic solution is extremely low in oxygenated freshwaters or seawater. Judging from laboratory culture experiments, iron should commonly limit productivity. The element is complexed by organic substances, however, and phytoplankton apparently can utilize colloidal particles of iron as well as the organic complexes. Gerloff and Skoog (1957) found 100 mg of Fe per liter and 4 mg of Mn per liter, on a fresh volumetric basis, in cultured blue-green algal cells at growth-limiting levels of these elements. Bloom cells from nature were much higher, suggesting luxury consumption of those elements.

The principal geochemical significance of iron concerns its role as an absorbent for other elements, including Mo, PO_4, and Mn, as discussed in the section on manganese.

MOLYBDENUM

Molybdenum is essential to the enzyme systems mediating nitrogen fixation. Goldman (1961) showed that nitrogen fixation by stands of alder trees contributed a substantial fraction of the fixed nitrogen (mainly nitrate) to Castle Lake, California, through runoff, springs, and leaf-fall into the lake. The same author (Goldman, 1960a) showed that Mo in the range of 1 to 50 ppb (25 ppb optimal) stimulated $^{14}CO_2$ fixation by Castle Lake phytoplankton. Nitrate was not a stimulator in the instances cited by Goldman. Therefore, we can postulate a role for Mo other than nitrogen fixation. In any event, Castle Lake seems to be one instance where productivity is limited by the availability of Mo. Sugawara et al. (1961) conducted experiments on the behavior of Mo during the oxidation-reduction cycle of lake-bottom waters (Figure 2). Suspensions of mud containing FeS and Mo in solution were permitted to stand for 109 days. The suspension pH was initially 8.1. As time passed, the pH dropped to 6.8, and iron in solution increased to 30 ppm at about 10 days, staying level until about the thirtieth day. FeS appeared on the fifteenth day. From 30 days to 100 days the iron rose to about 85 ppm. As iron rose, Mo fell until a minimum of less than 20 ppm was reached. On the 109th day the suspension was aerated and iron immediately fell to zero, with Mo rising in 1 day's time to about 65 ppm. After the aeration the system was allowed to stand for another 100 days, and as the system went anaerobic, iron rose to 60 ppm, with FeS appearing on the fortieth day. Mo fell gradually to less than 20 ppm.

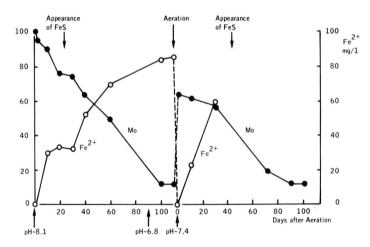

FIGURE 2   Behavior of Mo in redox changes in water in the presence of mud (Sugawara et al., 1961).

This experiment is interpreted as demonstrating that Mo is sorbed and removed to an extent by hydrous ferric oxide at pH 8 or higher and that Mo is also removed by ferrous sulfide or coprecipitated as the sulfide. This interpretation is consistent with the vertical distribution of Mo in stagnating lakes having sulfide in the deepest water. Molybdenum exists in lake water and in the sea at levels of a few parts per billion and is concentrated by a factor of 100–1,000 in sediments and plankton. In sediments from 27 Japanese lakes the range was 0.6 to 7.3 ppm, with a mean of 3.2 ppm and a single value of 18.6 ppm not included in the mean. In plankton (both phytoplankton and zooplankton), the range was found to be 0.2 to 4.0 ppm. But a single sample of plankton dominated by *Rhizosalenia* had 50 ppm.

ZINC

Bachman (1963) reviewed the geochemistry of zinc in fresh waters. Lake waters contained 5 to 100 ppb of Zn, but the equilibrium solubility of $Zn(OH)_2$ predicts 1.3×10^3 to 130×10^3 ppb over the pH range 8.0 to 9.0. Undoubtedly, there is a loss of zinc to sediments as sulfide; only 0.26 ppb would remain in solution in bottom waters saturated with FeS. Other mechanisms that remove zinc from solution are coprecipitation with $CaCO_3$ and $Fe(OH)_3$; ion-exchange on clays, silt, and humus; and the binding of zinc by microbes. Experiments by Bachman showed zinc to be absorbed by killed algae cells according to the same pattern as that of living cells, with "luxury" absorption occurring at high concentration of Zn. The sorption is increased by increasing pH or increased suspended matter and is decreased by an increase in total concentration of all other ions relative to zinc. Zinc should be similar to Mo in its vertical distribution in stagnant lakes, with an increase in deep waters but with a loss in deepest water containing sulfide. Zinc has never been shown to limit productivity in a lake; rainwater is reported by Bachman to contain 2.5 to 12 ppb of the element, which alone is probably sufficient for phytoplankton growth.

MANGANESE

The geochemistry of Mn relative to the fate of radioisotopes in the environment was reviewed by Wangersky (1963). The Mn contents of igneous rocks (0.01 percent) and soils (0.2 to 0.5 percent) are high relative to the needs of plants and animals. But Mn has several valence states, and it forms many insoluble compounds. Prominent among these is MnO_2, which will go into solution only under acidic or low redox conditions generally limiting to plants and animals. Manganese is oxidized and precipitated by so-called iron bacteria, and Wangersky cites the precipitation of manganese carbonate by sulfate-reducing bacteria. The literature on marine and freshwater manganese crusts and modules is extensive, but the mechanism of their formation is still unsettled. Wangersky stressed a mechanism involving the oxidation of lower-valence states of Mn released from the solution of $CaCO_3$, and he pointed out that the surfaces of $Mn(OH)_4$ and $Fe(OH)_3$ or oxides were catalytic.

Morgan and Stumm (1965) discussed transformations of iron and manganese that are important to the geochemistry of lake waters. Figure 3, from their paper, shows the effects of redox potential and pH on the state of iron and manganese in water. Also, it shows the behavior of iron and manganese compounds in the sorption of both cations and anions. The important general processes are the change of valence of both elements at redox-potential levels characteristic of molecular oxygen enhanced by increased pH, which results in (1) precipitation of both iron and manganese; (2) precipitation of ferrous and manganous carbonates at pH above 7.0; (3) the pH-dependent sorption of anions (low pH) and cations (including Mn^{2+} by hydrous iron oxide) at high pH; and (4) the stabilization of colloid at higher pH and flocculation of colloidal particles at lower pH. These processes have important relevance to water purification, waste treatment, and limnology. Equally important are the rates of the reactions involved and the

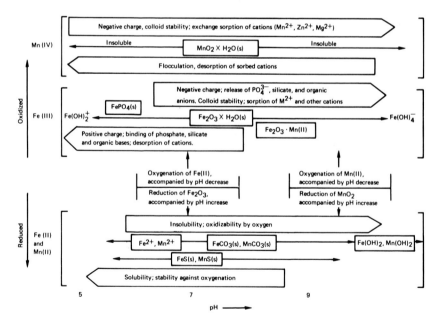

FIGURE 3   Limnological transformations of iron and manganese (Morgan and Stumm, 1965).

influence on the reaction rates of various factors, including temperature, total ionic strength, and hydrostatic pressure. Some of these factors, as they affect the rates of the reactions, are under study in J. J. Morgan's laboratory at the California Institute of Technology. The chemistry and mineralogy of iron compounds in the limnological thermodynamic realm form the subject of a recent Ph.D. thesis by Roger Doyle, from Prof. G. E. Hutchinson's laboratory at Yale.

COBALT

I have summarized the literature on analyses of freshwaters for cobalt and have given the results of my own analyses for Linsley Pond and four other Connecticut lakes (Benoit, 1957). Results for the Connecticut lakes were one or two orders of magnitude lower than the previously published values; that is, they were in the range of 0.02 to 0.2 ppb rather than 2 to 20 ppb. Carr and Turekian (1961) cite one additional report on Japanese mineral-spring waters falling in the range of 1 to 6 ppb. The same authors cited unpublished data of mine (a thesis that I completed in 1956) as reporting 2 to 4 ppb in solution from Linsley Pond. The citation is incorrect. Turekian and Kleinkopf (1958) determined several trace metals, including nickel but not cobalt, in 439 samples of stream and lake waters from Maine. Professor Turekian has pointed out to me in a personal communication, and it is obvious from other information in that paper, that the figures for average abundances given in Table 2 of the paper are too low by a factor of 10. The correct value for the average abundance of Ni in Maine surface waters is thus 0.2 ppb. On the basis of the general abundance of Ni and Co in lithosphere materials and on the basis of the general chemical similarities of the elements, I would expect the average abundance of Co to be 1/2 to 1/10 that of Ni, or somewhere between 0.02 and 0.1 ppb. Turekian has carried out extensive analyses on various Connecticut river waters and on the Neuse River in North Carolina. Professor Turekian was kind enough to permit me to cite some data from an unpublished progress report to the U.S. Atomic Energy Commission.* All data refer to samples of surface water collected in plastic bottles and filtered through a membrane filter having 0.45-micron pores. Freeze-dried evaporation residues were analyzed by neutron-activation analysis after the addition of specpure Na_2CO_3, which improves the handling characteristics of the residues.

*Turekian, K. K. 1966. *Trace elements in sea water and other natural waters.* Annual progress report to the Atomic Energy Commission for grant number AT (30-1)-2912, submitted August 31, 1966.

There is no better technique for analyzing Co, in my opinion, than the technique used by Turekian. The ranges, some individual values, and some mean values for Co in Connecticut waters as determined by Turekian are as follows:

Housatonic River:

One sample above the confluence of the Naugatuck River—0.34 ppb*

Two samples below the confluence of the Naugatuck River—0.11 and 0.78 ppb

Naugatuck River, a small river heavily polluted with industrial wastes, including wastes from various metal-working industries:

Three samples—6.14, 9.28, and 0.64 ppb (proceeding downstream)

Connecticut River between Hartford and the head of the estuary:

Eleven samples—0.036 to 0.18 ppb, with a mean of 0.103 ppb

Five tributaries of the Thames River:

Five samples—0.065 to 0.19 ppb, with a mean of 0.141 ppb

Neuse River, lightly contaminated in comparison with the Connecticut waters:

Twenty-five samples—0.026 to 0.223 ppb, with a mean of 0.078 ppb

The level of cobalt in the Neuse River is thus clearly of the same low order of magnitude as we found in Linsley Pond (Benoit, 1957). The vertical distribution of cobalt in Linsley Pond during stagnation resembles that of Mo as shown by Sugawara, Okabe, and Tanaka; that is, the element is low in oxygenated surface waters, increases in the anaerobic hypolimnion, and decreases again in the greatest depths if hydrogen sulfide is present. Cobalt in solution in the epilimnion of Linsley Pond was too low to measure in contrast to Mo in Japanese lakes (Sugawara *et al,* 1961). We suggest that Mo is not removed by sorption on iron oxide because, unlike Co, Mo is to an extent anionic under epilimnetic conditions.

CONCLUSION

Solubility equilibrium theory can account for the behavior of elements in weathering of rocks and for the general composition of surface waters. For

---

*This value should be compared with two values of 0.11 and 0.14 ppb *total* Co for Lake Zoar, an upstream impoundment of the Housatonic River (Benoit, 1957).

some elements, sorption phenomena seem to dominate their fate in water; the sorption processes are amenable to study by chemical methods as well as physical ones, and, in any event, they require careful quantitative study. There can be little doubt that nuisance species of algae evolved before man began to affect the environment significantly, as did agricultural weeds. But there can be little doubt that man is a dominating force in the geochemical economy of many lakes and rivers.

Both laboratory experiments on culture characteristics of planktonic algae and field experiments in which the chemistry of lakes is deliberately altered can shed light on the chemical factors that limit productivity or alter the trophic state of a lake enough for a nuisance to be created and a resource degraded.

REFERENCES

Bachman, R. W. 1963. Zn-65 in studies of the freshwater zinc cycle, p. 485–496 *In* V. Schultz and A. W. Klement [ed] Radioecology. Reinhold Publishing Corporation, New York, and AIBS.

Benoit, R. J. 1957. Preliminary observations on cobalt and vitamin B_{12} in fresh water. Limnol. Oceanogr. 2:233–240.

Carr, M. H., and K. K. Turekian. 1961. The geochemistry of cobalt. Geochim. Cosmochim. Acta 23:9–60.

Clarke, F. W. 1924. The data of geochemistry. U.S. Geol. Surv. Bull. 770, 5th ed. 841p.

Deevey, E. S. 1940. Limnological studies in Connecticut. V. A contribution to regional limnology. Amer. J. Sci. 238:717–741.

Englebrecht, R. S., and J. J. Morgan. 1959. Studies on the occurrence and degradation of condensed phosphate in surface waters. Sewage Ind. Wastes 31:458–478.

Gerloff, G. C., and F. Skoog. 1957. Nitrogen as a limiting factor for the growth of *Microcystis aeruginosa* in southern Wisconsin lakes. Ecology 38:556–561.

Goldman, C. R. 1960a. Molybdenum as a factor limiting primary productivity in Castle Lake, California. Science 132:1016–1017.

Goldman, C. R. 1960b. Primary productivity and limiting factors in three lakes of the Alaskan peninsula. Ecol. Monogr. 30:207–230.

Goldman, C. R. 1961. The contribution of alder trees (*Alnus tenuifolia*) to the primary productivity of Castle Lake, California. Ecology 42:282–288.

Goldman, C. R., and R. G. Wetzel. 1963. A study of primary productivity of Clear Lake, Lake County, California. Ecology 44:283–294.

Hsu, R. H. 1965. Fixation of phosphate by aluminum and iron acid soils. Soil Science 99:398–402.

Hutchinson, G. E. 1957. Treatise on limnology. Vol. 1. John Wiley & Sons, Inc., New York. 1015 p.

Kramer, J. R. 1964. Theoretical model for the chemical composition of fresh water with application to the Great Lakes, p. 147–160 *In* Pub. No. 11, Great Lakes Res. Div., Univ. Mich.

Kratz, W. A., and J. Myers. 1955. Nutrition and growth of several blue-green algae. Amer. J. Bot. 42:282–287.

Livingstone, D. A. 1963. The data of geochemistry. 6th ed., M. Fleischer [ed] Chapter G. Chemical composition of rivers and lakes. Geological Survey Professional Paper 440-G. U.S. Government Printing Office, Washington. 64 p.

Livingstone, D. A., and J. C. Boykin. 1962. Vertical distribution of phosphorus in Linsley Pond mud. Limnol. Oceanogr. 7:57–62.

Mackenthun, K. M. 1965. Nitrogen and phosphorus in water: an annotated selected bibliography of their biological effects. U.S. Public Health Service. U.S. Government Printing Office, Washington. 111 p.

Morgan, J. J., and W. Stumm. 1965. The role of multivalent metal oxides in limnological transmormations as exemplified by iron and manganese, p. 103–131 In Proc. Second Internat. Water Pollut. Res. Conf. (Tokyo), 1964. Pergamon Press, New York.

Provasoli, L. 1958. Nutrition and ecology of protozoa and algae. Ann. Rev. Microbiol. 12:279–308.

Provasoli, L., J. J. A. McLaughlin, and I. J. Pinter. 1954. Relative and limiting concentrations of major mineral constituents for the growth of algal flagellates. Trans. N.Y. Acad. Sci. (Ser. 2) 16:412–417.

Rodhe, W. 1949. The ionic composition of lake waters. Int. Assoc. Theor. and Appl. Limnology, Proc. V. 10:377–386.

Stumm, W. 1962. Discussion of a paper by G. A. Rohlich, p. 216–229 In Proc. First Int. Conf. Water Pollut. Res. (London), 1962; Pergamon Press, New York.

Stumm, W., and J. O. Leckie. 1967. Chemistry of ground waters: Models for their composition. Env. Sci. Tech. 1:298–302.

Sugawara, K., S. Okabe, and M. Tanaka. 1961. Geochemistry of molybdenum in natural waters (II). J. Earth Sci. 9:114–128 (Nagoya Univ.).

Sutherland, J. C., J. R. Kramer, L. Nichols, and T. D. Kurtz. 1966. Mineral-water equilibria, Great Lakes, silica and phosphate, p. 439–445 In Publ. No. 15, Great Lakes Res. Div., Univ. Mich.

Turekian, K. K., and M. D. Kleinkopf. 1958. Estimates of the average abundance of Cu, Mu, Pb, Ti, Ni, and Cr in surface waters of Maine. Bull. Geol. Soc. Amer. 67:1129–1132.

Wangersky, P. J. 1963. Manganese in ecology, p. 499–508 In V. Schultz and A. W. Kelment [ed] Radioecology. Reinhold Publishing Corporation, New York, and AIBS.

Weatherly, A. H. [ed]. 1967. Australian inland waters and their fauna: Eleven studies. Australian National University Press, Canberra.

Wetzel, R. G. 1966. Variations in productivity of Goose Lake and hypereutrophic Sylvan Lake, Indiana. Invest. Indiana Lakes and Streams 7:147–184.

FREDERICK E. SMITH

The University of Michigan, Ann Arbor

Effects of Enrichment in Mathematical Models

Ecosystems are at least as complex as the systems encountered in economics, industry, electronics, and aeronautics. Present technology permits the construction of complete models for any of these systems, so complete that their performance in the computer simulates precisely the performance of real systems. In ecology, however, such a goal is beyond both our present resources and our present level of interest.

Fortunately, systems analysis does not require complete models. Various degrees of simplification and abstraction are acceptable, provided the degree to which the model simulates the real world is allowed to be approximate rather than exact. Often a greatly simplified model will perform surprisingly well, simulating the major attributes of the real system and offering insight into the dominant forces operating in the real system.

The systems approach is not new in limnology. Energy flow is a systems analysis that has flourished since the inspirational study by Raymond Lindeman in 1941. The use of models as a means of simulating seasonal trends in plankton, and thereby evaluating the adequacy of the information going into the model, was developed rigorously in a remarkable series of papers by Gordon Riley, beginning in 1949.

To develop this field, it is necessary to become acquainted with the properties of mathematical systems. It often happens that a surprisingly large number of the attributes of systems can be determined by relatively few pieces of information. Some of these properties emerge from the model, instead of being put into it. Models seem to develop lives of their own, sometimes to the extent that the investigator shifts his area of study from the real world to the world of the computer.

The author is such an investigator, having spent the last year doing field research with an IBM 7090. These studies were not designed to explore eutrophication, but one of the major variables employed was the total nutrient density. The intent of this presentation is to suggest that some of the effects of eutrophication in lakes may be system effects, and do not depend upon whether species A or species B is present in a given trophic level. Secondarily, these results may challenge some of the accepted concepts of ecology.

THEORY OF MODELS

A discussion of enrichment in models requires first that some sort of general model be established. As a start, two very simple models will be presented, one based on three trophic levels and the other based on four, with one population at each level. In both, the mathematical expression for the feeding process will be greatly simplified, yet adequate for the gross aspects of the process. A general discussion of this expression follows; mathematical details are presented in a later section.

The rate at which food is consumed by a population shows a minimum of three major attributes: (1) the total rate of feeding increases with population density, (2) the total rate of feeding increases with food density, except that (3) the capacity of the population to feed is gradually satiated as the food density increases, the satiation producing an upper limit to the feeding rate no matter how much more abundant food becomes. These three statements can be shown graphically (Figure 1).

The best hope of describing this process accurately is through detailed systems analyses, such as those on the feeding of predators presented by C. S. Holling and Kenneth Watt in a series of papers beginning in 1959. A variety of simple statements will approximate the major attributes, however, and in the performance of a whole system it is doubtful that greater fidelity is needed.

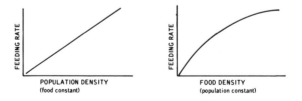

FIGURE 1   Empirical relations of feeding rates to variations in the densities of feeders and of food.

The enzyme-saturation model, the same that gives rise to the Michaelis-Menton equation, will be used here. It has appropriate properties and requires only two parameters, one expressing feeding ability and the other expressing anabolic capacity.

If the cycling of a mineral such as phosphate is considered in a system of three trophic levels, three feeding links of the two-step type can be used to connect the four components (Figure 2). In addition, inherent losses from each population (N) and inefficiencies of food conversion yield free resource (R). These links will also be simplified. Inherent losses from catabolism and natural death will be proportional to population density, and losses from the inefficiency of food conversion will be proportional to the rate of feeding.

This simple model has 12 parameters related to the feeding ability, growth capacity, inherent losses, and efficiency of conversion for each of the three populations. In addition, a value is needed for the total amount of mineral present. If all components are measured in units of the mineral contained, this thirteenth parameter is their sum.

Adding a fourth trophic level adds four more parameters, for a total of 17, producing the longer food chain (Figure 3).

By varying the values assigned to these parameters, an endless array of models can be generated, encompassing a wide range of real and imaginary situations. Some sets of parameters are incompatible. For example, the inherent loss rate of a population cannot exceed its capacity to grow, nor can an efficiency of food conversion be greater than unity. For each system, there is a minimal total density below which the system cannot operate. Within these restrictions, a wide range of systems can function.

If many systems are examined, the results are surprising. The two food chains produce consistently different results. If three levels are examined first, and then a fourth is added, the densities and dynamics of all the components are changed greatly. These changes are similar no matter what set of parameters is used.

Variations in the total density of a system has a strong effect on the performance of the system, but variations in the remaining parameters have less-striking effects. The rate constants of an alga can be replaced by those of a tree, or those of an aphid by those of a cow, without greatly changing the pattern of performance in the model. It appears that the structure of the model, and the total density of the system, are the two major determinants of the system, whereas the kinds of species in each trophic level are of less significance.

FIGURE 2   Flow chart of a closed system with three trophic levels.

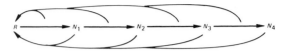

FIGURE 3   Flow chart of a closed system with four trophic levels.

If this is also true of natural systems, ecology can indeed develop a set of basic principles that is independent of species composition.

To observe the effect of enrichment in the model with three trophic levels, let us begin with its performance when the total density is not much above the minimum at which the system will operate. The following pattern is similar for most sets of parameters that are at all reasonable in comparison with nature:

1. The standing crop of plants is high, at least half the total. Free resources are moderately low, herbivores lower still, and predators very low.

2. The turnover rate of each population, relative to its maximum possible rate, is highest for herbivores, somewhat lower for plants, and minimal for predators. The turnover rate of free resources is moderately high.

As the total density of a given system is increased, the following results always occur, even in systems with unrealistic values for the parameters:

1. The standing crop of plants increases, accounting for most of the total increase. The density of free resource falls slowly, while the density of predators increases slowly.

2. The turnover rate of herbivores increases slowly toward its maximum, while that of the plants decreases. The turnover rate of free resources increases.

In brief, most of the enrichment is taken up in the plants, with a small increase in the density of predators.

A similar analysis can be made of the model with four trophic levels. It should be noted that the minimum total density at which four levels will operate is always well above the minimum at which the same system will operate if the fourth level is removed. The general results are as follows:

1. Free resources and plants are about equally abundant. Herbivores are moderately abundant. The abundance of predators is low, and that of secondary predators is lower still.

2. The turnover rate of each population, relative to its maximum, is very high for both plants and herbivores, intermediate for primary predators, and

minimal for secondary predators. The turnover rate of free resources is moderately high.

As the total density of these systems is increased, the following results always occur, for all sets of parameters:

1. The density of free resources increases, accounting for most of the total increase. The density of plants decreases slowly, while that of the herbivores and secondary predators increases very slowly.
2. The turnover rate of the plants increases slowly toward its maximum, while that of the herbivores decreases slowly. The turnover rate of free resources decreases.

In brief, most of the enrichment remains in free form, with small increases in herbivores and in secondary predators. Note that the effect of enrichment on free resources and plants is reversed completely by the presence of secondary predators. Also, only total nutrients (free plus plants plus animals) reflect the level of enrichment for both.

It is not intended that these models describe real systems. They are much too simple to have respresentations in nature. They are presented to illustrate how systems behave. Systems even as simplified as these show interesting properties and contain lessons for the study of real systems.

Suppose, for example, that one were interested in the dynamics of plants and made extensive measurements of their density, growth rate, metabolism, mineral absorption, losses to grazing, and so on. The same plant in the same habitat, with the same total nutrient density, would show one set of characteristics if secondary predators were present and a very different set if they were absent. The investigator may be able to describe the state of his plants with precision, but unless he takes into account the whole system in which the plants participate, he cannot possibly understand why those results are obtained.

As a step from these simple models toward real systems, the effects of complicating the food web can be considered. Several such complications have been explored.

In general, branches and convergences in the web have only minor effects on the major pattern. Branches tend to increase the total density of trophic levels above the branches, but convergences tend to have an opposite effect. It is difficult, however, to have branches survive at all. To provide sufficient freedom from competition, niche differentiation has to be provided for most cases of coexistence, and this is not easy to do in models. The same difficulties apply to convergences. One class of branches, however, allows coexistence without niche differentiation with respect to food. Several herbivores can coexist on the same food source if each has its own predator. Predator stress adjusts so as to balance exactly competitive inequalities in this system.

Mixing the trophic levels produces variable results. If a dominant path is present, the system follows close to the pattern set by the dominant path. Thus, minor exceptions in the diet can be ignored. If a species feeds abundantly on two trophic levels, the system as a whole displays an intermediate pattern that resembles more closely the chain having the shorter path. Thus, a hawk that eats both insectivorous and graminivorous birds will not function like a true fourth trophic level.

To the extent that the top trophic level is self-regulatory through territoriality or cannibalism, its role in the system as a whole is weakened. By including varying degrees of self-regulation in the fourth level, one may construct a whole array of systems intermediate in behavior between those with four and three trophic levels. A completely self-regulating population has no perceptible effect on the major aspects of the system to which it belongs.

A system with a partially self-regulating fourth trophic level responds in an interesting way to enrichment. At low nutrient densities, the system will exhibit some pattern intermediate between that for simple models of three trophic levels and that for simple models of four trophic levels. Since a low nutrient level for four trophic levels is well above the minimum for three, the system of three levels in this case is one that is very rich in plants. An intermediate pattern at a low total density, therefore, would be one in which plants account for the largest portion of nutrients, free nutrients and herbivores account for moderate portions, and smaller and smaller portions are present as primary and secondary predators.

As this system is enriched, the partial self-regulation of the secondary predators reduces their response; they will increase in density less than in a simple model of four trophic levels. Thus, the pattern shown will shift steadily toward the model with three trophic levels as enrichment continues. The largest increase will be in plant density, although moderate increases will occur in free nutrients. Herbivores will decrease and primary predators will increase.

It is tempting to relate this model to a limnetic system composed of nutrients, algae, zooplankton, forage fish, and piscivorous fish. The additional effect of mixed trophic levels among the two classes of fish would bring the performance of the model even closer to reality.

In the models discussed so far, resources removed from the populations return directly to the pool of free resources. There is, of course, an intermediate stage of decomposition that should be included for some of this flow. This stage is particularly significant in systems with high plant densities.

Very little of this plant crop is grazed, and the bulk of net primary production eventually forms detritus. This in turn serves as a food for microflora. Both the detritus and the microflora may be grazed to some extent by herbivores, increasing the variety of food available. Dr. Saunders has presented a system like this. (See page 556.) In such systems, the major path of the food web may be from free nutrients to plants to detritus to microflora to free nutrients, with only minor portions ever getting into the fauna.

In enriched aquatic systems, high plant densities can lead to difficulties with the oxygen budget. Free oxygen and organic matter are products of the same process and must be produced and used up together. When a lake turns green, and its total biomass is increased, it must have shown a net oxygen production during that period. Net oxygen production must also occur for every particle of organic matter lost to the sediment. Our best oxygen producers must be bogs, since they accumulate organic matter the most rapidly, while a healthy balanced lake must have almost no net oxygen production.

In a system rich in plants, photosynthesis during the day may be so intense that oxygen is lost from the water at a rate faster than it can re-enter at night. This loss imposes a physical limit to the respiratory rate of the system and may cause organic matter to accumulate whether organisms capable of utilizing the material are present or not. Organic matter will continue to accumulate as long as photosynthesis squeezes large amounts of oxygen out of the water. If nutrients and carbonate are steadily poured into the system, the only possible result is a thick soup of organic matter.

A system out of balance by this mechanism will show depressed oxygen levels late each night. Those species that are most sensitive to oxygen depletion will suffer first. If this sensitivity applies to secondary predators such as piscivorous fish, their role in the system will be reduced. This reduced role in turn will produce a still larger standing crop of plants, which will increase further the amplitude of the diurnal oxygen cycle. It is not difficult to imagine that such a process could rapidly lead to the extinction of the fourth trophic level.

Although some comment has been made on the effect of branches in the food web, the heterogeneity that becomes possible in complex systems has not been considered. The effect of heterogeneity within a trophic level is very interesting in systems where one trophic level is dense and lightly grazed while the next level above is sparse and heavily grazed. Such situations occur in those models having a high density of plants.

Let us consider a plant-rich system in which the plants are very diverse with respect to food quality. For terrestrial plants, the variation in food quality among pollen, fruits, seeds, young leaves, old leaves, bark, wood, and roots is a good example. In aquatic systems, the variations among nannoplankton, green algae, blue-green algae, and thorny and filamentous forms can be considered. In both systems, the additional diversity in detritus, bacteria, and fungi should be included. A third example is the aquatic rooted plants, which sometimes dominate aquatic habitats. In addition to a diversity in quality among plant parts, these plants tend to be covered with a film of microflora of excellent nutritional quality.

Let us assume, to avoid argument, that all plant materials are edible to one herbivore or another and that there is at least one herbivore capable of living on any one kind of plant material. Those eating the more nutritious or more easily digested material can grow rapidly and may show short generation times or high fecundities. Those eating the less nutritious material grow slowly and may show long generation times or low fecundities.

In a complex model with three trophic levels, each plant–herbivore path is able to support a given level of predation. The more productive herbivores can support more predators than the less productive herbivores. If this system contains generalized predators feeding on a variety of herbivores, their density will be set by the more productive of the herbivores.

This density of predators will generate a predation pressure on the less productive herbivores that could lead to their extinction. If the risk of predation per day is more or less the same for all herbivores, those with long generation times cannot make it, and those with low fecundities cannot compensate for mortality. Only those with sufficiently short generation times or sufficiently high fecundities can survive. They, of course, will continue to maintain the predators, preserving the system.

Thus, the less nutritious components among the plants may show little or no grazing. They are protected by the more nutritious components, which will show evidence of grazing.

Within this system, the accumulation of plant material is enhanced well beyond the expectations set by a simple model of a single food chain. Furthermore, the dominant features of the flora will tend to be those showing the least grazing. Obviously, this greater accumulation will increase even more the contribution of the plants to detritus.

Such a system, operating at a given total density, will have eliminated the slow-growing grazers on the poorer quality food up to that level of quality at which the herbivores just survive. If the system is now enriched, the intensity of predation will increase, eliminating another set of herbivores. Furthermore, grazing will be focused on a smaller segment of the flora, tending to keep its abundance down, and most of the enrichment will be taken up in the components of poorer quality. The end result is a preferential increase in the less nutritious plant materials.

This process may account for the generally intact appearance of terrestrial vegetation, for the accumulation of coarse algae in eutrophic systems, and for the remarkably intact appearance of aquatic rooted plants.

It must be emphasized, however, that this process can occur only in a system in which grazing pressures on the plants are already slight while those on the herbivores are strong. It is also necessary that the more nutritious components of the vegetation continue to be present so that their herbivores can continue to support many predators. With these restrictions, this mechanism can be powerful and effective. Several previous papers suggest that such a process may occur in real systems.

Diversity is a difficult aspect of systems to analyze since it is a property of complex models. A few efforts have been made in this direction, however, and some comments can be made on the relations between diversity and stability.

All the models that have been presented are unstable in the sense that if followed through time in the computer, the different components will oscillate with greater and greater amplitude around their steady-state values until the system crashes. This instability results from the very nature of biological growth: to the extent that population growth is a function of population size, positive gain is added to the system.

Adding branches or convergences (or both) to these models does not reduce their instability. With deterministic models, such diversity has no effect on stability. With stochastic processes added, the degree of instability is at least as bad with branches as without, and is possibly worse. Thus, the concept that food-web stability is derived from diversity cannot be supported in models.

Various modifications in the feeding reaction can make the systems stable. In general, these must be modifications whereby changes in food density have exaggerated effects on the feeding rate, or changes in the feeding rate have reduced effects on food densities.

Once a simple food chain has been made stable through an adequate degree of modification of the feeding links, an interesting result occurs. It is often possible to run several such chains simultaneously on a single limiting resource. Sometimes they can coexist indefinitely, even though they possess identical parameters. Thus, once a system is stabilized, it may accept diversity. Furthermore, it would appear that much less niche differentiation is needed to incorporate further diversity than was needed in the simpler models. It seems possible, therefore, that the relation between diversity and stability in natural systems is quite the reverse of current dogma.

The changes in diversity that were discussed earlier by Dr. Hooper (see page 225 are consistent with results from these models. As stated earlier, in systems with abundant plants, an increase in nutrients is taken up preferentially by the coarser plants. Such preferential shifts tend to reduce the diversity of plants. Similarly, a variety of slow-growing herbivores may be eliminated while rapidly growing forms carry more of the load of predation. This will tend to reduce the diversity of herbivores.

It seems to me, however, that many kinds of disturbances in natural systems are likely to reduce diversity and that this criterion alone is not diagnostic of eutrophication. It does have the important advantage of being independent of species composition and is applicable everywhere. Perhaps its diagnostic power can be improved if some of the kinds of changes involved in the decrease of diversity were considered. For example, in these models the plants shift in composition toward those of poor quality while the herbivores shift toward those of rapid growth capacities. Generally speaking, in limnetic systems these changes will be correlated with an increase in the average size of the phytoplankton and a decrease in the average size of the zooplankton. Such additional criteria still have the advantage of being species-independent, and they preserve the general utility of diversity as an index of eutrophication.

Probably any prolonged enrichment of an aquatic system leads unavoidably to a larger standing crop of plants. If one group of plants is removed, another more difficult to remove will take its place. Similarly, if predators are removed, others more difficult to remove will increase in abundance. Both of these forms of management operate against the natural performance of the system. The only management program that has the cooperation of the system is a reduction in nutrient supply.

Working with models is an excellent way of developing a sense of whole systems. The performance of these simple models suggests that it is foolhardy to try to understand any system unless the whole is studied. I an convinced, also, that ecosystems can be understood at levels far beyond the documentation of population dynamics, energy flow, and species diversity.

DESIGN OF MODELS

All the models used are similar in their basic structure. In any one model, all the components (resource, populations) are expressed as densities (area or volume), and the same unit of measurement is used in all estimates. Acceptable units are those that translate directly from one component to another, such as milligrams of phosphorous, or milligrams of protein-nitrogen, or calories. Units such as dry weight or numbers of individuals are not convenient; their composition varies among components and requires an additional set of translational parameters.

Most of the models discussed were considered to be mineral-limited. Phosphate is a good example to keep in mind, although any will do.

(Energy-limited systems will be mentioned at the end of this section.) The systems were considered to be closed in the sense that the input and output of the limiting resource from the system are very slow relative to its rate of cycling within the system. (Open systems will be mentioned at the end of this section.)

The basic notation is one in which:

$$S = R + N_1 + N_2 + N_3 + \dots$$
S = total density of the system
R = density of the free resource
N_i = density of the ith population
i = $1, 2, 3, \dots$

Much of the analysis concerns the densities and turnover rates of the components as a function of (1) the number of components, (2) their arrangement through linkage to each other, (3) the values assigned to parameters associated with the links, and (4) the total density of the system.

Each feeding link, unless otherwise modified, is expressed as a succession of two steps. The entire process from food capture through the synthesis of end products is compressed into these two steps. The resulting model differs considerably in its behavior from the one-step process used in most prey—predator models, probably shifting much more than half the distance to the performance of a realistic multistep model. The model is borrowed directly from enzyme kinetics:

$$Q + E \xrightarrow{x} (EQ) \xrightarrow{y} P + E .$$

Q = substrate concentration
E = concentration of active enzyme
EQ = concentration of enzyme-substrate complex
P = end product of the catalysis
x, y = rate constants

Reversibility has been eliminated from the model, since it is not a property of biological growth.

Biological growth differs further in being autocatalytic, so that the end product is more enzyme. Finally, biological growth cannot be viewed as a simple molecular process. Any reference to enzyme kinetics beyond the use of its algebra is improper. As used here, the "enzyme" becomes:

N = population density
A/N = degree of hunger in the population, zero to unity

A =   density of feeding portion of the population, equivalent to "active enzyme," but a conceptual, not a physical, category

B =   density of the anabolizing portion of the population, equivalent conceptually to the enzyme-substrate complex

N = $A + 2B$

The process of feeding and growth becomes:

$$F + A \xrightarrow{x} B \xrightarrow{y} 2A.$$

F =   food density

producing the following rate processes:*

$\dot{F}$ = $- xAF$.

$\dot{A}$ = $- xAF + 2yB$.

$\dot{B}$ = $+ xAF - yb$.

$\dot{N}$ = $+ xAF$.

This model must be modified further to include the inefficiency of food conversion and population losses.

Inefficiency implies less material gained by the population than is removed from the food source. The simplest way to incorporate this into a model is to set the rate of food removal at a level higher than the rate of population gain. That is, with $p > x$:

$\dot{F}$ = $- pAF$.

with the other equations unchanged. The ratio, x/p, is the efficiency with which the food removed becomes anabolic end products.

Inherent population losses are expressed in the simplest form as an exponential rate, $- cN$. The affected components are:

$\dot{A}$ = $- xAF + 2yB - cA$.

$\dot{B}$ = $+ xAF - yB - cB$.

$\dot{N}$ = $+ xAF - cN$.

These two sources of loss are balanced, in a total system, with a gain to the free resource (R):

$\dot{R}$ = $+ (p - x)AF + cN$.

*$F = dF/dt$.

These functions can now be assembled to form complete models. A system of three trophic levels, with one population at each level, takes the following form:

$$\dot{A_1} = -x_1 A_1 R + 2y_1 B_1 - c_1 A_1 - p_2 A_2 A_1.$$
$$\dot{B_1} = + x_1 A_1 R - y_1 B_1 - c_1 B_1 - p_2 A_2 B_1.$$
$$N_1 = A_1 + 2B_1.$$
$$\dot{A_2} = -x_2 A_2 N_1 + 2y_2 B_2 - c_2 A_2 - p_3 A_3 A_2.$$
$$\dot{B_2} = + x_2 A_2 N_1 - y_2 B_2 - c_2 B_2 - p_3 A_3 B_2.$$
$$N_2 = A_2 + 2B_2.$$
$$\dot{A_3} = - x_3 A_3 N_2 + 2y_3 B_3 - c_3 A_3.$$
$$\dot{B_3} = + x_3 A_3 N_2 - y_3 B_3 - c_3 B_3.$$
$$N_3 = A_3 + 2B_3.$$
$$\dot{R} = \text{by difference, since: } S = R + N_1 + N_2 + N_3$$
$$S = \text{fixed total density of the system}$$

The steady-state solution of this system, at which all rates of change are zero, can be obtained indirectly by working down from the top of the food chain, and solving for the total density that satisfies the system:

$$N_3 \quad \text{chosen between zero and } (y_2 - c_2)(y_3 + c_3)/p_3(y_3 - c_3).$$
$$A_3 = N_3 (y_3 - c_3)/(y_3 + c_3).$$
$$N_2 = c_3 N_3 / x_3 A_3.$$
$$A_2 = N_2 (y_2 - c_2 - p_3 A_3)/(y_2 + c_2 + p_3 A_3).$$
$$N_1 = N_2 (c_2 + p_3 A_3)/x_2 A_2.$$
$$A_1 = N_1 (y_1 - c_1 - p_2 A_2)/(y_1 + c_1 + p_2 A_2).$$
$$R = N_1 (c_1 + p_2 A_2)/x_1 A_1.$$
$$S = R + N_1 + N_2 + N_3.$$

By solving for several values of N_3, and graphing S against N_3, we can approximate the steady-state values of the system for any given level of S. For more precise solutions, the computer was programmed to follow a standard trial-and-error "half-interval" method until the derived value of S was within acceptable limits.

In some cases two sets of solutions exist for one total density. The set having the lower value of N_3 was discarded, since it was found to represent an unstable set of "hilltop" points rather than a set of "valley" points.

A model for four trophic levels is obtained by a simple extension of the above notation, both for the differential equations and for the steady-state solutions.

For simulation in time, the system was started very close to, but not exactly at, its steady-state solution. Stepwise integration was performed using a subroutine available in the computer, a fourth-order Rutte-Kunga method, equivalent to the first five terms of the Taylor expansion. The size of the step interval had to be very small for the integration to be accurate. If the system was unstable, so that oscillations were of greater and greater amplitude, integration eventually failed. The terminal results of such cases have not been included in the presentation.

The adequacy of methods of step-wise integration must be examined in any system, and especially in systems such as these. Their nonlinear, autocatalytic structure puts unusually heavy burdens on the process.

These models were stabilized by various modifications of the links. The two major classes are an exaggeration of the effect of food density and the addition of self-regulation. For the second trophic level, these are:

1. Change the notation $p_2 A_2 N_1$ and $x_2 A_2 N_1$ to $p_2' A_2 N_1^2$ and $x_2' A_2 N_1^2$
2. Add the term $- d_2 A_2 N_2$ to equation A_2, and $- d_2 B_2 N_2$ to equation B_2

In all comparative runs, d_2, $p_2{}'$, and $x_2{}'$ were chosen so as to produce the same steady-state densities as those of the basic model.

Detritus and decomposers can be added to the food chain in various ways. The following method allows the herbivores to feed on the decomposers, but not on the detritus.

D = density of detritus
N_4 = density of decomposers
c_i = $m_i + d_i$, $i = 1,2,3,4,$
m = losses, by catabolism, directly to free resource
d = losses of organic matter to detritus

Revisions and additions to the system for three trophic levels are:

$$\dot{D} = d_1 N_1 + d_2 N_2 + d_3 N_3 + d_4 N_4 - p_4 A_4 D .$$
$$\dot{A_4} = - x_4 A_4 D + 2y_4 B_4 - c_4 A_4 - p_2' A_2 A_4 .$$
$$\dot{B_4} = + x_4 A_4 D - y_4 B_4 - c_4 B_4 - p_2' A_2 B_4 .$$
$$N_4 = A_4 + 2B_4 .$$
$$\dot{A_2} = - x_2 A_2 N_1 - x_2' A_2 N_4 + 2y_2 B_2 - c_2 A_2 - p_3 A_3 A_2 .$$
$$\dot{B_2} = + x_2 A_2 N_1 + x_2' A_2 N_4 - y_2 B_2 - c_2 B_2 - p_3 A_3 B_2 .$$
$$N_2 = A_2 + 2B_2 .$$

$\dot{R}$ = by difference, or: $-\ p_1A_1R\ +\ m_1N_1\ +\ m_2N_2\ +\ m_3N_3\ +$
$m_4N_4\ +\ (p_1\ -\ x_1)A_1R\ +\ (p_2\ -\ x_2)A_2N_1\ +$
$(p_2'\ -\ x_2')A_2N_4\ + (p_3\ -\ x_3)A_3N_2.$

Stochastic processes were added in two ways. In some runs, the densities of the populations and of free resources were jiggled periodically; random normal numbers were drawn by the computer and added or subtracted among the components so that their sum (S) remained fixed. In most runs, the values of the parameters were jiggled; for each time period, each parameter was altered from its average value by using random normal deviates. The second procedure seems preferable as a means of simulating environmental variation. If the the parameters are adjusted too frequently, so that little population change can occur to any one set of values, the effect of jiggling tends to be averaged out. The effect is greater if the parameters are varied considerably and infrequently.

Adding flow through the system (Figure 4) or using calories as units of measurement (Figure 5) produces systems whose behavior is similar to that of systems already discussed. The only significant change is that the total density of the system (S) can vary as the parameters are altered, sometimes allowing a large storage at one trophic level to occur.

FIGURE 4   Flow chart of an open system with internal recycling.

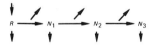

FIGURE 5   Flow chart of an open system with no internal recycling.

G. FRED LEE

University of Wisconsin, Madison

Analytical Chemistry of Plant Nutrients

If we are to manage water resources that are overgrown with aquatic weeds, it is essential that we understand the effect of plant nutrients on eutrophication of natural waters. This understanding must be based, to a large extent, on a thorough knowledge of the analytical chemistry of the nutrient elements. We have to know the aqueous environmental chemistry of these elements before we can place rational limits on one or more elements from a specific source. Although we have learned much about the analytical chemistry of plant nutrients during the past few years, it is clear that much additional work must be done in this area.

A study on the eutrophication of a particular body of water requires that the sources, types, forms, and amounts of the elements that may stimulate aquatic plant growth be determined. The investigator must ask the following questions in planning his analytical chemistry program:

What elements and forms of elements should be measured?
How should samples be taken? At what locations and at what frequency?
What type of sample preservation will be used?
What analytical methods will be used?
What accuracy is desired in the analytical results?
How will the data be processed and evaluated?

The answer to many of these questions requires careful attention to the planning part of the study dealing with analytical chemistry. Far too often, studies are conducted without proper attention to one or more of these questions. This paper will consider certain aspects of these questions and

646

emphasize areas that require additional study. For the purposes of this paper, only nitrogen and phosphorus will be considered. Other papers in this symposium have discussed the importance of other elements in the nutrition of aquatic plants. Much of what will be discussed on the analytical chemistry of nitrogen and phosphorus applies to other plant nutrients.

ANALYTICAL CHEMISTRY PROGRAM

SAMPLING

The first phase of a eutrophication study on a selected lake, stream, or estuary requires the delineation of the nutrient sources. Table 1 presents the results of an estimate of the nutrient (nitrogen and phosphorus) sources for Lake Mendota, Madison, Wisconsin. This table shows that 60 to 70 percent of the phosphorus is derived from sources that are associated with activities of man, whereas 60 to 70 percent of the nitrogen is derived from natural sources. It was not possible to estimate the nitrogen and phosphorus that may

TABLE 1   Estimated Nutrient Sources for Lake Mendota[a]

Source	Annual Contribution (lb)		Estimated Contribution (%)	
	Nitrogen	Phosphorus	Nitrogen	Phosphorus
Municipal and industrial waste water	47,000 (total)	17,000 (total)	10	36
Urban runoff	30,300 (soluble)	8,100 (soluble)	6	17
Rural runoff	52,000 (soluble)	20,000 (soluble)	11	42
Precipitation on lake surface	97,000	140–7,600	20	2[b]
Groundwater	250,000	600	52	2
Nitrogen fixation	2,000	–	0.4	–
Marsh drainage	(c)	(c)	–	–
Total	478,300	47,000[b]	–	–

[a]Source: Nutrient Sources Subcommittee (G. F. Lee, Chairman) of the Lake Mendota Problems Committee. 1966. *Report on Nutrient Sources of Lake Mendota.*
[b]These calculations are based on 1,000 lb per year of phosphorus in precipitation on lake surface.
[c]Information not available.

be derived from marshes. There is a need for study of drained marshes as a source of nitrogen and phosphorus.

Any study on nutrient sources should be preceded by an estimate of the expected contributions of each of the sources. Such an estimate is usually prepared from existing information and can be a useful guide to planning the sampling program.

If the basis for these estimates, as presented in Table 1, is examined, it is found that the spring runoff from the watershed contributes approximately 50 percent of the phosphorus contributed by rural runoff during the year. This phosphorus is derived from manure that is spread on frozen soil during the winter. The spring snow melt, coupled with rain, leaches large quantities of phosphorus from the winter's accumulation of manure. This phosphorus is carried by overland flow, because of frozen soil, to nearby streams that are tributary to the lake. Manure spread in the late spring, summer, and early fall probably contributes little phosphorus, since most of the rain carries leached phosphorus into the soil, where it is fixed by soil particles. A sampling program that attempts to measure accurately the amounts of nutrients contributed from manure must be set up so that frequent samples are taken during a relatively short period of time.

Studies on Black Earth Creek by the University of Wisconsin Water Chemistry Department have demonstrated the need for basing the sampling program on stream discharge rather than on arbitrary time frequency.

Figure 1 shows the discharge-conductance data for Black Earth Creek during the spring thaw in 1965 (Shannon and Lee, 1966). The stream had a base flow from groundwater of approximately 20 cfs for March 29 and 30. During the period of March 30 to April 2, the discharge increased to a peak of 150 cfs on April 1, then gradually receded. Flow during this period was primarily due to snow melt. On April 2 it started to rain; discharge increased to approximately 150 cfs on April 5. After this date, the discharge gradually decreased, approaching base flow after April 13. The specific conductance of the water showed the typical inverse relationship with discharge, with the minimum in specific conductance approximately corresponding to the maximum in discharge. The snow melt and precipitation runoff were low in dissolved salts and diluted the relatively hard groundwater base flow.

Figure 2, on the other hand, shows that maximum phosphate concentration corresponds to the maximum discharge (Shannon and Lee, 1966). As shown in this illustration, the snow melt and rainfall runoff contained more phosphorus than the base flow. This phosphorus is probably derived from manure. The condensed phosphates in the figure are composed of the nonfilterable phosphorus that is measured by mild acid hydrolysis. No known sources of domestic or industrial pollution are upstream from this sampling station; the condensed phosphates, therefore, are probably organic phos-

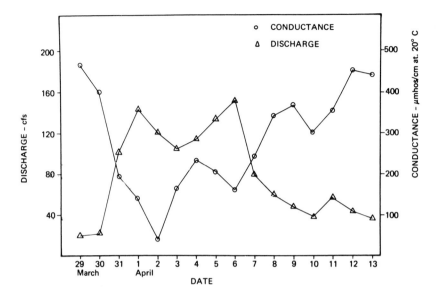

FIGURE 1   Conductance and discharge of Black Earth Creek during spring 1965.

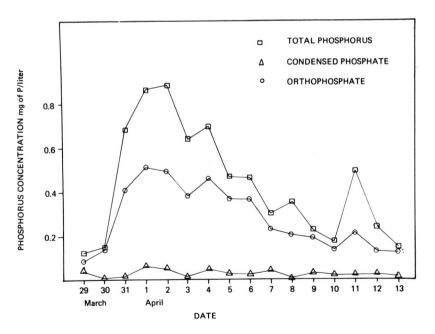

FIGURE 2   Phosphorus concentrations of Black Earth Creek during spring 1965.

phorus compounds. As expected, an appreciable amount of total phosphorus transported by the stream during spring thaw is particulate phosphate.

The same relation of phosphorus and nitrogen concentration to periods of high discharge has been found at all times of the year; the major proportion of these elements is transported during periods of high discharge. It is important, therefore, to establish a sampling program in which the frequency of sampling increases as the discharge increases. Although this kind of program is difficult to administer, it will yield the most meaningful results for estimating the amount of plant nutrients contributed by a stream.

These comments regarding sampling streams apply also to waste waters. Typical domestic waste waters show marked changes in flow and concentrations at different times of the day and on different days of the week. G. P. Fitzgerald (personal communication) has found that the *Cladophora* growing along the shores of Lake Mendota were nitrogen-starved until after a heavy rainfall. Evidently, the runoff from the urban area contributed sufficient nitrogen to change this algae from a starved to a surplus nitrogen status.

The practice of releasing stored waste waters during periods of high discharge may contribute to the nutrient load on a stream or lake. Some industries are given permits to store waste waters in lagoons during periods of low stream discharge and to release them during periods of high discharge. This practice can help alleviate problems such as oxygen demand and toxic chemicals. However, the lake to which the stream is a tributary will receive the same total load of nutrients over a year's time. In fact, the situation may worsen if nitrate is used to keep the storage lagoon from becoming anaerobic. A cheese plant in the Lake Mendota watershed used $NaNO_3$ in its lagoon to minimize odors. The plant was allowed to discharge the lagoon during periods of high stream discharge. It is reasonable to suspect that this practice contributed additional nitrogen to Lake Mendota.

FORMS OF THE ELEMENTS TO BE MEASURED

The investigator must decide what forms of the nutritive element are to be measured. Normally, total phosphate, nonfilterable o-phosphate, organic nitrogen, ammonia, nitrate, and nitrite are measured. Some investigators do not include nitrite, since its concentration is usually very low compared with concentrations of other forms of nitrogen. However, M. S. Nichols (personal communication) has found that some streams tributary to Lake Mendota may have several milligrams per liter of nitrite-N during late fall.

A major problem that requires much additional attention is the role of particulate forms of nitrogen and phosphorus in fertilizing lakes. In particular, what fraction of the nonfilterable organic nitrogen and total phosphate carried by a stream will become available in the lake? As shown in

Figure 2, a large part of the total phosphate present in Black Earth Creek during periods of high discharge is particulate, that is, will not pass through a 0.5-μ pore-size filter.

Except for the organic and colloidal phosphate measure in the o-phosphate test, as discussed below, it is probably reasonable to assume that essentially all nonfilterable nitrogen and phosphorus will become available for aquatic plant growth. However, it is impossible to define the fraction of particulate nitrogen and phosphorus that will become available. Additional research is needed on the aqueous environmental chemistry of plant nutrients with emphasis on sorption-desorption (both biotic and abiotic) reactions of nitrogen and phosphorus.

Rigler (1966) found that a substantial part of the o-phosphate measured in lakes by the molybdate procedure is not readily available for algal growth. Apparently, nonfilterable organic phosphorus compounds are present in natural waters that cannot be directly used by algae, yet are measured by the normal o-phosphate procedure.

Also deserving attention is the possible presence of large amounts of sorbed o-phosphate on clay minerals in streams during periods of high discharge. According to studies on sorption of o-phosphate on soil particles (R. B. Corey, personal communication), the concentrations of nonfilterable phosphate found in Black Earth Creek during period of high discharge are in excess of the equilibrium concentrations. Evidently, colloidal clay particles that contain substantial amounts of sorbed phosphate are present. This sorbed phosphate is released under the conditions of the o-phosphate test.

Many of the analytical procedures that are used apparently do not measure specific forms of elements. For example, the orthophosphate test measures some organic phosphorus and sorbed phosphorus on colloidal particles. These findings make very difficult the interpretation of the relationship between the measured quantity and the availability for plant growth. Additional research is needed on the forms of these elements in natural waters and measurement of the forms by operationally defined analytical procedures. It may be advisable to follow the practice of Strickland and Parsons (1965) of calling the results of the molybdate test for o-phosphate "reactive phosphate" until more is known about what is actually being measured.

PRESERVATION OF SAMPLES

Water samples can seldom be analyzed immediately after collection. Therefore, some provision must be made for preservation of samples. Frequently, chloroform is added in an attempt to preserve the forms of phosphorus, and H_2SO_4 is used to preserve the nitrogen compounds. Studies have shown that these compounds do not preserve the forms of these

TABLE 2 Phosphorus Released by Laboratory-Cultured Algae after Various Treatments[a]

Alga	Phosphorus Released[b]					
	Refrigerated[c]	Boiling Water[d]	Frozen[e]	Chloroform[f]	Algicide[g]	Copper Sulfate[h]
Scenedesmus	0.009	0.54	0.18	0.27	0.24	—
quadricauda	0.010	0.60	0.24	0.48	0.22	—
Chlorella	0.016	0.45	0.072	0.29	0.18	—
pyrenoidosa	0.013	—	0.046	0.25	0.039	0.002
Phormidium	0.015	0.38	0.11	0.33	0.20	0.082
inundatum	0.017	—	—	0.60	0.44	0.019

[a]Source: Fitzgerald and Faust (1967).
[b]Milligrams of ortho PO_4-P per 100 mg of algae.
[c]Overnight at 3 to 5°C.
[d]Overnight in refrigerator plus 60 min in boiling water bath.
[e]Overnight in freezer at −15°C.
[f]Saturated plus overnight in refrigerator.
[g]One mg of Algimycin-200 per mg of algae, plus overnight in refrigerator.
[h]One mg of $CuSO_4 \cdot 5H_2O$, plus 1 mg of citric acid (pH 7) per mg of algae, plus overnight in refrigerator.

elements. In fact, the addition of these compounds may alter the forms more than simple refrigeration. Fitzgerald and Faust (1967) found that the addition of chloroform to algal cultures resulted in cell rupture with the release of o-phosphate to the solution. Extrapolation of the results presented in Table 2 shows that o-phosphorus measurement in eutrophic lakes during periods of algal bloom may be meaningless if chloroform is used to preserve the sample. From these studies we may conclude that the only way to determine the nonfilterable "soluble" phosphate in a sample is to filter the sample immediately after collection with a minimum of vacuum to prevent cell rupture.

The recommended practice (American Public Health Association, 1965) of adding H_2SO_4 to preserve the forms of nitrogen has been found by Brezonik and Lee (1966) to result in denitrification of the nitrite present in the sample via the Van Slyke reaction. As shown in Figure 3, the addition of H^+ results in the rapid depletion of NO_2^- as compared with a sample to which no preservative was added. Jenkins (1965a, 1965b, 1967) has conducted extensive tests on the preservation of nitrogen and phosphorus compounds in fresh and estuarine waters. In general, he concludes that the addition of $HgCl_2$ and freezing to $-20°C$ is the best method of preserving the forms of these elements. If the samples cannot be preserved in this way, they should be filtered as soon after collection as possible, after which $HgCl_2$ should be added for nitrogen and chloroform should be added for phosphorus. The investigator should be aware that changes may occur in the "soluble" fraction of the element even after removal of the particulate matter. Hassenteufel *et al.* (1963) have reported significant loss of orthophosphate to the storage container (glass and plastic bottles), although others (see Strickland and Parsons, 1965) did not find similar losses on plastic containers. Studies in our laboratory have shown that approximately 50 percent of o-phosphate may be sorbed by glass bottles in a few days with concentrations of a few micrograms per liter. However, at 50 μg per liter or greater, the amount of loss was negligible.

SELECTION OF THE ANALYTICAL METHOD

One of the more important parts of the analytical chemistry of a eutrophication study is the selection of the analytical method. Ideally, the analytical procedure should measure a specific form of an element in all types of water without interference from other compounds. Few analytical methods meet these criteria. Therefore, the investigator has the responsibility of selecting the analytical method that measures the desired component in the water under study. Few investigators devote sufficient attention to this part of the study. Frequently, they will select an analytical method that is

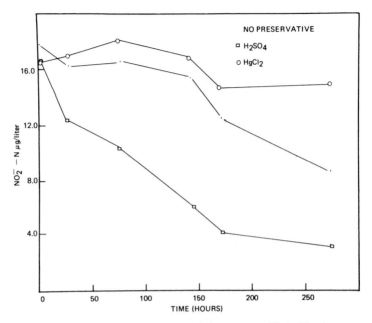

FIGURE 3   Effect of preservatives on nitrite content of Lake Mendota water.

published in a compilation of methods, such as *Standard Methods for the Examination of Water and Waste Water* (American Public Health Association, 1965). These methods have some legal status for state agencies; when used in research projects, however, they often leave much to be desired.

Some investigators exhibit the standard-methods syndrome in which they justify the use of these procedures in the belief that their results will be comparable with others if they all use the same method. Although standard methods usually provide a useful guide to the selection of an analytical method, they do not relieve the investigator of the responsibility for evaluating the analytical method. One of the primary problems with standard methods is that they tend to give the investigator a false sense of reliability. The same difficulty applies to attempts to determine whether the nutrient levels have changed in a lake during a period of time. Some investigators use the same analytical method as past investigators in the hope that they will obtain comparable results. The problem with the "same method yielding comparable results" is that interference may be present in the water at one time and not at another. For example, arsenic is measured as phosphate in most of the o-phosphate analytical procedures. The use of arsenite to control the growth of excessive amounts of aquatic plants has been a fairly common practice. It is impossible to determine o-phosphate in arsenite-treated waters unless a correction is made for arsenic or a modification of the procedure is used, such as suggested by Jenkins (1967).

The only rule that can be used in selecting an analytical method is to select one of the more promising methods and evaluate its reliability for use in the water under investigation.

In addition to the frequently used analytical methods for nitrogen and phosphorus, such as those listed in *Standard Methods* (American Public Health Association, 1965), consideration should be given to recently developed or modified procedures. Some of the methods that deserve particular attention are listed in Table 3. In general, these methods are reported to be easier to use, less subject to interference, and more reliable than the frequently used methods. A brief description of each of these methods follows.

Nitrate   The chromotropic acid-antimony method for nitrate developed by West and Ramachandran (1966) shows promise as a simple, reliable procedure. One problem with this method is that the minimum detectable level is 0.2 mg per liter of NO_3^-. Further work is needed to check for interferences in this method in natural waters.

Another very promising method for nitrate has been developed by Morris and Riley (1963). This method is based on the reduction of nitrate to nitrite in a cadmium-amalgam column. The nitrite is measured as the "azo" dye. It is receiving wide use in marine studies (Strickland and Parsons, 1965). Jenkins

TABLE 3   Analytical Methods for Nitrogen and Phosphorus in Natural Waters

Compound	Reagent or Principal	References
Nitrate	Chromotropic acid-antimony	West and Ramachandran (1966)
	Brucine	Jenkins and Medsker (1964)
	Cd-amalgam reduction	Morris and Riley (1963)
Nitrite	Sulfanilamide and n-(1-naphthyl)-ethylenediamine	Shinn (1941)
Ammonia	Oxidation to NO_2^-	Richards and Kletsch (1964)
o-Phosphate	Vanadomolybdophosphoric acid	Kitson and Mellon (1944)
	Extraction with benzene/isoamyl alcohol	Jenkins (1967)
	Ascorbic acid-antimony	Murphy and Riley (1962)
Total Phosphate	Persulfate oxidation	Menzel and Corwin (1965)
	Nitric- sulfuric or perchloric acid	Lee *et al.* (1965)

and Medsker (1964) have made significant improvements in the brucine method for nitrate. More consistent results were obtained by heating control, modified acid, and a change in the order of reagent addition than were obtained with the *Standard Methods* (American Public Health Association, 1965) brucine procedure.

Nitrite   The substitution of sulfanilamide for sulfanilic acid and n-(1-naphthyl)-ethylenediamine 2HC1 for naphthylene HCl in the determination of nitrite improves sensitivity and reliability (Richards and Kletsch, 1964; Shinn, 1941).

Ammonia   Richards and Kletsch (1964) have introduced the use of alkaline hypochlorite to oxidize ammonia to nitrite. The oxidation is 50 to 75 percent complete and is reported to be reproducible. Fitzgerald and Faust (1967) have found that this method did not yield satisfactory results for swimming pools. However, Strickland and Parsons (1965) state that the method is sensitive, convenient, and reliable. One of the problems with the method is that certain amino acids are partially measured as ammonia.

Organic nitrogen   Bremner (1965) has reviewed the determination of organic nitrogen in soils by the Kjeldahl procedure. His review should be consulted for further information.

Orthophosphate   D. A. Wentz (unpublished data)* has completed an evaluation of the vanadomolybdophosphoric yellow method (Jackson, 1958; Kitson and Mellon, 1944) for the determination of orthophosphate in lake sediments and water. He found that this is the method of choice with phosphate concentrations of 0.05 mg per liter of P or more. The method is simple and almost completely free of interferences.

The use of ascrobic acid-antimony as the reducing agent in the molybdate procedure for o-phosphate as developed by Murphy and Riley (1962) is reported to be simple, reliable, and superior to all other methods (Strickland and Parsons, 1965). This method is reported to have a low temperature coefficient, stable color, and no salt error.

Jenkins (1967) has reported that the interference from arsenic in the molybdate method for orthophosphate can be eliminated if the sample is extracted with benzene/isoamyl alcohol.

Total phosphate   The total phosphate procedure recommended by *Standard Methods* (American Public Health Association, 1965) has been found by Lee

Available Phosphorus in Lake Sediments. M.S. thesis, Univ. Wisconsin (1967).

et al. (1965) to measure only a small part of the total phosphate in some natural waters and algal cultures. The mild acid hydrolysis is not sufficiently rigorous to degrade many organic phosphorus compounds. Sulfuric-nitric or perchloric acid digestion yields complete recovery of the total phosphate. Menzel and Corwin (1965) and Gales *et al.* (1964) have reported that persulfate oxidation yields complete hydrolysis of organic phosphate compounds.

RELIABILITY OF DATA

The sampling of streams and lakes to determine the amounts of plant nutrients present requires that each investigator determine the reliability of his measurement. Frequently, arbitrary sampling schemes are established that call for collection of a grab sample at a fixed period of time. At the end of the period of study, the analytical results are tabulated and mean and statistical tests are performed. Ordinarily, no attempt is made to estimate how well the data describe the concentration of the element in the water under study. This approach is relatively easy to justify by lack of funds. However, it is highly debatable whether a few pieces of data of questionable reliability are better than no data.

Every investigator should determine the reliability of the sampling program and analytical methods. Much greater use should be made of statistical techniques in design of the analytical program and data evaluation. Future research on the eutrophication of natural waters requires that a substantial increase in funds will be needed for the part of the program involving analytical chemistry if significant progess is to be made in understanding the role that plant nutrients play in fertilization of natural waters.

This paper was supported by training grant number 5T1-WP-22-05 from the Federal Water Pollution Control Administration.

REFERENCES

American Public Health Association. 1965. Standard methods for the examination of water and waste water. 12th ed. American Public Health Association, New York. 769 p.

Bremner, J. M. 1965. Total nitrogen, inorganic forms of nitrogen, p. 1149–1286 *In* C. A. Black, D. D. Evans, L. E. Ensminger, J. L. White, and F. E. Clark [ed] Methods of soil chemical analysis. Part 2. Amer. Soc. Agron., Madison, Wis.

Brezonik, P. L., and G. F. Lee. 1966. Preservation of water samples for inorganic analysis with mercuric chloride. Air Water Pollut. 10:549–553.

Fitzgerald, G. P., and S. L. Faust. 1967. Effect of water samples preservation methods on the release of phosphorus from algae. Limnol. Oceanogr. 12:332–334.

Gales, M. E., E. C. Julian, and R. C. Kroner. 1966. Method for quantitative determination of total phosphorus in water. J. Amer. Water Works Ass. 58:1363–1368.

Hassenteufel, W. R., R. Jagitsch, and F. F. Koczy. 1963. Impregnation of glass surface against sorption of phosphate traces. Limnol. Oceanogr. 8:152–156.

Jackson, M. L. 1958. Soil chemical analysis. Prentice-Hall, Inc., Englewood Cliffs, N.J. 498 p.

Jenkins, D. 1965a. A study of methods suitable for the analysis and preservation of nitrogen forms in an estuarine environment. Rep. No. 65-13. Sanitary Engineering Research Laboratory, Univ. California, Berkeley. 50 p.

Jenkins, D. 1965b. A study of methods suitable for the analysis and preservation of phosphorus forms in an estuarine environment. Rep. No. 65-18. Sanitary Engineering Research Laboratory, Univ. California, Berkeley. 50 p.

Jenkins, D. 1967. Analysis of estuarine waters. J. Water Pollut. Control Fed. 39:159–180.

Jenkins, D., and L. L. Medsker. 1964. Brucine method for determination of nitrate in ocean, estuarine and fresh waters. Anal. Chem. 36:610–612.

Kitson, R. E., and M. G. Mellon. 1944. Colorimetric determination of phosphorus as molybdivanadosphosphoric acid. Ind. Eng. Chem. Anal. Edition 16:379–383.

Lee, G. F., N. L. Clesceri, and G. P. Fitzgerald. 1965. Studies on the analysis of phosphate in algal cultures. Air Water Pollut. 9:715–722.

Menzel, D. W., and N. Corwin. 1965. The measurement of total phosphorus in sea water based on the liberation of organically bound fractions by persulfate oxidation. Limnol. Oceanogr. 10:280–282.

Morris, A. W., and J. P. Riley. 1963. The determination of nitrate in sea water. Anal. Chim. Acta 29:272–279.

Murphy, J., and J. P. Riley. 1962. A modified single solution for the determination of phosphate in natural waters. Anal. Chim. Acta 27:31–36.

Richards, F. A., and R. A. Kletsch. 1964. The spectrophotometric determination of ammonia and labile amino compounds in fresh and sea water by oxidation to nitrate, p. 65–81 In Y. Miyake and T. Koyama [ed] Recent researches in the fields of hydrosphere, atmosphere and nuclear geochemistry. Maruzen Co., Tokyo.

Rigler, F. 1966. Radiobiological analysis of inorganic phosphorus in lake water. Verh. Internat. Verein. Limnol. 16:465–470.

Shannon, J. E., and G. F. Lee. 1966. Hydrolysis of condensed phosphates in natural waters. Air Water Pollut. 10:735–756.

Shinn, M. B. 1941. A colorimetric method for determination of nitrite. Ind. Engr. Chem. Anal. Edition 13:33–35.

Strickland, J. D. H., and T. R. Parsons. 1965. A manual of sea water analysis. 2nd ed. Fish. Res. Bd. Canada Bull. 125. 203 p.

West, P. W., and T. P. Ramachandran. 1966. Spectrophotometric determination of nitrate using chromotropic acid. Anal. Chim. Acta 35:317–324.

Contributors

A. M. Beeton, Associate Director, Center for Great Lakes Studies, University of Wisconsin—Madison

Richard J. Benoit, Chief, Marine Sciences Section, Research and Development Department, General Dynamics, Electric Boat Division, Groton, Connecticut

J. H. Beuscher, Professor of Law, University of Wisconsin—Madison*

J. W. Biggar, Department of Soils, University of Wisconsin—Madison

John Langdon Brooks, Department of Biology, Yale University, New Haven, Connecticut

J. H. Carpenter, Chesapeake Bay Institute and Department of Oceanography, The Johns Hopkins University, Baltimore, Maryland

Charles F. Cooper, School of Natural Resources, The University of Michigan, Ann Arbor

R. B. Corey, Department of Soils, University of Wisconsin—Madison

W. T. Edmondson, Department of Zoology, University of Washington, Seattle

David G. Frey, Department of Zoology, Indiana University, Bloomington

F. E. J. Fry, Department of Zoology, University of Toronto, Toronto, Ontario, Canada

G. C. Gerloff, Institute of Plant Development and Department of Botany, University of Wisconsin—Madison

Frank F. Hooper, School of Natural Resources, University of Michigan, Ann Arbor

*Deceased, July 12, 1967.

Shoji Horie, Otsu Hydrobiological Station, Kyoto University, Otsu, Shiga-Ken, Japan

G. E. Hutchinson, Department of Biology, Yale University, New Haven, Connecticut

H. B. N. Hynes, Chairman, Department of Biology, University of Waterloo, Ontario, Canada

Pétur M. Jónasson, Freshwater Biological Laboratory, University of Copenhagen, Hillerød, Denmark

Bostwick H. Ketchum, Woods Hole Oceanographic Institution, Woods Hole, Massachusetts

P. A. Larkin, Institute of Fisheries and Department of Zoology, The University of British Columbia, Vancouver, Canada

G. Fred Lee, Water Chemistry Laboratory, University of Wisconsin–Madison

D. F. Livermore, College of Engineering, University of Wisconsin–Madison

J. W. G. Lund, Freshwater Biological Association, Ambleside, England

Elizabeth McCoy, Department of Bacteriology, University of Wisconsin–Madison

C. H. Mortimer, Director, Center for Great Lakes Studies, University of Wisconsin–Milwaukee

Hugh F. Mulligan, Department of Agronomy, Cornell University, Ithaca, New York

T. G. Northcote, Institute of Fisheries and Department of Zoology, The University of British Columbia, Vancouver, Canada

R. T. Oglesby, Department of Civil Engineering, University of Washington, Seattle

Waldemar Ohle, Max Planck Institute for Limnology, August-Thienemann-Strabe, Germany*

D. W. Pritchard, Chesapeake Bay Institute and Department of Oceanography, The Johns Hopkins University, Baltimore, Maryland

Luigi Provasoli, Haskins Laboratories, New York, New York

Wilhelm Rodhe, Institute of Limnology, Uppsala, Sweden

Gerard A. Rohlich, Director, Water Resources Center, University of Wisconsin–Madison

William B. Sarles, Department of Bacteriology, University of Wisconsin–Madison

George W. Saunders, Jr., Department of Zoology and Great Lakes Research Division, Institute of Science and Technology, The University of Michigan, Ann Arbor

---

*Dr. Ohle presented a paper entitled "Chemical Indication of Eutrophication Processes in Lake Waters and Muds." The paper is omitted from the proceedings because a manuscript was not available.

Frederick E. Smith, Chairman, Department of Wildlife and Fisheries, School of Natural Resources, The University of Michigan, Ann Arbor

Milan Straškraba, Hydrobiological Laboratory, Czechoslovak Academy of Sciences, Prague, Czechoslovakia

Věra Straškrabová, Hydrobiological Laboratory, Czechoslovak Academy of Sciences, Prague, Czechoslovakia

Eugene A. Thomas, Zurich University and Cantonal Laboratory, Switzerland

R. A. Vollenweider, Italian Institute of Hydrobiology, Pallanza, Italy*

S. R. Weibel, Sanitary Engineer Director, U.S. Public Health Service†

R. C. Whaley, Chesapeake Bay Institute and Department of Oceanography, The Johns Hopkins University, Baltimore, Maryland

W. E. Wunderlich, U.S. Army Corps of Engineers, New Orleans District

---

*Dr. Vollenweider presented a paper entitled "Primary Production in Relation to Nutritional Factors with Special Reference to Eutrophication." The paper is omitted from the proceedings because a manuscript was not available.

†At the time of the symposium, Mr. Weibel was assigned to the Federal Water Pollution Control Administration, Cincinnati Water Research Laboratory, U.S. Department of the Interior.

DISCARD

COLLEGE
LIBRARY